PHYSICAL CHEMISTRY OF HIGH TEMPERATURE TECHNOLOGY

Physical Chemistry of High Temperature Technology

E. T. Turkdogan

UNITED STATES STEEL CORPORATION
RESEARCH LABORATORY
MONROEVILLE, PENNSYLVANIA

1980

ACADEMIC PRESS
A Subsidiary of Harcourt Brace Jovanovich, Publishers

New York London Toronto Sydney San Francisco

ACADEMIC PRESS, INC.
111 Fifth Avenue, New York, New York 10003

United Kingdom Edition published by
ACADEMIC PRESS, INC. (LONDON) LTD.
24/28 Oval Road, London NW1 7DX

Library of Congress Cataloging in Publication Data

Turkdogan, E T
Physical chemistry of high temperature technology.

Includes bibliographical references and index.
1. Thermochemistry. 2. High temperatures.
I. Title.
QD515.T87 541.3'687 79-26880
ISBN 0-12-704650-X

PRINTED IN THE UNITED STATES OF AMERICA

80 81 82 83 9 8 7 6 5 4 3 2 1

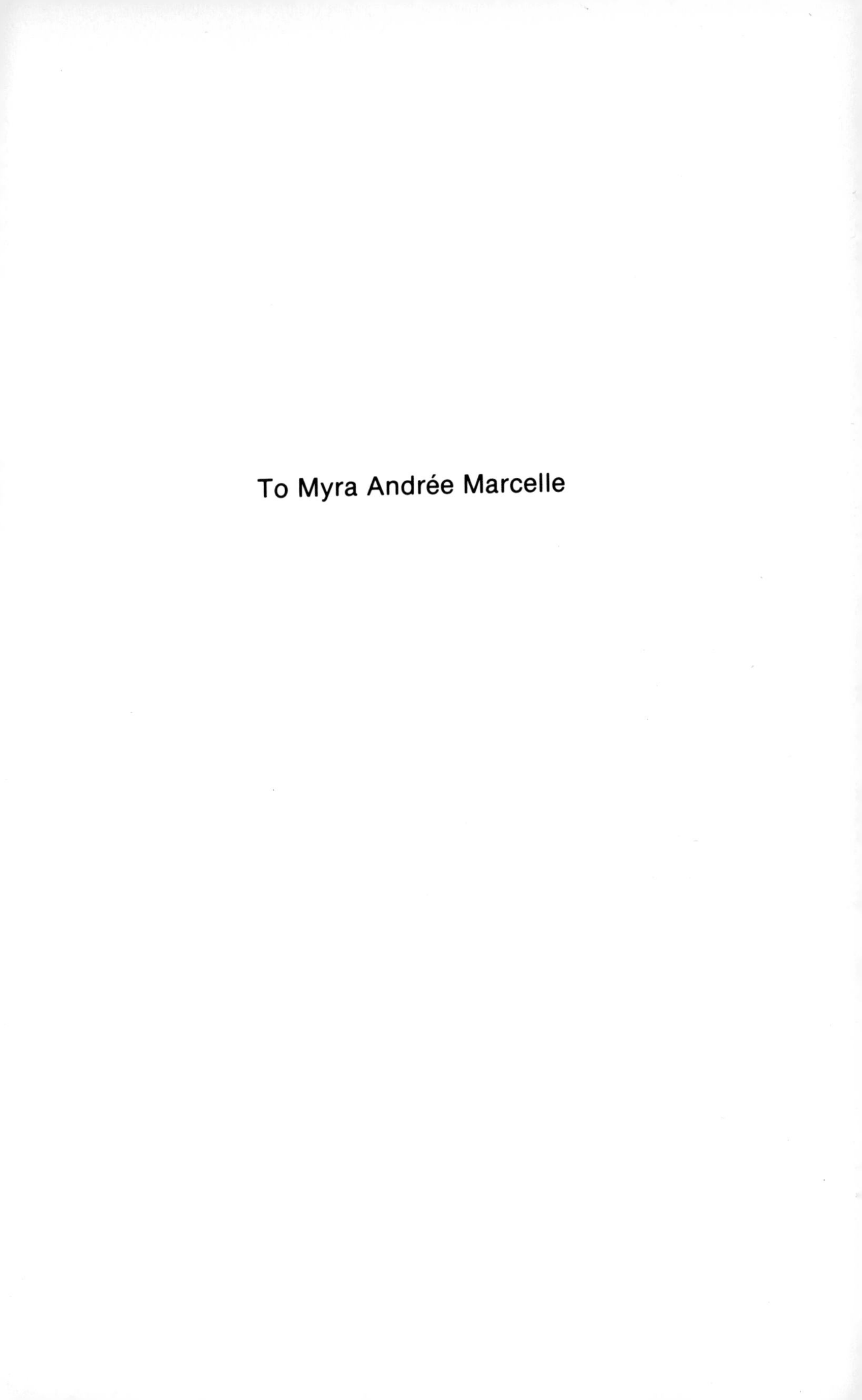

To Myra Andrée Marcelle

CONTENTS

PREFACE

High temperature technology encompasses diversified fields of applied sciences, incorporating a multitude of disciplines in mathematics, physical, chemical, and mechanical sciences, and chemical engineering. Most of the research efforts relating to high temperature technology in the first half of this century were in the areas of metallurgy, glass technology, ceramics, and refractory materials. The advent of nuclear energy since the 1950s, the search for special materials for a wide variety of applications from solid state electronic components to space rockets, and the ever-increasing demand for new energy conversion technology have broadened the scope of the high temperature technology that we have today. Increasing demands for exotic materials and processing techniques have opened up new avenues of research in high temperature chemical and physical phenomena in, for example, thermochemistry, solid state galvanic cells, mass spectrometry, spectroscopy, molecular beams, laser beams, and plasma arcs.

This book is intended for graduate students, research scientists, and technologists in the field of the high temperature technology of materials and their processing. An effort was made to present in a compact and comprehensive form our accumulated knowledge of the physical chemistry of materials and systems at elevated temperatures and pressures. The book is in two parts: Fundamentals and Applications.

Part I consists primarily of critical reviews and a compilation of data on the thermodynamic and physicochemical properties of materials and systems. The theoretical approaches presented are confined to those that are particularly useful as basic research tools and in correlating and forecasting the physicochemical properties of systems. Selected thermodynamic functions are given in Chapter 1 in a condensed form, keeping in mind their application to the study of systems and reactions at elevated temperatures. To facilitate thermodynamic calculations, the standard free-energy equations for reactions are tabulated at the end of Chapter 1, in the form of two-term equations. Since we are concerned mostly with heterogeneous reactions, Chapter 2 is devoted to heterogeneous phase equilibria, showing the effects of temperature, chemical potential, and pressure on phase relations, including correlations based on electron–atom ratios and metastable phase equilibria.

The theories and compilation of data given in Chapters 3–5 concern the physicochemical properties of (i) metals and alloys, (ii) molten glasses, slags, and mattes, and (iii) molten salts. The following properties are considered: thermodynamics, density and critical phemonena, surface tension, viscosity, diffusion, thermal conduction, and electrical conduction. The transport properties of gases, including gas diffusion in porous media, are discussed in Chapter 6.

The application of these physicochemical theories and concepts to high temperature technology is the subject of Part II. The rate phenomena presented in Chapter 7 are intended to provide guidance in understanding the workings of processes, with the ultimate objective of developing improved methods of control. All aspects of rate phenomena cannot be covered comprehensively in one chapter; the presentation is therefore confined to selected topics in rate phenomena: the kinetics of interfacial reactions, nucleation and growth, heat and mass transfer, reactions in porous media, and gas bubbles in liquids.

Because of the broad spectrum of subject matter, the discussion of high temperature processes had to be limited to a few chosen areas, with the specific purpose of demonstrating how our basic knowledge has been applied toward understanding the workings of processes and the behavior of materials at elevated temperatures. Instead of citing a multitude of processes in diversified areas of high temperature technology, with brief comments on their technical aspects, we proceed to an in-depth discussion of the state of reactions in certain types of processes. However, the topics chosen for discussion in Chapters 8–10 would also be pertinent to other high temperature processes.

The calcination, reduction, roasting, and sintering presented in Chapter 8 are typical examples of the processing of raw materials, primarily as feedstocks for the chemical industry and in smelting and metal refining processes.

Chapter 9 is devoted to the physical chemistry of reactions in ironmaking and steelmaking. The choice of this subject is partly personal, owing to the author's close association with research in the steel industry for 30 years. However, the primary reason for choosing this subject is that it provides an excellent means of demonstrating how diversified disciplines of science and technology have been utilized in the understanding of the state of reactions in complex processes. In addition, it is well to remember that much of the present knowledge of the physical chemistry of systems and reactions at elevated temperatures has originated from, or been stimulated by, research on materials and systems of interest to ferrous metallurgy.

Reactions in processes relating to energy conversion are discussed in Chapter 10 with particular emphasis on the thermochemical decomposition of water and coal conversion. Many of these processes are either in a conceptual or a pilot-plant demonstration stage; detailed technical information is therefore not readily available. The discussion in the latter half of Chapter 10 is on hot corrosion in process environments relating to coal gasifiers, petroleum refineries, jet engines, and nuclear reactors. As typical examples, the following cases of hot corrosion are considered: carbon deposition, oxidation and sulfidation of alloys and the effect of alkalies, and corrosion of alloys by molten fluorides and liquid sodium in fast breeder reactors.

Over 900 references are included to original studies, review papers, books, and data compilations without which this book could not have been written. In particular, the notable contributions of Wagner, Chipman, Darken, Richardson, and Kubaschewski, to whom reference is often made, have been of great value to the author in the preparation of this book.

All the quantities in this book are given in cgs units, calories, and atmospheres instead of SI units because almost all the available data in the literature on thermodynamic and physicochemical properties are tabulated or plotted in traditional units. For the reader's convenience, a table is given for conversion from traditional units to SI units.

ACKNOWLEDGMENTS

I am grateful to the United States Steel Corporation for the generous time allowed to me for the preparation of this book and for liberal use of the library and secretarial facilities of the Research Laboratory.

I am particularly grateful to Mrs. Louise Shimkus for her patience and care in typing the manuscript, to Mr. Arnold Cornelissen for his skillful drawing of the diagrams, and to the library staff of the Research Laboratory for their help in the search of the literature and the retrieval of publications.

I am indebted to the late Dr. L. S. Darken for his valuable written contribution to Section 2.3.1.

I would also like to thank Professors P. Grieveson and G. R. Belton for their encouragement and helpful discussions in the preparation of the manuscript.

CONVERSION TO SI UNITS

Quantity	*Traditional Units*	*Value in SI units*
Electron volt	eV	1.602×10^{-19} J
Energy, work, quantity of heat	cal $mole^{-1}$ (thermochemical)	4.184 J $mole^{-1}$
Entropy	cal $mole^{-1}$ deg^{-1}	4.184 J $mole^{-1}$ K^{-1}
Force	dyn (g cm sec^{-2})	10^{-5} N $\equiv 10^{-5}$ kg m sec^{-2}
Power	erg sec^{-1}	10^{-7} W $\equiv 10^{-7}$ J sec^{-1}
Pressure	atm	101.325 kPa
	bar	10^{5} Pa $\equiv 10^{5}$ N m^{-2}
	torr $\equiv$ mm Hg	133.322 Pa
Surface tension	dyn cm^{-1}	10^{-3} N $m^{-1} \equiv$ mN m^{-1}
Surface energy	erg cm^{-2}	10^{-3} J $m^{-2} \equiv$ mJ m^{-2}
Thermal conductivity	cal cm^{-1} sec^{-1} deg^{-1}	4.184×10^{2} W m^{-1} K^{-1}
Viscosity (dynamic)	poise (g cm^{-1} sec^{-1})	10^{-1} N sec $m^{-2} \equiv 10^{-1}$ kg m^{-1} sec^{-1}
Viscosity (kinematic)	stokes (cm^{2} sec^{-1})	10^{-4} m^{2} sec^{-1}

PART I

Fundamentals

CHAPTER 1

SELECTED THERMODYNAMIC FUNCTIONS

1.1 Introduction

The theme of this book is the application of our knowledge of physical chemistry and thermodynamics to a better understanding of the workings of processes in high temperature technology. With these objectives in mind, selected thermodynamic functions are presented in this chapter in a condensed form without dwelling on their derivations. It is not necessary, or even desirable, to recite here in detail the derivations of thermodynamic functions and concepts that are well documented in many textbooks. For in-depth study of thermodynamics, references may be made, for example, to the original teachings of Gibbs [1], the classic texts of Lewis and Randall [2], later rewritten by Pitzer and Brewer [3], and of Guggenheim [4], the textbook on the physical chemistry of metals by Darken and Gurry [5], and many others.

Selected thermodynamic functions are listed in Table 1.1.

1.2 Standard Free-Energy Data

Since the early pioneering work of Bichowsky and Rossini [6] and of Kelley [7], efforts have continued in updating the compilation of thermochemical data. In current use are the JANAF thermochemical tables [8–11], which contain critically evaluated data for C_p°, S°, $(F^\circ - H_{298}^\circ)/T$, and $(H^\circ - H_{298}^\circ)$ for numerous substances in gas, liquid, and solid form. In addition, the standard free energies and enthalpies of formation are also tabulated at temperature intervals of 100 °K.

In another compilation of thermochemical data, by Barin *et al.* [12, 13], an unconventional standard state has been used for the enthalpy; thus, in terms of the notations used here, their enthalpy H is equivalent to

$$H = \Delta H_{298}^\circ + (H^\circ - H_{298}^\circ). \tag{1.1}$$

TABLE 1.1

Selected Thermodynamic Functions

Function	Equations
1. Change in energy E of a homogeneous system in terms of state variables pressure P, volume V, and absolute temperature T.	$dE = \left(\frac{\partial E}{\partial P}\right)_V dP + \left(\frac{\partial E}{\partial V}\right)_P dV$ $dE = \left(\frac{\partial E}{\partial P}\right)_T dP + \left(\frac{\partial E}{\partial T}\right)_P dT$ $dE = \left(\frac{\partial E}{\partial V}\right)_T dV + \left(\frac{\partial E}{\partial T}\right)_V dT$
2. Heat capacity at constant volume C_V or constant pressure C_p.	$C_V = \left(\frac{\partial E}{\partial T}\right)_V, \quad C_P = \left(\frac{\partial(E + PV)}{\partial T}\right)_P = \left(\frac{\partial H}{\partial T}\right)_P$
3. Entropy change accompanying an isothermal and reversible process at constant volume or constant pressure.	$dS = (C_V/T)\,dT = C_V\,d(\ln T)$ $dS = (C_P/T)\,dT = C_P\,d(\ln T)$
4. Helmholtz free energy A and Gibbs free energy F.	$A = E - TS, \quad F = H - TS$
5. Maxwell's relations for combined statements of the first and second laws.	$\left(\frac{\partial T}{\partial V}\right)_S = -\left(\frac{\partial P}{\partial S}\right)_V, \quad \left(\frac{\partial S}{\partial V}\right)_T = \left(\frac{\partial P}{\partial T}\right)_V$ $\left(\frac{\partial T}{\partial P}\right)_S = \left(\frac{\partial V}{\partial S}\right)_P, \quad \left(\frac{\partial V}{\partial T}\right)_P = -\left(\frac{\partial S}{\partial P}\right)_T$
6. Statement of the third law for any homogeneous system of fixed composition at constant pressure.	$\frac{F - H_0}{T} = \int_0^T (H - H_0)\,d\left(\frac{1}{T}\right)$
7. Standard free-energy change as a function of temperature in terms of the standard enthalpy, entropy, and heat capacity changes.	$\Delta F^\circ = \Delta H^\circ_{298} + \int_{298}^T \Delta C^\circ_P\,dT - T\,\Delta S^\circ_{298} - T\int_{298}^T \frac{\Delta C^\circ_P}{T}\,dT$
8. Empirical relations for C°_p and ΔF° as a function of T.	$C^\circ_p = a + bT - cT^{-2}$, $\Delta F^\circ = A + BT\log T + CT$
9. Functions for the molar quantity G and the partial molar quantity $\bar{G}_i$ for 1 mole of solution at constant temperature and pressure.	$G = x_1\bar{G}_1 + x_2\bar{G}_2 + x_3\bar{G}_3 + \cdots$ $x_1\,d\bar{G}_1 + x_2\,d\bar{G}_2 + x_3\,d\bar{G}_3 + \cdots = 0$
10. Function for the partial molar quantity in a multicomponent system at constant relative amounts of all components except component 1.	$\bar{G}_1 = G + (1 - x_1)\left(\frac{\partial G}{\partial x_1}\right)_{x_2:x_3:x_4:\cdots}$
11. Relative partial molar quantity, relative to the molar quantity of the pure component i.	$\Delta\bar{G}_i = \bar{G}_i - G^\circ_i$
12. Chemical potential, or relative partial molar free-energy function in terms of fugacity p^*_i, activity a_i or enthalpy of solution $\bar{L}_i$, and entropy of solution $\Delta\bar{S}_i$.	$\mu_i = \mu^\circ_i + RT\ln p^*_i, \quad \bar{F}_i = F^\circ_i + RT\ln p^*_i$ $\Delta\bar{F}_i = \bar{F}_i - F^\circ_i = RT\ln a_i = \bar{L}_i - T\,\Delta\bar{S}_i$

TABLE 1.1 (continued)

Function	Equations
13. Molar quantities for an ideal solution.	$\Delta F^{M,id} = RT(x_1 \ln x_1 + x_2 \ln x_2 + x_3 \ln x_3 + \cdots)$ $\Delta S^{M,id} = -R(x_1 \ln x_1 + x_2 \ln x_2 + x_3 \ln x_3 + \cdots)$
14. Excess relative partial molar quantity $\bar{G}_i^{xs}$ relative to that for an ideal solution $\bar{G}_i^{id}$.	$\bar{G}_i^{xs} = \bar{G}_i - G_i^{id}$
15. Activity coefficient and excess free energy of solution of component i.	$\gamma_i = a_i/x_i, \quad \bar{F}_i^{xs} = RT \ln \gamma_i$ $\ln \gamma_i = (\bar{L}_i/RT) - (\bar{S}_i^{xs}/R)$
16. Free-energy change accompanying isothermal transfer of solute i from state (1) to state (2).	$\Delta F = F^\circ(1) - F^\circ(2) = RT \ln[(a_i)_2/(a_i)_1]$
For a 1 wt. % solution[a]	$\Delta F \simeq RT \ln((M_s/M_i)0.01\gamma_i^\circ)$,
17. Gibbs–Duhem integration for a binary system 1–2.	$\ln \gamma_2 = \ln \gamma_2' - \int_{\gamma_1'}^{\gamma_1} \frac{x_1}{x_2}\, d(\ln \gamma_1)$
18. Gibbs–Duhem integration for a ternary system 1–2–3 at a constant mole fraction ratio x_2/x_3.	$\left[\ln \gamma_2 = \ln \gamma_2' - \int_{\gamma_1'}^{\gamma_1} \frac{x_1}{x_2}\, d(\ln \gamma_1)\right]_{x_2/x_3}$
19. Equilibrium constant of a reaction and the accompanying change in the standard free energy.	$K = \frac{\prod a_i^{\nu_i}\text{(products)}}{\prod a_i^{\nu_i}\text{(reactants)}} \qquad \Delta F^\circ = -RT \ln K$
20. van't Hoff equation for the temperature dependence of the equilibrium constant.	$d(\ln K)/d(1/T) = -\Delta H^\circ/R$
21. Free-energy change in a galvanic-cell reaction.[b]	$\Delta F = -n\mathscr{F}\epsilon$

[a] Here γ_i° is the activity coefficient when % $i = 0$, and M_s and M_i are the atomic weights of solvent and solute, respectively.

[b] In the equation ϵ is emf in volts, $\mathscr{F}$ the Faraday constant, 23,061 cal V eq, and n the number of equivalents per gram-atom; the emf ϵ is positive if the reaction as written is a spontaneous one.

The heat of formation from elements in their standard states at temperature T is to be calculated from the H value for the pure substance minus the sum of the H values for the elements; thus

$$\Delta H^\circ = H \quad \text{(for product)} - \sum H \quad \text{(for elements)}. \tag{1.2}$$

Similarly, for the standard free energy of formation,

$$\Delta F^\circ = G \quad \text{(for product)} - \sum G \quad \text{(for elements)}. \tag{1.3}$$

Application of thermochemical data in the study of high temperature reactions is much facilitated by using free-energy equations instead of the tabulated data. Although Eq. (7) or (8) in Table 1.1 is an accurate description of the temperature dependence of the standard free energy over a wide range of temperatures, the linear equation with average values of ΔH° and ΔS° is adequate for most purposes because ΔC_p is small for most reactions.

Even when ΔC_p is not negligibly small, the linear equation can still be used, because for many reactions the changes in ΔH° and ΔS° with temperature are similar and tend to cancel each other, and consequently, the nonlinearity of the variation of ΔF° with temperature is minimized. Therefore, the best fit of the free-energy data to a linear equation by linear regression analysis is adequate for most thermodynamic calculations:

$$\Delta F^\circ = \Delta \tilde{H}^\circ - T\Delta \tilde{S}^\circ, \tag{1.4}$$

where a tilde designates the average value for the stated range of temperatures, which is usually taken to be the interval between the transition temperatures of the reactants or products.

The extent of error in ΔF° arising from the use of Eq. (1.4) is below 200 cal, which is much less than the accuracy of the tabulated data.

The standard free-energy equations are given in Table 1.2 for many reactions that are of interest in high temperature technology. This is not a critical evaluation of the thermochemical data. The values of $\Delta \tilde{H}^\circ$ and $\Delta \tilde{S}^\circ$ are derived by linear regression analysis of the tabulated values of ΔF°, using primarily the JANAF thermochemical tables [8–11], and the thermochemical data compiled by Barin *et al.* [12, 13], Kubaschewski *et al.* [14], and Hultgren *et al.* [15]. The limits of uncertainty assigned to the free-energy data in some compilations are much broader than those measured directly at elevated temperatures. In such cases, the preference is given to the results of direct measurements, to which references are given in Table 1.2.

Brackets are used to indicate the standard states for the pure reactants and products over the stated range of temperatures: $\langle$solid$\rangle$, $\{$liquid$\}$, and (gas). For the sake of brevity, the free-energy equations are given only for temperature ranges of greatest practical and experimental interest. By adding or subtracting the appropriate free-energy equations for phase transformations, these equations can be extended to other ranges of temperatures.

Small enthalpy changes for the allotropic phase transformations are not indicated in the free-energy equations for reactions. For example the values of $\Delta \tilde{H}^\circ$ and $\Delta \tilde{S}^\circ$ for the reaction $\langle FeO\rangle = \langle Fe\rangle + \frac{1}{2}(O_2)$ are for temperatures of 25–1371 °C, without making small adjustments for the α–γ and γ–δ phase transformations of iron: The resulting error in ΔF° is well below the limits of uncertainty of the thermochemical data. The standard state for carbon is β-graphite and, in most cases, the high temperature crystalline form is chosen as the standard state of the substance, e.g., α-Al_2O_3, α-Fe_2O_3, and β-cristobalite for SiO_2. The phase transformations are indicated by subscripts only in special cases. Whenever possible, the results of direct measurements are used for the free energies of multicomponent reactions such as carbonates, sulfates, etc. In such cases, thermochemical calculations are not made for critical evaluation of the experimental data. For multicomponent reactions such as silicates, aluminates, borates, etc., the free-energy equations are calculated for reactions involving the constituent oxides.

TABLE 1.2

Standard Free Energies for Reactions: $\Delta F^\circ = \Delta \tilde{H}^\circ - T\,\Delta \tilde{S}^\circ$

Reaction[a]	$\Delta \tilde{H}^\circ$ (cal)	$\Delta \tilde{S}^\circ$ (cal)	Error ($\pm$kcal)	Range[b] (°C)	Reference
$\langle Ag \rangle = \{Ag\}$	2700	2.19	0.1	961m	15
$\{Ag\} = (Ag)$	61,740	25.40	0.2	961–2163b	15
$\langle AgBr \rangle = \{AgBr\}$	2200	3.13	0.4	430m	12
$\{AgBr\} = (AgBr)$	45,900	25.0	4	1560b	12
$\{AgBr\} = \{Ag\} + \frac{1}{2}(Br_2)$	23,680	6.64	3	961–1560b	12
$\langle AgCl \rangle = \{AgCl\}$	3100	4.26	0.1	455m	12
$\{AgCl\} = (AgCl)$	42,500	23.1	4	1564b	12
$\{AgCl\} = \{Ag\} + \frac{1}{2}(Cl_2)$	24,970	5.50	1	961–1564b	12
$\langle AgF \rangle = \langle Ag \rangle + \frac{1}{2}(F_2)$	48,000	13.1	10	25–435m	12
$\langle AgI \rangle = \{AgI\}$	2250	2.71	0.5	558m	12
$\{AgI\} = (AgI)$	34,500	19.4	5	1505b	12
$\{AgI\} = \{Ag\} + \frac{1}{2}(I_2)$	18,430	5.46	2	961–1505b	12
$\langle Ag_2O \rangle = 2\langle Ag \rangle + \frac{1}{2}(O_2)$	7300	15.80	0.2	25	14
$\langle Ag_2S \rangle = \{Ag_2S\}$	1880	1.70	0.5	830m	13
$\langle Ag_2S \rangle = 2\langle Ag \rangle + \frac{1}{2}(S_2)$	38,540	40.29	2	25–830m	13
$\langle Al \rangle = \{Al\}$	2580	2.76	0.05	660m	15
$\{Al\} = (Al)$	72,810	26.17	0.5	660–2520b	15
$(AlBr) = \{Al\} + \frac{1}{2}(Br_2)$	5800	−14.4	5	660–2000	9
$(AlBr_3) = \{Al\} + \frac{3}{2}(Br_2)$	113,470	5.91	1	660–2000	9
$(AlCl) = \{Al\} + \frac{1}{2}(Cl_2)$	18,510	−13.90	1	660–2000	8
$(AlCl_3) = \{Al\} + \frac{3}{2}(Cl_2)$	143,910	16.24	2	660–2000	8
$(Al_2Cl_6) = 2\langle Al \rangle + 3(Cl_2)$	308,800	57.89	2	25–660	8
$(AlF) = \{Al\} + \frac{1}{2}(F_2)$	69,810	−12.96	1	660–2000	8
$\langle AlF_3 \rangle = \{Al\} + \frac{3}{2}(F_2)$	360,360	61.64	0.4	660–1276s	8
$(AlF_3) = \{Al\} + \frac{3}{2}(F_2)$	293,870	18.97	1	660–2000	8
$(AlI_3) = \{Al\} + \frac{3}{2}(I_2)$	75,140	17.63	3	660–2000	8
$\langle AlB_{12} \rangle = \{Al\} + 12\langle B \rangle$	52,580	1.78	2	660–2200m	13
$\langle Al_4C_3 \rangle = 4\{Al\} + 3\langle C \rangle$	63,330	22.72	2	660–2200m	8
$\langle AlN \rangle = \{Al\} + \frac{1}{2}(N_2)$	78,170	27.61	1	660–2000	8
$\langle AlP \rangle = \{Al\} + \frac{1}{2}(P_2)$	59,630	24.94	—	660–1700	13
$\langle Al_2O_3 \rangle = 2\{Al\} + \frac{3}{2}(O_2)$	403,260	78.11	0.5	660–2054m	11
$(Al_2O) = 2\{Al\} + \frac{1}{2}(O_2)$	40,800	−11.8	5	660–2000	11
$(Al_2O_2) = 2\{Al\} + (O_2)$	112,500	6.9	7	660–2000	11
$\langle AlPO_4 \rangle = \langle Al \rangle + \frac{1}{2}(P_2) + 2(O_2)$	430,720	107.54	2	25–660	13
$\langle 3Al_2O_3 \cdot 2SiO_2 \rangle = 3\langle Al_2O_3 \rangle + 2\langle SiO_2 \rangle$	−2055	−4.16	1	25–1750m	8
$\langle Al_2O_3 \cdot SiO_2 \rangle = \langle Al_2O_3 \rangle + \langle SiO_2 \rangle$	2106	0.93	0.5	25–1700	8
$\langle Al_2O_3 \cdot TiO_2 \rangle = \langle Al_2O_3 \rangle + \langle TiO_2 \rangle$	6040	0.94	—	25–1860m	13
$\langle As \rangle = \frac{1}{4}(As_4)$	8740	10.02	0.1	25–603s	15
$\langle Au \rangle = \{Au\}$	3000	2.25	0.1	1063m	15
$\{Au\} = (Au)$	81,400	26.06	0.3	1063–2857b	15
$\langle B \rangle = \{B\}$	12,000	5.22	0.8	2030m	15
$\langle B \rangle = (B)$	136,270	35.33	0.8	25–2030m	15
$\{B\} = (B)$	119,420	28.03	0.8	2030–4002b	15
$(BBr_3) = \langle B \rangle + \frac{3}{2}(Br_2)$	59,930	11.66	0.2	25–2030	8
$(BCl_3) = \langle B \rangle + \frac{3}{2}(Cl_2)$	96,560	12.35	0.2	25–2030	8
$(BF_3) = \langle B \rangle + \frac{3}{2}(F_2)$	272,610	15.56	0.3	25–2030	8
$\langle B_4C \rangle = 4\langle B \rangle + \langle C \rangle$	9920	1.33	2.5	25–2030	8

***TABLE 1.2* (*continued*)**

Reaction[a]	$\Delta\tilde{H}^\circ$ (cal)	$\Delta\tilde{S}^\circ$ (cal)	Error ($\pm$kcal)	Range[b] (°C)	Reference
$\langle BN\rangle = \langle B\rangle + \frac{1}{2}(N_2)$	59,900	20.94	0.5	25–2030	8
$\langle B_2O_3\rangle = \{B_2O_3\}$	5750	7.95	0.1	450m	9
$\{B_2O_3\} = 2\langle B\rangle + \frac{3}{2}(O_2)$	293,700	50.20	1.0	450–2043b	9
$(BO) = \langle B\rangle + \frac{1}{2}(O_2)$	909	−21.22	2	25–2030	8
$\langle Ba\rangle = \{Ba\}$	1850	1.85	0.5	729m	15
$\{Ba\} = (Ba)$	35,860	18.92	3	729–1622b	15
$\langle BaBr_2\rangle = \{BaBr_2\}$	7640	6.76	0.2	857m	11
$\{BaBr_2\} = \{Ba\} + (Br_2)$	178,200	28.4	3	857–1622	11
$\langle BaCl_2\rangle = \{BaCl_2\}$	3820	3.09	0.1	962m	10
$\{BaCl_2\} = \{Ba\} + (Cl_2)$	193,900	29.03	2	962–1622	10
$\langle BaF_2\rangle = \{BaF_2\}$	5580	3.40	0.2	1368m	10
$\{BaF_2\} = \{Ba\} + (F_2)$	275,900	30.85	1	1368–1622	10
$\langle BaC_2\rangle = \{Ba\} + 2\langle C\rangle$	21,400	0.5	1	729–1200	12
$\langle Ba_3N_2\rangle = 3\{Ba\} + (N_2)$	89,820	65.61	3	729–1000	13
$\langle BaO\rangle = \{Ba\} + \frac{1}{2}(O_2)$	133,170	24.54	1	729–1622	10
$\langle BaS\rangle = \{Ba\} + \frac{1}{2}(S_2)$	130,000	29.5	5	729–1622	13
$\langle 3BaO\cdot Al_2O_3\rangle = 3\langle BaO\rangle + \langle Al_2O_3\rangle$	50,700	−4.5	—	25–1750m	12
$\langle BaO\cdot Al_2O_3\rangle = \langle BaO\rangle + \langle Al_2O_3\rangle$	29,700	−1.6	—	25–1830m	12
$\langle BaCO_3\rangle = \langle BaO\rangle + (CO_2)$	59,930	35.15	0.3	800–1060m	16, 17
$\langle BaO\cdot HfO_2\rangle = \langle BaO\rangle + \langle HfO_2\rangle$	31,000	1.7	5	25–1100	13
$\langle BaSO_4\rangle = \langle BaO\rangle + (SO_2) + \frac{1}{2}(O_2)$	132,000	57.9	5	25–1350m	12
$\langle 2BaO\cdot SiO_2\rangle = 2\langle BaO\rangle + \langle SiO_2\rangle$	62,100	−1.4	3	25–1760m	12
$\langle BaO\cdot SiO_2\rangle = \langle BaO\rangle + \langle SiO_2\rangle$	35,600	−1.5	3	25–1605m	12
$\langle 2BaO\cdot TiO_2\rangle = 2\langle BaO\rangle + \langle TiO_2\rangle$	46,500	−1.2	4	25–1860m	18
$\langle BaO\cdot TiO_2\rangle = \langle BaO\rangle + \langle TiO_2\rangle$	37,400	3.75	3	25–1705m	18
$\langle BaO\cdot ZrO_2\rangle = \langle BaO\rangle + \langle ZrO_2\rangle$	30,000	3.8	5	25–1700	18
$\langle Be\rangle = \{Be\}$	2800	1.8	—	1287m	15
$\{Be\} = (Be)$	71,670	25.95	2	1287–2472b	15
$(BeBr_2) = \{Be\} + (Br_2)$	66,700	−1.3	4	1287–2000	11
$(BeCl_2) = \{Be\} + (Cl_2)$	90,720	−1.4	4	1287–2000	8
$(BeF_2) = \{Be\} + (F_2)$	195,670	0.56	2	1287–2000	8
$(BeI_2) = \{Be\} + (I_2)$	34,600	−2.0	8	1287–2000	11
$\langle Be_2C\rangle = 2\langle Be\rangle + \langle C\rangle$	22,300	3.3	4	25–1287	8
$\langle Be_2C\rangle = 2\{Be\} + \langle C\rangle$	27,500	6.8	4	1287–2100	8
$\langle Be_3N_2\rangle = 3\langle Be\rangle + (N_2)$	140,400	44.31	0.5	25–1287	8
$\langle Be_3N_2\rangle = 3\{Be\} + (N_2)$	147,300	48.57	0.5	1287–2200m	8
$\langle BeO\rangle = \langle Be\rangle + \frac{1}{2}(O_2)$	145,360	23.35	1	25–1287	11
$\langle BeO\rangle = \{Be\} + \frac{1}{2}(O_2)$	146,660	24.12	1	1287–2000	11
$\langle BeS\rangle = \langle Be\rangle + \frac{1}{2}(S_2)$	71,000	20.7	5	25–1287	13
$\langle BeO\cdot Al_2O_3\rangle = \langle BeO\rangle + \langle Al_2O_3\rangle$	1000	−0.7	5	25–1870m	8
$\langle 3BeO\cdot B_2O_3\rangle = 3\langle BeO\rangle + \{B_2O_3\}$	13,960	6.1	5	450–1495m	8
$\langle 2BeO\cdot SiO_2\rangle = 2\langle BeO\rangle + \langle SiO_2\rangle$	2600	1.7	7	25–1560m	8
$\langle BeSO_4\rangle_\beta = \langle BeO\rangle + (SO_2) + \frac{1}{2}(O_2)$	60,800	55.46	2	25–1287	8
$\langle Bi\rangle = \{Bi\}$	2700	4.96	0.1	272m	15
$\{Bi\} = (Bi)$	45,850	23.76	0.5	272–1564b	15
$\langle BiBr_3\rangle = \{BiBr_3\}$	5190	10.54	0.2	219m	13
$\{BiBr_3\} = \{Bi\} + \frac{3}{2}(Br_2)$	70,680	44.61	3	272–461b	13
$\langle BiCl_3\rangle = \{BiCl_3\}$	5650	11.15	0.2	234m	13
$\{BiCl_3\} = \{Bi\} + \frac{3}{2}(Cl_2)$	83,920	38.36	3	272–439b	13
$\langle BiF_3\rangle = \{BiF_3\}$	5160	5.60	1	649m	13
$\{BiF_3\} = (BiF_3)$	30,000	25.0	5	927b	13

TABLE 1.2 (continued)

Reaction[a]	$\Delta\tilde{H}^\circ$ (cal)	$\Delta\tilde{S}^\circ$ (cal)	Error ($\pm$kcal)	Range[b] (°C)	Reference
$\{BiF_3\}=\{Bi\}+\frac{3}{2}(F_2)$	204,000	43.7	5	649–927b	13
$\langle BiI_3\rangle=\{BiI_3\}$	9350	13.71	1	409m	13
$\{BiI_3\}=\{Bi\}+\frac{3}{2}(I_2)$	50,000	41.1	5	409–542b	13
$\langle Bi_2O_3\rangle=\{Bi_2O_3\}$	14,300	13.04	1	824m	13
$\{Bi_2O_3\} = 2\{Bi\} + \frac{3}{2}(O_2)$	106,400	38.14	3	824–1500	13
$\langle Bi_2S_3\rangle=\{Bi_2S_3\}$	18,700	17.81	3	777m	13
$\langle Bi_2S_3\rangle=2\{Bi\}+\frac{3}{2}(S_2)$	86,000	65.5	5	272–777m	13
graphite = diamond	345	−1.07	0.01	25–900	15
$\langle C\rangle=(C)$	170,520	37.16	1	1750–3800s	15
$(CH_4)=\langle C\rangle+2(H_2)$	21,760	26.45	0.1	500–2000	8
$(CCl_4)=\langle C\rangle+2(Cl_2)$	21,300	30.89	0.5	25–2000	8
$(CF_4)=\langle C\rangle+2(F_2)$	223,040	36.21	0.3	25–2000	8
$(CN)=\langle C\rangle+\frac{1}{2}(N_2)$	−103,600	−23.81	2.5	25–2000	8
$(CO) = \langle C\rangle + \frac{1}{2}(O_2)$	27,340	−20.50	0.1	500–2000	8
$(CO_2)=\langle C\rangle+(O_2)$	94,490	−0.13	0.02	500–2000	8
$(COS)=\langle C\rangle+\frac{1}{2}(O_2)+\frac{1}{2}(S_2)$	48,470	−2.38	0.3	500–2000	8
$(CS)=\langle C\rangle+\frac{1}{2}(S_2)$	−39,000	−21.0	5	25–2000	8
$(CS_2)=\langle C\rangle+(S_2)$	2730	−1.55	0.2	25–2000	8
$\langle Ca\rangle=\{Ca\}$	2040	1.84	0.1	839m	61
$\{Ca\}=(Ca)$	37,720	20.82	0.1	839–1491b	61
$\langle CaBr_2\rangle=\{CaBr_2\}$	6900	6.80	0.1	742m	10
$\{CaBr_2\}=\{Ca\}+(Br_2)$	160,000	25.0	5	839–1484	10
$\langle CaCl_2\rangle = \{CaCl_2\}$	6820	6.53	0.2	772m	8
$\{CaCl_2\}=(CaCl_2)$	56,200	25.44	1	1936b	8
$\{CaCl_2\}=\{Ca\}+(Cl_2)$	190,860	34.89	2	839–1484	8
$\langle CaF_2\rangle=\{CaF_2\}$	7100	4.20	0.1	1418m	8
$\{CaF_2\}=(CaF_2)$	73,780	26.29	1	2533b	8
$\langle CaF_2\rangle = \{Ca\} + (F_2)$	291,490	38.79	2	839–1484	8
$\langle CaI_2\rangle=\{CaI_2\}$	10,000	9.51	0.1	779m	10
$\{CaI_2\} = (CaI_2)$	42,880	21.14	1	1755b	10
$\{CaI_2\}=\{Ca\}+(I_2)$	130,000	23.0	5	839–1484	10
$\langle CaC_2\rangle=\{Ca\}+2\langle C\rangle$	14,400	−6.28	3	839–1484	12
$\langle Ca_3N_2\rangle=3\langle Ca\rangle+(N_2)$	104,000	47.5	5	25–839	13
$\langle CaAl_2\rangle=\{CaAl_2\}$	15,100	11.2	—	1080m	13
$\{CaAl_2\}=\{Ca\}+2\{Al\}$	45,200	0.8	—	1080–1484	13
$\langle CaAl_4\rangle=\{CaAl_4\}$	18,800	17.9	—	777m	13
$\{CaAl_4\}=\{Ca\}+4\{Al\}$	47,000	0.7	—	839–1484	13
$\langle Ca_2Si\rangle=2\langle Ca\rangle+\langle Si\rangle$	50,000	5.3	10	25–839	13
$\langle CaSi\rangle=\langle Ca\rangle+\langle Si\rangle$	36,000	3.7	5	25–839	13
$\langle CaSi_2\rangle=\langle Ca\rangle+2\langle Si\rangle$	36,000	6.8	5	25–839	13
$\langle Ca_3P_2\rangle=3\langle Ca\rangle+(P_2)$	155,000	51.7	10	25–839	13
$\langle CaO\rangle=\{CaO\}$	19,000	5.9	—	2927m	10
$\langle CaO\rangle=\{Ca\}+\frac{1}{2}(O_2)$	153,000	25.95	0.3	839–1484	10
$\langle CaS\rangle=\{Ca\}+\frac{1}{2}(S_2)$	131,000	24.82	1	839–1484	9
$\langle 3CaO\cdot Al_2O_3\rangle=3\langle CaO\rangle+\langle Al_2O_3\rangle$	3000	−5.9	1	500–1535m	18
$\langle CaO\cdot Al_2O_3\rangle=\langle CaO\rangle+\langle Al_2O_3\rangle$	4300	−4.5	0.5	500–1605m	18
$\langle CaO\cdot 2Al_2O_3\rangle=\langle CaO\rangle+2\langle Al_2O_3\rangle$	4000	−6.1	0.8	500–1750m	18
$\langle 3CaO\cdot B_2O_3\rangle=\{3CaO\cdot B_2O_3\}$	35,500	20.14	—	1490m	13
$\langle 3CaO\cdot B_2O_3\rangle=3\langle CaO\rangle+\{B_2O_3\}$	66,500	7.1	—	450–1490m	13
$\langle 2CaO\cdot B_2O_3\rangle=\{2CaO\cdot B_2O_3\}$	24,100	15.22	—	1310m	13
$\langle 2CaO\cdot B_2O_3\rangle=2\langle CaO\rangle+\{B_2O_3\}$	49,900	4.1	—	450–1310m	13
$\langle CaO\cdot B_2O_3\rangle=\{CaO\cdot B_2O_3\}$	17,700	12.35	—	1160m	13
$\langle CaO\cdot B_2O_3\rangle=\langle CaO\rangle+\{B_2O_3\}$	35,700	7.4	—	450–1160m	13

TABLE 1.2 (continued)

Reaction[a]	$\Delta\tilde{H}^\circ$ (cal)	$\Delta\tilde{S}^\circ$ (cal)	Error ($\pm$kcal)	Range[b] (°C)	Reference
$\langle CaO\cdot 2B_2O_3\rangle = \{CaO\cdot 2B_2O_3\}$	27,100	21.46	—	990m	13
$\langle CaO\cdot 2B_2O_3\rangle = \langle CaO\rangle + 2\{B_2O_3\}$	53,300	21.3	—	450–990m	13
$\langle CaCO_3\rangle = \langle CaO\rangle + (CO_2)$	38,560	32.8	0.3	700–1200	19, 20
$\langle 2CaO\cdot Fe_2O_3\rangle = 2\langle CaO\rangle + \langle Fe_2O_3\rangle$	12,700	−0.6	1	700–1450m	18
$\langle CaO\cdot Fe_2O_3\rangle = \langle CaO\rangle + \langle Fe_2O_3\rangle$	7100	−1.15	1	700–1216m	18
$\langle CaO\cdot HfO_2\rangle = \langle CaO\rangle + \langle HfO_2\rangle$	8000	−3.5	5	25–1700	13
$\langle 3CaO\cdot P_2O_5\rangle = 3\langle CaO\rangle + (P_2) + \frac{5}{2}(O_2)$	553,000	133.0	—	25–1730m	13
$\langle 2CaO\cdot P_2O_5\rangle = \{2CaO\cdot P_2O_5\}$	24,100	14.82	—	1353m	13
$\langle 2CaO\cdot P_2O_5\rangle = 2\langle CaO\rangle + (P_2) + \frac{5}{2}(O_2)$	523,200	140.0	—	25–1353m	13
$\langle CaSO_4\rangle_\beta = \langle CaO\rangle + (SO_2) + \frac{1}{2}(O_2)$	110,370	56.82	0.5	950–1195	21, 22
$\langle CaSO_4\rangle_\alpha = \langle CaO\rangle + (SO_2) + \frac{1}{2}(O_2)$	108,360	55.42	0.5	1195–1365m	21, 22
$\langle 3CaO\cdot SiO_2\rangle = 3\langle CaO\rangle + \langle SiO_2\rangle$	28,400	−1.6	3	25–1500	13
$\langle 2CaO\cdot SiO_2\rangle = 2\langle CaO\rangle + \langle SiO_2\rangle$	28,400	−2.7	3	25–2130m	13
$\langle 3CaO\cdot 2SiO_2\rangle = 3\langle CaO\rangle + 2\langle SiO_2\rangle$	56,600	2.3	3	25–1500	13
$\langle CaO\cdot SiO_2\rangle = \{CaO\cdot SiO_2\}$	13,400	7.9	—	1540m	13
$\langle CaO\cdot SiO_2\rangle = \langle CaO\rangle + \langle SiO_2\rangle$	22,100	0.60	3	25–1540m	13
$\langle 3CaO\cdot 2TiO_2\rangle = 3\langle CaO\rangle + 2\langle TiO_2\rangle$	49,500	−2.75	2.5	25–1400	18
$\langle 4CaO\cdot 3TiO_2\rangle = 4\langle CaO\rangle + 3\langle TiO_2\rangle$	70,000	−4.2	2	25–1400	18
$\langle CaO\cdot TiO_2\rangle = \langle CaO\rangle + \langle TiO_2\rangle$	19,100	−0.8	0.8	25–1400	18
$\langle CaO\cdot UO_3\rangle = \langle CaO\rangle + \langle UO_3\rangle$	32,000	−1.2	10	25–900	13
$\langle 3CaO\cdot V_2O_5\rangle = 3\langle CaO\rangle + \langle V_2O_5\rangle$	79,400	0.0	1.2	25–670	18
$\langle 2CaO\cdot V_2O_5\rangle = 2\langle CaO\rangle + \langle V_2O_5\rangle$	63,300	0.0	1.2	25–670	18
$\langle CaO\cdot V_2O_5\rangle = \langle CaO\rangle + \langle V_2O_5\rangle$	34,900	0.0	1.2	25–670	18
$\langle CaO\cdot ZrO_2\rangle = \langle CaO\rangle + \langle ZrO_2\rangle$	9400	0.1	—	25–2000	13
$\langle CaO\cdot MgO\cdot SiO_2\rangle = \langle CaO\rangle + \langle MgO\rangle + \langle SiO_2\rangle$	29,800	0.9	—	25–1200	13
$\langle 2CaO\cdot MgO\cdot 2SiO_2\rangle = 2\langle CaO\rangle + \langle MgO\rangle + 2\langle SiO_2\rangle$	49,000	0	—	25–1454m	13
$\langle CaO\cdot MgO\cdot 2SiO_2\rangle = \{CaO\cdot MgO\cdot 2SiO_2\}$	30,700	18.44	—	1392m	13
$\langle CaO\cdot MgO\cdot 2SiO_2\rangle = \langle CaO\rangle + \langle MgO\rangle + 2\langle SiO_2\rangle$	38,900	4.5	—	25–1392m	13
$\langle 3CaO\cdot Al_2O_3\cdot 3SiO_2\rangle = 3\langle CaO\rangle + \langle Al_2O_3\rangle + 3\langle SiO_2\rangle$	92,900	24.0	—	25–1400	13
$\langle 2CaO\cdot Al_2O_3\cdot SiO_2\rangle = 2\langle CaO\rangle + \langle Al_2O_3\rangle + \langle SiO_2\rangle$	40,900	2.1	—	25–1500	13
$\langle CaO\cdot Al_2O_3\cdot SiO_2\rangle = \langle CaO\rangle + \langle Al_2O_3\rangle + \langle SiO_2\rangle$	25,300	3.4	—	25–1400	13
$\langle CaO\cdot Al_2O_3\cdot 2SiO_2\rangle = \{CaO\cdot Al_2O_3\cdot 2SiO_2\}$	39,900	21.85	—	1553m	13
$\langle CaO\cdot Al_2O_3\cdot 2SiO_2\rangle = \langle CaO\rangle + \langle Al_2O_3\rangle + 2\langle SiO_2\rangle$	33,200	4.1	—	25–1553m	13
$\langle CaO\cdot TiO_2\cdot SiO_2\rangle = \langle CaO\rangle + \langle TiO_2\rangle + \langle SiO_2\rangle$	29,300	2.6	—	25–1400	13
$\langle Cd\rangle = \{Cd\}$	1480	2.49	0.05	321m	15
$\{Cd\} = (Cd)$	24,300	23.39	0.2	321–767b	15
$\langle CdBr_2\rangle = \{CdBr_2\}$	7970	9.48	0.3	568m	13
$\langle CdBr_2\rangle = \{Cd\} + (Br_2)$	82,200	36.17	3	321–568m	13
$\langle CdCl_2\rangle = \{CdCl_2\}$	7200	8.56	0.3	568m	13
$\langle CdCl_2\rangle = \{Cd\} + (Cl_2)$	93,120	36.56	3	321–568m	13
$\langle CdF_2\rangle = \{Cd\} + (F_2)$	167,900	41.3	4	321–1072m	13
$\langle CdI_2\rangle = \{CdI_2\}$	4950	7.45	0.4	388m	13
$\{CdI_2\} = \{Cd\} + (I_2)$	56,800	25.56	4	388–744b	13
$\langle CdSO_4\rangle = \langle CdO\rangle + (SO_2) + \frac{1}{2}(O_2)$	89,000	66.2	5	25–1000	12
$\langle CdO\cdot SiO_2\rangle = \langle CdO\rangle + \langle SiO_2\rangle$	5870	1.13	3	25–1241m	13
$\langle CdO\cdot TiO_2\rangle_\alpha = \langle CdO\rangle + \langle TiO_2\rangle$	6700	0.2	5	25–827	13
$\langle CdO\cdot TiO_2\rangle_\alpha = \langle CdO\cdot TiO_2\rangle_\beta$	3580	3.26	0.5	827	13
$\langle CdO\rangle = \{Cd\} + \frac{1}{2}(O_2)$	62,900	25.08	3	321–767	13
$\langle CdS\rangle = \{Cd\} + \frac{1}{2}(S_2)$	51,500	23.23	3	321–767	12
$\langle CdO\cdot Al_2O_3\rangle = \langle CdO\rangle + \langle Al_2O_3\rangle$	−4000	−5.5	4	25–900	13
$\langle Ce\rangle = \{Ce\}$	1310	1.22	0.05	798m	15
$\{Ce\} = (Ce)$	98,870	26.72	1	798–3426b	15
$\langle CeH_2\rangle = \langle Ce\rangle + (H_2)$	49,800	36.73	—	25–798	13

TABLE 1.2 (continued)

Reaction[a]	$\Delta\tilde{H}°$ (cal)	$\Delta\tilde{S}°$ (cal)	Error (±kcal)	Range[b] (°C)	Reference
$\langle CeBr_3\rangle = \{CeBr_3\}$	12,400	12.34	–	732m	13
$\langle CeBr_3\rangle = \langle Ce\rangle + \frac{3}{2}(Br_2)$	221,000	56.8	10	25–732m	13
$\langle CeCl_3\rangle = \{CeCl_3\}$	12,800	11.74	–	817m	13
$\langle CeCl_3\rangle = \langle Ce\rangle + \frac{3}{2}(Cl_2)$	251,000	57.29	4	25–798	13
$\langle CeF_3\rangle = \{CeF_3\}$	13,500	7.90	–	1437m	13
$\langle CeF_3\rangle = \langle Ce\rangle + \frac{3}{2}(F_2)$	424,000	60.2	10	25–798	13
$\langle CeI_3\rangle = \{CeI_3\}$	12,200	11.81	–	760m	13
$\langle CeI_3\rangle = \langle Ce\rangle + \frac{3}{2}(I_2)$	175,000	53.5	5	25–760m	13
$\langle CeB_6\rangle = \{Ce\} + 6\langle B\rangle$	90,000	13.2	10	798–2190m	13
$\langle Ce_2C_3\rangle = 2\{Ce\} + 3\langle C\rangle$	45,000	−3.5	5	798–1200	13
$\langle CeC_2\rangle = \{Ce\} + 2\langle C\rangle$	20,370	−6.45	1.5	798–2250m	23
$\langle CeN\rangle = \{Ce\} + \frac{1}{2}(N_2)$	116,700	42.33	1	2000–2575	24
$\langle CeAl_2\rangle = \{CeAl_2\}$	19,500	11.2	–	1465m	13
$\langle CeAl_2\rangle = \{Ce\} + 2\{Al\}$	51,000	12.1	10	798–1465m	13
$\langle CeAl_4\rangle = \langle Ce\rangle + 4\langle Al\rangle$	42,000	5.1	10	25–660	13
$\langle CeSi_2\rangle = \{Ce\} + 2\langle Si\rangle$	48,000	3.9	5	798–1427m	13
$\langle Ce_2O_3\rangle = 2\langle Ce\rangle + \frac{3}{2}(O_2)$	427,350	68.50	1	25–798	25
$\langle CeO_{1.72}\rangle = \langle Ce\rangle + 0.86(O_2)$	237,000	40.6	4	25–798	13
$\langle CeO_{1.83}\rangle = \langle Ce\rangle + 0.915(O_2)$	246,000	43.5	4	25–798	13
$\langle CeO_2\rangle = \langle Ce\rangle + (O_2)$	259,000	50.63	1	25–798	25
$\langle CeS\rangle = \{Ce\} + \frac{1}{2}(S_2)$	127,840	21.74	1.5	798–2450m	23
$\langle Ce_2O_2S\rangle = 2\{Ce\} + (O_2) + \frac{1}{2}(S_2)$	423,000	79.5	4	798–1500	26
$\langle Ce_2O_3\cdot Al_2O_3\rangle = \langle Ce_2O_3\rangle + \langle Al_2O_3\rangle$	19,000	−5.0	2.5	1100–1500	18
$\langle Co\rangle = \{Co\}$	3870	2.19	0.1	1495m	15
$\{Co\} = (Co)$	92,550	28.96	1	1495–2828b	15
$\langle CoBr_2\rangle = \langle Co\rangle + (Br_2)$	59,000	31.7	5	25–678m	13
$\langle CoCl_2\rangle = \{CoCl_2\}$	10,700	10.56	2	740m	10
$\langle CoCl_2\rangle = \langle Co\rangle + (Cl_2)$	73,400	31.0	0.5	25–740m	10
$\langle CoF_2\rangle = \langle Co\rangle + (F_2)$	164,000	33.8	5	25–1250m	13
$\langle CoI_2\rangle = \langle Co\rangle + (I_2)$	36,000	31.1	5	25–520m	13
$\langle Co_2B\rangle = 2\langle Co\rangle + \langle B\rangle$	31,000	2.6	10	25–1200	13
$\langle CoB\rangle = \langle Co\rangle + \langle B\rangle$	23,000	1.8	10	25–1460m	13
$\langle CoAl\rangle = \{CoAl\}$	15,000	7.82	—	1645m	13
$\langle CoAl\rangle = \langle Co\rangle + \{Al\}$	32,000	7.1	10	660–1495	13
$\langle Co_2Al_5\rangle = 2\langle Co\rangle + 5\{Al\}$	87,000	22.0	10	660–1170m	13
$\langle CoAl_3\rangle = \langle Co\rangle + 3\langle Al\rangle$	37,000	2.0	10	25–660	13
$\langle Co_2Al_9\rangle = 2\langle Co\rangle + 9\langle Al\rangle$	76,000	3.6	10	25–660	13
$\langle Co_2P\rangle = 2\langle Co\rangle + \frac{1}{2}(P_2)$	63,000	23.4	5	25–1100	13
$\langle CoP\rangle = \langle Co\rangle + \frac{1}{2}(P_2)$	47,000	20.7	5	25–1100	13
$\langle CoP_3\rangle = \langle Co\rangle + \frac{3}{2}(P_2)$	97,000	54.5	5	25–1100	13
$\langle CoO\rangle = \langle Co\rangle + \frac{1}{2}(O_2)$	58,700	18.80	3	25–1495	12
$\langle Co_3O_4\rangle = 3\langle Co\rangle + 2(O_2)$	228,800	109.21	3	25–700	12
$\langle Co_9S_8\rangle = 9\langle Co\rangle + 4(S_2)$	316,960	159.24	3	25–788	27
$\langle Co_4S_3\rangle = 4\langle Co\rangle + \frac{3}{2}(S_2)$	92,850	34.94	1	788–877	27
$\langle CoS_2\rangle = \langle Co\rangle + (S_2)$	67,000	43.6	4	25–600	13
$\langle CoO\cdot Cr_2O_3\rangle = \langle CoO\rangle + \langle Cr_2O_3\rangle$	14,200	2.0	1	700–1400	18
$\langle CoO\cdot Fe_2O_3\rangle = \langle CoO\rangle + \langle Fe_2O_3\rangle$	5900	−1.55	1	500–1200	18
$\langle CoSO_4\rangle = \langle CoO\rangle + (SO_2) + \frac{1}{2}(O_2)$	80,000	63.2	5	25–600	13
$\langle 2CoO\cdot SiO_2\rangle = \{2CoO\cdot SiO_2\}$	24,000	14.2	–	1417m	13
$\langle 2CoO\cdot SiO_2\rangle = 2\langle CoO\rangle + \langle SiO_2\rangle$	6700	1.0	3	25–1417m	13
$\langle 2CoO\cdot TiO_2\rangle = 2\langle CoO\rangle + \langle TiO_2\rangle$	5320	−0.26	3	25–1575m	13
$\langle CoO\cdot TiO_2\rangle = \langle CoO\rangle + \langle TiO_2\rangle$	5900	1.50	0.8	500–1400	18

TABLE 1.2 (continued)

Reaction[a]	$\Delta\tilde{H}°$ (cal)	$\Delta\tilde{S}°$ (cal)	Error ($\pm$kcal)	Range[b] (°C)	Reference
$\langle Cr\rangle=\{Cr\}$	4050	1.9	—	1857m	15
$\{Cr\}=(Cr)$	83,290	28.29	1	1857–2672b	15
$\langle CrBr_2\rangle=\langle Cr\rangle+(Br_2)$	79,000	29.4	10	25–842m	13
$\langle CrCl_2\rangle=\{CrCl_2\}$	7700	7.08	1.5	815m	13
$\langle CrCl_2\rangle=\langle Cr\rangle+(Cl_2)$	93,000	28.6	4	25–815m	13
$\langle CrCl_3\rangle=\langle Cr\rangle+\frac{3}{2}(Cl_2)$	131,000	51.5	5	25–945s	13
$\langle CrF_2\rangle=\langle Cr\rangle+(F_2)$	185,000	32.1	5	25–1100m	13
$\langle CrF_3\rangle=\langle Cr\rangle+\frac{3}{2}(F_2)$	265,000	53.9	5	25–1100m	13
$\langle CrI_2\rangle=\langle Cr\rangle+(I_2)$	52,000	24.7	5	25–793m	13
$\langle CrI_3\rangle=\langle Cr\rangle+\frac{3}{2}(I_2)$	70,000	47.4	5	25–600	13
$\langle CrB\rangle=\langle Cr\rangle+\langle B\rangle$	18,000	1.6	10	25–1000	13
$\langle CrB_2\rangle=\langle Cr\rangle+2\langle B\rangle$	22,000	2.0	10	25–1000	13
$\langle Cr_4C\rangle=4\langle Cr\rangle+\langle C\rangle$	23,000	−2.8	—	25–1520m	12
$\langle Cr_{23}C_6\rangle=23\langle Cr\rangle+6\langle C\rangle$	74,000	−18.5	10	25–1500	10
$\langle Cr_7C_3\rangle=7\langle Cr\rangle+3\langle C\rangle$	36,700	−8.9	5	25–1857	10
$\langle Cr_3C_2\rangle=3\langle Cr\rangle+2\langle C\rangle$	18,900	−4.22	3	25–1857	10
$\langle Cr_2N\rangle=2\langle Cr\rangle+\frac{1}{2}(N_2)$	23,710	11.23	0.2	1000–1400	28–30
$\langle CrN\rangle=\langle Cr\rangle+\frac{1}{2}(N_2)$	27,100	17.50	3	25–500	10
$\langle Cr_3Si\rangle=3\langle Cr\rangle+\langle Si\rangle$	25,400	0.81	3	25–1412	13
$\langle Cr_5Si_3\rangle=5\langle Cr\rangle+3\langle Si\rangle$	52,400	−3.72	3	25–1412	13
$\langle CrSi\rangle=\langle Cr\rangle+\langle Si\rangle$	12,800	−0.93	3	25–1412	13
$\langle CrSi_2\rangle=\langle Cr\rangle+2\langle Si\rangle$	19,000	0.47	3	25–1412	13
$\langle Cr_2O_3\rangle=2\langle Cr\rangle+\frac{3}{2}(O_2)$	265,330	59.11	0.2	900–1650	31–33
$\langle Cr_3O_4\rangle=3\langle Cr\rangle+2(O_2)$	323,900	63.25	0.2	1650–1665m	33
$\{CrO\}=\langle Cr\rangle+\frac{1}{2}(O_2)$	79,880	15.25	0.2	1665–1750	33
$\langle CrS\rangle=\langle Cr\rangle+\frac{1}{2}(S_2)$	48,400	13.4	2	1100–1300	34
$\langle Cr_2(SO_4)_3\rangle=\langle Cr_2O_3\rangle+3(SO_2)+\frac{3}{2}(O_2)$	213,120	196.0	0.3	600–800	35
$\langle Cs\rangle=\{Cs\}$	500	1.66	0.02	29m	15
$\{Cs\}=(Cs)$	17,140	18.21	0.1	29–771b	15
$\langle Cu\rangle=\{Cu\}$	3120	2.30	0.4	1083m	15
$\{Cu\}=(Cu)$	73,650	26.02	0.4	1083–2563b	15
$(CuBr)=\{Cu\}+\frac{1}{2}(Br_2)$	−28,000	−15.8	—	1083–2000	12
$(CuCl)=\{Cu\}+\frac{1}{2}(Cl_2)$	−15,120	−15.96	1	1083–2000	8
$(Cu_3Cl_3)=3\{Cu\}+\frac{3}{2}(Cl_2)$	74,990	11.24	1	1083–2000	8
$(CuF)=\{Cu\}+\frac{1}{2}(F_2)$	−5000	−15.8	10	1083–2000	8
$(CuI)=\{Cu\}+\frac{1}{2}(I_2)$	−48,000	−15.8	—	1083–2000	12
$\langle Cu_3P\rangle=3\langle Cu\rangle+\frac{1}{2}(P_2)$	45,000	20.1	5	25–1030m	13
$\langle CuP_2\rangle=\langle Cu\rangle+(P_2)$	57,000	37.0	5	25–800	13
$\langle Cu_2O\rangle=2\langle Cu\rangle+\frac{1}{2}(O_2)$	40,250	17.03	1	25–1083	8
$\langle Cu_2O\rangle=\{Cu_2O\}$	13,580	9.00	2	1236m	8
$\{Cu_2O\}=2\{Cu\}+\frac{1}{2}(O_2)$	28,380	9.43	1	1236–2000	8
$\langle CuO\rangle=\langle Cu\rangle+\frac{1}{2}(O_2)$	36,390	20.40	1	25–1083	8
$\langle Cu_2S\rangle_\gamma=2\langle Cu\rangle+\frac{1}{2}(S_2)$	33,630	10.36	0.3	25–435	36
$\langle Cu_2S\rangle_\mu=2\langle Cu\rangle+\frac{1}{2}(S_2)$	31,500	7.36	0.3	435–1129m	36
$\langle Cu_2S\rangle=\{Cu_2S\}$	2150	1.53	0.2	1129m	36
$\langle Cu_{1.738}S\rangle_\mu=1.738\langle Cu\rangle+\frac{1}{2}(S_2)$	27,160	6.35	0.3	435–620	36
$\langle CuS\rangle=\langle Cu\rangle+\frac{1}{2}(S_2)$	27,630	18.18	0.5	25–430	37
$\langle CuFeS_2\rangle_\alpha=\langle Cu\rangle+\langle Fe\rangle+(S_2)$	66,590	27.56	0.3	557–700	38
$\langle Cu_2O\cdot Fe_2O_3\rangle=\langle Cu_2O\rangle+\langle Fe_2O_3\rangle$	9000	4.5	4	25–1100	13

TABLE 1.2 (continued)

Reaction[a]	$\Delta\tilde{H}^\circ$ (cal)	$\Delta\tilde{S}^\circ$ (cal)	Error ($\pm$kcal)	Range[b] (°C)	Reference
$\langle CuSO_4\rangle = \frac{1}{2}\langle CuO\cdot CuSO_4\rangle + \frac{1}{2}(SO_2) + \frac{1}{4}(O_2)$	36,480	32.50	0.2	400–800	39, 40
$\langle CuO\cdot CuSO_4\rangle = 2\langle CuO\rangle + (SO_2) + \frac{1}{2}(O_2)$	70,890	59.80	0.2	500–900	39
$\langle Fe\rangle_\delta = \{Fe\}$	3300	1.82	0.2	1536m	15
$\{Fe\} = (Fe)$	86,900	27.78	0.3	1536–2862b	15
$\langle FeBr_2\rangle = \{FeBr_2\}$	12,000	12.45	3	691m	8
$\{FeBr_2\} = (FeBr_2)$	29,400	24.4	5	934b	8
$(FeBr_2) = \langle Fe\rangle + (Br_2)$	22,200	−10.54	3	934–2000	8
$\langle FeCl_2\rangle = \{FeCl_2\}$	10,280	10.82	0.05	677m	9
$\{FeCl_2\} = (FeCl_2)$	26,270	20.25	2	1074b	9
$(FeCl_2) = \langle Fe\rangle + (Cl_2)$	39,950	−6.00	1	1074–2000	9
$(FeCl_3) = \langle Fe\rangle + \frac{3}{2}(Cl_2)$	62,120	6.32	1	332–2000	8
$(Fe_2Cl_6) = 2\langle Fe\rangle + 3(Cl_2)$	156,600	43.46	3	332–2000	8
$\langle FeF_2\rangle = \langle Fe\rangle + (F_2)$	168,000	32.0	10	25–1100m	8
$(FeI_2) = \langle Fe\rangle + (I_2)$	−2000	−10.5	10	25–1536	8
$\langle Fe_2B\rangle = 2\langle Fe\rangle + \langle B\rangle$	21,000	4.4	5	25–1390m	13
$\langle FeB\rangle = \langle Fe\rangle + \langle B\rangle$	19,000	2.5	5	25–1650m	13
$\langle Fe_3C\rangle = 3\langle Fe\rangle_\alpha + \langle C\rangle$	−6940	−6.70	0.1	25–727	41
$\langle Fe_3C\rangle = 3\langle Fe\rangle_\gamma + \langle C\rangle$	−2685	−2.63	0.1	727–1137	41
$\langle Fe_4N\rangle = 4\langle Fe\rangle_\gamma + \frac{1}{2}(N_2)$	8000	16.68	0.3	400–680	42
$\langle Fe_{0.947}O\rangle = 0.947\langle Fe\rangle + \frac{1}{2}(O_2)$	63,030	15.38	0.2	25–1371m	8
$\langle Fe_{0.947}O\rangle = \{Fe_{0.947}O\}$	7490	4.54	0.2	1371m	8
$\{FeO\} = \{Fe\} + \frac{1}{2}(O_2)$	61,200	12.83	1	1371–2000	8
$\langle Fe_3O_4\rangle = 3\langle Fe\rangle + 2(O_2)$	263,430	73.46	0.5	25–1597m	8
$\langle Fe_2O_3\rangle = 2\langle Fe\rangle + \frac{3}{2}(O_2)$	194,580	59.91	0.5	25–1500	8
$\langle FeS\rangle = \langle Fe\rangle_\gamma + \frac{1}{2}(S_2)$	37,030	13.59	0.5	906–988	43
$\langle FeS\rangle = \{Fe\} + \frac{1}{2}(S_2)$	39,200	14.60	0.5	988–1195m	43
$\langle FeS_2\rangle = \langle FeS\rangle + \frac{1}{2}(S_2)$	43,500	44.85	1	630–760	27
$\langle FeO\cdot Al_2O_3\rangle = \{Fe_xO\} + (1-x)\{Fe\} + \langle Al_2O_3\rangle$	7920	1.46	1.2	1550–1750	44
$\langle FeO\cdot Cr_2O_3\rangle = \langle Fe\rangle + \frac{1}{2}(O_2) + \langle Cr_2O_3\rangle$	75,700	17.35	0.3	750–1536	45
$\langle FeO\cdot Cr_2O_3\rangle = \{Fe\} + \frac{1}{2}(O_2) + \langle Cr_2O_3\rangle$	79,000	19.20	0.3	1536–1700	45
$\langle Fe_2(SO_4)_3\rangle = \langle Fe_2O_3\rangle + 3(SO_2) + \frac{3}{2}(O_2)$	184,590	173.01	0.6	400–800	39, 40
$\langle Fe_2(SO_4)_3\rangle = 2\langle FeSO_4\rangle + (SO_2) + (O_2)$	94,560	84.04	0.2	430–630	40
$\langle FeSO_4\rangle = \frac{1}{2}\langle Fe_2O_3\rangle + (SO_2) + \frac{1}{4}(O_2)$	48,630	48.36	0.1	500–630	40
$\langle 2FeO\cdot SiO_2\rangle = \{2FeO\cdot SiO_2\}$	22,000	14.74	1	1220m	13
$\langle 2FeO\cdot SiO_2\rangle = 2\langle FeO\rangle + \langle SiO_2\rangle$	8660	5.04	1	25–1220m	13
$\langle 2FeO\cdot TiO_2\rangle = 2\langle FeO\rangle + \langle TiO_2\rangle$	8100	1.4	2	25–1100	18
$\langle FeO\cdot TiO_2\rangle = \langle FeO\rangle + \langle TiO_2\rangle$	8000	2.9	1	25–1300	18
$\langle FeO\cdot V_2O_3\rangle = \langle Fe\rangle + \frac{1}{2}(O_2) + \langle V_2O_3\rangle$	69,000	14.90	0.3	750–1536	45
$\langle FeO\cdot V_2O_3\rangle = \{Fe\} + \frac{1}{2}(O_2) + \langle V_2O_3\rangle$	72,000	16.73	0.3	1536–1700	45
$\langle FeO\cdot WO_3\rangle = \langle FeO\rangle + \langle WO_3\rangle$	13,100	−2.4	1.2	700–1150	18
$\langle Ga\rangle = \{Ga\}$	1340	4.11	0.04	30m	15
$\{Ga\} = (Ga)$	63,120	25.54	1	30–2205b	15
$\langle Ga_2O_3\rangle = 2\{Ga\} + \frac{3}{2}(O_2)$	260,500	77.34	3	30–1795m	13
$\langle GaS\rangle = \{Ga\} + \frac{1}{2}(S_2)$	66,000	26.5	5	30–960m	13
$\langle Ga_2S_3\rangle = 2\{Ga\} + \frac{3}{2}(S_2)$	172,000	76.1	5	30–1090m	13
$\langle Ge\rangle = \{Ge\}$	8830	7.30	0.5	937m	15
$\{Ge\} = (Ge)$	80,070	25.80	0.5	938–2834b	15
$\langle GeO_2\rangle = \langle Ge\rangle + (O_2)$	137,400	44.88	3	25–938	13

TABLE 1.2 (continued)

Reaction[a]	$\Delta\tilde{H}^\circ$ (cal)	$\Delta\tilde{S}^\circ$ (cal)	Error ($\pm$kcal)	Range[b] (°C)	Reference
$(HBr)=\frac{1}{2}(H_2)+\frac{1}{2}(Br_2)$	12,820	−1.64	0.2	25–2000	8
$(HCl)=\frac{1}{2}(H_2)+\frac{1}{2}(Cl_2)$	22,490	−1.53	0.2	25–2000	8
$(HF)=\frac{1}{2}(H_2)+\frac{1}{2}(F_2)$	65,600	−0.83	0.4	25–2000	8
$(HI)=\frac{1}{2}(H_2)+\frac{1}{2}(I_2)$	1000	−2.11	0.4	25–2000	8
$\{H_2O\}=(H_2O)$	9820	26.32	0.03	100b	8
$(H_2O)=(H_2)+\frac{1}{2}(O_2)$	59,150	13.35	0.3	25–2000	8
$(H_2S)=(H_2)+\frac{1}{2}(S_2)$	21,900	12.09	0.3	25–2000	8
$\langle Hf\rangle=\{Hf\}$	5800	2.3	–	2230m	15
$\{Hf\}=(Hf)$	137,240	28.14	1	2230–4600b	15
$\langle HfB_2\rangle=\langle Hf\rangle+2\langle B\rangle$	80,000	2.66	4	25–1200	13
$\langle HfC\rangle=\langle Hf\rangle+\langle C\rangle$	55,000	1.8	5	25–2000	12
$\langle HfO_2\rangle_\alpha=\langle Hf\rangle+(O_2)$	254,000	41.6	5	25–1700	12
$\langle HfO_2\rangle_\alpha=\langle HfO_2\rangle_\beta$	2500	1.27	–	1700	12
$\{Hg\}=(Hg)$	14,520	23.13	0.06	−39–357	15
$\langle In\rangle=\{In\}$	780	1.82	0.03	157m	15
$\{In\}=(In)$	56,150	23.97	0.3	157–2073b	15
$\langle In_2O_3\rangle=2\{In\}+\frac{3}{2}(O_2)$	219,600	73.95	3	157–1910m	13
$\langle InS\rangle=\{InS\}$	8600	8.91	–	692m	13
$\{InS\}=\{In\}+\frac{1}{2}(S_2)$	37,000	14.5	5	692–1500	13
$\langle In_2S_3\rangle=2\{In\}+\frac{3}{2}(S_2)$	130,000	68.4	5	157–900	13
$\langle Ir\rangle=\{Ir\}$	6250	2.3	–	2443m	15
$\{Ir\}=(Ir)$	146,930	31.29	2	2443–4428b	15
$(IrO_3)=\langle Ir\rangle+\frac{3}{2}(O_2)$	−4000	10.8	4	25–1500	13
$\langle IrO_2\rangle=\langle Ir\rangle+(O_2)$	56,000	40.5	4	25–1000	12
$\langle Ir_2S_3\rangle=2\langle Ir\rangle+\frac{3}{2}(S_2)$	102,600	71.4	4	25–1000	13
$\langle IrS_2\rangle=\langle Ir\rangle+(S_2)$	64,000	45.5	4	25–1000	13
$\langle K\rangle=\{K\}$	558	1.66	0.08	63m	15
$\{K\}=(K)$	20,190	19.60	0.1	63–759b	15
$\langle KBr\rangle=\{KBr\}$	6100	6.06	0.1	734m	8
$\{KBr\}=(K)+\frac{1}{2}(Br_2)$	107,480	31.43	0.3	734–1398b	8
$(KBr)=(K)+\frac{1}{2}(Br_2)$	68,500	8.42	0.5	1398–2000	8
$\langle KCl\rangle=\{KCl\}$	6280	6.02	0.1	771m	8
$\{KCl\}=(K)+\frac{1}{2}(Cl_2)$	113,300	31.51	0.1	771–1437b	8
$(KCl)=(K)+\frac{1}{2}(Cl_2)$	73,200	8.48	0.2	1437–2000	8
$\langle KF\rangle=\{KF\}$	6500	5.75	0.3	858m	8
$\{KF\}=(K)+\frac{1}{2}(F_2)$	145,190	33.97	0.1	858–1517b	8
$(KF)=(K)+\frac{1}{2}(F_2)$	100,000	9.01	0.5	1517–2000	8
$\langle K_3AlF_6\rangle=3\{K\}+\langle Al\rangle+3(F_2)$	794,000	130.2	5	63–660	8
$\langle KI\rangle=\{KI\}$	5740	6.02	0.1	681m	8
$\{KI\}=(K)+\frac{1}{2}(I_2)$	95,470	30.37	0.1	681–1345b	8
$(KI)=(K)+\frac{1}{2}(I_2)$	59,390	8.49	0.5	1345–2000	8
$\langle KCN\rangle=\{KCN\}$	3500	3.91	0.1	622m	8
$\{KCN\}=\frac{1}{2}(KCN)_2$	60,800	42.19	1.0	622–1132b	46
$\{KCN\}=(K)+\langle C\rangle+\frac{1}{2}(N_2)$	41,000	22.34	4	622–1132b	8
$\langle K_2O\rangle=2\{K\}+\frac{1}{2}(O_2)$	86,490	33.00	0.5	63–881d	8
$\langle K_2O_2\rangle=2\{K\}+(O_2)$	118,600	53.86	1	63–490m	8
$\langle KO_2\rangle=\{K\}+(O_2)$	67,300	35.84	3	63–402m	8
$\langle KOH\rangle=\{KOH\}$	2240	3.33	0.3	400m	8

TABLE 1.2 (continued)

Reaction[a]	$\Delta\tilde{H}^\circ$ (cal)	$\Delta\tilde{S}^\circ$ (cal)	Error ($\pm$kcal)	Range[b] (°C)	Reference
$\{KOH\}=(K)+\frac{1}{2}(H_2)+\frac{1}{2}(O_2)$	112,680	44.33	0.1	400–1327b	9
$\langle K_2CO_3\rangle=\{K_2CO_3\}$	6600	5.62	–	901m	8
$\{K_2CO_3\}=2(K)+\langle C\rangle+\frac{3}{2}(O_2)$	291,330	86.89	1	901–1700	8
$\langle K_2S\rangle=\{K_2S\}$	3860	3.16	–	948m	13
$\langle K_2S\rangle=2\{K\}+\frac{1}{2}(S_2)$	115,000	34.3	8	25–759	13
$\langle K_2O\cdot 2B_2O_3\rangle=\{K_2O\cdot 2B_2O_3\}$	24,900	22.89	1	815m	8
$\langle K_2O\cdot 2B_2O_3\rangle=\langle K_2O\rangle+2\{B_2O_3\}$	125,000	30.6	5	450–815m	8
$\langle K_2O\cdot 3B_2O_3\rangle=\langle K_2O\rangle+3\{B_2O_3\}$	142,000	45.4	5	450–825d	8
$\langle K_2O\cdot 4B_2O_3\rangle=\{K_2O\cdot 4B_2O_3\}$	29,900	26.46	1	857m	8
$\langle K_2O\cdot 4B_2O_3\rangle=\langle K_2O\rangle+4\{B_2O_3\}$	161,000	59.6	5	450–857m	8
$\langle K_2SO_4\rangle=\{K_2SO_4\}$	8480	6.32	0.1	1069m	9
$\langle K_2SO_4\rangle=\langle K_2O\rangle+(SO_2)+\frac{1}{2}(O_2)$	171,130	46.88	1	25–1069m	9
$\langle K_2O\cdot SiO_2\rangle=\{K_2O\cdot SiO_2\}$	12,000	9.61	–	976m	9
$\langle K_2O\cdot SiO_2\rangle=\langle K_2O\rangle+\langle SiO_2\rangle$	66,900	−0.11	3	25–976m	9
$\langle La\rangle=\{La\}$	1480	1.24	0.1	920m	15
$\{La\}=(La)$	99,500	26.96	1	920–3457b	15
$\langle LaBr_3\rangle=\{LaBr_3\}$	12,900	12.16	–	788m	13
$\langle LaBr_3\rangle=\langle La\rangle+\frac{3}{2}(Br_2)$	217,000	53.5	10	25–788m	13
$\langle LaCl_3\rangle=\{LaCl_3\}$	13,000	11.53	–	855m	12
$\langle LaCl_3\rangle=\langle La\rangle+\frac{3}{2}(Cl_2)$	254,000	53.9	5	25–855m	12
$\langle LaF_3\rangle=\{LaF_3\}$	12,000	6.80	–	1493m	12
$\langle LaF_3\rangle=\{La\}+\frac{3}{2}(F_2)$	426,000	57.7	5	25–920	13
$\langle LaI_3\rangle=\{LaI_3\}$	13,400	12.75	–	778m	13
$\langle LaI_3\rangle=\langle La\rangle+\frac{3}{2}(I_2)$	178,000	51.6	–	25–778m	13
$\langle LaN\rangle=\langle La\rangle+\frac{1}{2}(N_2)$	71,000	25.3	10	25–920	13
$\langle La_2O_3\rangle=2\langle La\rangle+\frac{3}{2}(O_2)$	427,000	66.51	3	25–920	12
$\langle LaS\rangle=\{La\}+\frac{1}{2}(S_2)$	126,000	24.9	10	920–1500	13
$\langle La_2S_3\rangle=2\{La\}+\frac{3}{2}(S_2)$	339,000	68.3	10	920–1500	13
$\langle Li\rangle=\{Li\}$	717	1.58	0.03	181m	15
$\{Li\}=(Li)$	36,160	22.43	0.3	181–1342b	15
$\langle LiBr\rangle=\{LiBr\}$	4220	5.13	0.2	550m	8
$\{LiBr\}=\{Li\}+\frac{1}{2}(Br_2)$	81,540	11.62	1	550–1289b	8
$(LiBr)=(Li)+\frac{1}{2}(Br_2)$	79,800	10.06	3	1289–2000	8
$\langle LiCl\rangle=\{LiCl\}$	4740	5.49	0.1	610m	8
$\{LiCl\}=\{Li\}+\frac{1}{2}(Cl_2)$	91,310	12.52	0.1	610–1383b	8
$(LiCl)=(Li)+\frac{1}{2}(Cl_2)$	86,100	10.09	3	1383–2000	8
$\langle LiF\rangle=\{LiF\}$	6470	5.77	0.01	848m	8
$\{LiF\}=\{Li\}+\frac{1}{2}(F_2)$	139,440	15.97	0.1	848–1342	8
$(LiF)=(Li)+\frac{1}{2}(F_2)$	119,300	10.28	2	1717b–2000	8
$\langle Li_3AlF_6\rangle=\{Li_3AlF_6\}$	20,600	19.47	1	785m	8
$\{Li_3AlF_6\}=3\{Li\}+\{Al\}+3(F_2)$	775,120	95.71	1.5	785–1342	8
$\langle LiI\rangle=\{LiI\}$	3500	4.72	0.1	449m	8
$\{LiI\}=\{Li\}+\frac{1}{2}(I_2)$	67,060	11.17	0.5	449–1176b	8
$(LiI)=(Li)+\frac{1}{2}(I_2)$	68,460	9.94	2	1342–2000	8
$\langle Li_3N\rangle=3\{Li\}+\frac{1}{2}(N_2)$	48,000	38.1	5	181–1000	8
$\langle Li_2O\rangle=\{Li_2O\}$	14,000	7.6	—	1570m	8
$\langle Li_2O\rangle=2\{Li\}+\frac{1}{2}(O_2)$	144,060	32.30	0.5	181–1342	8
$\langle Li_2S\rangle=2\{Li\}+\frac{1}{2}(S_2)$	123,000	29.0	5	181–1000	13
$\langle LiOH\rangle=\{LiOH\}$	5010	6.73	0.1	471m	8
$\{LiOH\}=\frac{1}{2}\langle Li_2O\rangle+\frac{1}{2}(H_2O)$	6680	5.22	0.5	471–1007d	8
$\langle Li_2O\cdot Al_2O_3\rangle=\{Li_2O\cdot Al_2O_3\}$	6000	3.19	—	1610m	18

TABLE 1.2 (continued)

Reaction[a]	$\Delta\tilde{H}°$ (cal)	$\Delta\tilde{S}°$ (cal)	Error ($\pm$kcal)	Range[b] (°C)	Reference
$\langle Li_2O\cdot Al_2O_3\rangle = \langle Li_2O\rangle + \langle Al_2O_3\rangle$	25,600	−2.53	1	500–1300	18
$\langle Li_2O\cdot B_2O_3\rangle = \{Li_2O\cdot B_2O_3\}$	8090	7.42	0.1	844m	8
$\{Li_2O\cdot B_2O_3\} = \langle Li_2O\rangle + \{B_2O_3\}$	24,000	−13.11	2	844–1570	8
$\langle Li_2O\cdot 2B_2O_3\rangle = \{Li_2O\cdot 2B_2O_3\}$	26,300	22.10	1.5	917m	8
$\{Li_2O\cdot 2B_2O_3\} = \langle Li_2O\rangle + 2\{B_2O_3\}$	36,000	−6.1	5	917–1570	8
$\langle Li_2O\cdot 3B_2O_3\rangle = \langle Li_2O\rangle + 3\{B_2O_3\}$	88,000	37.5	5	25–834d	8
$\langle Li_2CO_3\rangle = \{Li_2CO_3\}$	10,700	10.77	0.1	720m	8
$\{Li_2CO_3\} = \langle Li_2O\rangle + (CO_2)$	35,340	18.82	0.2	720–1570	8
$\langle Li_2O\cdot Fe_2O_3\rangle = \langle Li_2O\rangle + \langle Fe_2O_3\rangle$	9500	−6.05	1.2	25–700	18
$\langle Li_2SO_4\rangle_\alpha = \langle Li_2O\rangle + (SO_2) + \frac{1}{2}(O_2)$	123,000	62.1	5	25–386	13
$\langle Li_2SO_4\rangle_\beta = \langle Li_2O\rangle + (SO_2) + \frac{1}{2}(O_2)$	118,000	50.2	5	386–859m	13
$\langle Li_2SO_4\rangle_\beta = \{Li_2SO_4\}$	3300	2.92	—	859m	13
$\langle Li_2O\cdot SiO_2\rangle = \{Li_2O\cdot SiO_2\}$	6700	4.55	0.5	1201m	8
$\langle Li_2O\cdot SiO_2\rangle = \langle Li_2O\rangle + \langle SiO_2\rangle$	34,400	0.9	2	25–1201m	8
$\langle Li_2O\cdot 2SiO_2\rangle = \{Li_2O\cdot 2SiO_2\}$	12,900	9.84	—	1034m	8
$\langle Li_2O\cdot 2SiO_2\rangle = \langle Li_2O\rangle + 2\langle SiO_2\rangle$	35,000	0.0	2	25–1034m	8
$\langle Li_2O\cdot TiO_2\rangle = \langle Li_2O\rangle + \langle TiO_2\rangle$	31,000	−0.8	2.5	25–900	18
$\langle Mg\rangle = \{Mg\}$	2140	2.32	0.1	649m	15
$\{Mg\} = (Mg)$	30,970	22.74	0.4	649–1090b	15
$\langle MgBr_2\rangle = \{MgBr_2\}$	9400	9.55	1	711m	10
$\langle MgBr_2\rangle = \langle Mg\rangle + (Br_2)$	131,060	35.13	1	25–649	10
$\{MgBr_2\} = (Mg) + (Br_2)$	147,400	44.5	2	711–1158b	10
$\langle MgCl_2\rangle = \{MgCl_2\}$	10,300	10.44	0.1	714m	8
$\langle MgCl_2\rangle = \langle Mg\rangle + (Cl_2)$	152,320	37.14	0.5	25–649	8
$\{MgCl_2\} = (Mg) + (Cl_2)$	155,160	37.70	0.5	714–1437b	8
$\langle MgF_2\rangle = \{MgF_2\}$	14,030	9.13	0.1	1263m	11
$\langle MgF_2\rangle = \langle Mg\rangle + (F_2)$	268,000	40.91	0.5	25–649	11
$\{MgF_2\} = (Mg) + (F_2)$	280,200	51.52	0.5	1263–2260b	11
$\langle MgI_2\rangle = \langle Mg\rangle + (I_2)$	102,000	36.8	2	25–634m	11
$\langle MgB_2\rangle = \langle Mg\rangle + 2\langle B\rangle$	22,000	2.50	3	25–649	8
$\langle MgB_4\rangle = \langle Mg\rangle + 4\langle B\rangle$	25,800	2.67	3	25–649	8
$\langle Mg_3N_2\rangle = 3\langle Mg\rangle + (N_2)$	110,000	48.5	5	25–649	12
$\langle Mg_2Si\rangle = \{Mg_2Si\}$	20,500	14.93	2.5	1100m	12
$\langle Mg_2Si\rangle = 2\{Mg\} + \langle Si\rangle$	24,000	9.4	5	649–1090	12
$\langle MgO\rangle = \langle Mg\rangle + \frac{1}{2}(O_2)$	143,630	25.71	0.3	25–649	11
$\langle MgO\rangle = (Mg) + \frac{1}{2}(O_2)$	174,380	48.77	0.2	1090–2000	11
$\langle MgS\rangle = \langle Mg\rangle + \frac{1}{2}(S_2)$	97,900	22.56	3	25–649	13
$\langle MgS\rangle = (Mg) + \frac{1}{2}(S_2)$	129,000	46.14	3	1090–1700	13
$\langle MgO\cdot Al_2O_3\rangle = \langle MgO\rangle + \langle Al_2O_3\rangle$	8500	−0.5	0.8	25–1400	18
$\langle MgCO_3\rangle = \langle MgO\rangle + (CO_2)$	27,800	41.45	2	25–402d	8
$\langle MgO\cdot Cr_2O_3\rangle = \langle MgO\rangle + \langle Cr_2O_3\rangle$	10,250	1.7	1.2	25–1500	18
$\langle MgO\cdot Fe_2O_3\rangle = \langle MgO\rangle + \langle Fe_2O_3\rangle$	4600	−0.5	0.8	700–1400	18
$\langle MgO\cdot MoO_2\rangle = \langle MgO\rangle + \langle MoO_2\rangle$	−3300	−5.40	1.5	900–1200	18
$\langle MgO\cdot MoO_3\rangle = \langle MgO\rangle + \langle MoO_3\rangle$	12,900	−3.25	1	25–795	18
$\langle MgSO_4\rangle = \langle MgO\rangle + (SO_2) + \frac{1}{2}(O_2)$	88,670	62.31	0.5	670–1050	40, 47
$\langle 2MgO\cdot SiO_2\rangle = \{2MgO\cdot SiO_2\}$	17,000	7.8	5	1898m	8
$\langle 2MgO\cdot SiO_2\rangle = 2\langle MgO\rangle + \langle SiO_2\rangle$	16,060	1.03	1.5	25–1898m	8
$\langle MgO\cdot SiO_2\rangle = \{MgO\cdot SiO_2\}$	18,000	9.7	5	1577m	8
$\langle MgO\cdot SiO_2\rangle = \langle MgO\rangle + \langle SiO_2\rangle$	9830	1.46	1.5	25–1577m	8
$\langle 3MgO\cdot P_2O_5\rangle = \{3MgO\cdot P_2O_5\}$	29,000	17.9	10	1348m	8
$\langle 3MgO\cdot P_2O_5\rangle = 3\langle MgO\rangle + (P_2) + \frac{5}{2}(O_2)$	476,300	121.9	—	25–1348m	8
$\langle 2MgO\cdot TiO_2\rangle = 2\langle MgO\rangle + \langle TiO_2\rangle$	6100	0.30	0.5	25–1500	18

TABLE 1.2 (continued)

Reaction[a]	$\Delta\tilde{H}^\circ$ (cal)	$\Delta\tilde{S}^\circ$ (cal)	Error ($\pm$kcal)	Range[b] (°C)	Reference
$\langle MgO\cdot TiO_2\rangle=\langle MgO\rangle+\langle TiO_2\rangle$	6300	0.75	0.7	25–1500	18
$\langle MgO\cdot 2TiO_2\rangle=\langle MgO\rangle+2\langle TiO_2\rangle$	6600	0.15	0.8	25–1500	18
$\langle 2MgO\cdot V_2O_5\rangle=2\langle MgO\rangle+\langle V_2O_5\rangle$	20,900	0.0	1.5	25–670	18
$\langle MgO\cdot V_2O_5\rangle=\langle MgO\rangle+\langle V_2O_5\rangle$	12,750	2.0	1.5	25–670	18
$\langle MgO\cdot WO_3\rangle=\langle MgO\rangle+\langle WO_3\rangle$	17,700	2.6	0.8	25–1200	18
$\langle Mn\rangle=\{Mn\}$	2900	1.9	—	1244m	15
$\{Mn\}=(Mn)$	56,350	24.18	1	1244–2062b	15
$\langle MnBr_2\rangle=\{MnBr_2\}$	8000	8.24	—	698m	12
$\langle MnBr_2\rangle=\langle Mn\rangle+(Br_2)$	98,600	31.2	5	25–698m	12
$(MnBr_2)=\{Mn\}+(Br_2)$	67,300	3.6	5	1244–2062	12
$\langle MnCl_2\rangle=\{MnCl_2\}$	9000	9.75	0.2	650m	12
$\langle MnCl_2\rangle=\langle Mn\rangle+(Cl_2)$	114,300	30.52	3	25–650m	12
$(MnCl_2)=\{Mn\}+(Cl_2)$	72,100	2.34	2	1244–2062	12
$\langle Mn_3C\rangle=3\{Mn\}+\langle C\rangle$	3330	−0.26	3	25–1037m	12
$\langle Mn_7C_3\rangle=7\langle Mn\rangle+3\langle C\rangle$	30,500	5.04	3	25–1200	13
$\langle Mn_5Si_3\rangle=\{Mn_5Si_3\}$	41,300	26.26	—	1300m	13
$\langle Mn_5Si_3\rangle=5\langle Mn\rangle+3\langle Si\rangle$	49,000	−3.6	4	25–1244	13
$\langle MnSi\rangle=\{MnSi\}$	14,200	9.17	—	1275m	13
$\langle MnSi\rangle=\langle Mn\rangle+\langle Si\rangle$	14,700	1.5	4	25–1244	13
$\langle MnP\rangle=\langle Mn\rangle+\frac{1}{2}(P_2)$	40,000	20.8	5	25–1147m	13
$\langle MnP_3\rangle=\langle Mn\rangle+\frac{3}{2}(P_2)$	91,000	57.4	5	25–1244	13
$\langle MnO\rangle=\langle Mn\rangle+\frac{1}{2}(O_2)$	92,940	18.24	0.5	25–1244	48
$\langle Mn_3O_4\rangle=3\langle Mn\rangle+2(O_2)$	331,000	82.32	4	25–1244	12
$\langle Mn_3O_4\rangle=3\langle MnO\rangle+\frac{1}{2}(O_2)$	55,500	27.96	0.3	925–1540m	49, 50
$\langle Mn_2O_3\rangle=2\langle Mn\rangle+\frac{3}{2}(O_2)$	228,000	61.0	4	25–1244	12
$\langle Mn_2O_3\rangle=\frac{2}{3}\langle Mn_3O_4\rangle+\frac{1}{6}(O_2)$	8380	6.71	0.5	800–1000	51, 52
$\langle MnO_2\rangle=\langle Mn\rangle+(O_2)$	124,000	43.26	3	25–510d	12
$\langle MnS\rangle=\{MnS\}$	6240	3.46	—	1530m	13
$\langle MnS\rangle=\langle Mn\rangle+\frac{1}{2}(S_2)$	70,870	18.34	0.2	700–1200	53
$\langle MnO\cdot Al_2O_3\rangle=\langle MnO\rangle+\langle Al_2O_3\rangle$	11,500	1.75	1.5	500–1200	18
$\langle MnSO_4\rangle_\alpha=\frac{1}{3}\langle Mn_3O_4\rangle+(SO_2)+\frac{1}{3}(O_2)$	70,650	54.34	0.1	700–849	40
$\langle MnSO_4\rangle_\beta=\frac{1}{3}\langle Mn_3O_4\rangle+(SO_2)+\frac{1}{3}(O_2)$	59,800	44.73	0.2	849–1030	40, 53
$\langle 2MnO\cdot SiO_2\rangle=\{2MnO\cdot SiO_2\}$	21,420	13.24	—	1345m	13
$\langle 2MnO\cdot SiO_2\rangle=2\langle MnO\rangle+\langle SiO_2\rangle$	12,800	5.91	3	25–1345m	13
$\langle MnO\cdot SiO_2\rangle=\{MnO\cdot SiO_2\}$	16,000	10.23	—	1291m	13
$\langle MnO\cdot SiO_2\rangle=\langle MnO\rangle+\langle SiO_2\rangle$	6700	0.66	3	25–1291m	13
$\langle MnO\cdot MoO_3\rangle=\langle MnO\rangle+\langle MoO_3\rangle$	14,500	0.2	2.5	25–795	18
$\langle 2MnO\cdot TiO_2\rangle=2\langle MnO\rangle+\langle TiO_2\rangle$	9000	−0.4	5	25–1450m	12
$\langle MnO\cdot TiO_2\rangle=\langle MnO\rangle+\langle TiO_2\rangle$	5900	0.3	5	25–1360m	12
$\langle MnO\cdot V_2O_5\rangle=\langle MnO\rangle+\langle V_2O_5\rangle$	15,750	0.0	1.5	25–670	18
$\langle MnO\cdot WO_3\rangle=\langle MnO\rangle+\langle WO_3\rangle$	18,400	0.2	1.5	25–1100	18
$\langle Mo\rangle=\{Mo\}$	6650	2.30	1.5	2620m	15
$\{Mo\}=(Mo)$	141,200	28.73	1	2620–4640b	15
$\langle Mo_2C\rangle=2\langle Mo\rangle+\langle C\rangle$	10,900	−1.0	—	25–1100	12
$\langle MoC\rangle=\langle Mo\rangle+\langle C\rangle$	1800	−1.3	—	25–700	12
$\langle Mo_2N\rangle=2\langle Mo\rangle+\frac{1}{2}(N_2)$	14,500	3.5	5	25–500	13
$\langle Mo_3Si\rangle=3\langle Mo\rangle+\langle Si\rangle$	28,400	0.60	3	25–1412	13
$\langle Mo_5Si_3\rangle=5\langle Mo\rangle+3\langle Si\rangle$	74,400	0.98	3	25–1412	13
$\langle MoSi_2\rangle=\langle Mo\rangle+2\langle Si\rangle$	31,700	0.67	3	25–1412	13
$\langle MoO_2\rangle=\langle Mo\rangle+(O_2)$	138,200	39.80	3	25–2000	8
$(MoO_2)=\langle Mo\rangle+(O_2)$	4400	8.1	5	25–2000	8

TABLE 1.2 (continued)

Reaction[a]	$\Delta\tilde{H}°$ (cal)	$\Delta\tilde{S}°$ (cal)	Error ($\pm$kcal)	Range[b] (°C)	Reference
$\langle MoO_3\rangle=\{MoO_3\}$	11,400	10.8	—	795m	8
$\langle MoO_3\rangle=\langle Mo\rangle+\frac{3}{2}(O_2)$	176,900	58.97	3	25–795m	8
$(MoO_3)=\langle Mo\rangle+\frac{3}{2}(O_2)$	86,000	14.2	5	25–2000	8
$\langle Mo_2S_3\rangle=2\langle Mo\rangle+\frac{3}{2}(S_2)$	142,000	63.4	4	25–1200	13
$\langle MoS_2\rangle=\{MoS_2\}$	10,900	7.48	—	1185m	13
$\langle MoS_2\rangle=\langle Mo\rangle+(S_2)$	95,000	43.5	4	25–1185m	13
$(NH_3)=\frac{1}{2}(N_2)+\frac{3}{2}(H_2)$	12,840	27.85	0.1	25–2000	8
$(NO)=\frac{1}{2}(N_2)+\frac{1}{2}(O_2)$	−21,610	−3.03	0.1	25–2000	8
$(NO_2)=\frac{1}{2}(N_2)+(O_2)$	−7720	15.14	0.3	25–2000	8
$\langle Na\rangle=\{Na\}$	620	1.67	0.04	98m	15
$\{Na\}=(Na)$	24,220	21.01	0.2	98–883b	15
$\langle NaBr\rangle=\{NaBr\}$	6240	6.12	0.1	747m	8
$\langle NaBr\rangle=\{Na\}+\frac{1}{2}(Br_2)$	90,190	21.39	0.2	98–747m	8
$(NaBr)=(Na)+\frac{1}{2}(Br_2)$	64,490	9.21	0.5	1450–2000	8
$\langle NaCl\rangle=\{NaCl\}$	6730	6.27	0.04	801m	8
$\langle NaCl\rangle=\{Na\}+\frac{1}{2}(Cl_2)$	98,380	22.25	0.1	98–801m	8
$\{NaCl\}=(Na)+\frac{1}{2}(Cl_2)$	111,000	32.00	2	801–1465b	8
$\langle NaF\rangle=\{NaF\}$	7970	6.28	0.1	996m	8
$\langle NaF\rangle=\{Na\}+\frac{1}{2}(F_2)$	137,810	25.22	0.3	98–996m	8
$\{NaF\}=(Na)+\frac{1}{2}(F_2)$	149,200	35.43	3	996–1790b	8
$\langle Na_3AlF_6\rangle=\{Na_3AlF_6\}$	25,640	19.95	2	1012m	8
$\langle Na_3AlF_6\rangle=3\{Na\}+\langle Al\rangle+3(F_2)$	790,800	132.34	1.5	98–660	8
$\{Na_3AlF_6\}=3(Na)+\{Al\}+3(F_2)$	807,400	149.0	—	1012–2000	8
$\langle NaI\rangle=\{NaI\}$	5640	6.05	0.1	660m	8
$\langle NaI\rangle=\{Na\}+\frac{1}{2}(I_2)$	73,800	17.09	0.3	98–660m	8
$\{NaI\}=(Na)+\frac{1}{2}(I_2)$	91,400	31.67	3	660–1304b	8
$\langle NaCN\rangle=\{NaCN\}$	2100	2.5	—	562m	8
$\langle NaCN\rangle=\{Na\}+\langle C\rangle+\frac{1}{2}(N_2)$	21,640	8.16	1	98–562m	8
$\{NaCN\}=(Na)+\langle C\rangle+\frac{1}{2}(N_2)$	36,400	20.0	—	883–1530b	8
$\langle Na_2O\rangle=\{Na_2O\}$	11,400	8.11	1	1132m	8
$\langle Na_2O\rangle=2\{Na\}+\frac{1}{2}(O_2)$	100,760	33.78	2	98–1132m	8
$\{Na_2O\}=2(Na)+\frac{1}{2}(O_2)$	124,000	56.1	3	1132–1950d	8
$\langle Na_2O_2\rangle=2\{Na\}+(O_2)$	123,000	52.3	3	98–675m	8
$\langle Na_2S\rangle=\{Na_2S\}$	6300	5.04	—	978m	13
$\langle Na_2S\rangle=2\{Na\}+\frac{1}{2}(S_2)$	105,000	34.4	4	98–978m	13
$\langle NaOH\rangle=\{NaOH\}$	1520	2.58	0.1	229m	8
$\{NaOH\}=\frac{1}{2}\langle Na_2O\rangle+\frac{1}{2}(H_2O)$	17,700	6.73	3	229–1132	8
$\{NaOH\}=\frac{1}{2}\{Na_2O\}+\frac{1}{2}(H_2O)$	23,100	10.44	3	1132–1389b	8
$\langle Na_2O\cdot Al_2O_3\rangle=\langle Na_2O\rangle+\langle Al_2O_3\rangle$	44,150	−0.7	3	500–1132	18
$\langle Na_2O\cdot B_2O_3\rangle=\{Na_2O\cdot B_2O_3\}$	8660	6.99	1	966m	8
$\langle Na_2O\cdot B_2O_3\rangle=\langle Na_2O\rangle+\{B_2O_3\}$	70,800	5.67	3	450–966m	8
$\langle Na_2O\cdot 2B_2O_3\rangle=\{Na_2O\cdot 2B_2O_3\}$	19,400	19.09	1.5	743m	8
$\langle Na_2O\cdot 2B_2O_3\rangle=\langle Na_2O\rangle+2\{B_2O_3\}$	88,300	17.71	3	450–743m	8
$\langle Na_2O\cdot 3B_2O_3\rangle=\langle Na_2O\rangle+3\{B_2O_3\}$	101,000	26.6	4	450–766d	8
$\langle Na_2CO_3\rangle=\langle Na_2O\rangle+(CO_2)$	71,000	28.25	0.5	25–850m	8
$\{Na_2CO_3\}=\{Na_2O\}+(CO_2)$	75,610	31.27	0.5	1132–2000	8
$\langle Na_2O\cdot Cr_2O_3\rangle=\langle Na_2O\rangle+\langle Cr_2O_3\rangle$	48,600	−1.38	3	25–1132	13
$\langle Na_2O\cdot Fe_2O_3\rangle=\langle Na_2O\rangle+\langle Fe_2O_3\rangle$	21,000	−3.5	—	25–1132	12
$\langle Na_2SO_4\rangle=\{Na_2SO_4\}$	5670	4.90	0.1	884m	8
$\langle Na_2SO_4\rangle=\langle Na_2O\rangle+(SO_2)+\frac{1}{2}(O_2)$	155,700	56.72	3	250–884m	8

TABLE 1.2 (continued)

Reaction[a]	$\Delta\tilde{H}^\circ$ (cal)	$\Delta\tilde{S}^\circ$ (cal)	Error ($\pm$kcal)	Range[b] (°C)	Reference
$\langle Na_2O\cdot SiO_2\rangle = \{Na_2O\cdot SiO_2\}$	12,380	9.09	0.3	1089m	8
$\langle Na_2O\cdot SiO_2\rangle = \langle Na_2O\rangle + \langle SiO_2\rangle$	56,800	2.11	3	25–1089m	8
$\langle Na_2O\cdot 2SiO_2\rangle = \{Na_2O\cdot 2SiO_2\}$	8500	7.41	0.4	874m	8
$\langle Na_2O\cdot 2SiO_2\rangle = \langle Na_2O\rangle + 2\langle SiO_2\rangle$	55,800	−0.92	3	25–874m	8
$\langle Na_2O\cdot TiO_2\rangle = \{Na_2O\cdot TiO_2\}$	16,800	12.89	—	1030m	13
$\langle Na_2O\cdot TiO_2\rangle = \langle Na_2O\rangle + \langle TiO_2\rangle$	50,000	−0.3	5	25–1030m	13
$\langle Na_2O\cdot 2TiO_2\rangle = \{Na_2O\cdot 2TiO_2\}$	26,200	20.83	—	985m	13
$\langle Na_2O\cdot 2TiO_2\rangle = \langle Na_2O\rangle + 2\langle TiO_2\rangle$	55,000	−0.4	5	25–985m	13
$\langle Na_2O\cdot 3TiO_2\rangle = \{Na_2O\cdot 3TiO_2\}$	37,100	26.48	—	1128m	13
$\langle Na_2O\cdot 3TiO_2\rangle = \langle Na_2O\rangle + 3\langle TiO_2\rangle$	56,000	−2.8	5	25–1128m	13
$\langle 3Na_2O\cdot V_2O_5\rangle = 3\langle Na_2O\rangle + \langle V_2O_5\rangle$	172,500	0.0	5	25–670	18
$\langle 2Na_2O\cdot V_2O_5\rangle = 2\langle Na_2O\rangle + \langle V_2O_5\rangle$	128,000	−7.0	5	25–670	18
$\langle Na_2O\cdot V_2O_5\rangle = \langle Na_2O\rangle + \langle V_2O_5\rangle$	77,800	−3.6	4	25–670	18
$\langle Nb\rangle = \{Nb\}$	6430	2.34	0.3	2477m	10
$\{Nb\} = (Nb)$	164,890	32.11	1	2477–4863b	10
$\langle NbB_2\rangle = \langle Nb\rangle + 2\langle B\rangle$	60,000	2.2	—	25–1100	13
$\langle Nb_2C\rangle = 2\langle Nb\rangle + \langle C\rangle$	46,300	2.80	3	25–1500	12
$\langle NbC\rangle = \langle Nb\rangle + \langle C\rangle$	32,720	0.58	1	25–1500	10
$\langle Nb_2N\rangle = 2\langle Nb\rangle + \frac{1}{2}(N_2)$	60,000	19.9	—	25–2400m	12
$\langle NbN\rangle = \langle Nb\rangle + \frac{1}{2}(N_2)$	55,000	18.6	—	25–2050m	12
$\langle NbO\rangle = \{NbO\}$	20,000	9.2	5	1937m	10
$\langle NbO\rangle = \langle Nb\rangle + \frac{1}{2}(O_2)$	99,000	20.7	5	25–1937m	10
$\langle Nb_2O_5\rangle = \{Nb_2O_5\}$	24,920	13.96	0.5	1512m	10
$\langle Nb_2O_5\rangle = 2\langle Nb\rangle + \frac{5}{2}(O_2)$	451,300	100.31	3	25–1512m	10
$\langle NbO_2\rangle = \{NbO_2\}$	22,000	10.1	5	2150m	10
$\langle NbO_2\rangle = \langle Nb\rangle + (O_2)$	187,300	39.89	2.5	25–2150m	10
$\langle Ni\rangle = \{Ni\}$	4180	2.42	0.3	1453m	15
$\{Ni\} = (Ni)$	92,250	29.02	0.5	1453–2914b	15
$\langle NiBr_2\rangle = (NiBr_2)$	53,700	45.0	—	919s	13
$\langle NiBr_2\rangle = \langle Ni\rangle + (Br_2)$	57,000	30.9	5	25–919s	13
$\langle NiCl_2\rangle = (NiCl_2)$	53,800	42.7	—	987s	13
$\langle NiCl_2\rangle = \langle Ni\rangle + (Cl_2)$	73,000	35.0	5	25–987s	13
$\langle NiF_2\rangle = \langle Ni\rangle + (F_2)$	156,000	35.60	3	25–1474m	13
$\langle NiI_2\rangle = \langle Ni\rangle + (I_2)$	33,000	30.5	5	25–797m	13
$\langle NiB\rangle = \langle Ni\rangle + \langle B\rangle$	42,000	2.0	5	25–1000	13
$\langle Ni_4B_3\rangle = 4\langle Ni\rangle + 3\langle B\rangle$	76,000	7.9	5	25–1580m	13
$\langle Ni_3C\rangle = 3\langle Ni\rangle + \langle C\rangle$	−9500	−4.1	5	25–500	13
$\langle Ni_3Al\rangle = 3\langle Ni\rangle + \{Al\}$	42,000	7.7	—	660–1395m	13
$\langle NiAl\rangle = \langle Ni\rangle + \{Al\}$	33,000	6.7	—	660–1453m	13
$\langle NiSi\rangle = \{NiSi\}$	10,300	8.1	—	992m	13
$\langle NiSi\rangle = \langle Ni\rangle + \langle Si\rangle$	21,800	1.9	4	25–992m	13
$\langle Ni_7Si_{13}\rangle = 7\langle Ni\rangle + 13\langle Si\rangle$	151,000	4.7	4	25–972m	13
$\langle Ni_3Sn_2\rangle = 3\langle Ni\rangle + 2\{Sn\}$	53,000	19.0	5	232–1264m	13
$\langle Ni_3Ti\rangle = 3\langle Ni\rangle + \langle Ti\rangle$	35,000	6.3	5	25–1378m	13
$\langle NiTi\rangle = \langle Ni\rangle + \langle Ti\rangle$	16,000	2.8	5	25–1240m	13
$\langle NiO\rangle = \langle Ni\rangle + \frac{1}{2}(O_2)$	56,310	20.57	0.3	25–1984m	54
$\langle Ni_3S_2\rangle = 3\langle Ni\rangle + (S_2)$	79,240	39.01	2	25–790m	13
$\langle NiS\rangle = \langle Ni\rangle + \frac{1}{2}(S_2)$	34,980	17.20	1.5	25–500	13
$\langle NiO\cdot Al_2O_3\rangle = \langle NiO\rangle + \langle Al_2O_3\rangle$	1000	−3.0	1.5	700–1300	18
$(Ni(CO)_4) = \langle Ni\rangle + 4(CO)$	38,000	97.6	—	25–500	13
$\langle NiO\cdot Cr_2O_3\rangle = \langle NiO\rangle + \langle Cr_2O_3\rangle$	12,800	2.0	3	700–1200	18
$\langle NiO\cdot Fe_2O_3\rangle = \langle NiO\rangle + \langle Fe_2O_3\rangle$	4750	−0.9	1	582–1400	18
$\langle NiSO_4\rangle = \langle NiO\rangle + (SO_2) + \frac{1}{2}(O_2)$	83,040	70.08	0.2	600–860	40

TABLE 1.2 (continued)

Reaction[a]	$\Delta\tilde{H}^\circ$ (cal)	$\Delta\tilde{S}^\circ$ (cal)	Error ($\pm$kcal)	Range[b] (°C)	Reference
$\langle 2NiO\cdot SiO_2\rangle = 2\langle NiO\rangle + \langle SiO_2\rangle$	3700	2.2	4	25–1545m	13
$\langle NiO\cdot TiO_2\rangle = \langle NiO\rangle + \langle TiO_2\rangle$	4300	2.0	0.8	500–1400	18
$\langle NiO\cdot WO_3\rangle = \langle NiO\rangle + \langle WO_3\rangle$	11,000	0.0	1.5	500–1100	18
$\langle Os\rangle = \{Os\}$	7600	2.3	—	3030m	15
$\{Os\} = (Os)$	178,000	33.74	1	3030–5010b	15
$\langle OsO_2\rangle = \langle Os\rangle + (O_2)$	69,260	41.86	3	25–900	13
$(OsO_4) = \langle Os\rangle + 2(O_2)$	79,800	34.72	3	131b–70	13
$\langle OsS_2\rangle = \langle Os\rangle + (S_2)$	64,800	46.68	3	25–1100	13
$\langle P\rangle$white $= \{P\}$	157	0.49	0	44m	15
$\langle P\rangle$red $= \frac{1}{4}(P_4)$	7680	10.91	0.3	25–431s	15
$(P_4) = 2(P_2)$	51,900	33.22	0.5	25–1700	15
$(PH_3) = \frac{1}{2}(P_2) + \frac{3}{2}(H_2)$	17,100	25.86	—	25–1700	13
$(PBr_3) = \frac{1}{2}(P_2) + \frac{3}{2}(Br_2)$	62,500	28.8	—	25–1700	13
$(PCl_3) = \frac{1}{2}(P_2) + \frac{3}{2}(Cl_2)$	113,400	50.00	3	25–1300	13
$(PCl_5) = \frac{1}{2}(P_2) + \frac{5}{2}(Cl_2)$	100,300	66.75	3	25–1700	13
$(PF_3) = \frac{1}{2}(P_2) + \frac{3}{2}(F_2)$	246,000	32.8	5	25–1700	13
$(PF_5) = \frac{1}{2}(P_2) + \frac{5}{2}(F_2)$	396,800	73.24	3	25–1700	13
$(PO) = \frac{1}{2}(P_2) + \frac{1}{2}(O_2)$	18,600	−2.77	—	25–1700	13
$(PO_2) = \frac{1}{2}(P_2) + (O_2)$	92,200	14.4	—	25–1700	13
$(P_4O_{10}) = 2(P_2) + 5(O_2)$	754,300	241.6	—	358s–1700	13
$\langle Pb\rangle = \{Pb\}$	1150	1.91	0.1	328m	15
$\{Pb\} = (Pb)$	43,490	21.54	0.1	328–1750b	15
$\langle PbBr_2\rangle = \{PbBr_2\}$	3930	6.10	0.3	371m	10
$\{PbBr_2\} = \{Pb\} + (Br_2)$	66,800	24.54	1	371–912b	10
$(PbBr_2) = \{Pb\} + (Br_2)$	36,100	−1.73	3	912–1750	10
$\langle PbCl_2\rangle = \{PbCl_2\}$	5230	6.75	0.2	501m	10
$\{PbCl_2\} = \{Pb\} + (Cl_2)$	77,570	24.73	0.5	501–953b	10
$(PbCl_2) = \{Pb\} + (Cl_2)$	45,000	−1.8	10	953–1750	10
$\langle PbF_2\rangle = \{PbF_2\}$	3520	3.19	0.3	830m	10
$\{PbF_2\} = \{Pb\} + (F_2)$	150,000	86.08	3	830–1200	10
$\langle PbI_2\rangle = \{PbI_2\}$	5600	8.20	0.2	410m	10
$\{PbI_2\} = \{Pb\} + (I_2)$	49,100	24.14	1	410–832b	10
$\langle PbO\rangle = \{PbO\}$	6570	5.67	0.2	885m	9
$\{PbO\} = \{Pb\} + \frac{1}{2}(O_2)$	43,300	16.26	0.6	885–1535b	9
$\langle Pb_3O_4\rangle = 3\{Pb\} + 2(O_2)$	167,900	88.18	3	328–1200	13
$\langle PbO_2\rangle = \{Pb\} + (O_2)$	65,100	46.29	3	328–900	13
$\langle PbS\rangle = \{PbS\}$	4500	3.79	1.5	1113m	10
$\langle PbS\rangle = \{Pb\} + \frac{1}{2}(S_2)$	39,000	21.04	1	328–1113m	10
$\langle PbO\cdot B_2O_3\rangle = \{PbO\} + \{B_2O_3\}$	24,500	7.9	5	885–1535	8
$\langle PbO\cdot 2B_2O_3\rangle = \{PbO\} + 2\{B_2O_3\}$	39,800	19.0	5	885–1535	8
$\langle PbO\cdot MoO_3\rangle = \langle PbO\rangle + \langle MoO_3\rangle$	9800	−6.2	5	25–885	13
$\langle PbSO_4\rangle = \langle PbO\rangle + (SO_2) + \frac{1}{2}(O_2)$	95,900	62.5	—	25–1090m	12
$\langle 2PbO\cdot SiO_2\rangle = \{2PbO\cdot SiO_2\}$	12,200	12.01	—	743m	12
$\{2PbO\cdot SiO_2\} = 2\{PbO\} + \langle SiO_2\rangle$	8000	−1.6	5	885–1500	12
$\langle PbO\cdot SiO_2\rangle = \{PbO\cdot SiO_2\}$	6220	6.00	—	764m	12
$\{PbO\cdot SiO_2\} = \{PbO\} + \langle SiO_2\rangle$	6000	0.3	3	885–1500	12
$\langle PbO\cdot TiO_2\rangle = \langle PbO\rangle + \langle TiO_2\rangle$	7300	−1.1	—	25–885	12
$\langle 3PbO\cdot V_2O_5\rangle = 3\langle PbO\rangle + \langle V_2O_5\rangle$	42,500	0.0	2.5	25–670	18
$\langle 2PbO\cdot V_2O_5\rangle = 2\langle PbO\rangle + \langle V_2O_5\rangle$	35,600	0.0	2	25–670	18
$\langle PbO\cdot WO_3\rangle = \langle PbO\rangle + \langle WO_3\rangle$	14,300	−5.45	1.2	25–1100	18

TABLE 1.2 (continued)

Reaction[a]	$\Delta\tilde{H}^\circ$ (cal)	$\Delta\tilde{S}^\circ$ (cal)	Error ($\pm$kcal)	Range[b] (°C)	Reference
$\langle Pd \rangle = \{Pd\}$	4200	2.3	—	1552m	15
$\{Pd\} = (Pd)$	84,090	25.95	0.5	1552–2964b	15
$\langle PdO \rangle = \langle Pd \rangle + \frac{1}{2}(O_2)$	26,500	23.1	—	25–870m	13
$\langle Pd_4S \rangle = 4\langle Pd \rangle + \frac{1}{2}(S_2)$	32,000	20.58	3	25–761m	13
$\langle PdS \rangle = \langle Pd \rangle + \frac{1}{2}(S_2)$	32,000	21.8	5	25–970m	13
$\langle PdS_2 \rangle = \langle Pd \rangle + (S_2)$	48,000	40.0	5	25–972m	13
$\langle Pt \rangle = \{Pt\}$	4700	2.3	—	1769m	15
$\{Pt\} = (Pt)$	124,520	30.42	0.3	1769–3830b	15
$(PtO_2) = \langle Pt \rangle + (O_2)$	39,300	0.0	3	25–1700	13
$\langle PtS \rangle = \langle Pt \rangle + \frac{1}{2}(S_2)$	35,500	24.25	3	25–1200	13
$\langle PtS_2 \rangle = \langle Pt \rangle + (S_2)$	55,800	43.56	3	25–1200	13
$\langle Pu \rangle = \{Pu\}$	680	0.74	0.2	640m	15
$\{Pu\} = (Pu)$	80,000	22.9	4	640–3230b	15
$\langle PuBr_3 \rangle = \{PuBr_3\}$	14,000	14.68	—	681m	13
$\{PuBr_3\} = \{Pu\} + \frac{3}{2}(Br_2)$	192,000	35.2	5	681–1475b	13
$\langle PuCl_3 \rangle = \{PuCl_3\}$	15,200	14.62	—	767m	13
$\{PuCl_3\} = \{Pu\} + \frac{3}{2}(Cl_2)$	209,000	33.9	5	767–1790b	13
$\langle PuF_3 \rangle = \{PuF_3\}$	13,000	7.65	—	1426m	13
$\langle PuF_3 \rangle = \{Pu\} + \frac{3}{2}(F_2)$	368,000	54.6	5	640–1426m	13
$\langle PuI_3 \rangle = \{PuI_3\}$	12,000	11.43	—	777m	13
$\{PuI_3\} = \{Pu\} + \frac{3}{2}(I_2)$	161,000	35.2	10	777–1200	13
$\langle PuN \rangle = \{Pu\} + \frac{1}{2}(N_2)$	72,500	20.26	3	640–2000	13
$\langle Pu_2C_3 \rangle = 2\{Pu\} + 3\langle C \rangle$	31,600	−3.9	—	640–2000	13
$\langle PuC_2 \rangle = \{Pu\} + 2\langle C \rangle$	10,600	−5.4	—	640–2252m	13
$\langle PuO_2 \rangle = \{Pu\} + (O_2)$	250,000	42.52	3	640–2390m	13
$\langle PuS \rangle = \{Pu\} + \frac{1}{2}(S_2)$	122,000	24.2	10	640–1500	13
$\langle Pu_2S_3 \rangle = 2\{Pu\} + \frac{3}{2}(S_2)$	284,000	64.7	10	640–1200	13
$\langle Pu(SO_4)_2 \rangle = \langle PuO_2 \rangle + 2(SO_2) + (O_2)$	128,000	137.6	10	640–1200	13
$\langle Rb \rangle = \{Rb\}$	524	1.68	0.1	40m	15
$\{Rb\} = (Rb)$	18,240	19.04	0.1	40–688b	15
$\langle RbBr \rangle = \{RbBr\}$	3700	3.88	—	680m	12
$\{RbBr\} = (Rb) + \frac{1}{2}(Br_2)$	108,000	34.1	—	688–1352b	12
$\langle RbCl \rangle = \{RbCl\}$	4400	4.45	—	715m	12
$\{RbCl\} = (Rb) + \frac{1}{2}(Cl_2)$	114,000	35.2	—	715–1381b	12
$\langle RbF \rangle = \{RbF\}$	5490	5.24	1	775m	13
$\{RbF\} = (Rb) + \frac{1}{2}(F_2)$	140,000	34.3	4	775–1390b	13
$\langle RbI \rangle = \{RbI\}$	3000	3.29	—	640m	12
$\{RbI\} = (Rb) + \frac{1}{2}(I_2)$	96,000	32.9	—	688–1304b	12
$\langle Rb_2O \rangle = 2\{Rb\} + \frac{1}{2}(O_2)$	80,000	36.8	5	25–688	13
$\langle Rb_2CO_3 \rangle = \langle Rb_2O \rangle + (CO_2)$	96,000	34.8	5	25–600	13
$\langle Re \rangle = \{Re\}$	7940	2.3	—	3180m	15
$\{Re\} = (Re)$	168,570	28.71	2	3180–5600b	15
$\langle Re_5Si_3 \rangle = 5\langle Re \rangle + 3\langle Si \rangle$	37,000	−5.5	10	25–1412	13
$\langle ReSi \rangle = \langle Re \rangle + \langle Si \rangle$	12,000	−2.2	5	25–1412	13
$\langle ReSi_2 \rangle = \langle Re \rangle + 2\langle Si \rangle$	22,000	0.7	5	25–1412	13
$\langle ReO_2 \rangle = \langle Re \rangle + (O_2)$	102,400	40.57	3	25–1200	13
$\langle ReS_2 \rangle = \langle Re \rangle + (S_2)$	72,000	45.6	5	25–1200	13
$\langle Rh \rangle = \{Rh\}$	5140	2.3	—	1960m	15
$\{Rh\} = (Rh)$	120,700	30.44	1.5	1960–3700b	15

***TABLE 1.2* (continued)**

Reaction[a]	$\Delta\tilde{H}^\circ$ (cal)	$\Delta\tilde{S}^\circ$ (cal)	Error ($\pm$kcal)	Range[b] (°C)	Reference
$\langle Rh_2O_3\rangle = 2\langle Rh\rangle + \frac{3}{2}(O_2)$	90,000	63.57	3	25–1000	13
$(RhO_2) = \langle Rh\rangle + (O_2)$	47,600	−4.69	3	25–1200	13
$\langle Ru\rangle = \{Ru\}$	5800	2.3	—	2250m	15
$\{Ru\} = (Ru)$	144,500	32.70	1.5	2250–4150b	15
$\langle RuO_2\rangle = \langle Ru\rangle + (O_2)$	72,000	38.8	4	25–1200	13
$(RuO_3) = \langle Ru\rangle + \frac{3}{2}(O_2)$	19,000	14.5	5	25–1600	13
$(RuO_4) = \langle Ru\rangle + 2(O_2)$	43,000	34.1	4	25–1700	13
$\langle RuS_2\rangle = \langle Ru\rangle + (S_2)$	79,000	45.6	4	25–1200	13
$\langle S\rangle = \{S\}$	410	1.06	~0	115m	15
$\{S\} = \frac{1}{2}(S_2)$	14,000	16.32	0.5	115–445b	15
$(S_2) = 2(S)$	112,170	38.55	0.5	25–1700	13
$(S_4) = 2(S_2)$	15,000	27.6	5	25–1700	13
$(S_6) = 3(S_2)$	66,000	72.9	5	25–1700	13
$(S_8) = 4(S_2)$	95,000	107.1	5	25–1700	13
$(SO) = \frac{1}{2}(S_2) + \frac{1}{2}(O_2)$	13,810	−1.19	0.3	445–2000	8
$(SO_2) = \frac{1}{2}(S_2) + (O_2)$	86,440	17.37	0.1	445–2000	8
$(SO_3) = \frac{1}{2}(S_2) + \frac{3}{2}(O_2)$	109,440	39.04	0.3	445–2000	8
$\langle Sb\rangle = \{Sb\}$	4750	5.26	0.3	631m	15
$\{Sb\} = (Sb)$	20,380	10.89	0.5	631–1587b	15
$\langle Sb_2O_3\rangle = \{Sb_2O_3\}$	13,150	14.16	—	656m	13
$\{Sb_2O_3\} = 2\{Sb\} + \frac{3}{2}(O_2)$	157,900	47.35	3	656–1456b	13
$\langle Sc\rangle = \{Sc\}$	3370	1.86	—	1539m	15
$\{Sc\} = (Sc)$	78,200	25.25	1	1539–2830b	15
$\langle ScBr_3\rangle = \langle Sc\rangle + \frac{3}{2}(Br_2)$	179,000	52.1	10	25–929s	13
$\langle ScCl_3\rangle = \{ScCl_3\}$	16,100	12.98	—	967m	13
$\langle ScCl_3\rangle = \langle Sc\rangle + \frac{3}{2}(Cl_2)$	213,000	54.9	10	25–967m	13
$\langle ScF_3\rangle = \langle Sc\rangle + \frac{3}{2}(F_2)$	394,000	55.0	—	25–1000	13
$\langle ScN\rangle = \langle Sc\rangle + \frac{1}{2}(N_2)$	75,000	23.5	5	25–1539	13
$\langle Sc_2O_3\rangle = 2\langle Sc\rangle + \frac{3}{2}(O_2)$	454,600	69.52	3	25–1539	12
$\langle Se\rangle = \{Se\}$	1300	2.63	0.2	221m	12
$\{Se\} = \frac{1}{2}(Se_2)$	14,180	14.62	1	221–685b	12
$(SeO) = \frac{1}{2}(Se_2) + \frac{1}{2}(O_2)$	2200	−1.0	—	685–1700	13
$(SeO_2) = \frac{1}{2}(Se_2) + (O_2)$	42,500	15.80	3	685–1700	13
$\langle Si\rangle = \{Si\}$	12,080	7.17	0.4	1412m	15
$\{Si\} = (Si)$	94,500	26.62	1	1412–3280b	15
$(SiBr_4) = \langle Si\rangle + 2(Br_2)$	113,000	27.2	—	153b–900	13
$(SiCl_4) = \langle Si\rangle + 2(Cl_2)$	157,800	30.78	1	61b–1412	9
$(SiF_4) = \langle Si\rangle + 2(F_2)$	386,100	34.52	0.5	25–1412	11
$(SiI_4) = \langle Si\rangle + 2(I_2)$	55,300	27.6	4	301b–1412	11
$\langle Si_3N_4\rangle_\alpha = 3\langle Si\rangle + 2(N_2)$	173,000	75.3	1	25–1412	55
$\langle Si_3N_4\rangle_\alpha = 3\{Si\} + 2(N_2)$	209,000	96.8	1	1412–1700	55
$\langle SiC\rangle_\beta = \langle Si\rangle + \langle C\rangle$	17,460	1.83	2	25–1412	8
$\langle SiC\rangle_\beta = \{Si\} + \langle C\rangle$	29,300	8.85	2	1412–2000	8
$\langle SiP\rangle = \langle Si\rangle + \frac{1}{2}(P_2)$	31,200	−21.31	3	25–1410d	13
$(SiO) = \langle Si\rangle + \frac{1}{2}(O_2)$	24,900	−19.72	3	25–1412	8
$\langle SiO_2\rangle$quartz $= \{SiO_2\}$	1840	1.08	0.2	1423m	8
$\langle SiO_2\rangle$quartz $= \langle Si\rangle + (O_2)$	216,800	42.00	3	25–1412	8
$\langle SiO_2\rangle$cristobalite $= \{SiO_2\}$	2290	1.15	0.5	1723m	8
$\langle SiO_2\rangle$cristobalite $= \langle Si\rangle + (O_2)$	216,500	41.95	3	25–1412	8

***TABLE 1.2* (*continued*)**

Reaction[a]	$\Delta\tilde{H}°$ (cal)	$\Delta\tilde{S}°$ (cal)	Error ($\pm$kcal)	Range[b] (°C)	Reference
$(SiS) = \langle Si\rangle + \frac{1}{2}(S_2)$	−12,370	−19.50	0.5	700–1412	56, 57
$\langle SiS_2\rangle = \{SiS_2\}$	2000	1.47	1	1090m	9
$\langle SiS_2\rangle = \langle Si\rangle + (S_2)$	78,000	33.21	5	25–1090m	9
$\langle Sn\rangle = \{Sn\}$	1680	3.33	0.1	232m	15
$\{Sn\} = (Sn)$	70,740	24.59	0.8	232–2603b	15
$\{SnBr_2\} = (SnBr_2)$	23,300	25.49	—	641b	13
$(SnBr_2) = \{Sn\} + (Br_2)$	38,000	−7.8	5	641–1200	13
$\{SnCl_2\} = (SnCl_2)$	19,500	21.08	—	652b	13
$(SnCl_2) = \{Sn\} + (Cl_2)$	54,000	3.1	4	652–1200	13
$\{SnI_2\} = (SnI_2)$	24,000	24.22	—	718b	13
$(SnI_2) = \{Sn\} + (I_2)$	19,100	0.0	—	718–1200	13
$(SnO) = \{Sn\} + \frac{1}{2}(O_2)$	−1500	−12.17	3	232–1700	13
$\langle SnO_2\rangle = \{Sn\} + (O_2)$	137,400	47.41	3	232–1630m	12
$(SnS) = \{Sn\} + \frac{1}{2}(S_2)$	−6200	−11.80	3	232–1700	13
$\langle SnS_2\rangle = \{Sn\} + (S_2)$	68,000	46.8	5	232–765m	13
$\langle Sr\rangle = \{Sr\}$	2000	1.90	0.3	768m	9
$\{Sr\} = (Sr)$	33,820	20.49	1	768–1381b	9
$\langle SrBr_2\rangle = \{SrBr_2\}$	2420	2.60	0.1	657m	10
$\langle SrBr_2\rangle = \langle Sr\rangle + (Br_2)$	177,420	33.66	1	25–657m	10
$\langle SrCl_2\rangle = \{SrCl_2\}$	3880	3.38	0.2	874	10
$\langle SrCl_2\rangle = \langle Sr\rangle + (Cl_2)$	196,860	35.38	1	25–768	10
$\langle SrF_2\rangle = \{SrF_2\}$	7090	4.05	0.2	1477m	10
$\langle SrF_2\rangle = \langle Sr\rangle + (F_2)$	290,000	39.90	1	25–768	10
$\langle SrI_2\rangle = \{SrI_2\}$	4700	5.79	0.1	538m	10
$\langle SrI_2\rangle = \langle Sr\rangle + (I_2)$	147,860	33.60	1	25–538m	10
$\langle SrO\rangle = \{Sr\} + \frac{1}{2}(O_2)$	142,700	24.47	1	768–1377	10
$\langle SrS\rangle = \{Sr\} + \frac{1}{2}(S_2)$	124,000	23.0	5	768–1377	13
$\langle SrCO_3\rangle = \langle SrO\rangle + (CO_2)$	51,300	33.84	1	700–1243d	16, 58
$\langle SrO\cdot Al_2O_3\rangle = \langle SrO\rangle + \langle Al_2O_3\rangle$	17,000	−1.0	5	25–1300	13
$\langle SrO\cdot HfO_2\rangle = \langle SrO\rangle + \langle HfO_2\rangle$	19,000	0.6	5	25–1600	13
$\langle SrO\cdot MoO_2\rangle = \langle SrO\rangle + \langle MoO_2\rangle$	12,600	−6.4	1.5	950–1400	18
$\langle SrO\cdot MoO_3\rangle = \langle SrO\rangle + \langle MoO_3\rangle$	50,500	0.4	1.5	25–795	18
$\langle SrSO_4\rangle = \langle SrO\rangle + (SO_2) + \frac{1}{2}(O_2)$	131,000	65.2	5	25–1600m	12
$\langle 2SrO\cdot SiO_2\rangle = 2\langle SrO\rangle + \langle SiO_2\rangle$	51,000	1.4	—	25–900	13
$\langle SrO\cdot SiO_2\rangle = \langle SrO\rangle + \langle SiO_2\rangle$	32,000	1.0	—	25–900	13
$\langle SrO\cdot TiO_2\rangle = \langle SrO\rangle + \langle TiO_2\rangle$	32,800	0.5	2.5	25–900	13
$\langle Ta\rangle = \{Ta\}$	7560	2.30	1	3014m	13
$\{Ta\} = (Ta)$	176,810	30.57	0.6	3000–5513b	10
$\{TaCl_5\} = (TaCl_5)$	12,670	25.0	1	234b	11
$(TaCl_5) = \langle Ta\rangle + \frac{5}{2}(Cl_2)$	180,200	40.49	2	234–2000	11
$\{TaF_5\} = (TaF_5)$	12,100	24.10	—	224b	13
$(TaF_5) = \langle Ta\rangle + \frac{5}{2}(F_2)$	433,600	43.75	3	224–1700	13
$\{TaI_5\} = (TaI_5)$	15,500	23.13	2	397b	13
$(TaI_5) = \langle Ta\rangle + \frac{5}{2}(I_2)$	87,100	35.4	—	397–1700	13
$\langle TaB_2\rangle = \langle Ta\rangle + 2\langle B\rangle$	50,000	2.3	—	25–1700	12
$\langle Ta_2C\rangle = 2\langle Ta\rangle + \langle C\rangle$	48,000	0.5	4	25–1700	13
$\langle TaC\rangle = \langle Ta\rangle + \langle C\rangle$	34,000	0.29	1	25–1700	10
$\langle Ta_2N\rangle = 2\langle Ta\rangle + \frac{1}{2}(N_2)$	63,000	21.7	5	25–1700	13
$\langle TaN\rangle = \langle Ta\rangle + \frac{1}{2}(N_2)$	59,000	19.4	5	25–1700	13
$\langle Ta_2Si\rangle = 2\langle Ta\rangle + \langle Si\rangle$	30,000	−0.8	6	25–1412	12
$\langle Ta_5Si_3\rangle = 5\langle Ta\rangle + 3\langle Si\rangle$	80,000	−3.0	4	25–1412	12
$\langle TaSi_2\rangle = \langle Ta\rangle + 2\langle Si\rangle$	28,000	5.4	—	25–1412	12

TABLE 1.2 (continued)

Reaction[a]	$\Delta\tilde{H}^\circ$ (cal)	$\Delta\tilde{S}^\circ$ (cal)	Error ($\pm$kcal)	Range[b] (°C)	Reference
$(TaO)=\langle Ta\rangle+\frac{1}{2}(O_2)$	−45,000	−20.7	15	25–2000	10
$(TaO_2)=\langle Ta\rangle+(O_2)$	50,000	−4.9	15	25–2000	10
$\langle Ta_2O_5\rangle=\{Ta_2O_5\}$	36,100	16.80	—	1877m	13
$\langle Ta_2O_5\rangle=2\langle Ta\rangle+\frac{5}{2}(O_2)$	484,000	98.6	5	25–1877m	13
$\langle Te\rangle=\{Te\}$	4180	5.78	0.3	500m	15
$\{Te\}=\frac{1}{2}(Te_2)$	13,842	10.97	1	500–1009b	15
$(TeO)=\frac{1}{2}(Te_2)+\frac{1}{2}(O_2)$	1700	−1.44	—	1009–1700	13
$\langle Th\rangle=\{Th\}$	3900	1.9	—	1755m	15
$\{Th\}=(Th)$	126,000	24.94	0.5	1755–4790b	15
$\langle ThBr_4\rangle=\{ThBr_4\}$	13,000	13.66	—	679m	13
$\langle ThBr_4\rangle=\langle Th\rangle+2(Br_2)$	244,000	70.6	3	25–679m	13
$\langle ThCl_4\rangle=\{ThCl_4\}$	10,500	10.07	—	770m	12
$\langle ThCl_4\rangle=\langle Th\rangle+2(Cl_2)$	281,700	68.85	3	25–770m	12
$\langle ThF_4\rangle=\{ThF_4\}$	10,500	7.60	—	1110m	13
$\langle ThF_4\rangle=\langle Th\rangle+2(F_2)$	502,400	70.76	3	25–1110m	13
$\langle ThI_4\rangle=\{ThI_4\}$	11,500	13.64	—	570m	13
$\langle ThI_4\rangle=\langle Th\rangle+2(I_2)$	187,000	69.3	—	25–570m	13
$\langle ThC_2\rangle=\langle Th\rangle+2\langle C\rangle$	30,000	−2.0	5	25–1755	12
$\langle ThN\rangle=\langle Th\rangle+\frac{1}{2}(N_2)$	90,000	20.6	5	25–1755	13
$\langle Th_3N_4\rangle=3\langle Th\rangle+2(N_2)$	314,000	84.0	5	25–1755	13
$\langle Th_2N_2O\rangle=2\langle Th\rangle+(N_2)+\frac{1}{2}(O_2)$	306,000	62.4	5	25–1755	13
$\langle Th_3Si_2\rangle=3\langle Th\rangle+2\langle Si\rangle$	69,000	9.3	5	25–1412	13
$\langle ThSi\rangle=\langle Th\rangle+\langle Si\rangle$	31,000	3.7	5	25–1412	13
$\langle Th_3Si_5\rangle=3\langle Th\rangle+5\langle Si\rangle$	117,000	10.8	5	25–1412	13
$\langle ThSi_2\rangle=\langle Th\rangle+2\langle Si\rangle$	42,000	2.5	5	25–1412	13
$\langle ThP\rangle=\langle Th\rangle+\frac{1}{2}(P_2)$	100,000	20.6	10	25–1755	13
$\langle Th_3P_4\rangle=3\langle Th\rangle+2(P_2)$	339,000	76.4	10	25–1755	15
$(ThO)=\langle Th\rangle+\frac{1}{2}(O_2)$	16,000	−12.60	1	1600–2000	59
$\langle ThO_2\rangle=\langle Th\rangle+(O_2)$	292,000	43.81	1	25–2000	13
$\langle ThS\rangle=\langle Th\rangle+\frac{1}{2}(S_2)$	110,600	22.62	3	25–2000	13
$\langle Th_2S_3\rangle=2\langle Th\rangle+\frac{3}{2}(S_2)$	303,000	61.3	5	25–2000	13
$\langle ThS_2\rangle=\langle Th\rangle+(S_2)$	179,000	41.0	5	25–1915m	13
$\langle Ti\rangle=\{Ti\}$	3700	1.9	—	1670m	15
$\{Ti\}=(Ti)$	102,000	28.68	—	1670–3290b	15
$(TiBr_4)=\langle Ti\rangle+2(Br_2)$	146,900	29.47	3	25–1670	8
$(TiCl_4)=\langle Ti\rangle+2(Cl_2)$	182,600	29.03	3	25–1670	8
$(TiF_4)=\langle Ti\rangle+2(F_2)$	371,400	29.67	3	286b–1670	8
$(TiI_4)=\langle Ti\rangle+2(I_2)$	96,000	28.1	5	380b–1670	8
$\langle TiB\rangle=\langle Ti\rangle+\langle B\rangle$	39,000	1.4	10	25–1670	8
$\langle TiB_2\rangle=\langle Ti\rangle+2\langle B\rangle$	68,000	4.9	5	25–1670	8
$\langle TiC\rangle=\langle Ti\rangle+\langle C\rangle$	44,160	3.00	1.5	25–1670	8
$\langle TiN\rangle=\langle Ti\rangle+\frac{1}{2}(N_2)$	80,380	22.29	1.5	25–1670	8
$\langle TiO\rangle_\beta=\langle Ti\rangle+\frac{1}{2}(O_2)$	123,000	17.7	5	25–1670	10
$\langle TiO_2\rangle$rutile$=\langle Ti\rangle+(O_2)$	224,900	42.44	0.5	25–1670	10
$\langle Ti_2O_3\rangle=2\langle Ti\rangle+\frac{3}{2}(O_2)$	359,000	61.68	2.5	25–1670	10
$\langle Ti_3O_5\rangle=3\langle Ti\rangle+\frac{5}{2}(O_2)$	582,000	100.5	5	25–1670	10
$\langle Tl\rangle=\{Tl\}$	990	1.72	0.1	304m	15
$\{Tl\}=(Tl)$	40,570	23.32	0.6	304–1473b	15
$\langle U\rangle=\{U\}$	2040	1.45	0.7	1132m	15
$\{U\}=(U)$	114,000	25.2	5	1132–4130b	15
$\langle UBr_3\rangle=\{UBr_3\}$	10,500	10.50	—	727m	13

TABLE 1.2 (continued)

Reaction[a]	$\Delta\tilde{H}^\circ$ (cal)	$\Delta\tilde{S}^\circ$ (cal)	Error ($\pm$kcal)	Range[b] (°C)	Reference
$\langle UBr_3\rangle = \langle U\rangle + \frac{3}{2}(Br_2)$	182,000	50.6	5	25–727m	13
$\langle UCl_3\rangle = \{UCl_3\}$	11,100	10.00	—	837m	12
$\langle UCl_3\rangle = \langle U\rangle + \frac{3}{2}(Cl_2)$	212,000	50.2	—	25–837m	12
$\{UCl_4\} = (UCl_4)$	33,800	31.83	—	789b	12
$(UCl_4) = \{U\} + 2(Cl_2)$	202,800	19.70	3	1132–1700	12
$\{UCl_5\} = (UCl_5)$	18,000	22.50	—	527b	12
$(UCl_5) = \{U\} + \frac{5}{2}(Cl_2)$	230,000	41.3	—	1132–1700	12
$\{UCl_6\} = (UCl_6)$	12,000	18.05	—	392b	12
$(UCl_6) = \{U\} + 3(Cl_2)$	246,500	61.9	4	1132–1700	12
$\langle UF_4\rangle = \{UF_4\}$	10,200	7.79	—	1036m	12
$\langle UF_4\rangle = \langle U\rangle + 2(F_2)$	452,000	68.47	3	25–1036m	12
$\langle UF_6\rangle = (UF_6)$	11,520	34.91	—	57s	12
$(UF_6) = \langle U\rangle + 3(F_2)$	510,400	65.42	3	57–1132	12
$\langle UB_2\rangle = \{U\} + 2\langle B\rangle$	43,000	3.72	5	1132–2430m	13
$\langle UC\rangle = \{U\} + \langle C\rangle$	26,200	0.44	2	1132–252m	12
$\langle U_2C_3\rangle = 2\{U\} + 3\langle C\rangle$	53,000	0.60	3	1132–1700	13
$\langle UC_{1.93}\rangle = \{U\} + 1.93\langle C\rangle$	21,600	2.20	2	1132–1700	13
$\langle UN\rangle = \langle U\rangle + \frac{1}{2}(N_2)$	70,000	19.3	4	25–1132	12
$\langle UAl_2\rangle = \langle U\rangle + 2\{Al\}$	37,000	13.3	5	1132–1590m	13
$\langle USi\rangle = \langle U\rangle + \langle Si\rangle$	21,000	1.26	3	25–1132	13
$\langle UO_2\rangle = \{U\} + (O_2)$	259,700	41.18	3	1132–2000	12
$\langle U_4O_9\rangle = 4\langle U\rangle + \frac{9}{2}(O_2)$	1,073,600	176.91	3	25–600	12
$\langle U_3O_8\rangle = 3\langle U\rangle + 4(O_2)$	851,000	155.96	3	25–600	12
$\langle UO_3\rangle = \langle U\rangle + \frac{3}{2}(O_2)$	293,200	59.88	3	25–600	12
$\langle US\rangle = \{U\} + \frac{1}{2}(S_2)$	93,800	24.82	3	1132–2460m	13
$\langle U_2S_3\rangle = 2\{U\} + \frac{3}{2}(S_2)$	248,000	63.6	10	1132–2030m	13
$\langle US_2\rangle = \{U\} + (S_2)$	143,000	40.6	10	1132–1700	13
$\langle V\rangle = \{V\}$	5460	2.49	—	1920m	10
$\{V\} = (V)$	110,740	30.06	3	1920–3420b	10
$\langle VB\rangle = \langle V\rangle + \langle B\rangle$	33,000	1.4	—	25–2000	13
$\langle V_2C\rangle = 2\langle V\rangle + \langle C\rangle$	35,000	0.8	—	25–1700	12
$\langle VC\rangle = \langle V\rangle + \langle C\rangle$	24,400	2.29	3	25–2000	13
$\langle VN_{0.46}\rangle = \langle V\rangle + 0.23(N_2)$	31,000	10.6	2	25–1700	10
$\langle VN\rangle = \langle V\rangle + \frac{1}{2}(N_2)$	51,300	19.7	—	25–2346d	10
$\langle VO\rangle = \langle V\rangle + \frac{1}{2}(O_2)$	101,500	19.13	2	25–1800	10
$\langle V_2O_3\rangle = 2\langle V\rangle + \frac{3}{2}(O_2)$	287,500	56.77	2	25–2070m	10
$\langle VO_2\rangle = \langle V\rangle + (O_2)$	168,800	37.12	3	25–1360m	12
$\langle V_2O_5\rangle = \{V_2O_5\}$	15,400	16.33	0.8	670m	10
$\{V_2O_5\} = 2\langle V\rangle + \frac{5}{2}(O_2)$	345,930	76.86	2	670–2000	10
$\langle W\rangle = \{W\}$	8500	2.3	—	3400m	15
$\{W\} = (W)$	196,250	33.66	1	3400–5550b	15
$\langle W_2C\rangle = 2\langle W\rangle + \langle C\rangle$	7300	−0.56	0.1	1302–1400	60
$\langle WC\rangle = \langle W\rangle + \langle C\rangle$	10,100	1.19	0.1	900–1302	60
$\langle WO_2\rangle = \langle W\rangle + (O_2)$	138,900	41.07	3	25–2000d	8
$\langle WO_3\rangle = \{WO_3\}$	17,550	10.06	—	1472m	8
$\langle WO_3\rangle = \langle W\rangle + \frac{3}{2}(O_2)$	199,200	58.66	3	25–1472m	8
$\langle Y\rangle = \{Y\}$	2720	1.51	0.05	1526m	15
$\{Y\} = (Y)$	90,590	25.18	1	1526–3340b	15
$\langle YCl_3\rangle = \{YCl_3\}$	7500	7.55	—	721m	13
$\langle YCl_3\rangle = \langle Y\rangle + \frac{3}{2}(Cl_2)$	231,300	54.30	3	25–721m	13
$\langle YF_3\rangle = \{YF_3\}$	6690	4.68	—	1155m	13
$\langle YF_3\rangle = \langle Y\rangle + \frac{3}{2}(F_2)$	409,000	52.4	4	25–1155m	13

TABLE 1.2 (continued)

Reaction[a]	$\Delta\tilde{H}^\circ$ (cal)	$\Delta\tilde{S}^\circ$ (cal)	Error ($\pm$kcal)	Range[b] (°C)	Reference
$\langle YI_3\rangle = \langle Y\rangle + \frac{3}{2}(I_2)$	168,000	50.3	10	25–1000m	13
$\langle YN\rangle = \langle Y\rangle + \frac{1}{2}(N_2)$	71,000	23.8	5	25–1526	13
$\langle Y_2O_3\rangle = 2\langle Y\rangle + \frac{3}{2}(O_2)$	453,600	67.39	3	25–1526	12
$\langle Y_2O_3\cdot ZrO_2\rangle = \langle Y_2O_3\rangle + \langle ZrO_2\rangle$	5000	0.0	10	25–1200	13
$\langle Zn\rangle = \{Zn\}$	1750	2.53	0.05	420m	15
$\{Zn\} = (Zn)$	28,230	23.96	0.2	420–907b	15
$\langle ZnBr_2\rangle = \{ZnBr_2\}$	3740	5.54	—	402m	12
$\langle ZnBr_2\rangle = \langle Zn\rangle + (Br_2)$	85,000	35.0	4	25–402m	12
$(ZnBr_2) = (Zn) + (Br_2)$	84,400	25.9	4	907–1700	12
$\{ZnCl_2\} = (ZnCl_2)$	28,500	28.36	—	732b	12
$\{ZnCl_2\} = \{Zn\} + (Cl_2)$	96,200	31.4	4	420–732b	12
$(ZnCl_2) = (Zn) + (Cl_2)$	93,900	25.1	4	907–1700	12
$\langle ZnF_2\rangle = \{ZnF_2\}$	10,000	8.71	—	875m	13
$\langle ZnF_2\rangle = \{Zn\} + (F_2)$	183,700	41.76	3	420–875m	13
$\{ZnF_2\} = (Zn) + (F_2)$	198,800	54.37	3	907–1505b	13
$\langle ZnO\rangle = (Zn) + \frac{1}{2}(O_2)$	110,000	47.4	—	907–1700	12
$\langle ZnS\rangle = \{Zn\} + \frac{1}{2}(S_2)$	66,400	25.8	—	420–907	13
$(ZnS) = (Zn) + \frac{1}{2}(S_2)$	−1200	7.3	—	1182b–1700	13
$\langle ZnO\cdot Fe_2O_3\rangle = \langle ZnO\rangle + \langle Fe_2O_3\rangle$	2300	0.9	4	25–700	13
$\langle ZnSO_4\rangle = \langle ZnO\rangle + (SO_2) + \frac{1}{2}(O_2)$	78,300	63.81	3	25–1200	13
$\langle ZnSO_4\rangle = \frac{1}{3}\langle ZnO\cdot 2ZnSO_4\rangle + \frac{1}{3}(SO_2) + \frac{1}{6}(O_2)$	26,980	23.75	0.1	424–714	40
$\langle ZnO\cdot 2ZnSO_4\rangle = 3\langle ZnO\rangle + 2(SO_2) + (O_2)$	153,660	122.56	0.3	520–935	40
$\langle 2ZnO\cdot SiO_2\rangle = 2\langle ZnO\rangle + \langle SiO_2\rangle$	8700	0.68	3	25–1512m	13
$\langle 2ZnO\cdot TiO_2\rangle = 2\langle ZnO\rangle + \langle TiO_2\rangle$	200	−3.16	3	25–1700	13
$\langle Zr\rangle = \{Zr\}$	5000	2.35	—	1852m	12
$\{Zr\} = (Zr)$	138,550	29.59	1	1852–4410b	15
$(ZrBr_4) = \langle Zr\rangle + 2(Br_2)$	169,120	27.33	2	357s–1700	11
$\langle ZrCl_4\rangle = (ZrCl_4)$	26,420	43.38	0.2	336s	11
$(ZrCl_4) = \langle Zr\rangle + 2(Cl_2)$	208,190	27.80	0.5	336–2000	11
$\langle ZrF_4\rangle = (ZrF_4)$	56,800	48.18	—	903s	11
$(ZrF_4) = \langle Zr\rangle + 2(F_2)$	401,110	30.68	1	903–2000	11
$\langle ZrI_4\rangle = (ZrI_4)$	30,200	42.78	0.5	433s	11
$(ZrI_4) = \langle Zr\rangle + 2(I_2)$	116,730	26.96	2	433–2000	11
$\langle ZrB_2\rangle = \{ZrB_2\}$	25,000	7.52	—	3050m	12
$\langle ZrB_2\rangle = \langle Zr\rangle + 2\langle B\rangle$	78,400	5.6	4	25–1850	12
$\langle ZrC\rangle = \langle Zr\rangle + \langle C\rangle$	47,000	2.2	—	25–1850	12
$\langle ZrN\rangle = \langle Zr\rangle + \frac{1}{2}(N_2)$	86,900	22.0	4	25–1850	12
$\langle ZrSi\rangle = \langle Zr\rangle + \langle Si\rangle$	38,000	1.0	10	25–1412	13
$(ZrO) = \langle Zr\rangle + \frac{1}{2}(O_2)$	−13,040	−16.04	1	1300–1850	59
$\langle ZrO_2\rangle = \{ZrO_2\}$	20,800	7.05	—	2680m	12
$\langle ZrO_2\rangle = \langle Zr\rangle + (O_2)$	261,000	43.9	4	25–1850	12
$(ZrS) = \langle Zr\rangle + \frac{1}{2}(S_2)$	−56,700	−18.7	5	25–1850	13
$\langle ZrS_2\rangle = \langle Zr\rangle + (S_2)$	167,000	42.6	5	25–1550m	13
$\langle ZrO_2\cdot SiO_2\rangle = \langle ZrO_2\rangle + \langle SiO_2\rangle$	6400	3.0	5	25–1707m	13

[a] ⟨Solid⟩, {liquid}, (gas).

[b] m denotes melting, b boiling, s sublimation, and d decomposition temperature.

References

1. "The Collected Works of J. Willard Gibbs," Vols. I and II. Longmans, Green, New York, 1928.
2. G. N. Lewis and M. Randall, "Thermodynamics and the Free Energy of Chemical Substances." McGraw-Hill, New York, 1923.
3. K. S. Pitzer and L. Brewer, "Thermodynamics—Lewis & Randall." McGraw-Hill, New York, 1961.
4. E. A. Guggenheim, "Thermodynamics." Wiley (Interscience), New York, 1949.
5. L. S. Darken and R. W. Gurry, "Physical Chemistry of Metals." McGraw-Hill, New York, 1953.
6. F. R. Bichowsky and F. D. Rossini, "The Thermochemistry of Chemical Substances." Rheinhold, New York, 1936.
7. K. K. Kelley, *U.S. Bur. Mines, Bull.* **371** (1934); **406** and **407** (1937); **476** (1949); **584** (1960).
8. D. R. Stull and H. Prophet, "JANAF Thermochemical Tables," NSRDS-NBS37. U.S. Dept. Commer., Washington, D.C., 1971.
9. M. W. Chase, J. L. Curnutt, A. T. Hu, H. Prophet, A. N. Syverud, and L. C. Walker, *J. Phys. Chem. Ref. Data* **3**, 311 (1974).
10. M. W. Chase, J. L. Curnutt, H. Prophet, R. A. McDonald, and A. N. Syverud, *J. Phys. Chem. Ref. Data* **4**, 1 (1975).
11. M. W. Chase, J. L. Curnutt, R. A. McDonald, and A. N. Syverud, *J. Phys. Chem. Ref. Data* **7**, 793 (1978).
12. I. Barin and O. Knacke, "Thermochemical Properties of Inorganic Substances." Springer-Verlag, Berlin and New York, 1973; *Metall. Trans.* **5** 1769 (1974).
13. I. Barin, O. Knacke, and O. Kubaschewski, "Thermochemical Properties of Inorganic Substances, Supplement." Springer-Verlag, Berlin and New York, 1977.
14. O. Kubaschewski, E. L. Evans, and C. B. Alcock, "Metallurgical Thermochemistry." Pergamon, Oxford, 1967.
15. H. Hultgren, P. D. Desai, D. T. Hawkins, M. Gleiser, K. K. Kelley, and D. D. Wagman, "Selected Values of the Thermodynamic Properties of the Elements." Am. Soc. Met., Metals Park, Ohio, 1973.
16. J. J. Lander, *J. Am. Chem. Soc.* **73**, 5794 (1951).
17. E. H. Baker, *J. Chem. Soc.* p. 699 (1964).
18. O. Kubaschewski, *High Temp.—High Pressures* **4**, 1 (1972).
19. E. H. Baker, *J. Chem. Soc.* p. 462 (1962).
20. A. W. D. Hills, *Inst. Min. Metall., Trans., Sect. C* **76**, 241 (1967).
21. E. W. Dewing and F. D. Richardson, *Trans. Faraday Soc.* **55**, 611 (1959).
22. E. T. Turkdogan, B. B. Rice, and J. V. Vinters, *Metall. Trans.* **5**, 1527 (1974).
23. K. A. Gschneidner and N. Kippenhan, "Thermochemistry of the Rare Earth Carbides, Nitrides, and Sulfides for Steelmaking," IS-RIC-5. Rare Earth Inf. Cent., Iowa State Univ., Ames (1971).
24. K. D. O'Dell and E. B. Hensley, *J. Phys. Chem. Solids* **33**, 443 (1972).
25. F. B. Baker and C. E. Holley, *J. Chem. Eng. Data* **13**, 405 (1968).
26. R. J. Fruehan, *Metall. Trans. B* **10**, 143 (1979).
27. T. Rosenqvist, *J. Iron Steel Inst., London* **176**, 37 (1954).
28. A. U. Seybolt and R. A. Oriani, *Trans. AIME* **206**, 556 (1956).
29. I. A. Tomilin and N. A. Savostyanova, *Issled. p. Zharoproch. Splavam, Akad. Nauk SSSR, Inst. Metall.* **10**, 283 (1963).
30. K. Schwerdtfeger, *Trans. Metall. Soc. AIME* **239**, 1432 (1967).
31. Y. Jeannin, C. Mannerskantz, and F. D. Richardson, *Trans. Metall. Soc. AIME* **227**, 300 (1963).
32. F. N. Mazandarany and R. D. Pehlke, *J. Electrochem. Soc.* **121**, 711 (1974).
33. N. Toker, Ph.D. Thesis, Pennsylvania State Univ., University Park (1977).

34. J. P. Hager and J. F. Elliott, *Trans. Metall. Soc. AIME* **239**, 513 (1967).
35. K. T. Jacob, D. B. Rao, and H. G. Nelson, *Metall. Trans. A* **10**, 327 (1979).
36. H. H. Kellogg, *Can. Metall. Q.* **8**, 3 (1969).
37. E. G. King, A. D. Mah, and L. B. Pankratz, "The Metallurgy of Copper," INCRA Monograph II. U.S. Bur. Mines, Washington, D.C. (1973).
38. J. P. Pemsler and C. Wagner, *Metall. Trans. B* **6**, 311 (1975).
39. H. H. Kellogg, *Trans. Metall. Soc. AIME* **230**, 1622 (1964).
40. J. M. Skeaff and A. W. Espelund, *Can. Metall. Q.* **12**, 445 (1973).
41. J. Chipman, *Metall. Trans.* **3**, 55 (1972).
42. M. Hillert and M. Jarl, *Metall. Trans. A* **6**, 553 (1975).
43. M. Hillert and L.-I. Staffansson, *Metall. Trans. B* **6**, 37 (1975).
44. A McLean and R. G. Ward, *J. Iron Steel Inst., London* **204**, 8 (1966).
45. K. T. Jacob and C. B. Alcock, *Metall. Trans. B* **6**, 215 (1975).
46. L. L. Simmons, L. F. Lowden, and T. C. Ehlert, *J. Phys. Chem.* **81**, 709 (1977).
47. E. T. Turkdogan and B. B. Rice, *Metall. Trans.* **5**, 1537 (1974).
48. C. B. Alcock and S. Zador, *Electrochim. Acta* **12**, 673 (1967).
49. W. C. Hahn and A. Muan, *Am. J. Sci.* **258**, 66 (1960).
50. K. Schwerdtfeger and A. Muan, *Trans. Metall. Soc. AIME* **239**, 1114 (1967).
51. E. M. Otto, *J. Electrochem. Soc.* **111**, 88 (1964).
52. T. R. Ingraham, *Can. Metall. Q.* **5**, 109 (1966).
53. E. T. Turkdogan, R. G. Olsson, and B. B. Rice, *Metall. Trans. B* **8**, 59 (1977).
54. H. H. Kellogg, *J. Chem. Eng. Data* **14**, 41 (1969).
55. R. D. Pehlke and J. F. Elliott, *Trans. Metall. Soc. AIME* **215**, 781 (1959).
56. T. Rosenqvist and K. Tungesvik, *Trans. Faraday Soc.* **67**, 2945 (1971).
57. G. R. Belton, R. J. Fruehan, and E. T. Turkdogan, *Metall. Trans.* **3**, 596 (1972).
58. E. H. Baker, *J. Chem. Soc.* p. 339 (1963).
59. R. J. Ackermann and E. G. Rauh, *J. Chem. Phys.* **60**, 2266 (1974).
60. D. K. Gupta and L. L. Seigle, *Metall. Trans. A* **6**, 1939 (1975).
61. E. Schurmann and R. Schmid, Arch. Eisenhuettenwes **46**, 773 (1975).

CHAPTER 2

HETEROGENEOUS PHASE EQUILIBRIA

2.1 Phase Rule

In most processes we are concerned with heterogeneous reactions involving more than one phase, often of variable composition. Gibbs [1] begins his treatment of the thermodynamics of heterogeneous equilibrium and phases of variable composition by writing the energy of a homogeneous phase at internal equilibrium as a function of independent state variables. Let us consider a single homogeneous phase which contains $x_1, x_2, x_3, \ldots$ atom fractions of constituents 1, 2, 3, ... and is free of gravitational, electric, and magnetic fields, surface tension effects, etc. From the fundamental theorem of partial differentiation, the total derivative of the free energy in terms of pressure, temperature, and composition is given by

$$dF = \left(\frac{\partial F}{\partial P}\right) dP + \left(\frac{\partial F}{\partial T}\right) dT + \left(\frac{\partial F}{\partial x_1}\right) dx_1 + \left(\frac{\partial F}{\partial x_2}\right) dx_2 + \cdots. \tag{2.1}$$

Substituting for $V = (\partial F/\partial P)$, $-S = (\partial F/\partial T)$, and the partial molar free energies for $(\partial F/\partial x)$,

$$dF = V\,dP - S\,dT + \bar{F}_1\,dx_1 + \bar{F}_2\,dx_2 + \cdots. \tag{2.2}$$

Integration and total differentiation and combining with Eq. (2.1) gives what is known as "Gibbs equation 97" [1, p. 88],

$$0 = S\,dT - V\,dP + x_1\,d\bar{F}_1 + x_2\,d\bar{F}_2 + \cdots. \tag{2.3}$$

This is another general form of the combined statement of the first and second laws and applies to any partial molar quantity. For the special case of constant temperature and pressure, this equation is reduced to that given in Table 1.1 for function 9.

Of the n number of components in the single phase considered, $n - 1$ numbers are independent variables. With the additional two variables temperature and pressure, the state of a single-phase system is defined by $n + 1$ independent variables. The number of independent variables is called the "variance" or the number of "degrees of freedom" designated by $f = n + 1$.

For a system of n number of components with p number of coexistent phases at equilibrium (in the absence of gravitational, electric, and magnetic fields, surface tension effects, etc.), there are p number of equations of the type (2.3), one for each phase. Because of the equilibrium between coexistent phases, temperature, pressure, and chemical potentials are the same in all phases; hence of the $n + 2$ total number of derivatives, there are $n + 2 - p$ independent variables. Therefore, the (Gibbs) phase rule is defined by the relation

$$f = n + 2 - p. \tag{2.4}$$

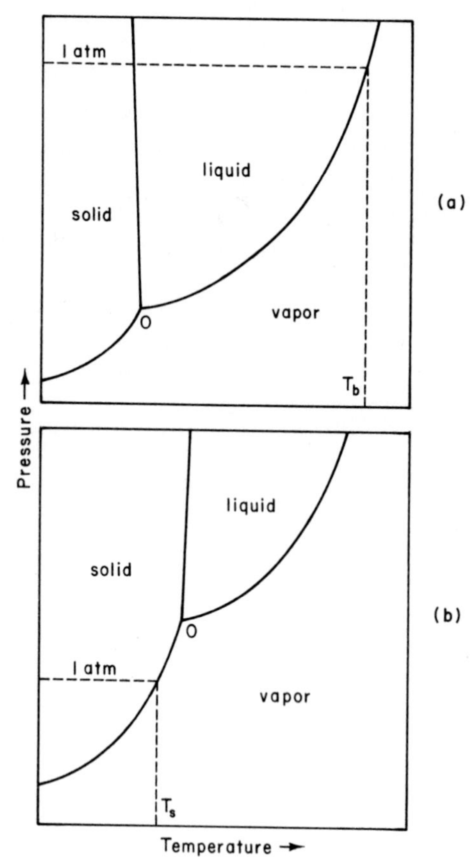

FIG. 2.1 Pressure–temperature phase equilibrium diagram for a single-component system which (a) has a normal boiling temperature and (b) sublimes at atmospheric pressure.

For definition of certain parameters, let us consider a phase equilibrium diagram (Fig. 2.1) for a system of a single component such as H_2O or CO_2. Each curve is for coexistent two-phase equilibrium, solid–liquid, solid–vapor, and liquid–vapor for which the number of degrees of freedom is 1. The point of intersection of three univariant curves for a single-component system is the invariant point at which the variance is 0. Within a single-phase region the variance is 2; for this bivariant equilibrium, both temperature and pressure can be set or changed arbitrarily. When the invariant (triple) point is at a pressure below atmospheric (Fig. 2.1a), the system has a normal boiling point. When the pressure for the invariant point is above atmospheric, the substance sublimes, and the melting temperature under pressure is higher than the boiling (sublimation) temperature at atmospheric pressure. For example, the sublimation temperature of solid carbon dioxide is $-78.5\,^{\circ}C$ at 1 atm CO_2. When the pressure is raised to 5.11 atm, the carbon dioxide melts at $-56.4\,^{\circ}C$.

Next, let us consider the dissociation of calcium carbonate represented by the equilibrium

$$CaCO_3 = CaO + CO_2. \tag{2.5}$$

Although there are three atomic species, this system has two components, because of the stoichiometric composition of the three phases involved. The pressure–temperature phase equilibrium diagram, as determined by Baker [2], is shown in Fig. 2.2. The eutectic invariant is at 1242°C and 39.5 atm CO_2 at which four coexistent phases are $CaCO_3$, CaO, liquid, and CO_2 gas. Each univariant curve, i.e., $f = 1$, depicts the state of equilibrium between two condensed phases and a gas phase.

As indicated by Eq. (2.4), for a two-component system, there should be four univariant curves in the pressure versus temperature plot. Yet in Fig. 2.2 we see only three univariant curves; the missing fourth curve, almost vertical to the temperature axis, is, for the univariant equilibrium, between three condensed phases: $CaCO_3$, CaO, and liquid. At pressures much below the kilobar range, the state of equilibrium between condensed phases is not much affected by pressure, therefore the variance can be reduced by 1 and the phase rule for ordinary pressures is stated by the relation

$$f = n + 1 - p. \tag{2.4a}$$

With this form of the phase rule for ordinary pressures (more exactly for constant pressure), it is seen that for a system with n number of components, there are $n + 1$ univariant equilibrium curves intersecting at an invariant point when $n + 1$ condensed phases are at equilibrium for which the variance is zero. Each one of the $n + 1$ univariant curves, in the plot of chemical potential versus temperature, represents the state of equilibrium between n coexistent condensed phases at constant pressure. The effect of total pressure on phase equilibria is discussed in Sec. 2.5.

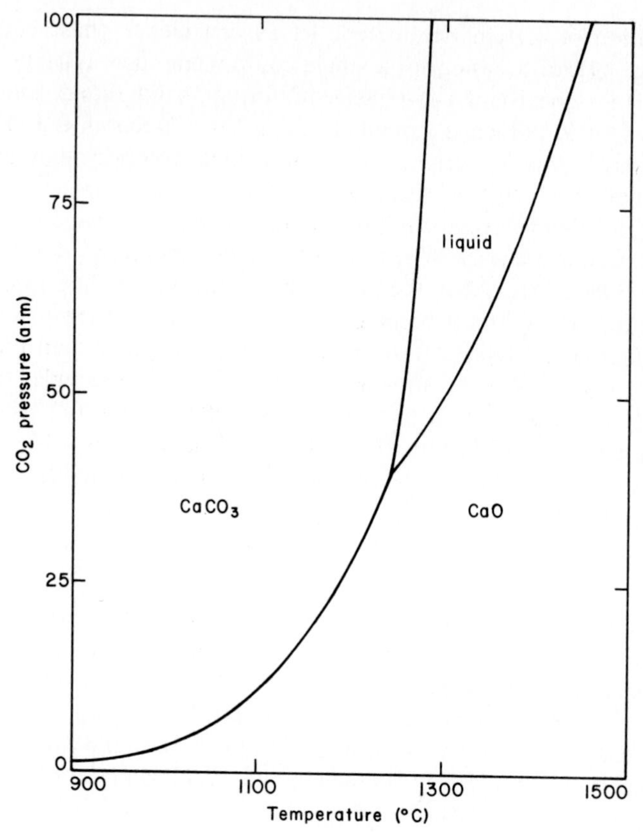

FIG. 2.2 Pressure–temperature phase equilibrium diagram for $CaCO_3$–CaO system. After Baker [2].

2.2 Intersection of the Univariant Curves at an Invariant Point

Based on the fundamental concept of Gibbs on coexistent phases of matter, a theorem was developed by Roozeboom and Schreinemakers [3, 4] on the relation of univariant curves intersecting at an invariant point in a P–T plot. They showed that there are only three types of composition diagrams for an invariant equilibrium in a ternary system, as shown in Fig. 2.3 with the corresponding P–T plots for univariant curves intersecting at an invariant point. With respect to the positions of phases A, B, and C in the composition diagram, forming the base triangle ABC, the positions of the other two phases D and E determine the geometric shape of the two-dimensional composition diagram as a pentagon (I), tetragon (II), and triangle (III).

In the P–T plots, the univariant curves are labeled with the missing phase. With C designating the gas phase, the line C which is almost vertical to the temperature axis represents the univariant equilibrium for coexistent condensed

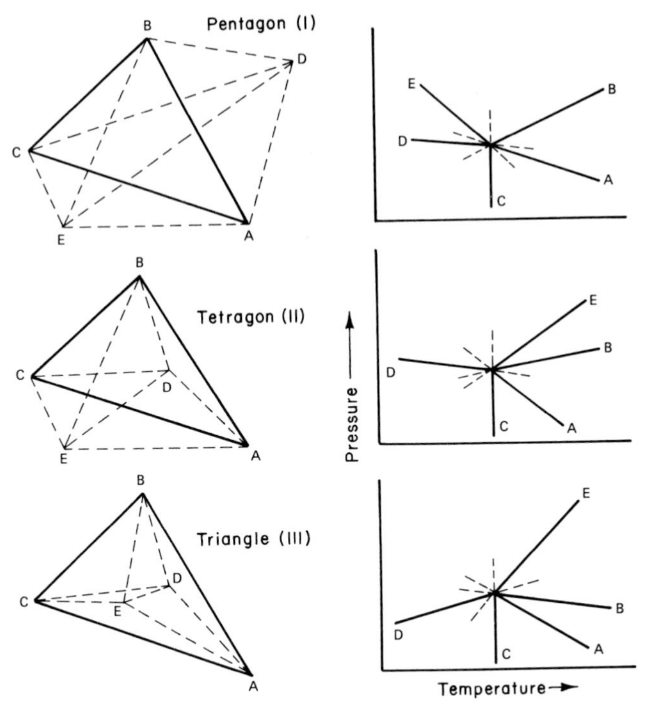

FIG. 2.3 Classification of composition diagrams for an invariant equilibrium in a ternary system, and the corresponding P–T plots for univariant curves intersecting at an invariant point. After Roozeboom and Schreinemakers [3].

phases A, B, D, and E. Let us consider univariants D and E: starting with the pentagon arrangement of phases in the composition diagram, when the composition of phase D is imagined to move closer to the line joining A and B, the slope of the univariant E in the P–T plot (I) increases. When the position of D in the composition diagram is collinear with AB, the univariant E coincides with the metastable extension of C. When the position of D is imagined to move to the left of the line AB as in the tetragon arrangement, the univariant E tilts clockwise closer to the univariant B. Similarly, we see how the slope of the univariant D would change when the position of phase E in the composition diagram is imagined to move across the line AC as from the tetragon to the triangle arrangement.

The following statements summarize the essential features of the theorem on the interrelation of univariant curves intersecting at an invariant point as applied to a ternary system.

(1) In the composition diagram, if the positions of three phases are collinear, two univariant curves containing these three phases are identical, i.e., the curves meet at the invariant point without a discontinuity.

(2) Two univariant curves are adjacent if the three phases common to them have compositions such that collinearity may be achieved by an imaginary movement of their positions in the composition diagram without crossing any of the interconnecting lines. For example, in the tetragon arrangement, D can be made collinear only with AB, AC, and BE by an imaginary movement of its position without crossing any of the interconnecting lines; therefore, the adjacent univariants in II, all containing phase D, are C–E, B–E, and A–C.

(3) The stable part of the univariant with the missing phase C is adjacent to the metastable extension of the univariant with the missing phase E if the positions of C and E in the composition diagram are both on the same side of the lines joining the compositions of the other three phases A, B, D as in the pentagon, tetragon, and triangle arrangements.

(4) The stable part of the univariant with the missing phase B is adjacent to the stable part of the univariant with the missing phase E if the positions of B and E are on the opposite sides of the lines joining the compositions of the other three phases A, C, D as in the tetragon arrangement.

With this theorem it is possible to predict the relative positions of the univariant curves in the vicinity of an invariant point from the compositions of the phases involved. Roozeboom and Schreinemakers derived equations for quantitative application of their theorem to binary and ternary systems. A more general form of this theorem was later developed by Morey and Williamson [5, 6] for multicomponent systems. Before discussing the quantitative application of this theorem, let us consider the extension of the theorem to quaternary systems.

When the theorem is applied to quaternary systems, the statements given above for ternary systems involving the concept of collinearity of the positions of three phases in the composition diagram would apply to a four-phase plane for the quaternary system. In this case, the relation between the adjacent univariant curves is derived from the relative positions of the other two phases with respect to the four-phase plane.

For a quaternary system the compositions of six phases of an invariant equilibrium are given in a three-dimensional diagram, i.e., a tetrahedron, each corner representing one of the phases. Depending on the composition of the phases, the polyhedron constructed for an invariant equilibrium may be a hexahedron (type I), an octahedron (type II), or a tetrahedron (type III). These are shown in Fig. 2.4 with the corresponding relations between the univariant curves in the P–T plots. Of the six phases, ABCD form the base tetrahedron. In type I (hexahedron) phases E and F are situated outside the base tetrahedron; in type II (octahedron) E is outside and F inside the base tetrahedron; in type III (tetrahedron) phases E and F are inside the base tetrahedron. In all cases, there are 20 planes; ten planes merge at the composition of every phase and four planes intersect at a line joining the compositions of two phases.

With C taken to be the gas phase, the univariant C is drawn normal to the temperature axis. In type I, for example, the univariants B and F adjacent to C

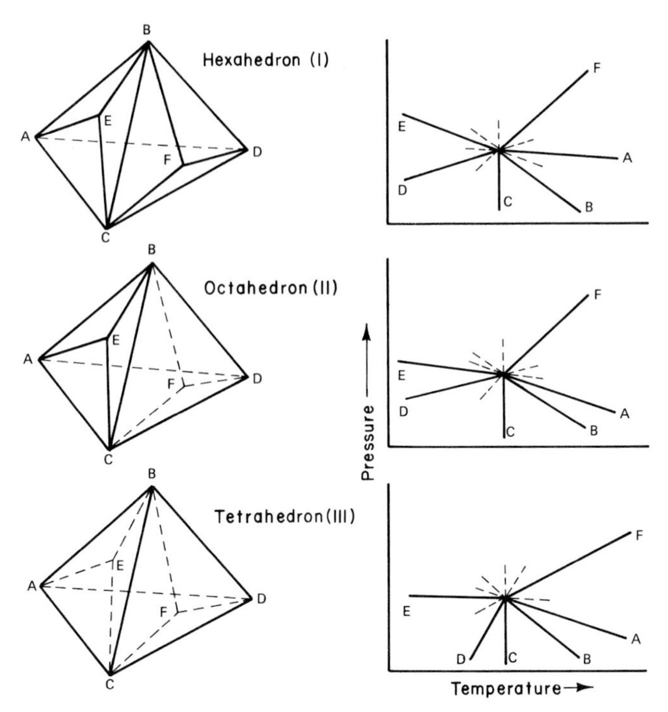

FIG. 2.4 Classification of composition diagrams for an invariant equilibrium in a quaternary system, and the corresponding P–T plots for univariant curves intersecting at an invariant point.

are found by seeking the formation of coplanar planes by changing the position of one of the condensed phases. It is seen that the phases ABDE can be made coplanar without crossing any of the three-phase planes. Since C and F both are on the same side of this plane, the stable part of the univariant C is adjacent to the metastable part of F. Another four-phase plane that can be formed in type I is ADFE; in this case B and C are on the opposite sides, and therefore the stable parts of the corresponding univariants are adjacent. Following this procedure, the sequence of the remainder of the univariant curves in the P–T plots are obtained, as shown in Fig. 2.4.

Needless to say, the application of the theorem on univariant curves to quaternary systems is complex and cumbersome. However, the problem may be overcome by considering a pseudoternary section of the system where there are two condensed phases common to all the univariant equilibria.

2.3 Univariant Equilibria at Constant Pressure

In a generalized treatment of the theorem on univariant equilibria, Morey and Williamson [5, 6] derived equations for dP/dT as a function of independent variables entropy S, volume V, and composition in the form of a determinant.

With the exception of geological systems, heterogeneous phase equilibria and reaction equilibria of experimental and practical interest are for pressures well below the kilobar range, and mostly at atmospheric pressure. To facilitate the use of the theorem in studies of phase equilibria, it is more meaningful to describe the univariant equilibrium in terms of chemical potential, or activity, at constant pressure as a function of temperature, enthalpy, and composition.

Darken [7] gave a constructive illustration of the application of this theorem to ternary systems at constant pressure, with particular reference to the iron–silicon–oxygen system, by converting the determinants of Morey and Williamson for P–T curves about an invariant equilibrium to the thermodynamic activity versus temperature relations. Although not given in his paper [7], Darken derived the following relation between the variations in activity and temperature for any univariant equilibrium in a ternary system at constant pressure.

$$\left(\frac{d\ln a_3}{d(1/T)}\right)_p = \left(\frac{1}{R}\right)\begin{vmatrix} x_1' & x_2' & \Delta H' \\ x_1'' & x_2'' & \Delta H'' \\ x_1''' & x_2''' & \Delta H''' \end{vmatrix} \Bigg/ \begin{vmatrix} x_1' & x_2' & 1 \\ x_1'' & x_2'' & 1 \\ x_1''' & x_2''' & 1 \end{vmatrix}, \tag{2.6}$$

where a_3 is the activity of component 3, the primes indicate coexistent three condensed phases, and ΔH is the enthalpy of the phase indicated by the prime. In the determinant the units for concentration and enthalpies of phases must be interconsistent. A convenient form for concentration is the atom ratio x_2/x_1, x_3/x_1, and enthalpy is that for per mole of component 1.

With the kind consent of Dr. Darken, his derivation of Eq. (2.6) is given in the next section.

2.3.1 A Note by Darken on the Univariant Curves for a Ternary System Involving Three Condensed Phases at Constant Pressure

Let us consider a single phase region (here designated by a prime) of a ternary system. Clearly, from the phase rule, the variance is 4. At internal equilibrium and constant pressure then, there remain three independent variables that establish any (and all) thermodynamic property. If we select $(\mu_1' - \mu_1^\circ)/T$ as the property to be considered and $1/T$, x_2', and x_3' as the independent variables, we may thus write the differential equation

$$d\,\frac{\mu_1' - \mu_1^\circ}{T} = \left[\frac{\partial(\mu_1' - \mu_1^\circ)/T}{\partial(1/T)}\right] d\,\frac{1}{T} + \left[\frac{\partial(\mu_1' - \mu_1^\circ)/T}{\partial x_2'}\right] dx_2' + \left[\frac{\partial(\mu_1' - \mu_1^\circ)/T}{\partial x_3'}\right] dx_3', \tag{2.6a}$$

or alternatively, noting that the terms on the left are equal to $R\,d\ln a_1$ and that the first partial derivative is equal to $\bar{H}_1' - H_1^\circ$, and multiplying through by x_1',

$$Rx_1'\left[\frac{d\ln a_1}{d(1/T)}\right] = x_1'(\bar{H}_1' - H_1^\circ) + \frac{x_1'}{T}\left[\frac{\partial(\mu_1' - \mu_1^\circ)}{\partial x_2'}\right]dx_2' + \frac{x_1'}{T}\left[\frac{\partial(\mu_1' - \mu_1^\circ)}{\partial x_3'}\right]dx_3'. \tag{2.6b}$$

Clearly, the following similar equations may also be written:

$$Rx_2'\left[\frac{d\ln a_2}{d(1/T)}\right] = x_2'(\bar{H}_2' - H_2^\circ) + \frac{x_2'}{T}\left[\frac{\partial(\mu_2' - \mu_2^\circ)}{\partial x_2'}\right]dx_2' + \frac{x_2'}{T}\left[\frac{\partial(\mu_2' - \mu_2^\circ)}{\partial x_3'}\right]dx_3', \tag{2.6c}$$

$$Rx_3'\left[\frac{d\ln a_3}{d(1/T)}\right] = x_3'(\bar{H}_3' - H_3^\circ) + \frac{x_3'}{T}\left[\frac{\partial(\mu_3' - \mu_3^\circ)}{\partial x_2'}\right]dx_2' + \frac{x_3'}{T}\left[\frac{\partial(\mu_3' - \mu_3^\circ)}{\partial x_3'}\right]dx_3', \tag{2.6d}$$

On addition, of Eqs. (2.6b)–(2.6d) we note that

$$x_1'(\bar{H}_1' - H_1^\circ) + x_2'(\bar{H}_2' - H_2^\circ) + x_3'(\bar{H}_3' - H_3^\circ) = \Delta H',$$

which is the enthalpy of formation of one mole of the phase (′) from the pure components, and that the sum of each set of partial derivatives is zero by virtue of the (isothermal) Gibbs–Duhem relation, Table 1.1; hence,

$$x_1'\left[\frac{d\ln a_1}{d(1/T)}\right] + x_2'\left[\frac{d\ln a_2}{d(1/T)}\right] + x_3'\left[\frac{d\ln a_3}{d(1/T)}\right] = \frac{\Delta H'}{R}. \tag{2.6e}$$

If we now consider two other condensed phases (″), (‴) simultaneously in equilibrium with the phase (′)—noting that temperature and each activity must be the same for all phases at equilibrium, i.e., the standard states must be the same for all phases—we may write the following analogous equations:

$$x_1''\left[\frac{d\ln a_1}{d(1/T)}\right] + x_2''\left[\frac{d\ln a_2}{d(1/T)}\right] + x_3''\left[\frac{d\ln a_3}{d(1/T)}\right] = \frac{\Delta H''}{R}, \tag{2.6f}$$

$$x_1'''\left[\frac{d\ln a_1}{d(1/T)}\right] + x_2'''\left[\frac{d\ln a_2}{d(1/T)}\right] + x_3'''\left[\frac{d\ln a_3}{d(1/T)}\right] = \frac{\Delta H'''}{R}. \tag{2.6g}$$

Simultaneous solution of Eqs. (2.6e)–(2.6g) yields the values of the total derivatives; for example,

$$\left[\frac{d\ln a_3}{d(1/T)}\right]_P = \frac{1}{R}\begin{vmatrix} x_1' & x_2' & \Delta H' \\ x_1'' & x_2'' & \Delta H'' \\ x_1''' & x_2''' & \Delta H''' \end{vmatrix} \Bigg/ \begin{vmatrix} x_1' & x_2' & x_3' \\ x_1'' & x_2'' & x_3'' \\ x_1''' & x_2''' & x_3''' \end{vmatrix} = \frac{1}{R}\begin{vmatrix} x_1' & x_2' & \Delta H' \\ x_1'' & x_2'' & \Delta H'' \\ x_1''' & x_2''' & \Delta H''' \end{vmatrix} \Bigg/ \begin{vmatrix} x_1' & x_2' & 1 \\ x_1'' & x_2'' & 1 \\ x_1''' & x_2''' & 1 \end{vmatrix}, \tag{2.6h}$$

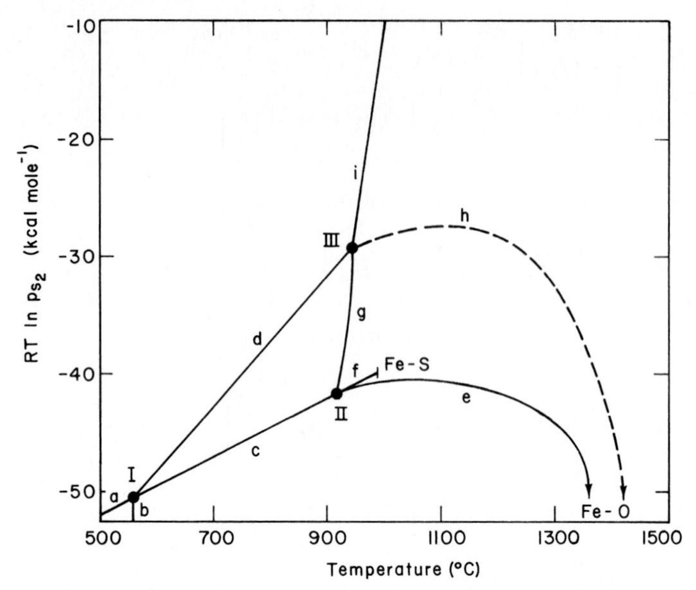

FIG. 2.5 Sulfur potentials for univariant equilibria in Fe–S–O system. W, wustite; M, magnetite; P, pyrrhotite; *l*, liquid oxysulfide. Univariants: (a) Fe,M,P; (b) Fe,W,M; (c) Fe,W,P; (d) W,M,P; (e) Fe,W,l; (f) Fe,P,l; (g) W,P,l; (h) W,M,l; (i) P,M,l. After Turkdogan and Kor [8].

where the second form follows from the first by virtue of the properties of determinants and noting that for each phase $x_1 + x_2 + x_3 = 1$.

Note that if there exists a maximum (or minimum) temperature for the three-phase equilibrium, then the determinator of either expression must be zero at this temperature and fixed pressure. This condition corresponds to collinearity of the three composition points on the isothermal isobaric composition plot.

2.3.2 Application of the Theorem

To demostrate the application of the theorem, let us consider the phase equilibria in the Fe–S–O system shown in Fig. 2.5 in terms of the sulfur chemical potential as a function of temperature [8, 9]. There are three stable invariant equilibria within the oxide–sulfide region of this system:

(I) iron, wustite, magnetite, pyrrhotite, gas;
(II) iron, wustite, pyrrhotite, liquid, gas;
(III) wustite, magnetite, pyrrhotite, liquid, gas.

Because of the negligible oxygen solubility in pyrrhotite, the compositions of iron, pyrrhotite, and sulfur are collinear in the composition plane; hence the univariants *a*, *c*, and *f* are essentially linear extensions of one another. The liquid compositions near the eutectic invariant II (915 °C) and the peritectic invariant III (942 °C) are nearly collinear with those of wustite and pyrrhotite. Consequently, the univariant *g* either passes through a maximum temperature or approaches the invariant III with almost infinite slope. The univariant curves terminate at the Fe–S and Fe–O binary systems as indicated in Fig. 2.5.

The univariant curves shown by full lines are calculated from the available thermodynamic data. The position of the univariant *h*, shown by dotted lines, near the invariant point III is calculated by using Eq. (2.6). First, the enthalpy of the liquid oxysulfide at the invariant (III) temperature is evaluated from the slope of the univariant *i* and Eq. (2.6); this gives $\Delta H_l = -36.4$ kcal g atom^{-1} Fe. Noting that the activity of sulfur is proportional to $\sqrt{P_{S_2}}$, the slope of the univariant *h* at III is now obtained from the determinant.

$$\left[\frac{d\ln a_S}{d(1/T)}\right]_P = \frac{1}{R}\,\frac{\begin{vmatrix} 1 & 1.13 & -72.09 \\ 1 & 1.333 & -87.00 \\ 1 & 0.388 & -36.4 \end{vmatrix}\begin{matrix}\text{wustite}\\ \text{magnetite}\\ \text{liquid}\end{matrix}}{\begin{vmatrix} 1 & 1.13 & 0 \\ 1 & 1.333 & 0 \\ 1 & 0.388 & 0.654 \end{vmatrix}\begin{matrix}\text{wustite}\\ \text{magnetite}\\ \text{liquid}\end{matrix}} \tag{2.7}$$

If chemical potential is plotted against temperature, instead of $\ln a$ vs $1/T$, the slope of one can be obtained from the slope of the other by making use of the equality $T^2/dT = -1/d(1/T)$. Thus

$$\frac{d\ln a}{d(1/T)} = T\ln a - \frac{Td(T\ln a)}{dT}. \tag{2.8}$$

From Eq. (2.7), $d\ln a_S/d(1/T)$ is equal to $-29.4/R\,°\text{K}$ which, in turn, gives $dRT\ln P_{S_2}/dT = 0.024$ kcal mole^{-1} deg^{-1} for the slope of the univariant *h* near the invariant point III. The remainder of this univariant is not known; however, as is indicated in Fig. 2.5, it has to pass through a maximum and then approach the Fe–O invariant (1424 °C) asymptotically.

The isothermal sections can readily be constructed from the available data on phase equilibria in the Fe–S–O system. Figure 2.6a is for 1100 °C showing composition ranges for univariants *e* and *h* (three condensed phases), four two-phase equilibria, and one liquid phase. Figure 2.6b is for 1400 °C showing regions of the miscibility gap and other phase equilibria.

As another example, the SO_2 chemical potentials for univariant equilibria are shown in Fig. 2.7 for the Pb–S–O system from the work of Kellogg and Basu [10]. In this system there are two invariant points at 616 and 733 °C, and eight stable univariant equilibria described in terms of SO_2 chemical potentials. The fourth univariant curve to intersect at 733 °C invariant, not shown in Fig. 2.7, is for four phases—$PbSO_4 \cdot PbO$, $PbSO_4 \cdot 2PbO$, $PbSO_4 \cdot 4PbO$, and gas of composition SO_3. Since the compositions of these four phases are collinear, this is a special invariant equilibrium for the PbO–SO_3 binary system; therefore, it does not appear in Fig. 2.7.

Areas between the lines for univariant equilibria in Fig. 2.7 are two-phase regions. Another useful form of the chemical potential diagrams for ternary systems is the isothermal section for bivariant equilibria showing the phase stability regions in terms of two isothermal chemical potentials. Such equilibria involving two condensed phases and a gas phase are shown in Fig. 2.8 for the

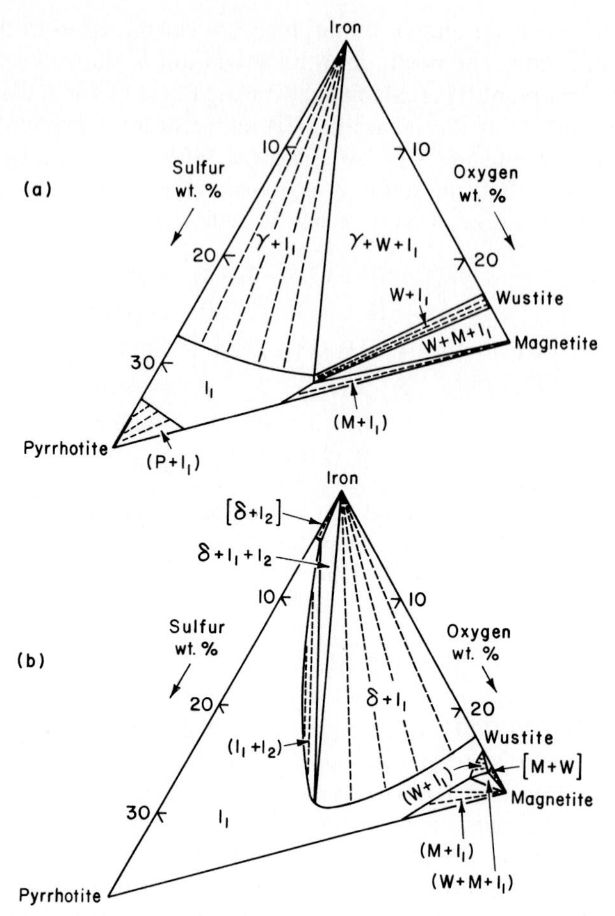

FIG. 2.6 Isotherms for Fe–S–O system for (a) 1100 °C and (b) 1400 °C. γ, gamma iron; W, wustite; M, magnetite; P, pyrrhotite; l_1, liquid oxysulfide; l_2, liquid metal. After Turkdogan and Kor [8].

Pb–S–O system at 1100 °K as given by Kellogg and Basu. Intersection of three bivariant equilibria corresponds to the univariant equilibrium. Areas between the lines for bivariant equilibria are regions of single condensed phases. The diagrams of this type are most helpful in the study of reactions occurring in the roasting of sulfide ores, calcination of sulfates and carbonates, and many smelting processes, for which examples are given in other chapters.

Similar phase diagrams are drawn to describe the regions of phase stability in aqueous solutions, by plotting the half-cell potential E_h against the pH of the aqueous solution. The E_h–pH diagrams, also known as Pourbaix diagrams, are used extensively in electrochemical, hydrometallurgical, and geological applications [11, 12].

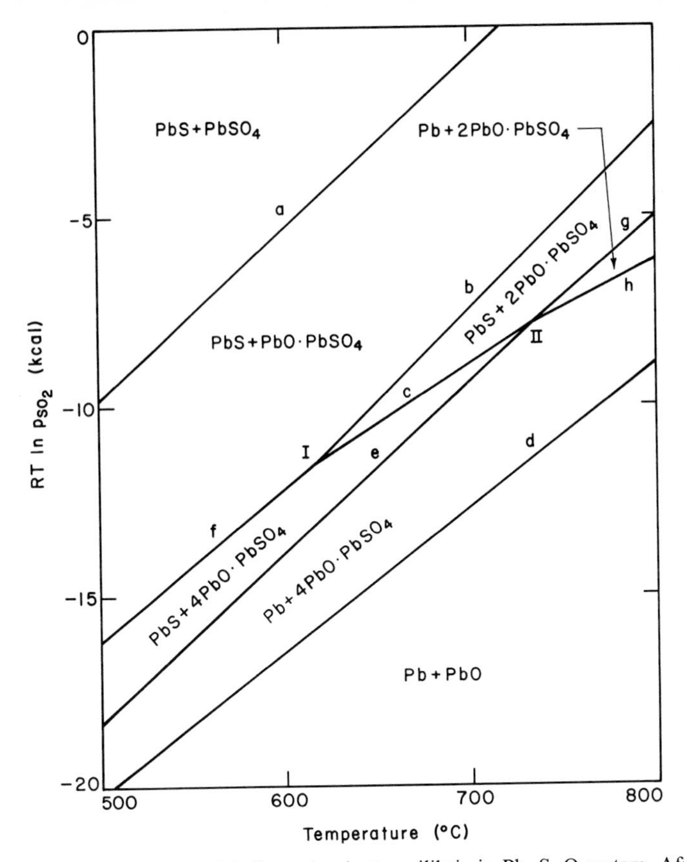

FIG. 2.7 SO_2 chemical potentials for univariant equilibria in Pb–S–O system. After Kellogg and Basu [10].

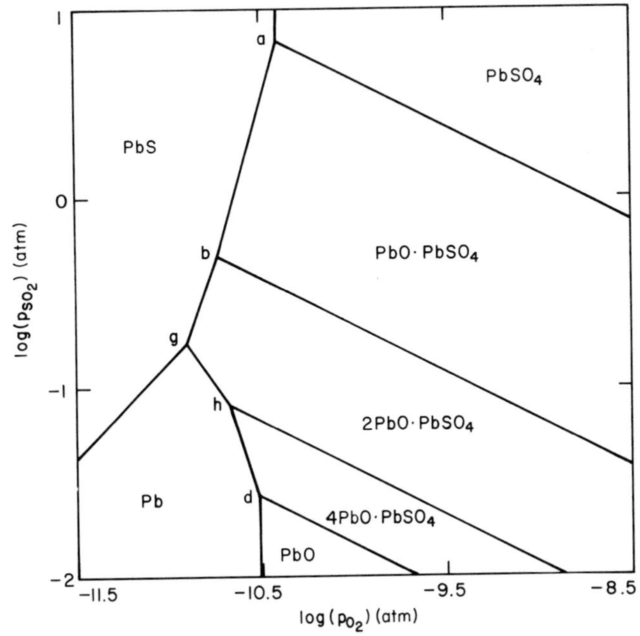

FIG. 2.8 Bivariant equilibria in Pb–S–O system at 1100°K. After Kellogg and Basu [10].

2.4 Interrelations between Phase Diagrams and Thermodynamic Functions

Since the early work of van Larr [13, 14], many studies have been made of theoretical relations between thermodynamic functions and the phase equilibrium diagrams. With the accumulation of experimental data on the thermodynamic properties of materials, much use has now been made of the thermodynamic functions for more accurate evaluation of the phase diagrams.

Schottky *et al.* [15] derived the following thermodynamic functions for the temperature dependence of the composition of coexisting phases in binary systems at constant pressure:

$$\frac{dx_2'}{dT} = -\frac{(1 - x_2'')\Delta H_1 + x_2''\Delta H_2}{(x_2'' - x_2')T(d^2F'/dx_2'^2)}, \tag{2.9}$$

$$\frac{dx_2''}{dT} = -\frac{(1 - x_2')\Delta H_1 + x_2'\Delta H_2}{(x_2'' - x_2')T(d^2F''/dx_2''^2)}, \tag{2.10}$$

where x_2 is the atom fraction of component 2 in coexisting phases (′) and (″), ΔH_1 the enthalpy change accompanying the reversible transfer of component 1 from phase (′) to phase (″), similarly ΔH_2 for component 2, and F', F'' are the integral molar free energies for phases (′) and (″). For some applications it is convenient to substitute the first derivatives of the partial molar free energies for the second derivatives of the integral molar free energies. Thus, from the equation for function 10 in Table 1.1, and noting that $dx_1 = -dx_2$,

$$d\bar{F}_1/dx_2 = -x_2(d^2F/dx_2{}^2), \tag{2.11a}$$

$$d\bar{F}_2/dx_2 = (1 - x_2)(d^2F/dx_2{}^2). \tag{2.11b}$$

As a special case, let us consider the equilibrium distribution of solute 2 between two coexistent phases in which the solute concentration is low. For $x_2' \ll x_2'' \ll 1$, Eq. (2.10) is simplified to

$$\frac{dx_2''}{dT} = -\frac{\Delta H_{t1}}{(x_2'' - x_2')T(d^2F''/dx_2''^2)} \tag{2.12}$$

where ΔH_{t1} is the heat of transformation of one mole of the solvent from phase (′) to phase (″). For the dilute solutions considered the laws of ideal solutions may be assumed; therefore,

$$\frac{d\bar{F}_2}{dx_2''} = RT\frac{d\ln x_2''}{dx_2''} = \frac{RT}{x_2''}. \tag{2.13}$$

Substituting Eqs. (2.11b) and (2.13) in Eq. (2.12),

$$dT/dx_2'' = (RT^2/\Delta H_{t1})[1 - (x_2'/x_2'')]. \tag{2.14}$$

In ideal dilute solutions, the solute distribution ratio is constant $x_2'/x_2'' = k$; inserting this in Eq. (2.14) and integrating for small values of x_2'' gives for the depression of the transformation temperature ΔT

$$\Delta T = (RT_t^2/\Delta H_{t1})(1 - k)x_2'', \tag{2.15}$$

where T_t is the transformation temperature of the pure component 1.

For a eutectic system with negligible terminal solid solution, i.e., $x_2' \simeq 0$, Eq. (2.10) is simplified to

$$\frac{dx_2''}{dT} = -\frac{\Delta H_m}{x_2''T_m(d^2F''/dx_2''^2)},$$

where $\Delta H_m = \Delta H_1$ is the enthalpy of fusion and T_m is the melting temperature of pure component 1. Substituting for $d^2F/dx_2{}^2$ from Eq. (2.11a), and assuming that the difference in the specific heats of solid and liquid is small, integration gives for the liquidus temperature T_l

$$T_l - T_m = (T_m/\Delta H_m)\,\Delta\bar{F}_1, \tag{2.16}$$

where $\Delta\bar{F}_1$ is the partial molar free energy of solution of component 1 in the liquid phase. Noting that the entropy of fusion ΔS_m is equal to $\Delta H_m/T_m$, gives for the liquidus temperature

$$T_l = (\Delta H_m + \bar{L}_1)/(\Delta S_m + \Delta\bar{S}_1), \tag{2.17}$$

where $\bar{L}_1$ and $\Delta\bar{S}_1$ are the partial molar enthalpy and entropy of solution of component 1 in the liquid phase at the liquidus temperature T_l.

The liquidus curve for the zinc–tin system calculated by Kubaschewski and Chart [16] (Fig. 2.9), using Eq. (2.17) gives results in good agreement with the direct measurements of Heycock and Neville [17].

Equation (2.16) can be used also to calculate the activity of component 1 in the liquid along the liquidus curve from the phase diagram. For pure liquid as the standard state,

$$\Delta\bar{F}_1 = RT\ln a_1 = (\Delta H_m/T_m)(T_l - T_m), \tag{2.18}$$

$$\ln a_1 = \Delta S_m(T_l - T_m)/RT, \tag{2.19}$$

where a_1 is the activity. With the specific heats of solid and liquid component 1 included, the exact form of Eq. (2.18) becomes [18]

$$RT\ln a_1 = \frac{\Delta H_m}{T_m}(T_l - T_m) + T\int_{T_m}^{T_l}\frac{dT}{T^2}\int_{T_m}^{T_l}\Delta C_{p1}\,dT, \tag{2.20}$$

where ΔC_p is the difference between the specific heats of liquid and solid phases. For metals ΔC_p is less than 0.7 cal g atom^{-1} deg^{-1}; therefore, Eq. (2.19) is adequate. For oxides ΔC_p is much higher, e.g., 2–10 cal mole^{-1} deg^{-1}, therefore, it would be necessary to use Eq. (2.20). It should be noted that the activities obtained are for varying temperatures along the liquidus curve; if the enthalpy of solution is known, conversion may be made to isothermal activity.

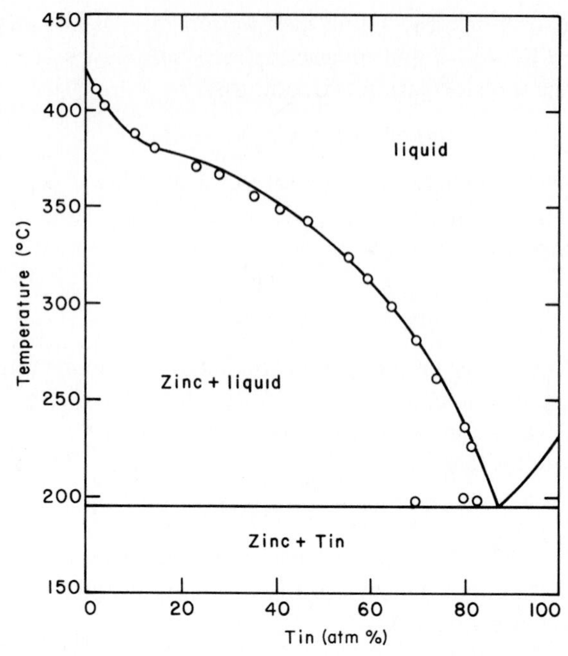

FIG. 2.9 Calculated liquidus curve for zinc–tin system compared with experimental results (○) of Heycock and Neville [17]. After Kubaschewski and Chart [16].

Above equations are for negligible solid solution of component 2 in 1. For small ranges of solid solubility, the Raoult's law may be assumed for the solid solution and Eq. (2.19) is modified to

$$\ln a_1 = \ln x_1\,(\text{solid}) + \Delta S_{\mathrm{m}}(T_1 - T_{\mathrm{m}})/RT. \tag{2.21}$$

In metallic systems where there is complete solid and liquid solubility, the liquidus–solidus gap is narrow and cannot be measured with sufficient accuracy by usual experimental techniques. Realizing this problem, Wagner [19] derived an equation for the width of the temperature gap $\Delta\theta$ for binary metallic solutions which are completely miscible, and are assumed to obey the laws of regular solutions. The following equation was derived for the equiatomic composition:

$$\Delta\theta = \frac{x_1{}'\,\Delta H_{\mathrm{m}1} + x_2{}'\,\Delta H_{\mathrm{m}2}}{RT^2/x_1''x_2'' + T(d^2F^{x''}/dx_2''^2)}\left(\frac{dT}{dx_2''}\right), \tag{2.22}$$

where the primes (′) and (″) indicate solid and liquid phases, respectively, $F^{x''}$ is the excess integral molar free energy of the liquid, and $\Delta H_{\mathrm{m}1}$ and $\Delta H_{\mathrm{m}2}$ are enthalpies of fusion of pure components 1 and 2. In carefully conducted experiments with silver–gold alloys by White [20], and with an iron–chromium (equiatomic) alloy by Cook and Hume-Rothery [21], the liquidus–solidus gaps measured were found to be in close agreement with those calculated from Eq. (2.22).

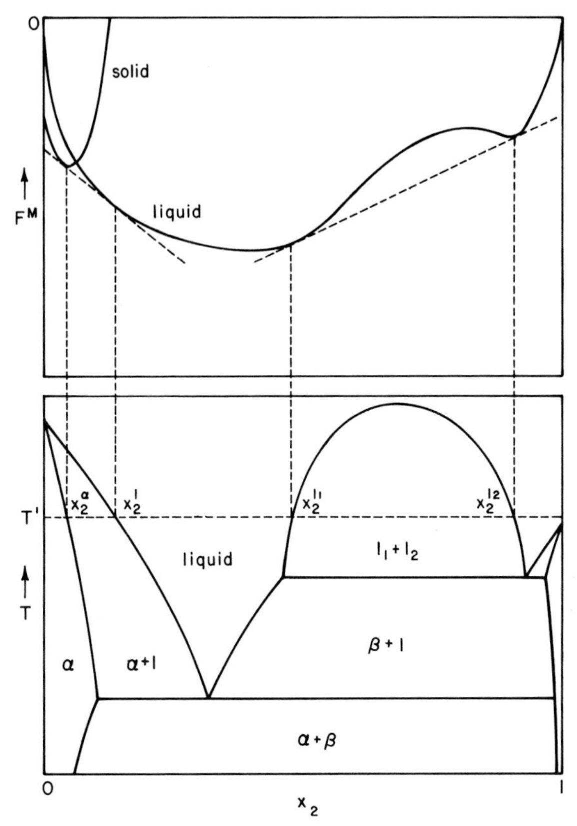

FIG. 2.10 Free-energy and temperature-composition diagrams illustrating graphical method of determining phase diagram from free-energy curves for solid and liquid solutions.

When a miscibility gap forms, the partial molar free energies are independent of composition. That is, within the range of composition of the miscibility gap for a given temperature $d\bar{F}_2/dx_2$ is equal to zero, and by virtue of the relation in Eq. (2.11), the second derivative $d^2F/dx_2{}^2$ is also zero. Inserting this in Eq. (2.10) gives dx_2/dT equal to zero within the range of the miscibility gap.

2.4.1 Graphical Method

If partial molar free energies are known as functions of temperature and composition for all the phases present in a binary system, the phase equilibrium diagram may be constructed graphically by drawing common tangents to the free-energy curves. As an example, let us consider a binary system with partial terminal solid solutions and a liquid miscibility gap as shown in Fig. 2.10. The integral molar free energies of solution are given in the top diagram for the α solid solution and the liquid phase at temperature T'. The tangent that is common to both free-energy curves gives the partial molar free energies in coexistent phases. Therefore, for alloy compositions within the range $x_2{}^\alpha$ and $x_2{}^1$, the solid solution of composition $x_2{}^\alpha$ is in equilibrium with liquid of composition $x_2{}^1$. The hump on the hypothetical free-energy curve for the liquid phase indicates immiscibility. The common tangent $x_2^{1_1}x_2^{1_2}$ depicts the composition range of the miscibility gap at T'. Since the free energy of mixing of liquids 1_1 and 1_2 must ride along the common tangent, the hump on the integral molar free energy is for the metastable liquid.

Graphical methods of deriving binary phase diagrams from free-energy curves have been described in detail by Darken and Gurry [22], and in other textbooks. Numerical examples are given by Kubaschewski and Chart [16] for several binary metallic systems, and Hillert [23] demonstrated the use of free-energy functions and common tangent methods in describing the role of interfaces in phase transformations.

2.4.2 Computer Calculations

Since the mid 1960s, there has been a rapid growth of interest in the calculation of phase diagrams from compiled thermochemical data using computer techniques. Basically there are two methods of approach: equality of chemical potential and minimization of the integral free energy of solution.

In a multicomponent system containing n components, there may be, at constant pressure, $\phi \leqslant (n+1)$ number of coexisting phases in which the chemical potential of each component is the same. This equilibrium condition for component i in coexisting phases, $\alpha, \beta, \gamma, \ldots, \psi$ at constant temperature and pressure is represented by

$$\begin{aligned}\Delta\bar{F}_{i\alpha} &= \Delta\bar{F}_{i\beta} + F_i{}^\circ(\beta-\alpha) = \Delta\bar{F}_{i\gamma} + \Delta F_i{}^\circ(\gamma-\alpha)\\ &= \Delta\bar{F}_{i\psi} + \Delta F_i{}^\circ(\psi-\alpha),\end{aligned} \tag{2.23}$$

where $\Delta\bar{F}_i$ is the chemical potential of i in the phase indicated by the subscript, and $\Delta F_i{}^\circ$ denotes the change in the free energy of the pure component i accompanying the phase transformation. The equilibrium composition that satisfies the equality of chemical potential in ϕ number of phases is obtained by simultaneous solution of n number of equations of type (2.23).

In the second method of approach, the equilibrium composition is calculated by minimizing the sum of the integral free energies of solution in ϕ number of possible phases,

$$\Delta F^{\mathrm{M}} = x_\alpha \Delta F_\alpha{}^{\mathrm{M}} + x_\beta \Delta F_\beta{}^{\mathrm{M}} + x_\gamma \Delta F_\gamma{}^{\mathrm{M}} + \cdots + x_\psi \Delta F_\psi{}^{\mathrm{M}}, \tag{2.24}$$

where x_α, $x_\beta \ldots$ are the mole fractions of the phases so that

$$x_\alpha + x_\beta + x_\gamma + \cdots + x_\psi = 1. \tag{2.25}$$

In order to perform such computer calculations, it is necessary to represent the free-energy functions in suitable mathematical forms using quasi-chemical solution models and/or empirical relations, some of which are discussed in

Chapter 3. The papers cited [24–33] give numerous examples of calculations of phase equilibrium diagrams by both of these methods for binary, ternary, and quaternary systems. As examples, the results of calculations of Counsell *et al.* [33] for isothermal sections of Fe–Cu–Ni and Fe–Cr–Ni are reproduced in Figs. 2.11 and 2.12.

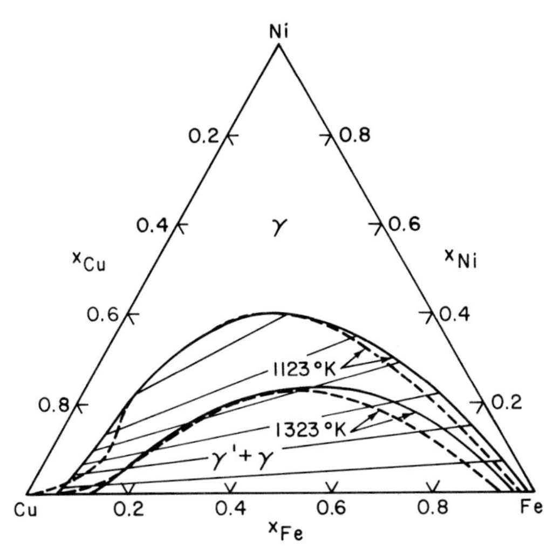

FIG. 2.11 Calculated phase boundaries for a two-phase region in Fe–Cu–Ni system at 1123 and 1323°K compared with experimental boundaries determined by Bradley *et al.* [34]. After Counsell *et al.* [33]. Dashed line—experimental; straight line—calculated.

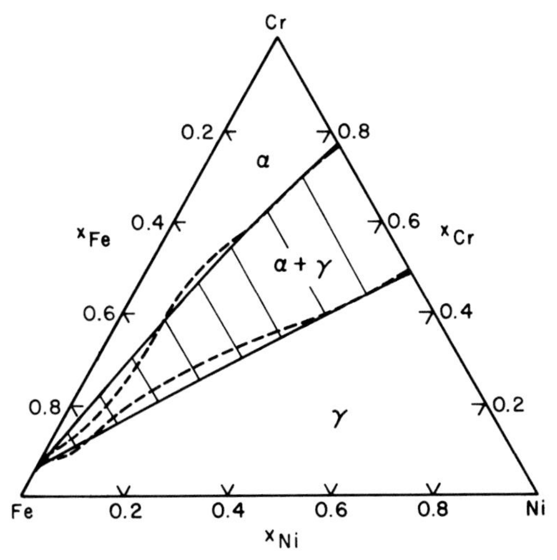

FIG. 2.12 Calculated phase boundaries of two-phase region in Fe–Cr–Ni system at 1550°K compared with experimental boundaries determined by Jenkins *et al.* [35]. After Counsell *et al.* [33]. Dashed line—experimental; straight line—calculated.

2.5 Effect of Pressure on Phase Equilibria

Although modest changes in total pressure have a negligible effect on the temperature of phase transformation, or the temperature dependence of the composition of coexistent phases, the phase equilibrium diagrams are much altered at very high pressures in the kilobar range. The effect of pressure on phase equilibria becomes even more significant at elevated temperatures.

Much of the early studies of high pressure effects on metallurgical systems are attributed to the pioneering work of Bridgman [36]. In the 1960s, there was a rapid growth of research in the field of high temperature–high pressure effect on phase equilibria. This is well documented in a review paper by Klement and Jayaraman [37].

For any univariant equilibrium, the pressure dependence of the temperature of coexisting phases is given by the Clausius–Clapeyron equation, which may be derived also from one of Maxwell's relations, Table 1.1,

$$dT/dP = \Delta V/\Delta S = T\,\Delta V/\Delta H \tag{2.26}$$

where ΔV, ΔS, and ΔH are changes in volume, entropy, and enthalpy of the system accompanying the phase transformation at the temperature T and pressure P. With the units of ΔV in cm^3 g atom^{-1}, ΔS in cal g atom^{-1} deg^{-1} and P in kbar, the change of transition temperature with pressure is given by the relation

$$dT/dP = 23.9(\Delta V/\Delta S). \tag{2.27}$$

Because the variations of ΔV and ΔS with temperature, pressure, and composition are not usually known, only the initial slope dT/dP can be estimated at atmospheric pressure for which the appropriate data may be available.

In a study of entropy and volume changes accompanying melting of metals, Kubaschewski [38] noted that for hexagonal close-packed (hcp) metals the product $V_s(\Delta S/\Delta V)_{s\to l}$, where V_s is the molar volume of the solid, is about 57 cal g atom^{-1} deg^{-1}. Subsequent study of the available data [39] indicated that for most hcp metals the approximate relation

$$V_s\left(\frac{\Delta S}{\Delta V}\right)_{s\to l} \sim 60 \quad \text{cal g atom}^{-1}\ \text{deg}^{-1} \tag{2.28}$$

may be used to estimate the initial slope dT/dP for $P \to 1$ atm. There are other semiempirical relations which have been postulated [40–42] for the estimation of $(\Delta S/\Delta V)_{s\to l}$.

Most of the available data on phase transformations of metals at elevated pressures have been compiled by Klement and Jayaraman [37]; only a few examples need be given here. The phase relations for bismuth is shown in Fig. 2.13. The bismuth is often used for pressure calibration of the apparatus from the known pressure-induced polymorphs.

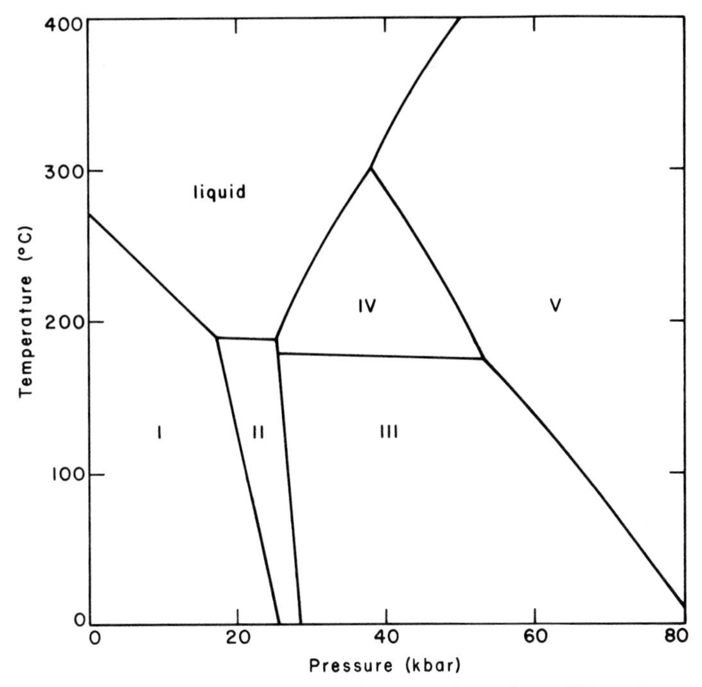

FIG. 2.13 Effect of pressure on phase transformations of bismuth.

The effect of pressure on phase transformations of iron has been studied extensively, particularly for the α–γ transformation; compiled data [28, 37] are shown graphically in Fig. 2.14. The invariant point for the α (bcc)–γ (fcc)–ε (hcp) equilibrium at about 500 °C and 110 kbar was determined by Johnson *et al.* [43] with the shock-wave (dynamic-pressure) measurements. The effect of pressure on the melting temperature of iron was determined by Strong [44]; the position of the invariant point for the δ–γ–liquid equilibrium has not yet been evaluated. The tangents drawn to the experimental phase boundaries at $P \to 1$ atm agree well with those calculated from Eq. (2.27) using the known values of ΔS and ΔV.

The isothermal change in composition as a function of pressure for a univariant equilibrium in binary systems, with $x_2' \ll x_2'' \ll 1$, is derived by combining Eqs. (2.12) and (2.26).

$$\left(\frac{dx_2''}{dP}\right)_T = \frac{x_2''(1 - x_2'')}{x_2' - x_2''}\left[RT\left(1 + \frac{\partial \ln \gamma_2''}{\partial \ln x_2''}\right)\right]^{-1} \Delta V, \tag{2.29}$$

where $\Delta V = V' - (1 - x_2'')\bar{V}_2'' - x_2''\bar{V}_2''$, V' is the molar volume of phase ($'$), $\bar{V}''$ the partial molar volume of phase ($''$), x_2 the atom fraction of component 2, and γ_2'' the activity coefficient of component 2. With this equation and the available thermodynamic data on the Fe–C system, Hilliard calculated

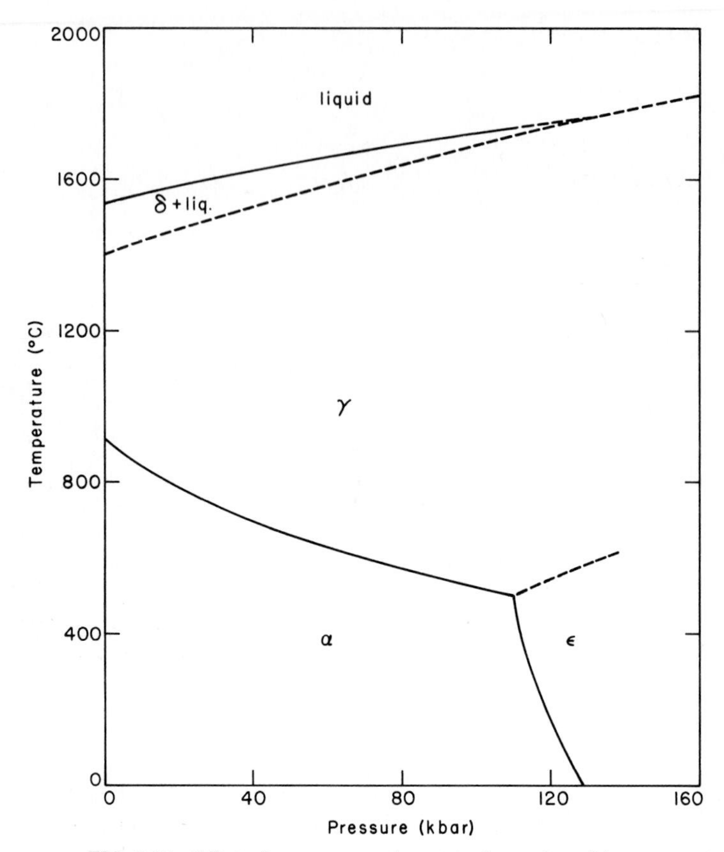

FIG. 2.14 Effect of pressure on phase transformation of iron.

the following limiting slopes. For (γ) iron—Fe_3C univariant at 727 °C, $(\partial \log x_C/\partial P)_T = -9.5 \times 10^{-3}\ \text{kbar}^{-1}$; for ($\gamma$) iron—($\alpha$) iron univariant at 727 °C, $(\partial x_C/\partial P)_T = -1.3 \times 10^{-3}\ \text{kbar}^{-1}$.

The pronounced effect of pressure on the phase boundaries of the Fe–C system measured by Hilliard is shown in Fig. 2.15. From these studies, it was concluded that Fe_3C coexisting with saturated austenite will be stable with respect to both diamond and graphite at all pressures above 3.5 kbar, which is reverse of the condition prevailing at atmospheric pressure.

As another example, let us consider the SiO_2–H_2O system; the univariant equilibrium for this system, as determined by Kennedy *et al.* [46] and others [47–49], is shown in Fig. 2.16. There are two invariant equilibria: (I) at about 1470 °C and 0.4 kbar for cristobalite–tridymite–liquid I–liquid II equilibrium, and (II) at 1160°C and 1.5 kbar for tridymite–quartz–liquid I–liquid II equilibrium. There is a critical end point at 1080 °C and 9.7 kbar above which only

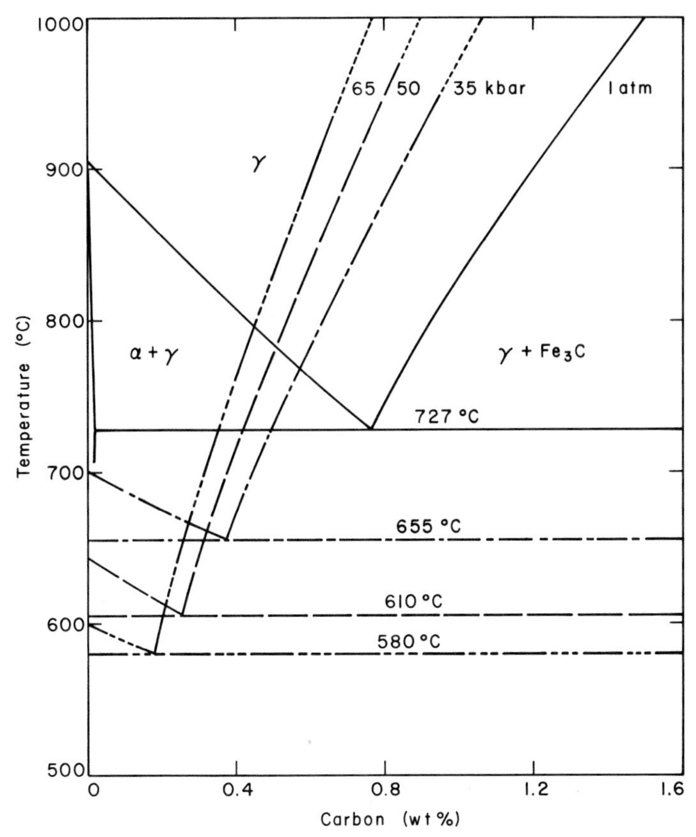

FIG. 2.15 Effect of pressure on positions of phase boundaries in iron–carbon system. After Hilliard [45].

quartz and a liquid phase coexist. The miscibility gap univariant in Fig. 2.17 shows the compositions of the H_2O-rich and SiO_2-rich fluids in equilibrium with silica as a function of pressure. The SiO_2 content of the fluid phase increases from 1.8 mole % at 2 kbar to about 48 mole % at the critical endpoint. A small initial H_2O pressure, for example, of 0.4 kbar sharply decreases the melting point of silica from 1720 to 1470 °C and introduces about 10 mole % H_2O into molten silica.

The knowledge of pressure effect on phase equilibria in mineralogical systems has been of much assistance in understanding the major geological problems and metamorphism. For example, the possibility of the formation of silica-rich critical fluid at terrestrial pressures and temperatures, and the precipitation of silica at lower pressures may explain the occurrences of quartz veins in many mineral deposits and metamorphic terranes.

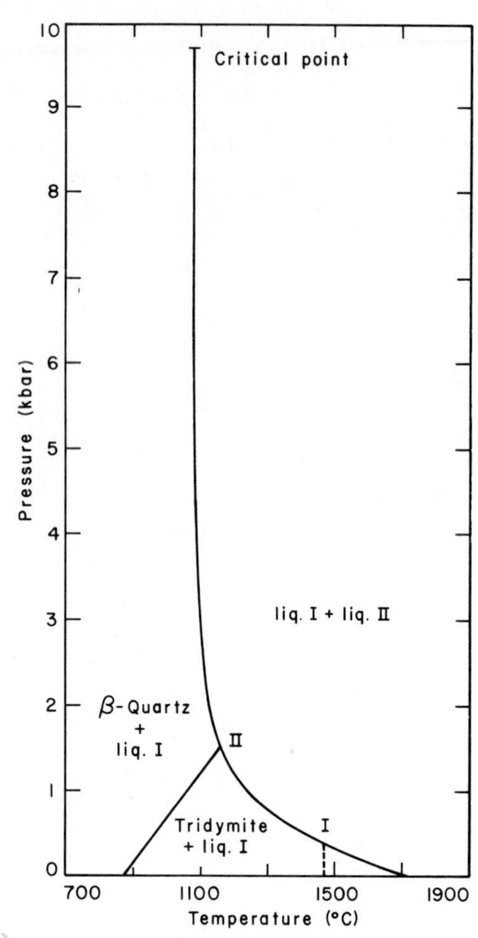

FIG. 2.16 Univariant equilibrium curves for SiO_2–H_2O system. After Kennedy *et al.* [46].

2.6 Correlations Based on Electron–Atom Ratios

Much has been learned about the thermodynamic properties of metallic solid solutions and intermetallic phases from concepts based on the electron theory of metals; notable contributions made by Hume-Rothery were particularly outstanding. There are many books and proceedings of symposia devoted to this subject which cannot be covered adequately here. However, when discussing phase equilibria in metallic systems, some consideration should be given to correlations based on electron–atom ratios.

From a study of the relative effects of various alloying elements on the liquid$\rightleftharpoons$solid and δ/γ transition points in iron-rich binary alloys, Buckley and Hume-Rothery [50] showed the importance of melting point and size factor differences between solute and solvent elements in determining the effects on

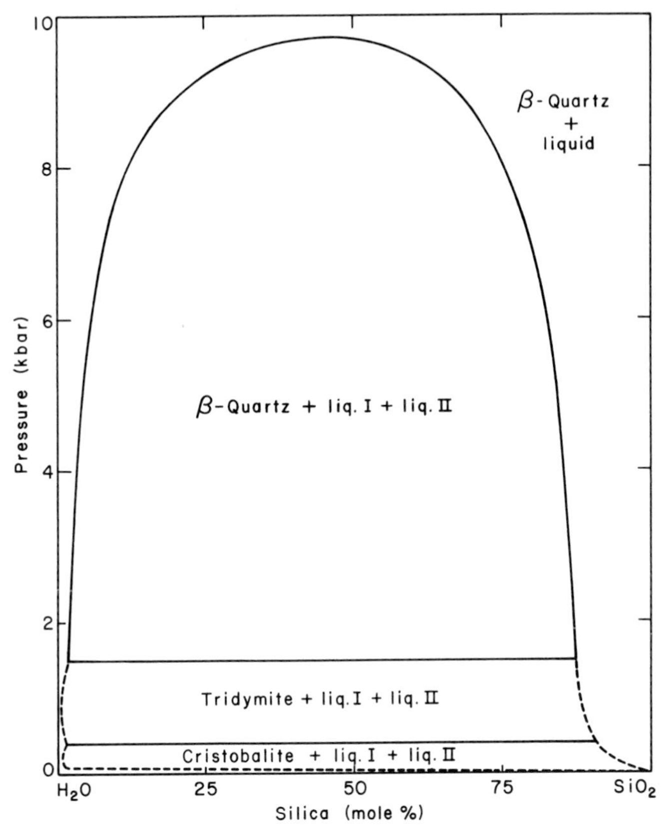

FIG. 2.17 Miscibility–gap univariant for SiO_2–H_2O system. After Kennedy *et al.* [46].

the liquid⇌solid equilibrium. The free-energy change accompanying the transfer of solute from solid to liquid was used as a measure of the liquid stabilizing power of the element, and was considered to be the sum of melting point effect ΔF_{m} and lattice-disturbing effects ΔF_{d} due to electronic and size effects.

$$\Delta F = \Delta F_{\mathrm{m}} + \Delta F_{\mathrm{d}}. \tag{2.30}$$

At infinitely dilute solutions, i.e., $x_2 \to 0$, combined form of Eqs. (2.9) and (2.10) is reduced to

$$\Delta F = -RT \ln[(dx_2/dT)_{\mathrm{l}} - (dx_2/dT)_{\mathrm{s}}]. \tag{2.31}$$

The melting point effect is obtained from the entropy of fusion of the solute element ΔS_{m},

$$\Delta F_{\mathrm{m}} = -\Delta S_{\mathrm{m}} \Delta T, \tag{2.32}$$

where ΔT is the difference between the liquidus temperature and the melting point of the pure solutes.

From these equations, and known values of ΔS_m and measured values of $(dx/dT)_{l,s}$

$$\Delta F_d = -RT \ln\left[\left(\frac{dx_2}{dT}\right)_l - \left(\frac{dx_2}{dT}\right)_s\right] + \Delta S_m \Delta T. \tag{2.33}$$

The results obtained by Buckley and Hume-Rothery are reproduced in Fig. 2.18; the size factor in the lower diagram is the percent difference between the atom size of the solute and the solvent (iron). As they pointed out, the ΔF values for solutes from the first long period are less than those in the second and third long periods. With the exception of Zr, the observed trend emphasizes the size factors and electronic factors as major contributors to the lattice disturbing effects.

The effects of different solutes on the $\delta \rightleftharpoons \gamma$ transformation were derived in a similar way, and the results (Fig. 2.19) were interpreted by Buckley and Hume-Rothery in terms of s and d hybridization of the bonding electrons in the solute

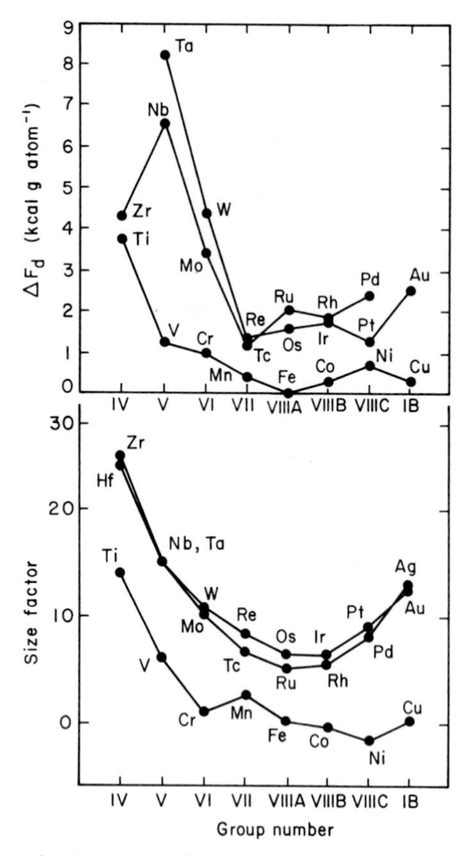

FIG. 2.18 Variation of ΔF_d and size factor with group number for transition elements in solution in iron. After Buckley and Hume-Rothery [50].

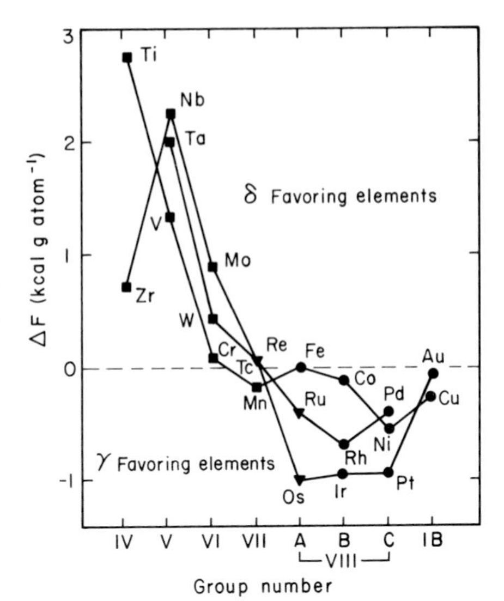

FIG. 2.19 Solute free-energy difference ΔF for the $\gamma \rightarrow \delta$ transition of iron. Legend: ■, bcc; ●, fcc; ▲, cph.

atoms, and pointed out that the electronic factors are more important than size factors in this case of phase stability.

From experimental observations, Engel [51, 52] postulated a correlation between the valence state of electrons and the crystal structure of metal atoms as an extension of the $(8 - N)$ rule proposed by Hume-Rothery [53, 54].* According to the Engel correlation, the body-centered-cubic (bcc) structure is formed when the valence state corresponds to one s electron ($d^{n-1}s$); when the valence state corresponds to one s electron and one p electron ($d^{n-2}sp$), the hexagonal close-packed (hcp) structure is formed; and when the valence state corresponds to one s electron and two p electrons ($d^{n-3}sp^2$), the face-centered-cubic (fcc) structure is expected.

In subsequent studies of many metallic systems, Brewer [55–58] refined and broadened Engel's concept. In the application of Engel–Brewer correlation, the promotion energies required to obtain the various unpaired electronic configurations are derived from the spectroscopic data. The bonding energy that results when the gaseous atoms are condensed to the solid metal is given by the sum of enthalpy of sublimation and the promotion energy for a given electronic configuration. The valence state that yields the lowest bonding energy determines the structure that is stable at high temperatures.

By recognizing nonintegral electron distributions in the subshells, the following characteristic electron–atom ratios have been postulated: for the bcc, <1–1.75 electrons/atom; for the hcp, 1.8–2.2 electrons/atom; and for the fcc, 2.25–3 electrons/atom. As shown by Brewer [56], other factors such as atom size and internal pressure differences have to be recognized in fixing the stability

* In the crystal of the atom of group N in the periodic table of elements, each atom has $8 - N$ nearest neighbors.

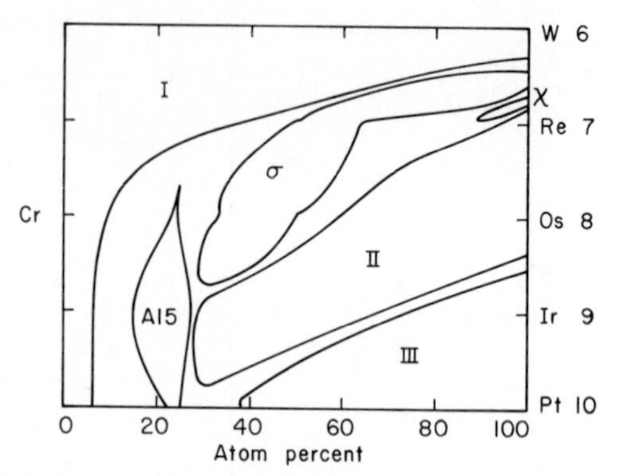

FIG. 2.20 Phase stability ranges in binary and multicomponent alloys of Cr with W, Re, Os, Ir, and Pt. After Brewer [56].

ranges of the various phases. Also, depending on the atom size ratios, there are other phases to be considered: *L* (Laves phases),* *A*15 (cubic Cr_3Si type), σ (tetragonal β-Mn type), ϕ (tetragonal HfRe type), χ (cubic α-Mn type), etc. In further development of these ideas, Brewer and Wengert [59] arrived at a generalized description of Lewis' acid–base reaction by considering the transfer of d electrons from transition metal atoms that have empty d orbitals, e.g., Nb, to other transition metal atoms that have internally paired d electrons, e.g., Ru.

The application of Engel–Brewer correlation in the calculation of multicomponent phase diagrams, by using heats of sublimation, solubility parameters, and spectroscopic data, has been described by Brewer [56]. Of the 30 diagrams thus derived for the transition metals, Fig. 2.20 is reproduced here as an example for binary and multicomponent alloys of Cr with W, Re, Os, Ir, and Pt. The phase diagram is the projection along the temperature axis, so that the composition ranges given are maximum ranges at the optimum temperature for each phase. The number after each element is the number of valence electrons per atom. The top horizontal line represents the binary Cr–W system; a horizontal line at the level of the element Os represents the binary Cr–Os system, and so on. The horizontal line halfway between Re and Os is for the ternary Cr–Re–Os system with the atom ratio Re/Os = 1. The same horizontal line also shows phases formed when Cr is alloyed with an equimolal mixture of W and Ir, or any mixture of W, Re, Os, Ir, and Pt which has an average 7.5 valence electrons/atom. A single diagram of this type gives information on composition ranges of all possible phases of high thermodynamic stability in a total of 25 systems with Cr as one of the components.

Despite some misgivings about the theoretical aspects of the concept [60, 61], the Engel–Brewer correlation has been substantiated by data on many metallic

* Laves phases are intermetallic compounds AB_2 crystallizing with $MgCu_2$, $MgZn_2$, or $MgNi_2$ type structure in which metallic bonding predominates.

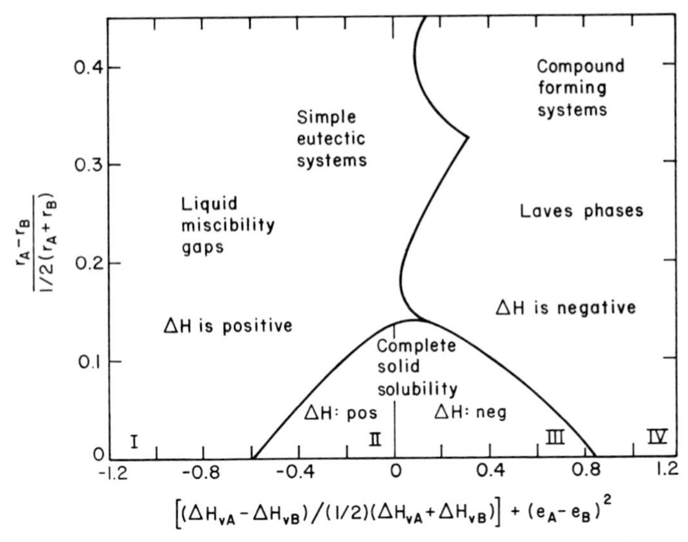

FIG. 2.21 Regions of preference for the formation of certain types of binary equilibrium diagrams as depicted by atomic radius, heat of vaporization, and electronegativity of the component metals. After Kubaschewski [62].

systems and found to be helpful in predicting high temperature stable phases in multicomponent systems of transition metals.

Hume-Rothery held the view that the type of phase equilibrium diagram formed in binary metallic systems depends primarily on the atom size factor, the electron concentration, and the electrochemical factor. Based on these thoughts, Kubaschewski [62] developed a correlation for binary equilibrium diagrams involving average atom size and average heat of vaporization of component metals. This is shown in Fig. 2.21, where the size factor $(r_A - r_B)/\frac{1}{2}(r_A + r_B)$ is plotted against the vaporization energy factor $[(\Delta H_{vA} - \Delta H_{vB})/\frac{1}{2}(\Delta H_{vA} + \Delta H_{vB})] + (e_A - e_B)^2$, where r is atomic radius ($r_A > r_B$), ΔH_v heat of vaporization at 25 °C, and e the electronegativity as defined by Pauling [63]. Since the metals tend to be electropositive, the electronegativity term is smaller than the vaporization energy term. The boundaries drawn by Kubaschewski indicate regions of transition from one type of phase diagram to another. Of the 350 binary systems considered, about 20 do not fit the pattern in Fig. 2.21, e.g. Cr–W, Mg–Ni, Ca–Na, and systems involving Mn and U whose atomic diameters are difficult to assess.

2.7 Metastable Phase Equilibria

The coexisting phases can be in a state of stable or metastable equilibrium, depending on the constraint imposed on the system by the kinetics of the phase transformation involving nucleation and diffusion, and the contribution to the free energy of the system arising from the interface around the new phase initially formed. The metastable phases are common occurrences in nature and in man-made materials, e.g., glass, ceramics, and metal alloys, many of which have unusual properties of practical importance. An effective means of obtaining metastable phases is by rapidly changing the temperature of the system at rates of the order of 10^5 °C sec^{-1} as in splat cooling [64, 65]. The metastable phases are also formed in metal–film deposition by vacuum evaporation or cathodic sputtering [66, 67] and in catalyzed electrodeposition [68, 69].

Many studies have been made of the thermodynamic interpretation of the formation of metastable phases by rapid cooling of liquids [70–73]. This is illustrated by considering a simple binary eutectic system *A–B* with limited terminal solid solubilities as shown in Fig. 2.22. In the lower diagram, the free-energy curves are given for: (1) the liquid supercooled to the indicated temperatures $F_l(T)$, at which the solidification begins, and (2) the solid solution with a weak F_s or strong F_s' departure from Raoult's law. For simplicity, the

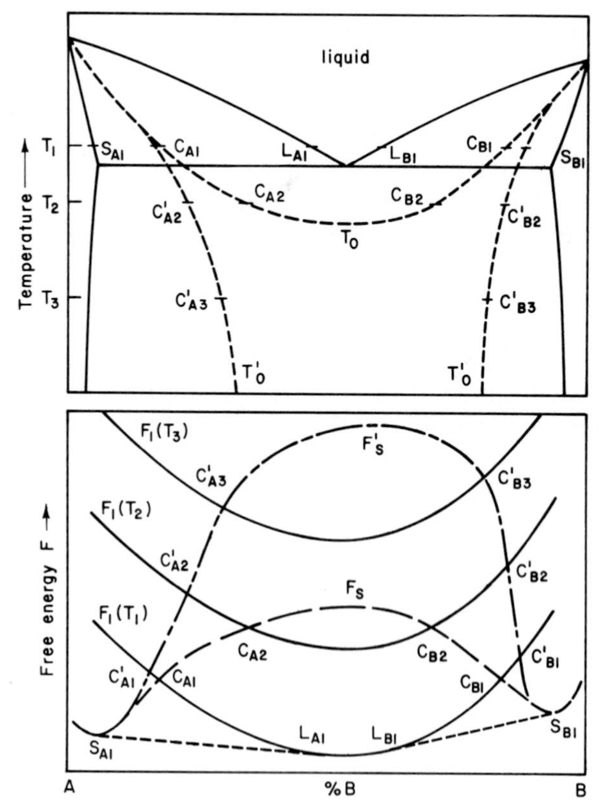

FIG. 2.22 Schematic representation of extended metastable solid solutions formed upon quenching to undercooling temperatures T_1, T_2, and T_3, and construction of T_0 curves from the free-energy curves for liquid and solid phases.

F_s curves are assumed to be independent of temperature. The common tangents to the free-energy curves at T_1 give the compositions of the coexisting phases S_{A1}, L_{A1}, for the conditions of equilibrium freezing of the supercooled liquid. For melt compositions below C_{A1}, the free energy $F_l(T_1) > F_s$. The melt crystallizes, therefore, to a single phase without change of composition, known as diffusionless solidification. For melt compositions between C_{A1} and L_{A1}, the solidification results in a two-phase structure containing mixed crystals with the concentration of B varying between S_{A1} and C_{A1}. Hence, the crossover point C_{A1} of the F_l and F_s curves gives the limit of metastable solid solubility C_{A1} of component B in A; similarly, C_{B1} is the extended limiting solubility of A in B.

The free energy of the liquid increases with decreasing supercooling temperature. Therefore, lowering of the supercooling temperature increases the limit of metastable solid solubility of B in A, and A in B. The *loci* of the crossover points, designated by T_0 and T_0', are shown by the dotted curves in Fig. 2.22 for the free-energy curves F_s and F_s', respectively. At the supercooling temperature T_3, the free-energy curve $F_l(T_3)$ lies above the F_s curve but intersects the F_s' curve at C'_{A3} and C'_{B3}. This indicates complete metastable solid solution for F_s at all compositions, and limited solubilities for F_s'. In fact, for the case of F_s' considered, there can be no continuous series of metastable solid solutions, no matter how low the supercooling temperature. For example, as shown by Duwez *et al.* [74], complete metastable solid solution is possible in the Ag–Cu system by rapid quenching of the melt to 0 °C; on the other hand, in the Cd–Zn system the extent of metastable solid solution is limited to the terminal regions [75].

A criterion for the formation of a metastable phase was estimated by Baker and Cahn [73] through thermodynamic calculations based on a Carnot cycle and experimentally observed limits of supercooling to suppress nucleation. If the free-energy increase in the formation of the metastable phase is more than about one-third the heat of fusion of the metal, i.e., $\Delta F > 0.3\,\Delta H_m$, it is impossible to produce a metastable solid by quenching the melt. For vapor quenching on a very cold substrate, the limit is much higher, i.e., $\Delta F \sim \Delta H_v$, where ΔH_v is the heat of vaporization. Therefore, vapor quenching on a very cold substrate offers a greater potential for the formation of metastable phases.

For information on experimental data for metastable equilibria in metallic systems and their properties, reference may be made to the review papers cited [72, 76–78].

References

1. "The Collected Works of J. Willard Gibbs," Vol. 1. Longmans, Green, New York, 1928.
2. E. H. Baker, *J. Chem. Soc.* p. 464 (1962).
3. H. W. B. Roozeboom and F. A. H. Schreinemakers, "Die Heterogenen Gleichgewichte," Vol. 3, Part I. Wieweg, Braunschweig, 1911.

4. F. A. H. Schreinemakers, *Proc. K. Ned. Akad. Wet.* **18**, 116 (1916).
5. G. W. Morey and E. D. Williamson, *J. Am. Chem. Soc.* **40**, 59 (1918).
6. G. W. Morey, "Commentary on the Scientific Writings of J. Willard Gibbs," Vol. 1. Yale Univ. Press, New Haven, Connecticut, 1936.
7. L. S. Darken, *J. Am. Chem. Soc.* **70**, 2046 (1948).
8. E. T. Turkdogan and G. J. W. Kor, *Metall. Trans.* **2**, 1561 (1971).
9. G. J. W. Kor and E. T. Turkdogan, *Metall. Trans.* **3**, 1269 (1972).
10. H. H. Kellogg and S. K. Basu, *Trans. Metall. Soc. AIME* **218**, 70 (1960).
11. M. Pourbaix, "Atlas of Electrochemical Equilibria." Pergamon, Oxford, 1966.
12. R. M. Garrels and C. L. Christ, "Solutions, Minerals and Equilibria." Harper, Row, New York, 1965.
13. van J. J. Larr, *Versl. Gewone Vergad. Wis-Natvvrkd. Afd., K. Akad. Wet. Amsterdam* **11**, 478 (1903); **12**, 25 (1904).
14. van J. J. Larr, *Z. Phys. Chem., Stoechiom. Verwandschaftsl.* **63**, 216 (1908); **64**, 257 (1908).
15. W. Schottky, H. Ulich, and C. Wagner, "Thermodynamik." Springer-Verlag, Berlin, 1929.
16. O. Kubaschewski and T. G. Chart, *J. Inst. Met.* **93**, 329 (1964–1965).
17. C. T. Heycock and F. H. Neville, *J. Chem. Soc.* **71**, 383 (1897).
18. C. Wagner, "Thermodynamics of Alloys." Addison-Wesley, Reading, Massachusetts, 1952.
19. C. Wagner, *Acta Metall.* **2**, 242 (1954).
20. J. L. White, *Trans. Metall. Soc. AIME* **215**, 178 (1959).
21. C. J. Cooke and W. Hume-Rothery, *Acta Metall.* **9**, 982 (1961).
22. L. S. Darken and R. W. Gurry, "Physical Chemistry of Metals." McGraw-Hill, New York, 1953.
23. M. Hillert, *in* "The Mechanism of Phase Transformations in Crystalline Solids," Monogr. Rep. Ser. No. 33, p. 231. Inst. Met., London, 1969.
24. J. L. Meijering, *Acta Metall.* **5**, 257 (1957).
25. D. T. J. Hurle and E. R. Pike, *J. Mater. Sci.* **1**, 399 (1966).
26. N. J. Olson and G. W. Toop, *Trans. Metall. Soc. AIME* **245**, 905 (1969).
27. I. Ansara, P. Desre and E. Bonnier, *J. Chim. Phys.* **66**, 297 (1969).
28. L. Kaufman and H. Bernstein, "Computer Calculation of Phase Diagrams." Academic Press, New York, 1970.
29. H. Gaye and C. H. P. Lupis, *Scr. Metall.* **4**, 685 (1970).
30. J. F. Counsell, E. B. Lees, and P. J. Spencer, *Met. Sci. J.* **5**, 210 (1970).
31. L. Kaufman and H. Nesor, *Metall. Trans.* **5**, 1617, 1623 (1974); *Metall. Trans. A* **6**, 2115, 2123 (1975).
32. T. G. Chart, J. F. Counsel, G. P. Jones, W. Slough, and P. J. Spencer, *Int. Metall. Rev.* **20**, 57 (1975).
33. J. F. Counsell, E. B. Lees, and P. J. Spencer, *in* "Metallurgical Chemistry" (O. Kubaschewski, ed.), p. 451. HM Stationery Off., London, 1972.
34. A. J. Bradley, W. F. Cox, and H. J. Goldschmidt, *J. Inst. Met.* **67**, 189 (1941).
35. C. H. M. Jenkins, E. H. Bucknall, C. R. Austin, and G. A. Mellor, *J. Iron Steel Inst., London* **136**, 215 (1937).
36. H. Brooks, F. Birch, G. Holton, and W. Paul, "Collected Experimental Papers of P. W. Bridgman." Harvard Univ. Press, Cambridge, Massachusetts, 1964.
37. W. Klement, Jr. and A. Jayaraman, *Prog. Solid State Chem.* vol. 3 p. 289 (1967).
38. O. Kubaschewski, *Trans. Faraday Soc.* **45**, 931 (1949).
39. A. Schneider and G. Heymer, *in* "The Physical Chemistry of Metallic Solutions and Intermetallic Compounds," Pap. 4A. HM Stationery Off., London, 1959.
40. J. Lumsden, "Thermodynamics of Alloys." Inst. Met. London, 1952.
41. F. Simon, *Z. Elektrochem.* **35**, 618 (1929); *Trans. Faraday Soc.* **33**, 65 (1937).
42. J. Gilvarry, *Phys. Rev.* **102**, 308, 317, 325 (1956).
43. P. C. Johnson, B. A. Stein, and R. S. Davis, *J. Appl. Phys.* **33**, 557 (1962).

44. H. M. Strong, *in* "Progress in Very High Pressure Research" (F. P. Bundy, W. R. Hibbard, Jr., and H. M. Strong, eds.), p. 182. Wiley, New York, 1961.
45. J. E. Hilliard, *Trans. Metall. Soc. AIME* **227**, 429 (1963).
46. G. C. Kennedy, G. J. Wasserburg, H. C. Heard, and R. C. Newton, *Am. J. Sci.* **260**, 501 (1962).
47. O. F. Tuttle and J. L. England, *Geol. Soc. Am., Bull.* **66**, 149 (1955).
48. O. F. Tuttle and N. L. Bowen, *Geol. Soc. Am. Mem.* **74**, 153 (1958).
49. I. A. Ostrovskii, G. P. Mishina, and V. M. Povilaitis, *Dokl. Akad. Nauk SSSR* **126**, 645 (1959).
50. R. A. Buckley and W. Hume-Rothery, *J. Iron Steel Inst., London* **201**, 227 (1963).
51. N. Engel, *Kem. Maanedsbl. Nord. Handelsbl. Kem. Ind.* **30**, 53, 75, 97, 105, 114 (1949).
52. N. Engel, *Powder Metall. Bull.* **7**, 8 (1954).
53. W. Hume-Rothery, "The Structure of Metals and Alloys," Monogr. Rep. Ser., No. 1. Inst. Met., London, 1936.
54. W. Hume-Rothery, "Atomic Theory for Students of Metallurgy," Monogr. Rep. Ser., No. 3. Inst. Met., London, 1946.
55. L. Brewer, *in* "Electronic Structure and Alloy Chemistry of the Transition Elements" (P. A. Beck, ed.), p. 221. Wiley (Interscience), New York, 1963.
56. L. Brewer, *in* "High-Strength Materials" (V. F. Zackay, ed.), p. 12. Wiley, New York, 1965.
57. L. Brewer, *Acta Metall.* **15**, 553 (1967).
58. L. Brewer, *in* "Phase Stability in Metals and Alloys" (P. S. Rudman, J. Stringer, and R. I. Jaffee, eds.), p. 39. McGraw-Hill, New York, 1967.
59. L. Brewer and P. R. Wengert, *Metall. Trans.* **4**, 83 (1973).
60. Discussion, *in* "Phase Stability in Metals and Alloys" (P. S. Rudman, J. Stringer, and R. I. Jaffee, eds.), p. 39. McGraw-Hill, New York, 1967.
61. W. Hume-Rothery, *Prog. Mater. Sci.* **13**, 231 (1967).
62. O. Kubaschewski, *in* "Phase Stability in Metals and Alloys" (P. S. Rudman, J. Stringer, and R. I. Jaffee, eds.), p. 63. McGraw-Hill, New York, 1967.
63. L. Pauling, "The Nature of the Chemical Bond." Cornell Univ. Press, Ithaca, New York, 1960.
64. P. Duwez, *Prog. Solid State Chem.* **3**, 377 (1967); *Trans. Am. Soc. Met.* **60**, 607 (1967).
65. P. Duwez, *in* "Intermetallic Compounds" (J. H. Westbrook, ed.), p. 340. Wiley, New York, 1970.
66. L. I. Maissel and R. Glang (eds.), "Handbook of Thin-Film Technology." McGraw-Hill, New York, 1970.
67. K. L. Chopra, "Thin-Film Phenomena." McGraw-Hill, New York, 1969.
68. G. C. Cargill, III, *J. Appl. Phys.* **41**, 12 (1970).
69. J. P. Marton and M. Schlesinger, *J. Electrochem. Soc.* **115**, 16 (1968).
70. I. S. Misoschnichenko and I. V. Salli, *Izv. V.U.Z. Chernaya Metall.* **8**, 104 (1960).
71. H. Biloni and B. Chalmers, *Trans. Metall. Soc. AIME* **233**, 373 (1965).
72. B. C. Giessen and R. H. Willens, *in* "Phase Diagrams: Materials Science and Technology" (A. M. Alper, ed.), Vol. III, p. 103. Academic Press, New York, 1970.
73. J. C. Baker and J. W. Cahn, *in* "Solidification," p. 23, ASM, Metals Park, Ohio, 1970.
74. P. Duwez, R. H. Willens, and W. Klement, *J. Appl. Phys.* **31**, 1136 (1960).
75. J. C. Baker and J. W. Cahn, *Acta Metall.* **17**, 575 (1969).
76. H. Jones, in *Rep. Prog. Phys.* **36**, 1425 (1973).
77. A. K. Sinha, B. C. Giessen, and D. E. Polk, *in* "Treatise on Solid State Chemistry" (N. B. Hannay, ed.), p. 1. Plenum, New York, 1976.
78. N. J. Grant and B. C. Giessen (eds.), "Rapidly Quenched Metals." MIT Press, Cambridge, Massachusetts, 1977.

CHAPTER 3

PHYSICOCHEMICAL PROPERTIES OF METALS AND ALLOYS

3.1 Thermodynamic Properties

The development of new materials, novel methods of processing, and the mere want of knowledge have led to much being known about the thermodynamic properties of metals and alloys. Reference may be made to the compilations of Hultgren *et al.* [1, 2] on selected thermodynamic properties of elements and binary metal alloys, and to the compilations of Hansen and Anderko [3] and Shunk [4] on phase diagrams of binary metal systems. Many other sources of information are, of course, referred to in the following discussion of the subject with particular emphasis on liquid metal alloys. The theoretical approaches presented are confined to those which have been found to be particularly useful in correlating or forecasting physicochemical properties of metallic solutions.

3.1.1 Binary Systems

In an attempt to rationalize the variation of activity coefficient with composition, several solution models have been developed over the years. The first attempt was that made by Margules [5], who suggested a power series for the composition dependence of the activity coefficients γ_1 and γ_2 of a binary solution; thus for a given temperature*

$$\ln \gamma_1 = \alpha_1 x_2 + \tfrac{1}{2}\alpha_2 x_2{}^2 + \tfrac{1}{3}\alpha_3 x_2{}^3, \qquad \ln \gamma_2 = \beta_1 x_1 + \tfrac{1}{2}\beta_2 x_1{}^2 + \tfrac{1}{3}\beta_3 x_1{}^3. \quad (3.1)$$

* It should be noted that Margules did not use the term activity coefficient which was first introduced by Lewis in 1907 (see this volume, Chapter 1 [8]). Margules used the ratio p/x which is, of course, proportional to γ.

For these equations to hold over the entire composition range, the Gibbs–Duhem relation indicates that the first coefficients α_1 and β_1 be zero. Also, if the system can be represented by the quadratic terms only, then α_2 is equal to β_2. Subsequently, Porter [6] showed that the quadratic approximation

$$\ln \gamma_1 = \alpha x_2^2 \qquad \text{and} \qquad \ln \gamma_2 = \alpha x_1^2 \tag{3.2}$$

holds for several binary nonelectrolyte solutions. On the basis of these observations, Hildebrand [7] developed the concept of regular solution on the assumption of random distribution of constituents with the entropy of mixing equal to the ideal configurational entropy. The regular solution behavior, or its modification through quasi-chemical considerations [8–10], does not hold for many solutions because (i) the entropy of mixing differs from the ideal configurational entropy, (ii) the enthalpy of mixing of one atom of component 1 in pure component 2 may not be equal to that for the transfer of one atom of 2 to pure 1, and (iii) the coordination number may not remain constant throughout the concentration range.

To obtain relations of wider applicability, Guggenheim [11] has suggested a series expansion for the concentration dependence of the excess integral molar quantities, i.e., excess free energy, enthalpy, and entropy.

$$G^{xs} = x_1 x_2 \sum_{\nu = 0,1,\ldots} A_\nu (x_1 - x_2)^\nu, \tag{3.3}$$

where $A_0, A_1, A_2, \ldots$ are coefficients to be evaluated from experimental data. In terms of activity coefficients

$$\begin{aligned} \ln \gamma_1 &= x_2^2 \left\{ \alpha_0 + \sum_{\nu = 1,2,\ldots} \alpha_\nu [(2\nu + 1)x_1 - x_2](x_1 - x_2)^{\nu - 1} \right\}, \\ \ln \gamma_2 &= x_1^2 \left\{ \alpha_0 + \sum_{\nu = 1,2,\ldots} \alpha_\nu [x_1 - (2\nu + 1)x_2](x_1 - x_2)^{\nu - 1} \right\}. \end{aligned} \tag{3.4}$$

If the terms of the series expansion are omitted, these equations are reduced to that for the regular solution, Eq. (3.2). Equation (3.4) with one or two terms in the series expansion represents activity data well in nonelectrolyte solutions and metallic solutions with moderate departures from ideality.

Because of the limitations of Eq. (3.4), Darken [12] has suggested that the entire concentration range may be divided into three regions: two terminal regions and a central region. He proposed a quadratic formalism for the concentration dependence of the activity coefficient in the two terminal regions. For region (I), $0 < x_2 < x_2'$,

$$\log(\gamma_2/\gamma_2^\circ) = \alpha_{12}(x_1^2 - 1) = \alpha_{12}(-2x_2 + x_2^2), \tag{3.5}$$

$$\log \gamma_1 = \alpha_{12}(1 - x_1)^2. \tag{3.6}$$

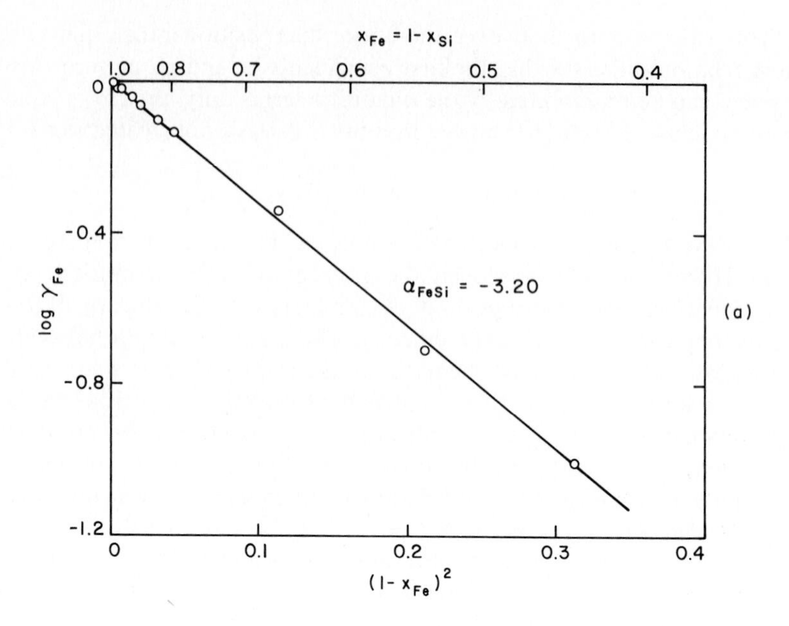

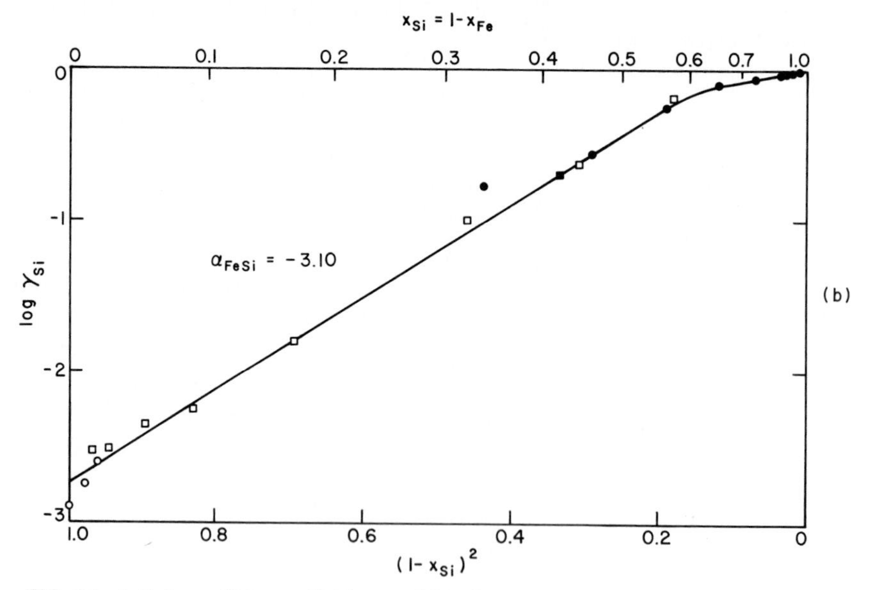

FIG. 3.1 Activity coefficients of (a) iron and (b) silicon in liquid iron–silicon alloys at 1600 °C. After Darken [12].

For region (II), $x_2'' < x_2 < 1$,

$$\log \gamma_2 = \alpha_{21}(1 - x_2)^2, \tag{3.7}$$

$$\log(\gamma_1/\gamma_1^\circ) = \alpha_{21}(x_2^2 - 1) = \alpha_{21}(-2x_1 + x_1^2), \tag{3.8}$$

where the superscript $^\circ$ indicates activity coefficient at infinite dilution and α's are temperature-dependent parameters to be determined from experimental data. It should be noted that these equations satisfy the limiting solution laws (Raoult's and Henry's) at infinite dilution.

The applicability of the quadratic formalism is well substantiated by the available activity data of which those in Figs. 3.1–3.3 are typical examples. The extent of terminal region varies from one system to another, commonly from 10 to 70 at. %. Consistency of independently measured activities of both components in a given binary system is readily checked by comparing the slopes α_{12} and α_{21}. For example, the value of α_{FeSi} from the slope in Fig. 3.1a is -3.20, which compares well with the corresponding value -3.10 from the slope in Fig. 3.1b. Similarly, the slopes of the lines in Fig. 3.2 for the potassium amalgams at 300 °C show consistencies between independent measurements of the activities of potassium [14, 15] and of mercury [16].

In the intermediate concentration range, there is either a sharp change in the slope $d\log\gamma_2/d(1 - x_2)^2$, as in the Fe–Si and Hg–K systems, or a stepwise displacement of the curve as in Fig. 3.3 for the Bi–Mg system at 700 °C [17]. The thermodynamic behavior in binary systems in the intermediate concentration range is described by Darken in terms of the second derivative of the molar free energy with mole fraction $d^2F/dx_2^2 \equiv d^2F/dx_1^2$, which Darken called the *stability*. The stability is evaluated by differentiating the isothermal and isobaric thermodynamic relation (10) in Table 1.1

$$\bar{F}_2 = F + (1 - x_2)(dF/dx_2), \tag{3.9}$$

where $\bar{F}_2$ is the partial molar free energy (chemical potential) of component 2 and F is the molar free energy. Differentiation and rearrangement gives

$$\text{stability} = \frac{d^2F}{dx_2^2} = \frac{1}{(1 - x_2)}\frac{d\bar{F}_2}{dx_2} = -2\frac{d\bar{F}_2}{d(1 - x_2)^2}, \tag{3.10}$$

$$\text{excess stability} = \frac{d^2F^{xs}}{dx_2^2} = -2\frac{d\bar{F}_2^{xs}}{d(1 - x_2)^2}, \tag{3.11}$$

In terms of activity and activity coefficient

$$\text{stability} = -4.605RT\frac{d\log a_2}{d(1 - x_2)^2}, \tag{3.12}$$

$$\text{excess stability} = -4.605RT\frac{d\log \gamma_2}{d(1 - x_2)^2}. \tag{3.13}$$

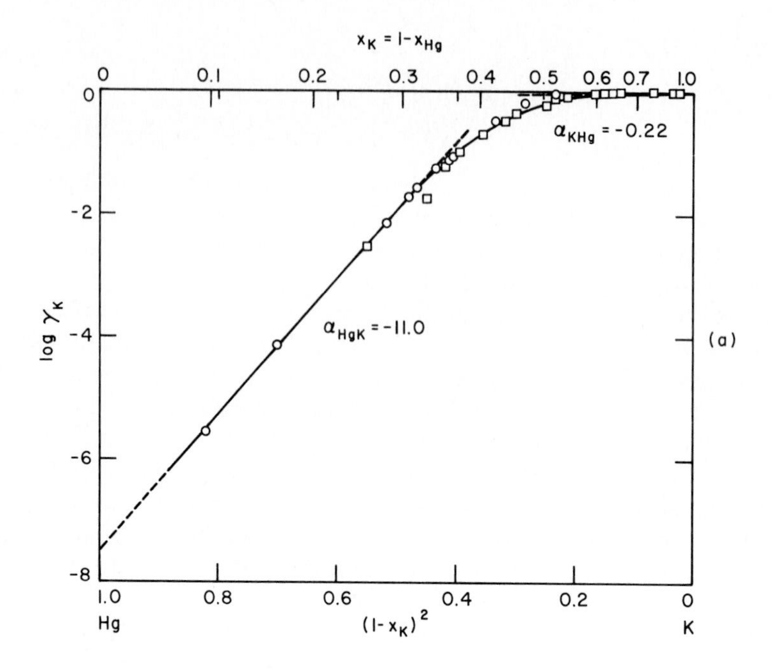

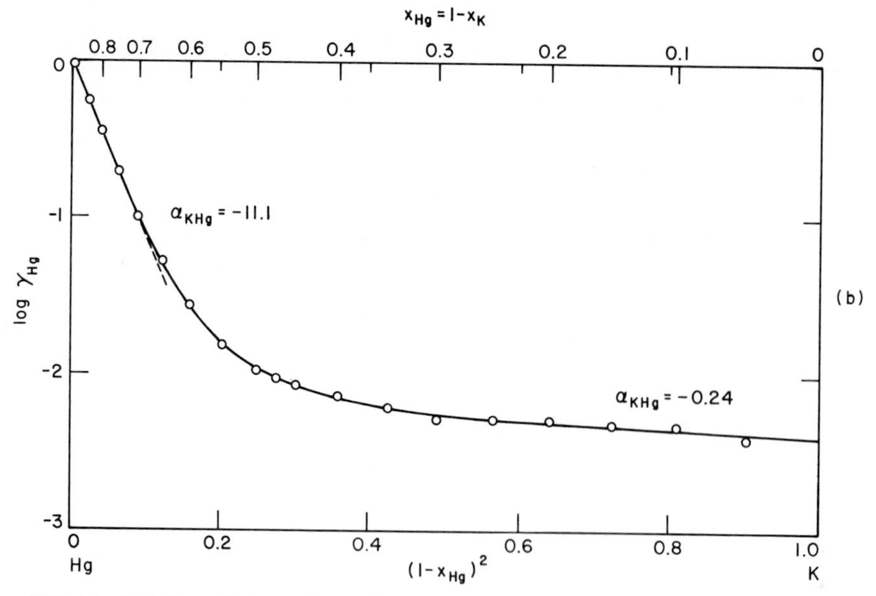

FIG. 3.2 Activity coefficients of potassium and mercury in potassium amalgams. (a) $\gamma_{K'}$, experimental data of Vierk and Hauffe (□) for 325 °C [14] and of Morachevskii (○) for 280 °C [15]. (b) γ_{Hg}, experimental data of Roeder and Morawietz for 300°C [16]. After Turkdogan and Darken [13].

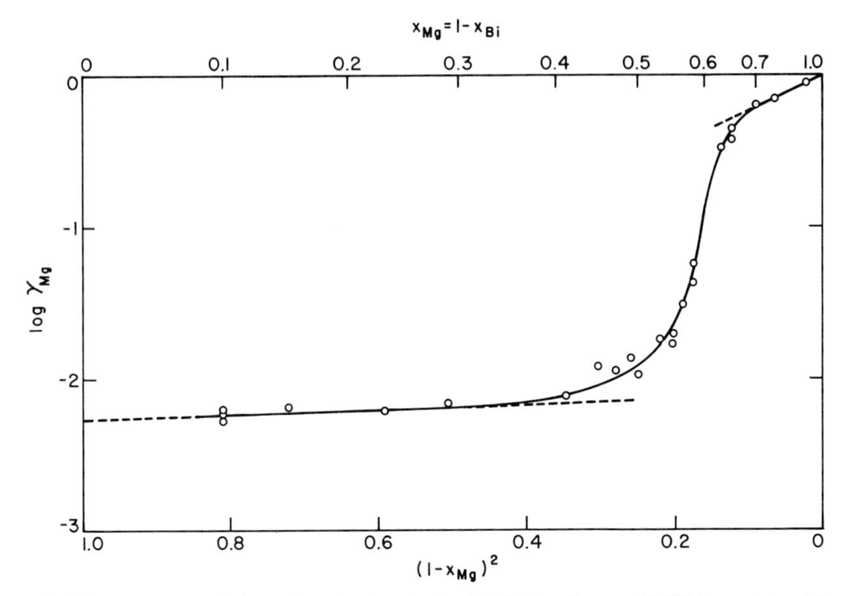

FIG. 3.3 Activity coefficient of magnesium in liquid Bi–Mg alloys at 700°C from data of Egan [17]. After Darken [12].

It is seen that the excess stability is evaluated by multiplying the slope of the curve in the log γ_2 versus $(1 - x_2)^2$ plot by the factor $(-4.605RT)$. In the Bi–Mg system the excess-stability peak occurs at the composition corresponding to that of the solid phase Mg_3Bi_2 (Fig. 3.4a). In the Fe–Si and Fe–Ni, and many other systems, no stability peak is observed (Fig. 3.4b); the flat excess-stability curves of the two terminal regions are connected by a stepwise transition.

Because of the constancy of the excess stability in the terminal regions and the symmetry of this function with respect to the two components, Gibbs–Duhem integration is much facilitated, as indicated by the following equality derived by differentiating Eqs. (3.5) and (3.6) with respect to x_1^2 and x_2^2, respectively.

$$\frac{d \log \gamma_2}{d(x_1^2)} = \frac{d \log \gamma_1}{d(x_2^2)}. \tag{3.14}$$

The partial molar enthalpy of solution $\bar{L}_2$ and $\bar{L}_2^\circ$ (the value of $\bar{L}_2$ at infinite dilution) are related to the activity coefficients by the thermodynamic relation (1.64)

$$\frac{d \log(\gamma_2/\gamma_2^\circ)}{d(1/T)} = \frac{\bar{L}_2 - \bar{L}_2^\circ}{2.303R}. \tag{3.15}$$

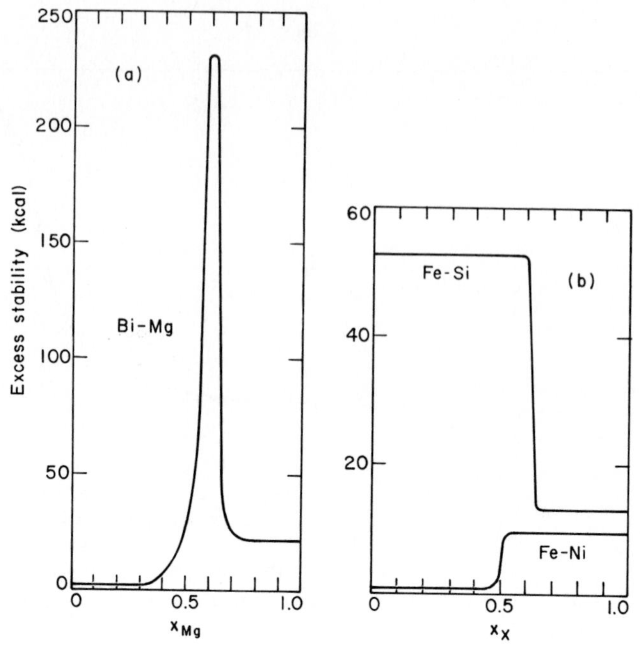

FIG. 3.4 Excess stability of (a) liquid Bi–Mg alloys at 700°C and (b) Fe–Si and Fe–Ni liquid alloys at 1600 °C. After Darken [12].

Differentiating Eq. (3.5) with respect to $1/T$ and substituting the above thermodynamic relation gives for region (I)

$$\bar{L}_2 - \bar{L}_2^{\circ} = 2.303R(x_1^{\,2} - 1)\frac{d\alpha_{12}}{d(1/T)}. \tag{3.16}$$

It is reasonable to assume that over a temperature range of experimental interest, the enthalpy of mixing is essentially independent of temperature. On the basis of this argument, the temperature dependence of α has the form

$$\alpha = (\mathscr{L}/2.303RT) - (\mathscr{S}/2.303R), \tag{3.17}$$

where $\mathscr{L}$ and $\mathscr{S}$ have the units of enthalpy and entropy, respectively; they are characteristic parameters for a terminal region of a binary system, e.g., for region (I), $\mathscr{L}_{12}$, $\mathscr{S}_{12}$, and for region (II), $\mathscr{L}_{21}$, $\mathscr{S}_{21}$. This equation gives for $d\alpha_{12}/d(1/T)$

$$\frac{d\alpha_{12}}{d(1/T)} = \frac{\mathscr{L}_{12}}{2.303R} = \left(\alpha_{12} + \frac{\mathscr{S}_{12}}{2.303R}\right)T. \tag{3.18}$$

On the basis of the above argument, Turkdogan and Darken [13] extended the quadratic formalism to the enthalpy and entropy of mixing. For example, for region (I),

$$\bar{L}_2 = \bar{L}_2^\circ + \mathscr{L}_{12}(x_1^{\,2} - 1), \tag{3.19}$$

$$\bar{L}_1 = \mathscr{L}_{12}x_2^{\,2}, \tag{3.20}$$

$$\Delta H^{\mathrm{M}}/x_2 = \bar{L}_2^\circ - \mathscr{L}_{12}x_2, \tag{3.21}$$

$$\bar{S}_2^{\mathrm{xs}} = S_2^{\mathrm{xs}\circ} + \mathscr{S}_{12}(x_1^{\,2} - 1), \tag{3.22}$$

$$\bar{S}_1^{\mathrm{xs}} = \mathscr{S}_{12}x_2^{\,2}, \tag{3.23}$$

$$\Delta S^{\mathrm{xs}}/x_2 = \bar{S}_2^{\mathrm{xs}\circ} - \mathscr{S}_{12}x_2, \tag{3.24}$$

where ΔH^{M} is the integral molar enthalpy and ΔS^{xs} the integral excess molar entropy. Similar equations are derived for region (II).

The calorimetric heat of solution data of Kleppa [18] for the liquid cadmium–lead alloys are plotted in Fig. 3.5 in accord with Eq. (3.21). It should be noted that because of the symmetry of the functions with respect to the two components, the tangent drawn to the $\Delta H^{\mathrm{M}}/x_{\mathrm{Cd}}$ curve at $x_{\mathrm{Cd}} = 1$ intercepts the ordinate at $x_{\mathrm{Cd}} = 0$ at value corresponding to $\bar{L}_{\mathrm{Pb}}^\circ$; this must coincide with the linear extension of the curve $\Delta H^{\mathrm{M}}/x_{\mathrm{Pb}}$ to $x_{\mathrm{Pb}} = 0$. This tangent–intercept relation is helpful in the extrapolation of the data to infinite dilution.

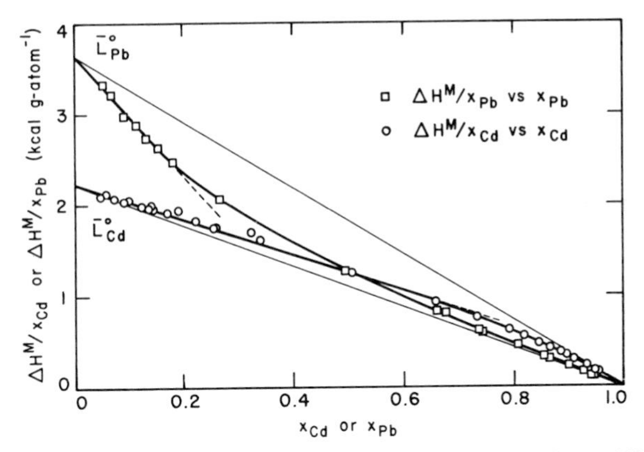

FIG. 3.5 Heats of solution of liquid Cd–Pb alloys at temperatures of 350–450 °C from data of Kleppa [18]. After Turkdogan and Darken [13].

In some thermodynamic measurements, the ratio γ_1/γ_2 is determined directly from the experiment. The treatment of such data is much simplified by using the quadratic formalism. From the isothermal and isobaric thermodynamic relation in (10) in Table 1.1, Redlich and Kister [19] derived the relation

$$\int_0^1 \log(\gamma_1/\gamma_2)\,dx_2 = 0. \tag{3.25}$$

That is, in a plot of $\log(\gamma_1/\gamma_2)$ versus x_2, the areas under the curve on either side of the horizontal line for $\log(\gamma_1/\gamma_2) = 0$ are equal. Using the quadratic formalism, it follows from Eq. (3.5) and (3.6) that for the terminal region (I)

$$\log(\gamma_1/\gamma_2) = 2\alpha_{12}x_2 - \log\gamma_2{}^\circ, \tag{3.26}$$

and similarly, for the terminal region (II)

$$\log(\gamma_1/\gamma_2) = -2\alpha_{21}(1 - x_2) + \log\gamma_1{}^\circ, \tag{3.27}$$

Belton and Fruehan [20] have shown that the thermodynamic properties of solutions can be related to the ratio of ion currents of vapor species in equilibrium with a condensed phase as measured by a mass spectrometer. Since the ratio of ion currents $I_2{}^+/I_1{}^+$ is proportional to the ratio of partial pressures of the components, the following relation is obtained:

$$\log(I_2{}^+/I_1{}^+) - \log(x_2/x_1) = \log(\gamma_2/\gamma_1) + C, \tag{3.28}$$

where C is a constant for a given temperature. The experimental results of Belton and Fruehan are plotted in Fig. 3.6. The position of the dot–dash horizontal line, corresponding to the constant $C = 1.085$, is so chosen that the shaded areas are equal in accord with Eq. (3.25); intercepts with the ordinate give $\log\gamma_{\mathrm{Ni}}^\circ = -0.15$ and $\log\gamma_{\mathrm{Fe}}^\circ = -0.40$ at 1600 °C.

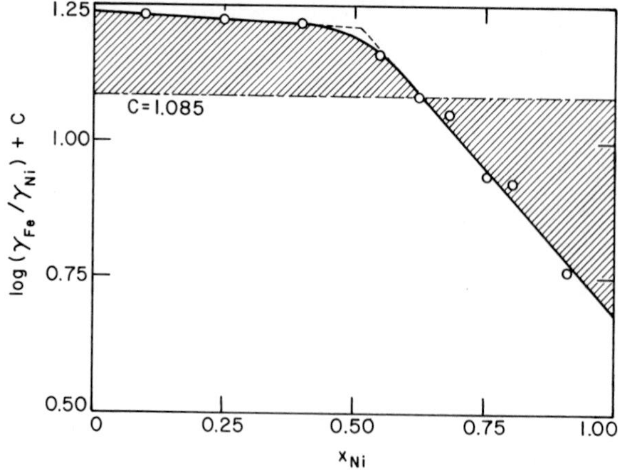

FIG. 3.6 Activity coefficient ratio for liquid iron–nickel alloys at 1600 °C. After Belton and Fruehan [20].

Another empirical correlation is that proposed by Wilson [21] for the excess free energy as a logarithmic function of solute concentration over the whole range of concentrations; thus for a binary system at constant temperature

$$\begin{aligned} \ln \gamma_1 &= -\ln(x_1 + \Lambda_{12}x_2) + x_2\left[\frac{\Lambda_{12}}{x_1 + \Lambda_{12}x_2} - \frac{\Lambda_{21}}{x_2 + \Lambda_{21}x_1}\right], \\ \ln \gamma_2 &= -\ln(x_2 + \Lambda_{21}x_1) - x_1\left[\frac{\Lambda_{12}}{x_1 + \Lambda_{12}x_2} - \frac{\Lambda_{21}}{x_2 + \Lambda_{21}x_1}\right], \end{aligned} \tag{3.29}$$

where the parameters Λ_{12} and Λ_{21}, characteristic of the system, are positive and evaluated from the experimental values of γ_1 and γ_2 at several concentrations. This is the modified form of the Wilson equation as proposed by Orye and Prausnitz [22]. For an ideal system $\Lambda_{12} = \Lambda_{21} = 1$ and γ_i approaches a finite value as x_i approaches zero. For positive departure from ideality, $0 < \Lambda < 1$, and for negative departure, $\Lambda > 1$.

The Wilson equation holds well for many organic solutions [21–24] and also for some metallic solutions [25]. This equation, involving only two temperature-dependent parameters Λ_{12} and Λ_{21} over the entire concentration range, represents the composition dependence of γ_1 and γ_2 well only when there is moderate departure from ideality, and more staisfactorily for positive departures.

In an attempt to represent the composition dependence of activity coefficient over the entire range of concentrations, Krupkowski [26] proposed the following formalism for binary solutions.

$$\ln \gamma_1 = \omega(T)(1 - x_1)^m \tag{3.30}$$

$$\ln \gamma_2 = \omega(T)\left[(1 - x_1)^m - \frac{m}{m-1}(1 - x_1)^{m-1} + \frac{1}{m-1}\right]. \tag{3.31}$$

Equation (3.30) is the assumed relation from which Eq. (3.31) is derived through the Gibbs–Duhem integration. The function $\omega(T)$ is temperature dependent and, for many systems, has the same form as Eq. (3.17). The exponent m is usually, but not necessarily independent of temperature; m is evaluated from the slope of the line $\ln(\ln \gamma_1)$ versus $\ln(1 - x_1)$, and the intercept gives the value of $\omega(T)$ for a given temperature. Moser [27, 28] explored the extent of the applicability of Eq. (3.30) to binary metallic solutions. This formalism generally applies to systems with positive departure from Raoult's law; for such systems the value of m is within 1 and 2. For systems with negative deviations from ideality, m has values greater than 2. For systems with strong departure from ideality, as in Bi–Mg, this formalism does not hold well.

There are numerous other empirical correlations based on power series representation of the composition dependence of excess free energy and enthalpy of mixing for binary and multicomponent systems, as, for example, those proposed by Wohl [29], Redlich and Kister [19], Scatchard [30], Williams [31], and Bale and Pelton [32]. In the polynomials for the series representations, as many parameters are used as are required to fit the data over the entire concentration range. The coefficients α_{ij}, β_{ij}, γ_{ij}, etc., obtained by curve fitting using computer techniques, have no physical significance and cannot be used for intercomparison of the thermodynamic properties of a series of binary systems, for example. The power series with two to four parameters represents the excess free energy over the entire concentration range only when departure from ideality is moderate as in many nonelectrolyte solutions.

There are also quasi-chemical approximations of various complexity [10, 33–36], and the equations derived therefrom again are of limited applicability, particularly for metallic solutions with strong negative departures from ideality.

All attempts to date to derive thermodynamic properties of metal alloys, solid or liquid, through the application of fundamental theories of quantum mechanics have failed because of the complexity of the bond mechanism. Nevertheless, some progress has been made in relating thermodynamic properties to various models of bond mechanisms. The discussion of this subject is beyond the scope of this book; the subject is well covered in review papers; see, for example, the references cited in [37, 38].

Since thermodynamic quantities are functions of similar state properties, one thermodynamic quantity may be related to another for certain types of solutions. For example, as is shown in Fig. 3.7, Kubaschewski [39] found that

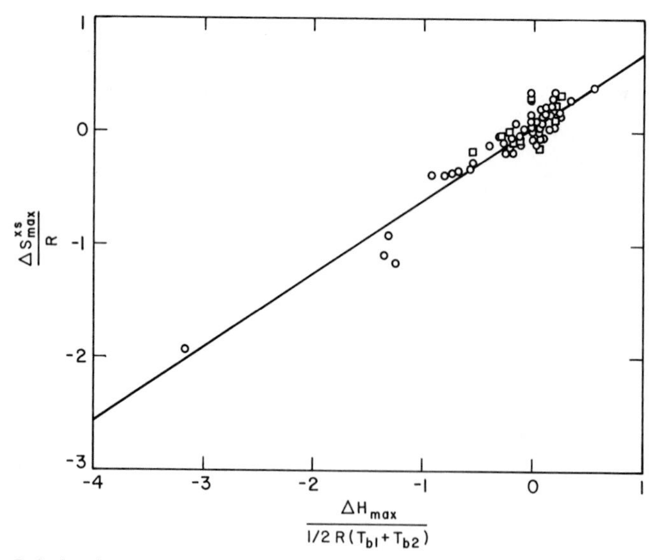

FIG. 3.7 Relation between maximum and minimum heats and excess entropies of mixing of of solid (□) and liquid (○) solutions. After Kubaschewski [39].

there is a linear relationship between $\Delta S^{xs}_{max}/R$ and $2\,\Delta H^{M}_{max}/R(T_{b1}+T_{b2})$, where ΔS^{xs}_{max} and ΔH^{M}_{max} are the integral maximum or minimum excess entropy and enthalpy of mixing per gram atom, respectively; and T_{b1} and T_{b2} are the boiling points of component metals. The scatter in the data corresponds to the limit of accuracy of ± 0.2 cal g atom^{-1} deg^{-1} in the excess entropy of mixing. The straight line in Fig. 3.7 may be represented by

$$\Delta S^{xs}_{max} = 1.28\,\Delta H^{M}_{max}/(T_{b1}+T_{b2}). \tag{3.32}$$

With the assumption that the maxima or minima in ΔH^M and ΔS^{xs} curves occur at the same composition for a given binary system, Kubaschewski derived the following relation from Eq. (3.32).

$$\Delta S^{xs}_{max} = \Delta F^{xs}_{max}/[0.78(T_{b1}+T_{b2}) - T], \tag{3.33}$$

where T is the absolute temperature at which the free energy has been measured.

In a study of the applicability of the quadratic formalism to binary metallic solutions, Turkdogan and Darken [13] observed two significant correlations between the thermodynamic quantities pertaining to the terminal regions. As is shown in Fig. 3.8, the quantity $(\mathscr{L}_{12}-\mathscr{L}_{21})$ for liquid and solid solutions is directly proportional to the difference $(\bar{L}_2^\circ - \bar{L}_1^\circ)$. Similarly, Fig. 3.9 shows

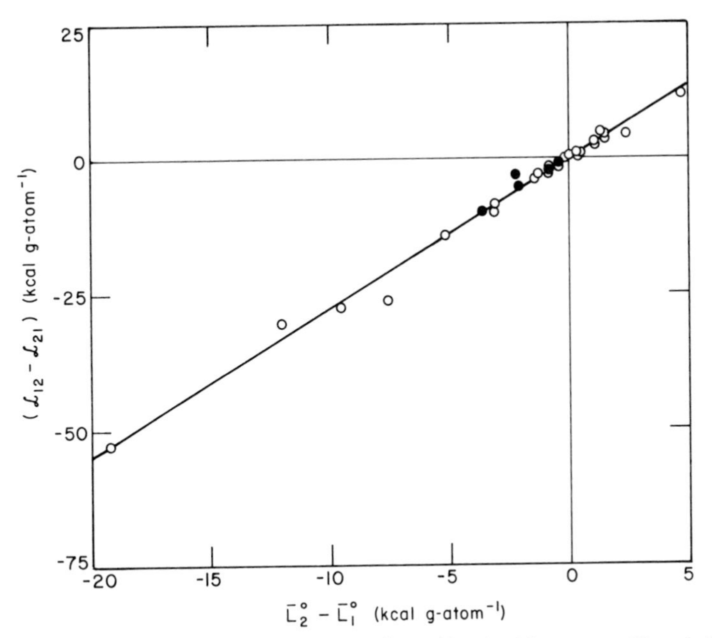

FIG. 3.8 Relation between $(\mathscr{L}_{12}-\mathscr{L}_{21})$ and $\bar{L}_2^\circ - \bar{L}_1^\circ$ for binary metallic solutions. (●) solid; (○) liquid. After Turkdogan and Darken [13].

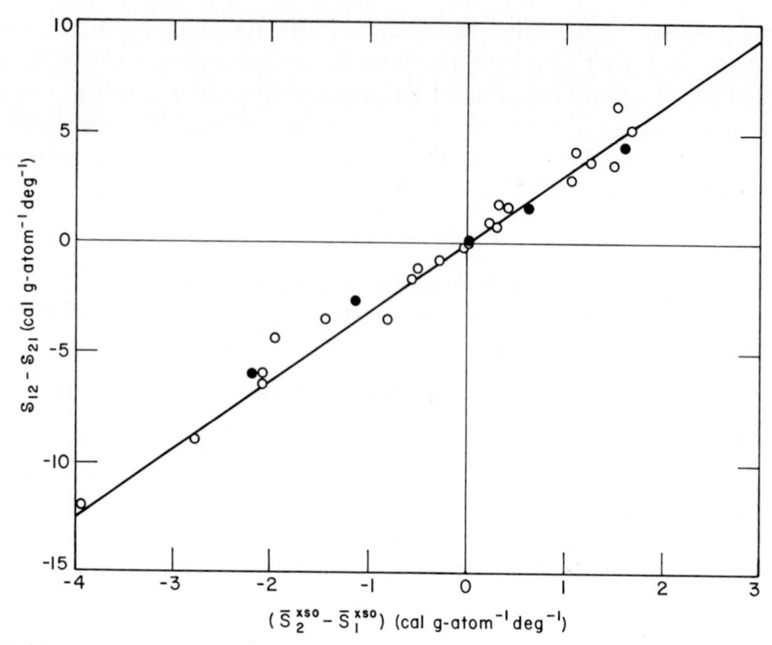

FIG. 3.9 Relation between $(\mathscr{S}_{12} - \mathscr{S}_{21})$ and $(\Delta\bar{S}_2^{xs\circ} - \Delta\bar{S}_1^{xs\circ})$ for binary metallic solutions. (●) solid; (○) liquid. After Turkdogan and Darken [13].

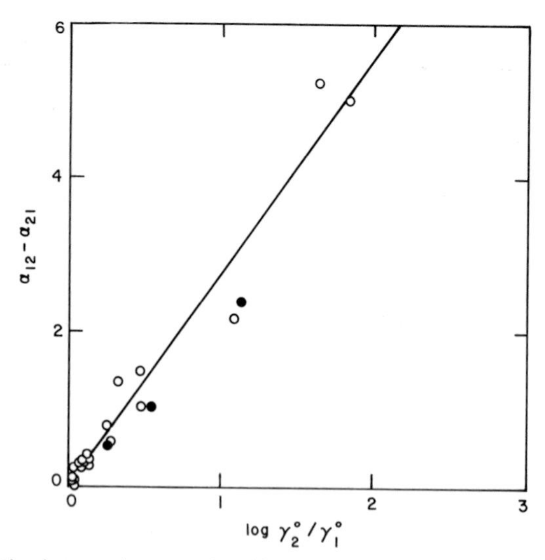

FIG. 3.10 Relation between $(\alpha_{12} - \alpha_{21})$ and $\log \gamma_2{}^\circ/\gamma_1{}^\circ$ for binary liquid metallic solutions. (○) nonferrous alloys, 300–1000 °C; (●) ferrous alloys, 1600 °C. After Turkdogan *et al.* [40].

the direct proportionality of $(\mathscr{S}_{12} - \mathscr{S}_{21})$ to the difference $(\bar{S}_2^{xs\circ} - \bar{S}_1^{xs\circ})$. These empirical relations may be represented by the following equations, in which the units of $\mathscr{S}$ and $\bar{L}^\circ$ are kcal g atom^{-1}, and $\mathscr{S}$ and $\mathscr{S}^{xs\circ}$ are in cal g atom^{-1} deg^{-1}:

$$\mathscr{L}_{12} - \mathscr{L}_{21} = 2.74(\bar{L}_2{}^\circ - \bar{L}_1{}^\circ), \tag{3.34}$$

$$\mathscr{S}_{12} - \mathscr{S}_{21} = 3.11(\bar{S}_2^{xs\circ} - \bar{S}_1^{xs\circ}). \tag{3.35}$$

Combining these with Eq. (3.17) and omitting the resulting small term $0.08(\bar{S}_2^{xs\circ} - \bar{S}_1^{xs\circ})$, the following relation is obtained.

$$\alpha_{12} - \alpha_{21} = 2.74 \log(\gamma_2{}^\circ/\gamma_1{}^\circ). \tag{3.36}$$

The plot in Fig. 3.10 [40] substantiates the general validity of this relation.

Binary alloys considered for the correlations in Figs. 3.7–3.10 are of different bond mechanisms, e.g., Na–Hg, Fe–Si, Ag–Au, Tl–Sb, Cr–Mo, yet the data fit the same pattern; this observation indicates that for any given binary system, the fundamental causes for the values of excess thermodynamic quantities are essentially the same.

In a study of the thermodynamics of copper-base binary liquid alloys, Azakami and Yazawa [41] found that the activity coefficient of solute at infinite dilution varies in a systematic manner, as shown in Fig. 3.11, with the group

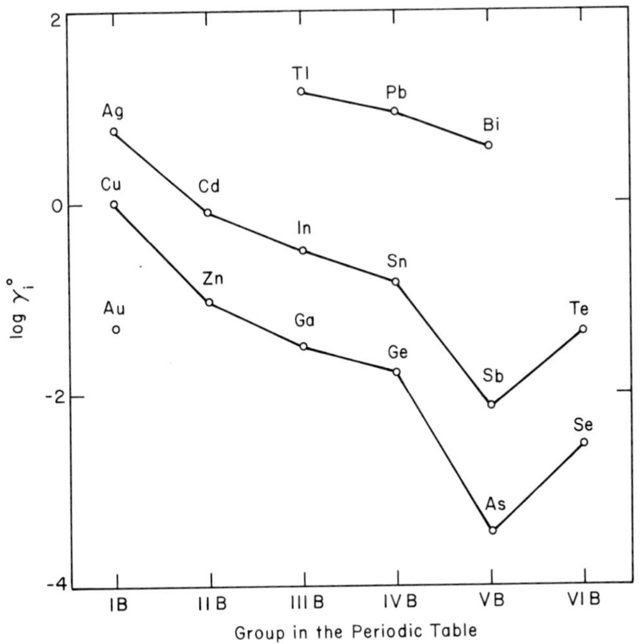

FIG. 3.11 Relation between $\gamma_1{}^\circ$ and the periodic table group of the alloying element for binary liquid copper alloys at infinite dilution at 1100 °C. After Azakami and Yazawa [41].

and number of the element in the periodic table, except for Cu–Au alloys. This bears close resemblance to the relation found by Buckley and Hume-Rothery [42] for the variation of the free-energy change for the transfer of solute from one phase to another at infinite dilution (Figs. 2.18 and 2.19).

3.1.2 Ternary Systems—High Solute Concentrations

It follows from the general thermodynamic relation (9) in Table 1.1 for isothermal isobaric conditions that a thermodynamic quantity change dG per mole of solution caused by changing the atom fractions of constituents by dx_1, $dx_2, \ldots$ equals

$$dG = \bar{G}_1\,dx_1 + \bar{G}_2\,dx_2 + \bar{G}_3\,dx_3 + \cdots. \tag{3.37}$$

From this general relation, Darken [43] derived the following thermodynamic relation for ternary systems

$$(1 - x_3)\frac{\partial \bar{G}_2}{\partial x_3} + x_3\frac{\partial \bar{G}_3}{\partial x_3} = x_2\frac{\partial \bar{G}_2}{\partial x_2} + (1 - x_2)\frac{\partial \bar{G}_3}{\partial x_2}. \tag{3.38}$$

In terms of activity coefficients γ_2 and γ_3 of solutes 2 and 3, component 1 being the solvent.

$$(1 - x_3)\frac{\partial \ln \gamma_2}{\partial x_3} + x_3\frac{\partial \ln \gamma_3}{\partial x_3} = x_2\frac{\partial \ln \gamma_2}{\partial x_2} + (1 - x_2)\frac{\partial \ln \gamma_3}{\partial x_2}. \tag{3.39}$$

This relation states the condition of thermodynamic consistency between the activity coefficients of solutes. Any proposed relations for the composition dependence of γ_2 and γ_3 must satisfy the above relation for thermodynamic consistency.

From Eqs. (3.5) and (3.6) the following is obtained for the excess integral molar free energy in terminal region (I) of a binary system, by making use of the thermodynamic relation in Eq. (1.35).

$$F^{\text{xs}} = 2.303RT(x_2 \log \gamma_2{}^\circ - \alpha_{12}x_2{}^2). \tag{3.40}$$

As proposed by Darken [43], a quadratic equation for ternary solutions corresponding to Eq. (3.40) is

$$\begin{aligned} F^{\text{xs}}/2.303RT = {} & x_2 \log \gamma_2{}^\circ + x_3 \log \gamma_3{}^\circ - \alpha_{12}x_2{}^2 \\ & - \alpha_{13}x_3{}^2 - (\alpha_{12} + \alpha_{13} - \alpha_{23})x_2x_3. \end{aligned} \tag{3.41}$$

The coefficients α_{12} and α_{13} are obtained from the binary systems 1–2 and 1–3; the third coefficient α_{23} is obtainable only from measurements on the ternary system and is a characteristic of a hypothetical system 2–3. From Eq. (3.41), Darken derived the following relations of activity coefficients.

$$\begin{aligned}
\log\gamma_1 &= \alpha_{12}x_2^{\,2} + \alpha_{13}x_3^{\,2} + (\alpha_{12} + \alpha_{13} - \alpha_{23})x_2x_3, \\
\log(\gamma_2/\gamma_2^{\circ}) &= -2\alpha_{12}x_2 + (\alpha_{23} - \alpha_{12} - \alpha_{13})x_3 \\
&\quad + \alpha_{12}x_2^{\,2} + \alpha_{13}x_3^{\,2} + (\alpha_{12} + \alpha_{13} - \alpha_{23})x_2x_3, \\
\log(\gamma_3/\gamma_3^{\circ}) &= -2\alpha_{13}x_3 + (\alpha_{23} - \alpha_{12} - \alpha_{13})x_2 \\
&\quad + \alpha_{12}x_2^{\,2} + \alpha_{13}x_3^{\,2} + (\alpha_{12} + \alpha_{13} - \alpha_{23})x_2x_3.
\end{aligned} \tag{3.42}$$

These relations reduce to the expressions for the binaries when x_2 or x_3 is set equal to zero.

When the activity of one of the components, e.g., component 2, is held constant, such as graphite saturation, the following is derived from Eq. (3.42) for the solubility of x_2

$$\begin{aligned}
&\log(x_2^{*}/x_2') + \alpha_{12}[x_2'(2 - x_2') - x_2^{*}(2 - x_2^{*})] - \alpha_{13}x_3^{\,2} \\
&\quad = (\alpha_{23} - \alpha_{12} - \alpha_{13})x_3(1 - x_2'),
\end{aligned} \tag{3.43}$$

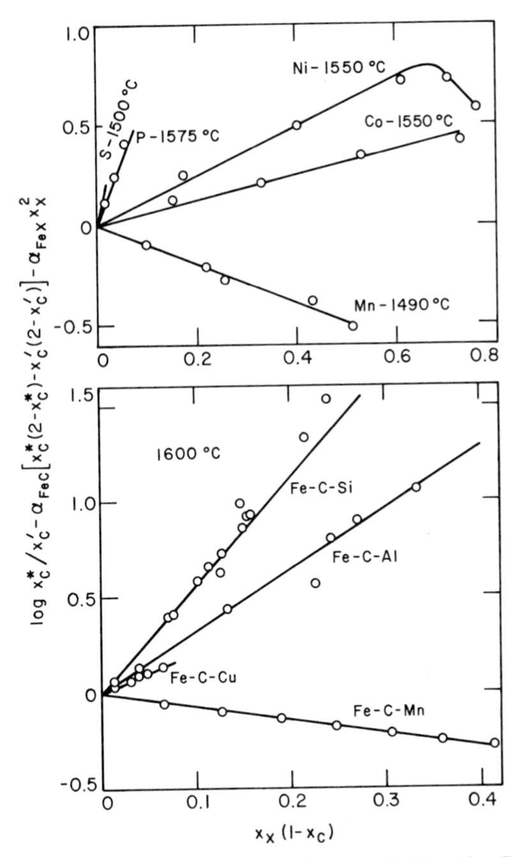

FIG. 3.12 Solubility of graphite in iron alloys at 1600 °C. After Darken [43].

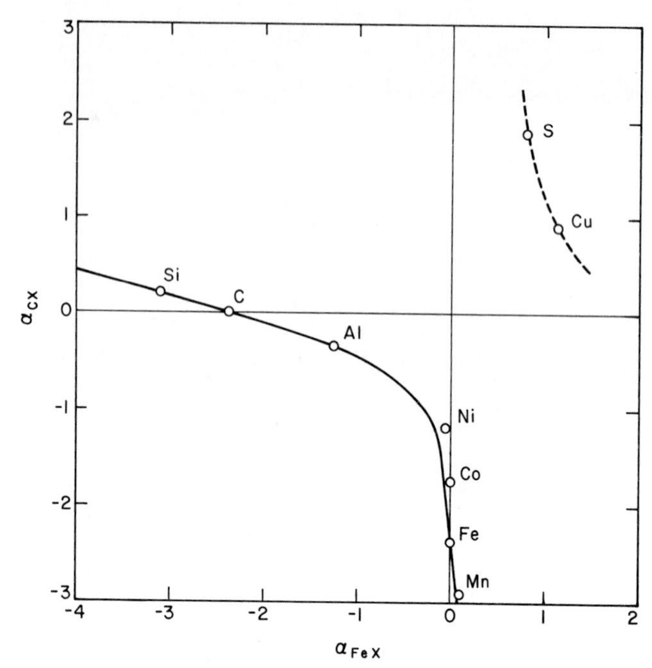

FIG. 3.13 A correlation of binary and ternary constants for Fe–X and Fe–C–X systems at 1600 °C. After Darken [43].

where x_2^* is the solubility of component 2 in the binary 1–2 and x_2' in the ternary 1–2–3. When the coefficients α_{12} and α_{13} for the binaries are known, the parameter α_{23} can be evaluated from the solubility data by using Eq. (3.43).

Available data on the solubility of graphite in Fe–Si, Fe–Al, and Fe–Mn alloys from the work of Chipman *et al.* [44, 45]; in Fe–Cu alloys from the work of Schenck and Perbix [46]; and in several ferrous alloys from the work of Turkdogan *et al.* [47–49] are plotted in Fig. 3.12 in accord with Eq. (3.43). The terminal lines (vertical) correspond to the binary systems X–C. It is seen that Eq. (3.43) holds over substantial ranges of solute concentration x_X with x_C within the range 0.05–0.2. The slopes of the lines give $(\alpha_{23} - \alpha_{12} - \alpha_{13})$. Since the parameters α_{12} and α_{13} (α_{FeX}) are known for the iron-rich region of several Fe–X alloys, the parameter α_{23} (α_{CX}) can be evaluated from the data in Fig. 3.12.

As is seen from Fig. 3.13, α_{CX} varies in a systematic manner with α_{FeX}, except for copper and sulfur. The curve as drawn appears to have asymptotes; if so, then another branch might be expected as shown by the dotted curve. It is interesting that the systems on this branch of the curve exhibit miscibility gaps.

3.1.3 Ternary Systems—Dilute Solutions

In many metal-refining processes, we are concerned with purification of metals to low, or even trace, levels of impurities. For this reason, a great deal of work has been done on solute activities at low concentrations in base metals such as iron and copper, following the earlier observations of Chipman [50] that for adequate representation of reaction equilibria in steelmaking, effects of solutes on activity coefficients must be known.

Using a Taylor series expansion for the excess partial molar free energy, Wagner [51] derived the following expression for the activity coefficient of component 2 in a multicomponent solution:

$$\ln \gamma_2 = \ln \gamma_2{}^\circ + \left[x_2 \frac{\partial \ln \gamma_2}{\partial x^2} + x_3 \frac{\partial \ln \gamma_2}{\partial x_3} + \cdots \right] + \left[\frac{1}{2} x_2{}^2 \frac{\partial^2 \ln \gamma_2}{\partial x_2{}^2} + x_2 x_3 \frac{\partial^2 \ln \gamma_2}{\partial x_2\, \partial x_3} + \cdots \right], \tag{3.44}$$

where the derivatives are for the limiting case of zero concentration of all solutes. For dilute solutions, the second and higher terms may be neglected to give a simplified linear equation.

$$\ln f_2 = x_2 \varepsilon_2^{(2)} + x_3 \varepsilon_2^{(3)} + x_4 \varepsilon_2^{(4)} + \cdots, \tag{3.45}$$

where f_2 is the ratio $\gamma_2/\gamma_2{}^\circ$, a measure of departure from Henry's law, and $\varepsilon_2^{(2)}$, $\varepsilon_2^{(3)}$, etc., are coefficients defined as

$$\varepsilon_2^{(2)} = \lim_{x_1 \to 1} \frac{\partial \ln f_2}{\partial x_2}, \qquad \varepsilon_2^{(3)} = \lim_{x_1 \to 1} \frac{\partial \ln f_2}{\partial x_3}. \tag{3.46}$$

For infinitely dilute solutions, i.e., $x_2 \to 0$, $x_3 \to 0$,

$$\varepsilon_2^{(3)} = \varepsilon_3^{(2)}. \tag{3.47}$$

Simplification of the quadratic formalism for infinite dilution gives the interrelation between ε and α, thus,

$$\begin{aligned} \varepsilon_2^{(2)}/2.303 &= -2\alpha_{12}, \qquad \varepsilon_3^{(3)}/2.303 = -2\alpha_{13}, \\ \varepsilon_2^{(3)}/2.303 &= \varepsilon_3^{(2)}/2.303 = (\alpha_{23} - \alpha_{12} - \alpha_{13}). \end{aligned} \tag{3.48}$$

For convenience, the solute concentration is often given in weight percent for which Eq. (3.45) for a ternary system is written as

$$\log f_i = e_i{}^i[\%\, i] + e_i{}^j[\%\, j], \tag{3.49}$$

where the coefficient e, known as the interaction coefficient, is related to ε by the relation

$$\begin{aligned} \varepsilon_i{}^i &= 230(M_i/M_1)e_i{}^i + (M_1 - M_i)/M_1, \\ \varepsilon_i{}^j &= 230(M_j/M_1)e_i{}^j + (M_1 - M_j)/M_1, \end{aligned} \tag{3.50}$$

where M is the atomic weight of the component indicated by the subscript, 1 being the solvent.

Sigworth and Elliott [52–54] compiled the available data on interaction coefficients in liquid iron, cobalt, and copper alloys. The values of e_C^j, e_H^j, e_N^j, e_O^j, and e_S^j taken from these compilations (with minor adjustments of some values of e) are listed in Tables 3.1–3.3 for selected alloys of practical interest.

As suggested by Lupis and Elliott [55], the temperature dependence of ε_i^j may be represented by the relation

$$\varepsilon_i^j = (\eta_i^j/RT) - (\sigma_i^j/R), \tag{3.51}$$

where η and σ are the enthalpy- and entropy-like terms. Chipman and Corrigan [56] have found that in Fe–N–X alloys, η_N^j is a linear function of the interaction coefficient ε_N^j, as shown in Fig. 3.14. A similar relation was found by Jacob and Jeffes [57] for Cu–O–X alloys; this is shown in Fig. 3.15.

In calculating the equilibrium constants for metallurgical reactions involving dilute solutions, it is convenient to change the standard state from pure component to that for 1 wt.% in solution for which the free-energy change is given by relation (16) in Table 1.1. Free energies of solution ΔF_s of selected elements in liquid iron, cobalt, and copper are listed in Table 3.4.

TABLE 3.1

Interaction Coefficients in Ternary Iron-Base Alloys for C, H, N, O, *and* S *at* 1600 °C[a]

Element j	e_C^j	e_H^j	e_N^j	e_O^j	e_S^j
Al	0.043	0.013	−0.028	−3.9	0.035
B	0.24	0.05	0.094	−2.6	0.13
C	0.14	0.06	0.13	−0.13	0.11
Co	0.008	0.002	0.011	0.008	0.003
Cr	−0.024	−0.002	−0.047	−0.04	−0.011
Cu	0.016	<0.001	0.009	−0.013	−0.008
Mn	−0.012	−0.001	−0.02	−0.021	−0.026
Mo	−0.008	0.002	−0.011	0.004	0.003
N	0.11	—	0	0.057	0.01
Nb	−0.06	−0.002	−0.06	−0.14	−0.013
Ni	0.012	0	0.01	0.006	0
O	−0.34	−0.19	0.05	−0.20	−0.27
P	0.051	0.011	0.045	0.07	0.29
S	0.046	0.008	0.007	−0.133	−0.028
Si	0.08	0.027	0.047	−0.131	0.063
Sn	0.041	0.005	0.007	−0.011	−0.004
Ti	—	−0.019	−0.53	−0.31	−0.072
V	−0.077	−0.007	−0.093	−0.14	−0.016
W	−0.006	0.005	−0.002	−0.009	0.01
Zr	—	—	−0.63	(−3.0)	−0.052

[a] From [52].

TABLE 3.2

Interaction Coefficients in Ternary Cobalt-Base Alloys for H, N, O, *and* S *at* 1600 °C[a]

Element j	e_H^j	e_N^j	e_O^j	e_S^j
Al	0.014	0.04	—	—
Au	—	—	0.003	—
B	0.09	—	—	—
Cr	0	−0.043	−0.07	—
Cu	−0.003	−0.009	−0.009	−0.016
Fe	0.001	−0.01	−0.019	−0.003
Mn	−0.003	—	−0.20	—
Mo	0.004	−0.008	−0.001	—
Nb	0	−0.042	—	—
Ni	−0.002	0.024	—	−0.007
S	—	—	−0.133	—
Si	0.021	0.12	−0.28	—
Sn	0.006	—	—	—
Ta	0.005	−0.029	—	—
Ti	—	−0.45	—	—
V	—	−0.1	—	—
W	0.006	−0.007	—	—

[a] From [53].

TABLE 3.3

Interaction Coefficients in Ternary Copper-Base Alloys for H, O, *and* S *at Temperatures* 1100–1200 °C[a]

Element j	e_H^j	e_O^j	e_S^j
Ag	0.001	−0.001	—
Al	0.006	—	—
Au	0	0.005	0.053
Co	0.015	−0.32	−0.046
Cr	0.009	—	—
Fe	−0.015	−3.25	−0.094
Mn	−0.006	—	−0.082
Ni	−0.026	−0.04	0.095
P	0.088	−0.74	—
Pb	0.031	−0.007	—
Pt	−0.008	0.008	—
S	0.073	−0.19	—
Sb	0.031	—	—
Si	0.042	—	0.055
Sn	0.016	−0.01	—
Te	−0.012	—	—
Zn	0.029	—	—

[a] From [54].

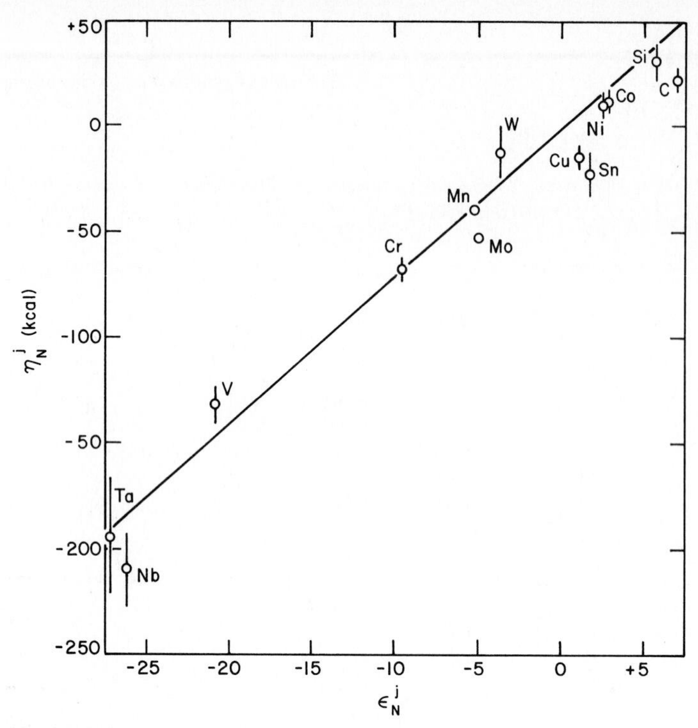

FIG. 3.14 Enthalpy parameter η as a function of interaction coefficient ε for oxygen in liquid iron alloys. After Chipman and Corrigan [56].

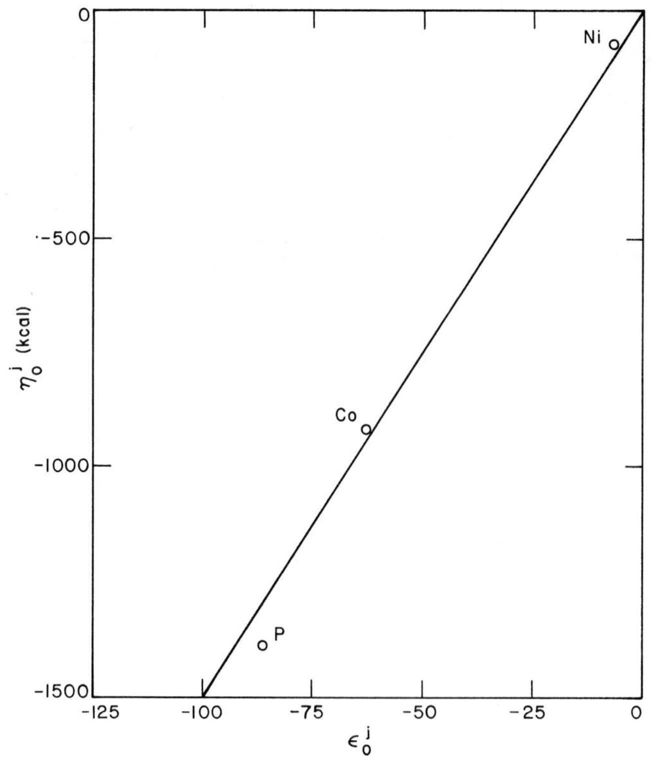

FIG. 3.15 Enthalpy parameter η as a function of interaction coefficient ε for oxygen in liquid copper alloys. After Jacob and Jeffes [57].

TABLE 3.4

Free Energies of Solution in Liquid Iron, Cobalt and Copper at 1 wt. %[a]

Element i	γ_i°	ΔF_s (cal g atom^{-1})
	M_s, liquid iron at 1600 °C[b]	
Al (l)	0.029	$-15{,}100 - 6.67T$
C (g)	0.57	$5400 - 10.10T$
Co (l)	1.07	$240 - 9.26T$
Cr (s)	1.14	$4600 - 11.20T$
Cu (l)	8.6	$8000 - 9.41T$
$\frac{1}{2}H_2$ (g)	—	$8720 + 7.28T$
Mn (l)	1.3	$976 - 9.12T$
$\frac{1}{2}N_2$ (g)	—	$860 + 5.71T$
Ni (l)	0.66	$-5000 - 7.42T$
$\frac{1}{2}O_2$ (g)	—	$-28{,}000 - 0.69T$
$\frac{1}{2}P_2$ (g)	—	$-29{,}200 - 4.60T$
$\frac{1}{2}S_2$ (g)	—	$-32{,}280 + 5.60T$
Si (l)	0.0013	$-31{,}430 - 4.12T$
Ti (s)	0.038	$-7440 - 10.75T$
V (s)	0.1	$-4950 - 10.90T$
W (s)	1.2	$7500 - 15.20T$
Zr (s)	0.043	$-8300 - 11.95T$
	M_s, liquid cobalt at 1600 °C[c]	
Al (l)	0.005	$-21{,}800 - 6.49T$
C (gr)	6.7	$4990 - 4.88T$
Fe (l)	1.6	$-2880 - 6.57T$
$\frac{1}{2}H_2$ (g)	—	$8500 + 7.53T$
Mn (l)	1.0	$- 9.01T$
$\frac{1}{2}N_2$ (g)	—	$10{,}000 + 5.24T$
Ni (l)	0.53	$300 - 10.55T$
$\frac{1}{2}O_2$ (g)	—	$-14{,}730 - 2.32T$
$\frac{1}{2}S_2$(g)	—	$-32{,}300 + 7.23T$
Si (l)	0.0017	$-35{,}000 - 1.68T$
Ti (s)	0.0009	$-22{,}480 - 10.62T$
V (s)	0.06	$-6140 - 11.14T$
	M_s, liquid copper at 1200 °C[d]	
Ag (l)	3.23	$3900 - 10.52T$
Al (l)	0.0028	$-8630 - 13.84T$
Au (l)	0.14	$-4630 - 12.09T$
Bi (l)	1.25	$5960 - 15.10T$
Fe (s)	24.1	$12{,}970 - 11.34T$
$\frac{1}{2}H_2$ (g)	—	$10{,}400 + 7.50T$
Mg (l)	0.044	$-8670 - 7.53T$

TABLE 3.4 **(*continued*)**

Element i	γ_i°	ΔF_s (cal g atom^{-1})
$\frac{1}{2}O_2$ (g)	—	$-20{,}400 + 4.43T$
Pb (l)	5.27	$8620 - 14.01T$
$\frac{1}{2}S_2$ (g)	—	$-28{,}600 + 6.03T$
Sb (l)	0.014	$-12{,}500 - 10.40T$
Se (g)	0.002	$-18{,}200 - 9.58T$
Si (s)	0.01	$-2900 - 14.68T$
Sn (l)	0.048	$-8900 - 10.40T$
Te (g)	0.033	$-10{,}000 - 10.53T$
Tl (l)	8.5	$6730 - 11.74T$
Zn (l)	0.146	$-5640 - 9.17T$

[a] i (s, l, g) = i (1 wt. %) for which $\Delta F_s = RT\ \ln[M_s/M_i)0.01\gamma_i^\circ]$.
[b] See Sigworth and Elliott [52].
[c] See Sigworth and Elliott [53].
[d] See Sigworth and Elliott [54].

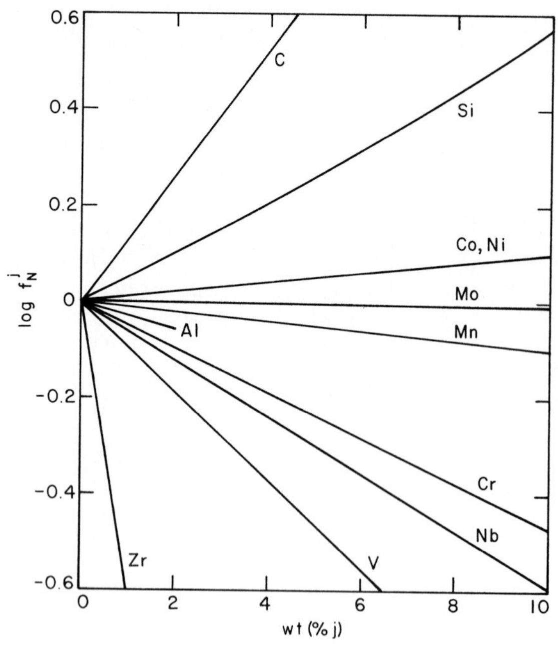

FIG. 3.16 Effect of alloying elements on activity coefficient of nitrogen in liquid iron alloys at 1600 °C.

Effects of alloying elements on activity coefficients of nitrogen, oxygen, and sulfur in liquid iron at 1600 °C are shown in Figs. 3.16–3.18: references to experimental data are given in a paper by Sigworth and Elliott [52]. The data of Jacob and Jeffes [57] in Fig. 3.19 show how the alloying elements affect the activity coefficient of oxygen in liquid copper at 1200 °C.

Although Eq. (3.49) is for infinitely dilute solutions, as is seen from the data in Figs. 3.16–3.19, the linearity extends to substantial concentrations in many alloys. However, in some alloys, $\partial \log f_i^j / \partial(\% j)$ increases or decreases with increasing concentration of the alloying element j. For these alloys the quadratic formalism may be more appropriate for analytical represention of the activity coefficient.

Another method of approach is to include the second-order term of Eq. (3.44) in representing the concentration dependence of $\ln \gamma_i$; thus for an n component system

$$\ln \gamma_i = \ln \gamma_i^\circ + \sum_{j=2}^{n} \varepsilon_i^j x_j + \sum_{j=2}^{n} \rho_i^j x_j^2, \tag{3.52}$$

where ρ_i^j is the second-order interaction coefficient.

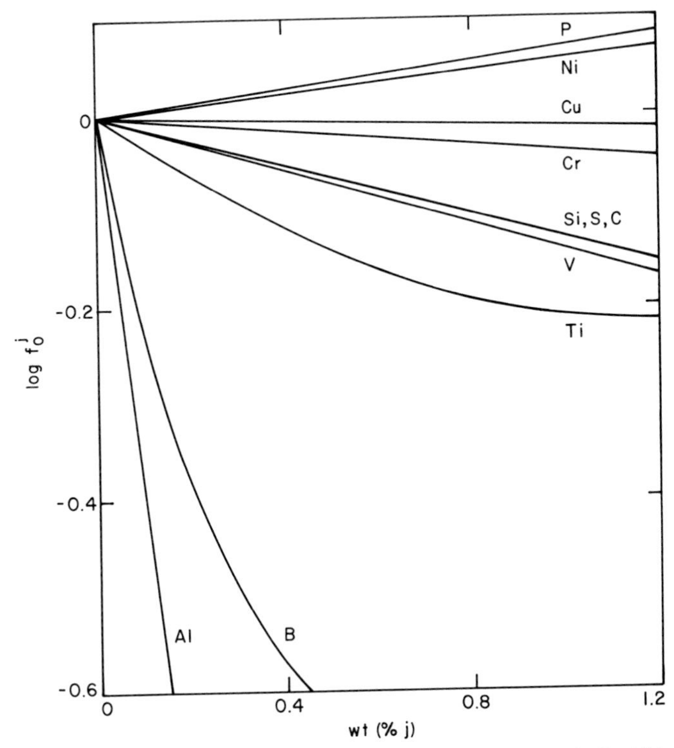

FIG. 3.17 Effect of alloying elements on activity coefficient of oxygen in liquid iron alloys at 1600 °C.

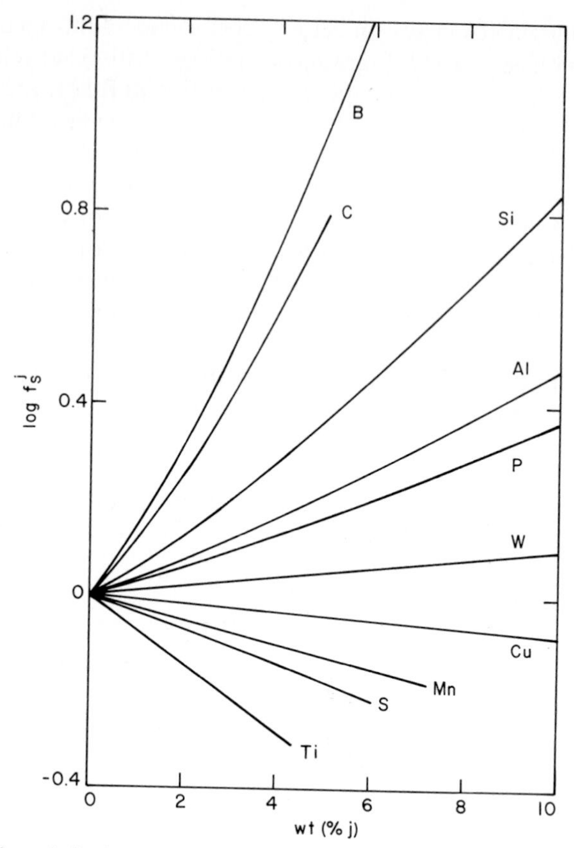

FIG. 3.18 Effect of alloying elements on activity coefficient of sulfur in liquid iron alloys at 1600 °C.

Following the suggestion of Lupis and Elliott [58], this equation has been used often to linearize the composition dependence of $\ln \gamma_i$.

3.1.4 Ternary Systems—Dilute Solutions (Quasi-Chemical Model)

Thermodynamics of metallic solutions consisting of two major constituents A and B and a solute X at low concentration has been analyzed by Alcock and Richardson [59] on the basis of a quasi-chemical solution model. The concept of regular solution was extended to ternary systems by Alcock and Richardson to describe the thermodynamics of a dilute solute X present in an alloy consisting of the major constituents A and B. The following equation was derived on the assumption that (1) the distribution of atoms in the solution is random; (2) the coordination number of all three types of atoms is equal; and (3) the energy of interaction between atom pairs is independent of concentration.

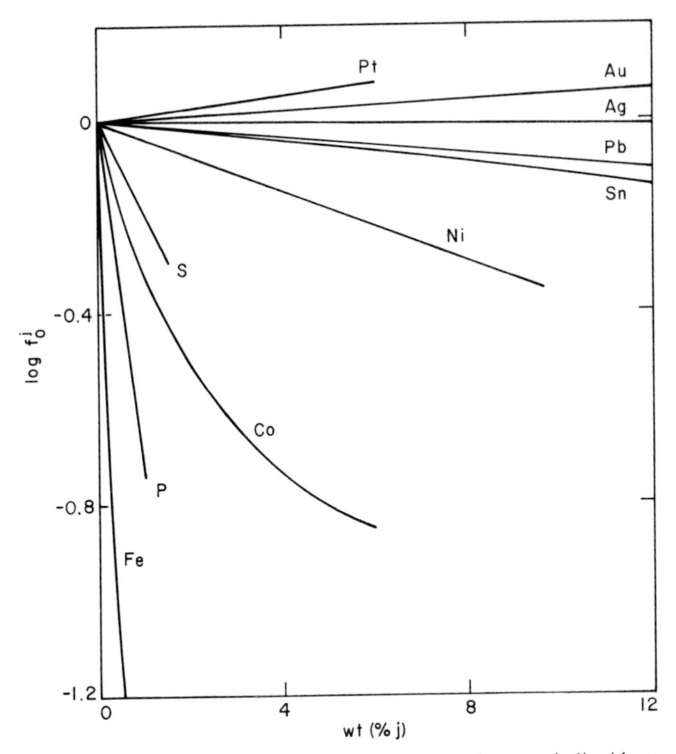

FIG. 3.19 Effect of alloying elements on activity coefficient of oxygen in liquid copper alloys at 1200 °C; for alloys with Ag, Pb, and Sn 1100 °C. After Jacob and Jeffes [57].

$$\ln \gamma_{X(AB)} = x_A \ln \gamma_{X(A)} + x_B \ln \gamma_{X(B)} - x_A \ln \gamma_A - x_B \ln \gamma_B, \tag{3.53}$$

where γ_X is the activity coefficient of the solute at infinite dilution in the alloy (AB) and pure metals (A) and (B) as indicated by the subscripts; γ_A and γ_B are the activity coefficients of components in the alloy (AB) with atom fraction x_A and x_B.

Because of limited applicability of Eq. (3.53), Alcock and Richardson [60] modified their solution model by assuming that atoms of solute X are surrounded preferentially by that type of metal atoms for which the energy of interaction with solute atoms is high. With this modification of the model, the following relation was obtained.

$$(1/\gamma_{X(AB)})^{1/Z} = x_A(\gamma_A/\gamma_{X(A)})^{1/Z} + x_B(\gamma_B/\gamma_{X(B)})^{1/Z}, \tag{3.54}$$

where Z is the coordination number.

For the special case of sparingly soluble X in an ideal solution A–B, Wagner [51] derived the following relation for the solubility of X.

$$(x_{X(AB)})^{1/Z} = x_A(x_{X(A)})^{1/Z} + x_B(x_{X(B)})^{1/Z}. \tag{3.55}$$

This equation may be obtained also from Eq. (3.54) by setting γ_A and γ_B equal to 1, and putting γ_X equal to the reciprocal of the solubility of X.

Because of systematic deviations of experimental data from Eq. (3.54), Wagner [61] proposed another quasi-chemical model. The basic assumptions in Wagner's model are (1) the concentration of the nonmetallic solute is sufficiently low for Sievert's law to be obyed, (2) the solute atoms occupy quasi-interstitial sites with a coordinate number Z equal to 6, and (3) binomial distribution of complexes which consist of an atom of solute X surrounded by varying amounts of constituents A and B.

On the basis of this model, Wagner derived the following relation for the solubility of solute X, such as the oxygen or sulfur, in a binary alloy $A_{1-x}B_x$:

$$x_{X(AB)} = C\sqrt{p_{x_2}} \sum_{i=0}^{i=Z} \binom{Z}{i} (1 - x_B)^{Z-i} x_B{}^i \exp\left(-\frac{\Delta H_i}{RT}\right), \tag{3.56}$$

where x_B is the atom fraction of B such that $x_A + x_B = 1$, C is the solubility constant, and p_{x_2} is the partial pressure of the dissolved gas, e.g., O_2, S_2, N_2. The enthalpy of dissolution of the gas ΔH_i is for the reaction

$$\tfrac{1}{2}X_2(g) + V(A_{z-i}B_i) = X(A_{z-i}B_i), \tag{3.57}$$

where $X(A_{Z-i}B_i)$ is an X atom in a quasi-interstitial site surround by a number $(Z - i)$ of A atoms and a number i of B atoms; $V(A_{Z-i}B_i)$ denotes a corresponding vacant quasi-interstitial site.

The transfer of atom X from a quasi-interstitial site with iB atoms as nearest neighbors to a site with $(i + 1)$B atoms as nearest neighbors is represented by

$$X(A_{Z-i}B_i) + V(A_{Z-i-1}B_{i+1}) = V(A_{Z-i}B_i) + X(A_{Z-i-1}B_{i+1}). \tag{3.58}$$

The enthalpy change for the exchange reaction (3.58) is the difference of the enthalpy changes ΔH_i for consecutive values of i,

$$\Delta H_{i+1} - \Delta H_i = \Delta H_i(i \to i + 1).$$

Wagner solved Eq. (3.56) for the special case of constant difference between the consecutive exchange reactions: $\Delta H(i + 1 \to i + 2) - \Delta H(i \to i + 1) = \Delta H(i + 2 \to i + 3) - \Delta H(i + 1 \to i + 2) = \cdots = h =$ constant, and obtained the following relation.

$$Q = \frac{\gamma_{X(AB)}}{(\gamma_{X(A)}\,\gamma_{X(B)})^{1/2}} = \left\{\sum_{i=0}^{i=Z} \binom{Z}{i} \left[\frac{(1 - x_B)}{\phi^{1/2Z}}\right]^{Z-i} [x_B \phi^{1/2Z}]^i \exp\left[\frac{(Z - i)ih}{2RT}\right]\right\}^{-1}, \tag{3.59}$$

where $\phi = [x_{X(B)}/x_{X(A)}]_{p_{x_2}} = \gamma_{X(A)}/\gamma_{X(B)}$. For a given value of ϕ from the data on the A–X and B–X binaries, and assuming $Z = 6$ for example, the value of h may be derived by fitting experimental data for A–B–X to Eq. (3.59). Comparison of available experimental data with Eq. (3.59) indicates that h has a positive value.

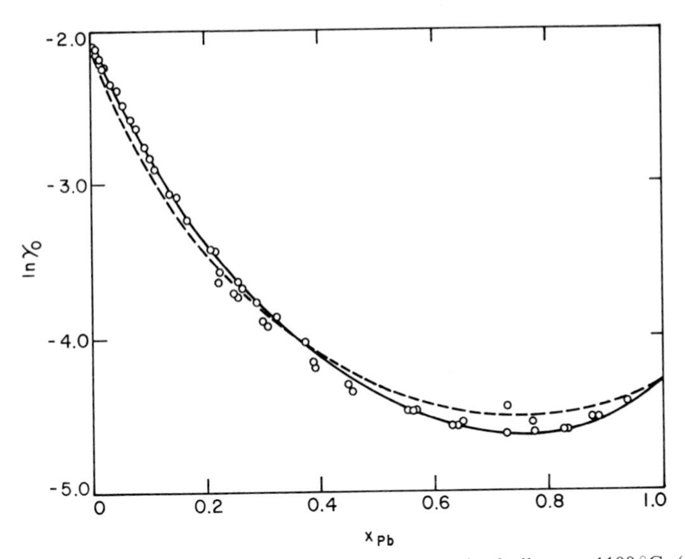

FIG. 3.20 Activity coefficient of oxygen in liquid copper–lead alloys at 1100 °C; (○) data of Jacob and Jeffes [63] compared with data calculated from Eqs. (3.59) (– – –) and (3.63) (———). After Chiang and Chang [62].

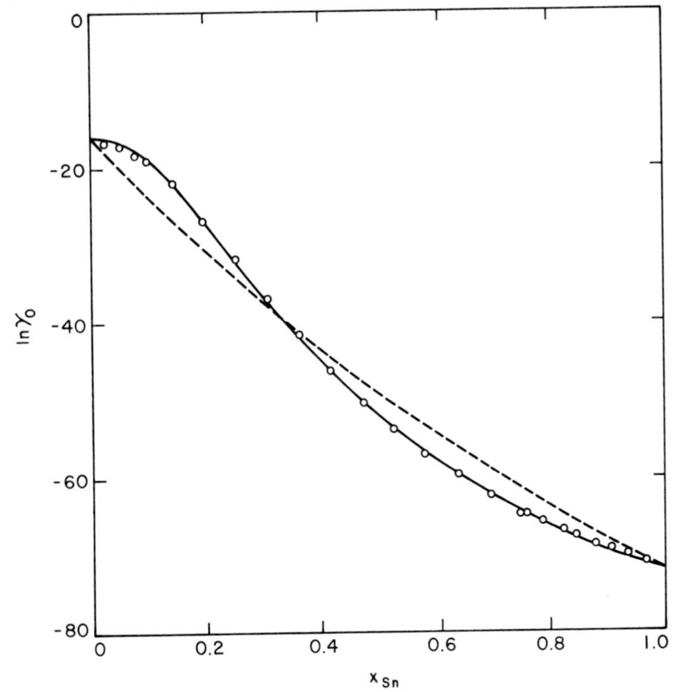

FIG. 3.21 Activity coefficient of oxygen in liquid copper–tin alloys at 1200 °C; (○) data of Block and Stuwe [64] compared with data calculated from Eqs. (3.59) (– – –) and (3.63) (———). After Chiang and Chang [62].

For the limiting case of $h = 0$, i.e., a linear dependence of ΔH_i on i, Eq. (3.59) is reduced to

$$Q = \frac{\gamma_{X(AB)}}{[\gamma_{X(A)}\,\gamma_{X(B)}]^{1/2}} = [(1 - x_B)\phi^{-1/2Z} + x_B\phi^{1/2Z}]^{-Z} \tag{3.60}$$

or

$$\gamma_{X(AB)} = \left[\frac{(1 - x_B)}{(\gamma_{X(A)})^{1/Z}} + \frac{x_B}{(\gamma_{X(B)})^{1/Z}}\right]^{-Z}. \tag{3.61}$$

In a subsequent study, Chiang and Chang [62] noted that Eq. (3.59) for constant h does not fit experimental data well for the activity coefficient of oxygen in ternary A–B–O systems. They modified Wagner's equation by assuming h is a linear function of i.

$$\Delta H_{i+2} - 2\,\Delta H_{i+1} + \Delta H_i = h_1 + h_2 i \tag{3.62}$$

and derived the following relation

$$\frac{\gamma_{X(AB)}}{[\gamma_{X(A)}\,\gamma_{X(B)}]^{1/2}} = \left\{\sum_{i=0}^{i=Z}\binom{Z}{i}\left[\frac{1 - x_B}{\phi^{1/2Z}}\right]^{Z-i}\right.$$
$$\left.\times\,[x_B\phi^{1/2Z}]^i \exp\left[\frac{(Z - i)i}{2RT}(h_1 - h_2) + \frac{(Z^2 - i^2)i}{6RT}h_2\right]\right\}. \tag{3.63}$$

For $h_2 = 0$, this equation is reduced to Wagner's equation (3.59). Chiang and Chang showed that with this modification, Wagner's equation represents the experimental data well as demonstrated by the examples in Figs. 3.20 and 3.21 for liquid Cu–Pb–O [63] and Cu–Sn–O [64] alloys.

3.2 Density and Critical Phenomena

The molar volume is an intensive state property of a system, and as an independent state variable, the molar volume often appears in many thermodynamic and kinetic functions. For convenience, the available data on molar volumes of solid and liquid elements at their melting points are listed in Table 3.5. References to sources of data on densities are given in review papers by Lucas [65] and Allen [66].

With the exception of Hg, Ga, In, and most of the transition elements, for about 22 elements the temperature coefficient of molar volume in terms of $(1/V)(dV/dT)$ is close to being inversely proportional to the melting temperature; the proportionality factor is found to be about 0.09. Another correlation is that shown by Lucas [65]: For a given group in the periodic table, the temperature coefficient of liquid density $d\rho/dT$ increases linearly with increasing melting temperature.

TABLE 3.5

Molar Volumes of Elements at Their Melting Temperatures (mp)[a]

Element	mp (°K)	V (cm^3 g atom^{-1}) solid	V (cm^3 g atom^{-1}) liquid	$\frac{1}{V}\frac{dV}{dT} \times 10^6$ (°K^{-1})
Ag	1234	11.16	11.54	97
Al	932	10.59	11.33	114
Au	1336	10.76	11.35	70
B	2448	4.96	5.20	—
Ba	1000	—	41.34	83
Be	1556	4.98	5.33	69
Bi	544	21.59	20.75	125
Ca	1106	—	28.86	162
Cd	594	13.49	13.94	134
Ce	1071	21.11	21.02	34
Co	1768	7.27	7.37	160
Cr	2130	—	8.11	—
Cs	302	—	72.36	283
Cu	1356	7.61	7.91	100
Fe	1809	7.66	7.96	119
Ga	303	11.82	11.47	100
Ge	1210	13.93	13.22	94
Hf	2503	—	14.87	—
Hg	234	14.17	14.64	174
In	430	15.96	16.35	115
Ir	2716	—	9.61	—
K	336	46.01	47.18	268
La	1193	—	23.32	40
Li	453	13.29	13.66	164
Mg	922	14.85	15.29	167
Mn	1517	—	9.41	127
Mo	2893	—	10.26	—
Na	371	—	24.81	237
Nb	2741	—	11.86	—
Nd	1297	—	21.56	—
Ni	1726	7.11	7.56	142
Os	3303	—	9.46	—
Pb	600	18.70	19.41	123
Pd	1825	9.58	10.14	121
Pr	1208	—	21.32	—
Pt	2042	9.68	10.32	153
Pu	913	14.89	14.52	91
Rb	312	—	56.67	272
Re	3433	—	9.96	—
Rh	2239	—	9.27	—
Ru	2523	—	9.27	—
Sb	904	18.63	18.83	85
Se	494	16.82	19.68	274
Si	1685	12.29	11.12	145

TABLE 3.5 (continued)

		$V(\text{cm}^3 \text{ g atom}^{-1})$		$\frac{1}{V}\frac{dV}{dT} \times 10^6$
Element	mp (°K)	solid	liquid	(°K^{-1})
Sn	505	16.58	16.96	88
Sr	1041	—	35.25	110
Ta	3287	—	12.07	—
Te	773	20.98	22.01	96
Th	2028	—	22.10	—
Ti	1943	—	11.65	170
Tl	577	—	18.11	115
U	1405	—	13.30	58
V	2175	—	9.18	—
W	3673	—	10.51	—
Zn	693	9.55	9.94	177
Zr	2123	—	16.58	—

[a] References to sources of data are given in review papers by Lucas [65] and Allen [66].

The densities of saturated liquid and vapor phases of mercury and alkali metals have been measured at elevated temperatures and pressures close to the critical point. As an example, the results of Dillon *et al.* [67] for densities of saturated liquid and vapor rubidium are shown in Fig. 3.22. The curves for the liquid and vapor phases merge at the critical point such that $(\partial P/\partial V)_{T_c} = 0$ and $(\partial \rho/\partial T)_{p_c} = \infty$. The average density of the saturated liquid and the vapor phase obeys a so-called "law of rectilinear diameters." The point of intersection of this line with the density curve gives the values of critical density ρ_c and temper-

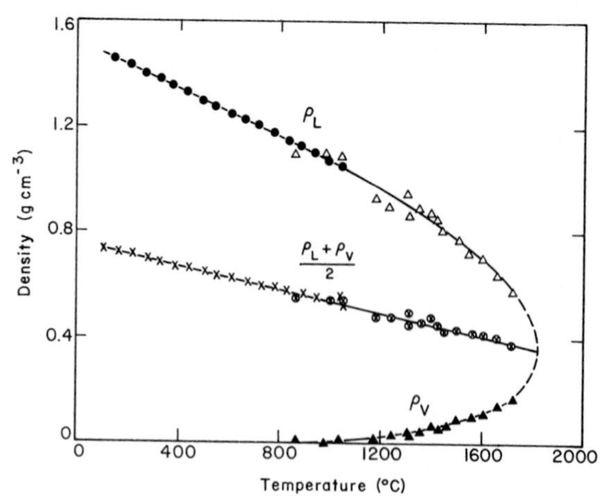

FIG. 3.22 Densities of saturated liquid and vapor rubidium. After Dillon *et al.* [67].

ature T_c; extrapolation of measured vapor pressure of saturated liquid to the critical temperature gives the critical pressure p_c.

The critical properties of alkali metals determined by Dillon *et al.* are listed in Table 3.6, together with the critical values for mercury, an average of two independent determinations [68, 69]. The compressibility factor at the critical point $p_c V_c / R T_c$ is not constant, indicating that the principle of corresponding states for p–V–T properties may not apply to metals. The principle of corresponding states for p–V–T properties, as originally stated by van der Waals, is based on the reduction of the variables using the critical constants. The reduced variables are defined by the ratios

$$p_r = p/p_c, \qquad V_r = V/V_c, \qquad T_r = T/T_c. \tag{3.64}$$

If the principle of corresponding states applies, the equation of state in terms of the reduced variables should be the same for all substances. Therefore, the compressibility factor at the critical point should be a universal constant. For simple, nonpolar molecules $p_c V_c / R T_c$ is about 0.29 and for hydrocarbons 0.27. For alkali metals the compressibility factor at the critical point increases from 0.17 to 0.25 with increasing atomic weight of the metal. Despite this departure from the principle of corresponding states, Dillon *et al.* found that in a plot of ρ/ρ_c against T/T_c, the data points for all the alkali metals fall on the same curve.

TABLE 3.6

Critical Constants[a]

Element	T_c (°K)	ρ_c (g cm^{-3})	p_c (atm)	$p_c V_c / R T_c$
Li	3223 ± 600	0.120 ± 0.033	680	0.17 ± 0.12
Na	2573 ± 350	0.206 ± 0.041	350	0.20 ± 0.12
K	2223 ± 330	0.194 ± 0.037	160	0.21 ± 0.13
Rb	2093 ± 35	0.346 ± 0.009	157	0.22 ± 0.02
Cs	2057 ± 40	0.428 ± 0.012	145	0.25 ± 0.02
Hg	1758 ± 12	5.0 ± 0.3	1515	0.42 ± 0.03

[a] Data of Dillon *et al.* [67] for alkali metals. Data of Hensel and Franck [68] and Kikoin and Senchenkov [69] for mercury.

It should be noted that the derivation of critical constants from p–V–T data is subject to interpretation of the experimental data. For example, the critical constants derived by Ross and Greenwood [70] from a reevaluation of the experimental data differ somewhat from those given in Table 3.6.

Several attempts have been made to estimate critical temperatures of metals by semiempirical relations based on the application of the principle of corresponding states. From estimates of critical temperatures by assuming the validity of the principle of corresponding states, as applied to entropies of vaporization relative to data for mercury and alkali metals, Grosse [71] obtained the following approximate proportionality of the critical temperature to the melting temperature:

$$T_c \sim 6.6 T_m. \tag{3.65}$$

For many binary liquid alloys, the molar volume is a linear function of composition, indicating ideal molar volume. For most liquid alloys, there are only modest departures from ideal molar volumes, therefore, the molar volumes of alloys may be estimated by assuming additivity of molar volumes of the components.

$$V \sim \sum n_i V_i. \tag{3.66}$$

3.3 Surface Tension

The plane of separation of two phases is known as a surface or interface. For a given system at constant temperature and pressure, the increase in free energy per unit increase in surface area is the surface tension (surface energy) or interfacial tension σ,

$$\sigma = \left(\frac{\partial F}{\partial A_s}\right)_{T,P,n_i}, \tag{3.67}$$

where A_s is the surface area and n_i is the number of moles of the ith component. With the inclusion of the surface-tension effect, the differential free-energy Eq. (2.3) for a k-component system with a flat surface now becomes

$$dF = -S\,dT + V dP + \sigma\,dA_s + \sum_i^k \mu_i\,dn_i, \tag{3.68}$$

where μ_i is the chemical potential of the ith component. Gibbs' exact treatment of surface thermodynamics gives, for fixed unit surface area and constant pressure [72],

$$d\sigma = -S_s\,dT - \sum_i^k \Gamma_i\,d\mu_i, \tag{3.69}$$

where S_s is the surface entropy and Γ_i is the surface excess concentration of the ith component. For alternative approaches to the thermodynamics of surfaces, reference may be made to the writings of Guggenheim [73]and of Wagner [74].

3.3.1 Surface Tensions of Pure Liquid Metals

For a single-component system, the summation in Eq. (3.69) is zero, and the surface entropy is therefore given by

$$S_s = -d\sigma/dT. \tag{3.70}$$

Combining this with Eq. (3.68) for constant pressure and volume gives the total surface energy E_s, which is also equal to the surface enthalpy H_s.

$$E_s = H_s = \sigma - T\,d\sigma/dT \tag{3.71}$$

For numerical evaluation of molar surface quantities, an estimate is made of the molar surface area A_s by assuming spreading of a mole of atoms into a monolayer for which

$$A_s = bN^{1/3}V^{2/3}, \tag{3.72}$$

where N is the Avogadro's number, V is the molar volume of the substrates, and the factor b accounts for the atomic arrangement in the monolayer: $b = 1.09$ for hexagonal close packing (hcp) and $b = 1.12$ for bcc packing. For $b = 1.12$, A_s in units of $\mathrm{cm^2\,g\,atom^{-1}}$ and σ in $\mathrm{erg\,cm^{-2}}$ ($\equiv \mathrm{dyn\,cm^{-1}} \equiv \mathrm{g\,sec^{-2}}$), the molar surface free energy and enthalpy in units of $\mathrm{cal\ g\ atom^{-1}}$ are given by

$$F_s^{\mathrm{M}} = A_s\sigma = 2.26V^{2/3}\sigma, \tag{3.73}$$

$$H_s^{\mathrm{M}} = A_sH_s = 2.26V^{2/3}(\sigma - T\,d\sigma/dT). \tag{3.74}$$

For curved surfaces, Eq. (3.68) has to be modified to account for the change of A_s with the radius of curvature of the surface. Thomson (Lord Kelvin) [75] showed that the vapor pressure of a liquid droplet is greater than that of the same liquid with a plane surface; when the surface is concave, the vapor pressure is lowered. From thermodynamic considerations, the escaping tendency from a curved surface is given by the relation

$$\mu_i - \mu_i^{\mathrm{P}} = (2\sigma/r)\bar{V}_i, \tag{3.75}$$

where μ_i is the chemical potential over a curved surface and μ_i^{P} is that for a plane surface, r is the radius of curvature, and $\bar{V}_i$ is the relative partial molar volume of the ith component. In terms of fugacity, or partial pressure for an ideal vapor,

$$\ln(P_i/P_i^{\mathrm{P}}) = (2\sigma/r)(\bar{V}_i/RT). \tag{3.76}$$

This thermodynamic relation can be anticipated directly from the Laplace equation derived in 1805 by considering a simple physical model [76]. For a spherical bubble or a drop of radius r, the excess pressure in the interior of the drop is

$$\Delta P = 2\sigma/r \tag{3.77}$$

and for an ellipsoidal drop with major and minor axes r_1 and r_2,

$$\Delta P = \sigma(1/r_1 + 1/r_2). \tag{3.78}$$

These equations form the basis of experimental determinations of surface tension of liquids with techniques such as capillary rise, maximum bubble pressure, pendant drop, and sessile drop.

Available data on surface tensions of liquid metals at their melting temperatures and temperature coefficients of surface tension are listed in Table 3.7; references to sources of data are given in review papers by Allen [66]and Grosse [77].

TABLE 3.7

Surface Tensions of Liquid Metals at Their Melting Temperatures[a]

Element	σ (dyn cm^{-1})	$-d\sigma/dT$ (dyn cm^{-1} deg^{-1})	Element	σ (dyn cm^{-1})	$-d\sigma/dT$ (dyn cm^{-1} deg^{-1})
Ag	903	0.16	Na	191	0.10
Al	914	0.35	Nb	1900	0.24
Au	1140	0.52	Nd	689	0.09
B	1070	—	Ni	1778	0.38
Ba	277	0.08	Os	2500	—
Be	1390	—	Pb	468	0.13
Bi	378	0.07	Pd	1500	—
Ca	838	0.10	Pt	1800	—
Cd	570	0.26	Pu	550	—
Ce	740	0.33	Rb	85	0.06
Co	1873	0.49	Re	2700	—
Cr	1700	0.32	Rh	2000	—
Cs	70	0.06	Ru	2250	—
Cu	1360	0.21	Sb	367	0.05
Fe	1872	0.49	Se	106	0.10
Ga	718	0.10	Si	865	—
Gd	810	—	Sn	544	0.07
Ge	621	0.26	Sr	303	0.10
Hf	1630	—	Ta	2150	—
Hg	498	0.20	Te	180	0.06
In	556	0.09	Th	978	—
Ir	2250	—	Ti	1650	—
K	115	0.08	Tl	464	0.08
La	720	0.32	U	1550	0.14
Li	398	0.14	V	1950	0.31
Mg	559	0.35	W	2500	—
Mn	1090	0.20	Zn	782	0.17
Mo	2250	—	Zr	1480	—

[a] References to sources of data are given in review papers by Allen [66] and by Grosse [77].

The surface tension of a pure liquid decreases with increasing temperature and becomes zero at the critical temperature. On this theoretical basis, Eötvös [78] proposed the following relation for the temperature dependence of liquids:

$$\sigma V^{2/3} = K(T_c - T), \tag{3.79}$$

where the proportionality factor K is about 2.1 erg deg^{-1} for most organic liquids. Subsequently, van der Waals [79] introduced another semiempirical relation for the dependence of surface tension on reduced temperature T/T_c

$$\sigma = \sigma^\circ[1 - (T/T_c)]^n, \tag{3.80}$$

where σ° is a constant that can be related with critical properties of the liquid. This equation holds well for many organic liquids for which the exponent n is $\frac{11}{9}$ [80].

Lennard-Jones and Corner [81, 82] have developed a theory of surface tension based on Lennard-Jones intermolecular potential function and the concept of free volume proposed by Lennard-Jones and Devonshire [83]. For nonpolar molecules the intermolecular potential energy function $\varphi(r)$ is usually represented by the Lennard-Jones (6–12) potential

$$\varphi(r) = 4\epsilon[(\delta/r)^{12} - (\delta/r)^6], \tag{3.81}$$

where r is the distance of separation of molecular or atomic species, ϵ the maximum energy of attraction, i.e., potential well in the $\varphi(r)$ versus r plot, and δ the collision diameter for low-energy collisions, i.e., the value of r at $\varphi(r) = 0$. The force constants δ and ϵ/k, where k is Boltzmann constant, as characteristic of molecular or atomic species, are independent variables and describe many fundamental properties of gases and liquids.

By applying the principle of corresponding states with reduced variables in terms of force constants, Lennard-Jones and Corner have shown that the reduced surface tension is a function of reduced volume and temperature,

$$\sigma^* = f(V^*, T^*), \tag{3.82}$$

where

$$\sigma^* = \sigma\delta^2/\epsilon, \tag{3.83}$$

$$V^* = V/N\delta^3, \tag{3.84}$$

$$T^* = kT/\epsilon. \tag{3.85}$$

From the values of σ^*, T^*, and V^* calculated by Corner [82] for simple liquids, such as liquid inert gases, a corresponding state plot of σ^*/T^* versus $1/(V^*)^2$ is obtained. As is shown in Sec. 3.4, transport properties of liquid metals may be described approximately in terms of a rigid-sphere model for which the atomic packing factor at the melting point gives $V^* = 1.10$. For this value of the reduced volume at the melting point, σ^*/T^* is equal to 3.1; that is, at the melting point

$$\sigma\delta^2/\epsilon = 3.1T^* \qquad \text{or} \qquad \sigma\delta^2 = 3.1kT_\mathrm{m}. \tag{3.86}$$

If a rigid-sphere model is assumed for the structure of liquid metals, the collision diameter may be estimated from the molar volume of the liquid metal at its melting point, thus $\delta = 1.146 \times 10^{-8} V^{1/3}$ cm (see Sec. 3.4 on viscosity). With this estimate of δ, the following relation is obtained from Eq. (3.86).

$$\sigma = 3.26 T_\mathrm{m} V^{-2/3} \quad \mathrm{dyn\,cm^{-1}}. \tag{3.87}$$

This semitheoretical relation is in general accord with the experimental data in Fig. 3.23, reproduced from Allen's paper [66] with the addition of the lower line for Eq. (3.87). The upper line with a slope of 3.6 was chosen by Allen from empirical considerations involving estimated critical temperatures.

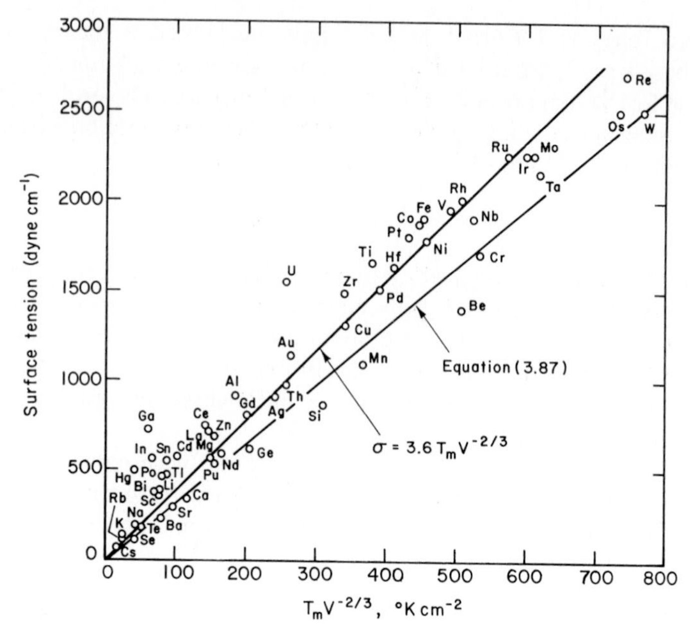

FIG. 3.23 Surface tension of liquid metals at their melting points related to their melting temperatures and molar volumes. After Allen [66], by courtesy of Marcel Dekker, Inc., with the addition of lower line representing Eq. (3.87).

There are other theoretical equations for surface tension. In all cases, however, approximations have to be made to estimate interatomic energy functions and force constants for numerical evaluation of the surface tension.

Since vaporization involves breaking of interatomic bonds, the surface tension is expected to be related to the heat of vaporization. In fact, Strauss [84] showed that at melting points of metals, $\log \sigma$ is a linear function of $\log(\Delta H_v/V)$, where ΔH_v is the heat of vaporization. In a subsequent study Grosse [77] showed that for liquid metals solidifying with cubic or tetragonal crystallographic modifications, the following relation holds well at the melting point; with ΔH_v in calories per mole,

$$\sigma = 0.066\,\Delta H_v V^{-2/3}. \tag{3.88}$$

For metals crystallizing in hexagonal or rhombohedral forms, e.g., Hg, Cd, Mg, Zn, surface tensions are a little higher than those given by Eq. (3.88).

3.3.2 Surface Tensions of Binary Alloys

As in seen from typical examples [85–88] in Fig. 3.24,* surface tensions of binary liquid alloys vary nonlinearly with composition; departure from linearity

* The value of 1730 dyn cm^{-1} for nickel determined by Fesenko *et al.* [85] is much lower than 1934 dyn cm^{-1} determined subsequently by Kozakevitch and Urbain [90] for high purity nickel.

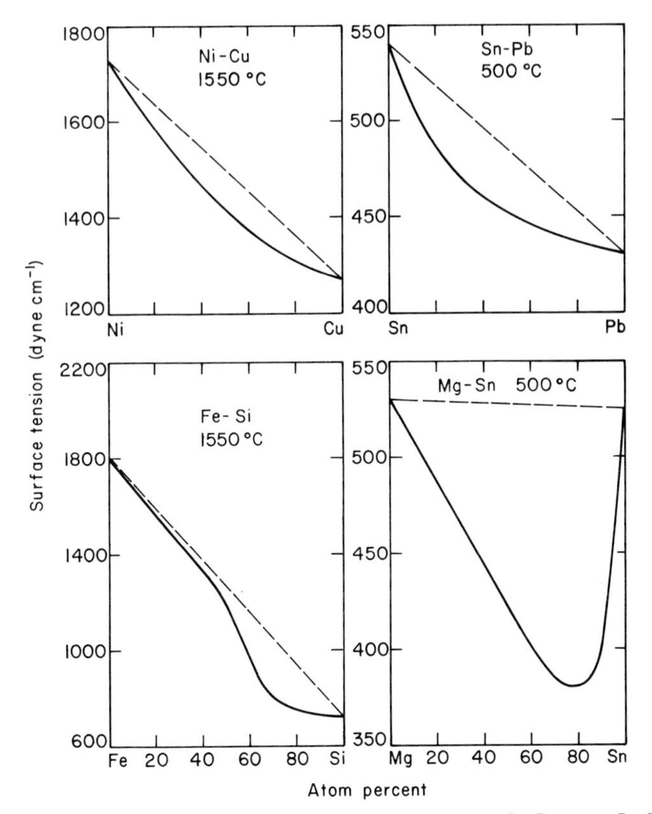

FIG. 3.24 Surface tensions of selected binary liquid alloys: Ni–Cu [85], Sn–Pb [86], Fe–Si [87], Mg–Sn [88].

is more pronounced in systems showing strong negative deviations from ideal thermodynamic behavior, e.g., Fe–Si, Mg–Sn.

Although many attempts have been made, no general relation could be found to describe the composition dependence of surface tension. However, Eberhart [89] has shown that for organic liquids, molten salts, and for some metallic solutions with modest departure from ideal behavior, the following relation may be used for binary mixtures 1–2:

$$\sigma = (\phi x_1 \sigma_1 + x_2 \sigma_2)/(\phi x_1 + x_2), \tag{3.89}$$

where σ_1 and σ_2 are surface tensions of pure liquid components 1 and 2; the coefficient ϕ, characteristic of the system, is determined by fitting the surface tension–composition curve to this equation.

Effects of various solutes on surface tensions of liquid iron [90], copper [91], and aluminum [92] are shown in Fig. 3.25.

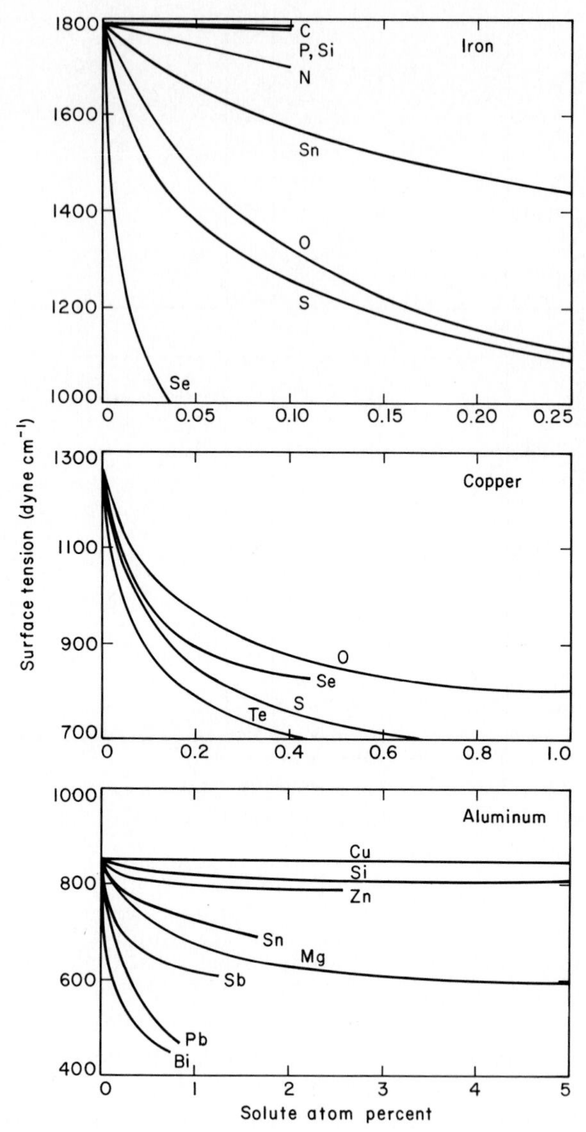

FIG. 3.25 Effect of solutes on surface tensions of liquid iron at 1550 °C [90], liquid copper at 1150 °C [91], and liquid aluminum at about 700 °C [92].

3.3.3 Surface Tensions of Ternary Alloys

Consistent with Gibbs' concept of the mathematical surface separating two adjacent phases, the surface excess concentration of the solute is relative to the position of the surface chosen such that for the solvent, component 1, Γ_1 equals zero. With this frame of reference, for a ternary system at constant temperature, Eq. (3.69) is reduced to

$$-d\sigma = \Gamma_2\, d\mu_2 + \Gamma_3\, d\mu_3. \tag{3.90}$$

To evaluate Γ_2 and Γ_3 from the surface tension data for a ternary system, Eq. (3.90) must be transformed to two independent equations. Such a transformation made by Whalen *et al.* [93] gives the following thermodynamic relations:

$$\Gamma_2 = (\alpha_2 + \beta_{23}\alpha_3)/(1 - \beta_{23}\beta_{32}), \qquad \Gamma_3 = (\alpha_3 + \beta_{32}\alpha_2)/(1 - \beta_{23}\beta_{32}) \tag{3.91a}$$

where

$$\begin{aligned} \alpha_2 &= -(\partial\sigma/\partial\mu_2)_{x_1/x_3}, & \alpha_3 &= -(\partial\sigma/\partial\mu_3)_{x_1/x_2}, \\ \beta_{23} &= (\partial x_2/\partial x_3)_{\mu_2, x_1}, & \beta_{32} &= (\partial x_3/\partial x_2)_{\mu_3, x_1}. \end{aligned} \tag{3.91b}$$

The parameters α and β are determined from data on surface tension and chemical potentials of components 2 and 3 plotted in a tenary composition diagram.

Whalen *et al.* measured the surface tension of Fe–Cr–3%C and Fe–Si–3%C liquid alloys. Because their measurements are confined to essentially constant atom fraction of carbon, Eq. (3.91) cannot be solved with sufficient accuracy to evaluate Γ_2 and Γ_3. An expression derived by Belton [94] from Eq. (3.90) is more readily applicable for correct interpretation of the surface-tension data of Whalen *et al.* By differentiating Eq. (3.90) with respect to $\ln x_2$ and by substituting $\mu_i = \mu_i^\circ + RT \ln \gamma_i x_i$ and the interaction coefficients $\varepsilon_i^{(j)} = \partial \ln \gamma_i/\partial x_j$, Belton obtained the following equation for constant concentration of component 3:

$$-\frac{1}{RT}\left(\frac{\partial\sigma}{\partial \ln x_2}\right)_{x_3} = \Gamma_2[1 + x_2\varepsilon_2^{(2)}] + \Gamma_3 x_2 \varepsilon_3^{(2)}. \tag{3.92}$$

The surface tension data of Whalen *et al.* plotted in Fig. 3.26 show the existence of a minimum surface tension at $x_{Cr} = 0.18$ in alloys containing 0.125 atom fraction of carbon. By solving Eq. (3.92) for $(\partial\sigma/\partial \ln x_{Cr})_{x_C}$ equal zero with the known values of $\varepsilon_i^{(j)}$, Belton found that the ratio of relative adsorption Γ_{Cr}/Γ_C is unity at about x_{Cr} equal to 0.18, corresponding to the position of the minimum in Fig. 3.26. This finding is indicative of associative adsorption of CrC.

Although carbon has virtually no effect on the surface tension of liquid iron (Fig. 3.25a), as shown by Kozakevitch [90a], the addition of carbon to iron–sulfur alloys lowers the surface tension (Fig. 3.27a). In Fig. 3.27b the surface tension plotted against the activity of sulfur illustrates that the apparent effect of carbon on the surface tension results from the effect of carbon on the activity coefficient of sulfur in iron.

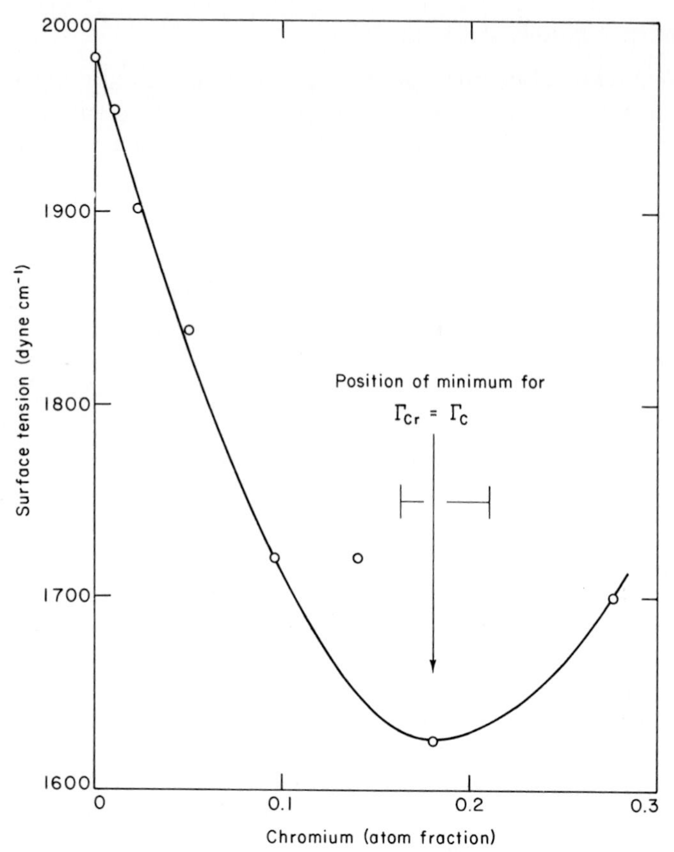

FIG. 3.26 Surface tensions of Fe–Cr–3%C liquid alloys at 1350°C measured by Whalen *et al.* [93] and the position of minmum σ computed by Belton [94] for $\Gamma_{Cr}/\Gamma_C = 1$. After Belton [94].

3.3.4 Chemisorption

With the substitution of $RT \ln a_i$ for the chemical potential in Eq. (3.69), the differential surface tension at constant temperature is given by the thermodynamic relation

$$d\sigma = -RT \sum_i^k \Gamma_i \, d \ln a_i. \tag{3.93}$$

For a binary system, surface excess concentration is then given by

$$\Gamma_2 = -(1/RT)(d\sigma/d \ln a_2), \tag{3.94}$$

which is known as the *Gibbs adsorption equation*. In dilute solutions, the surface excess concentration is much greater than that in the substrate: Γ_2 is therefore essentially equal to the actual surface concentration. At low concentrations of

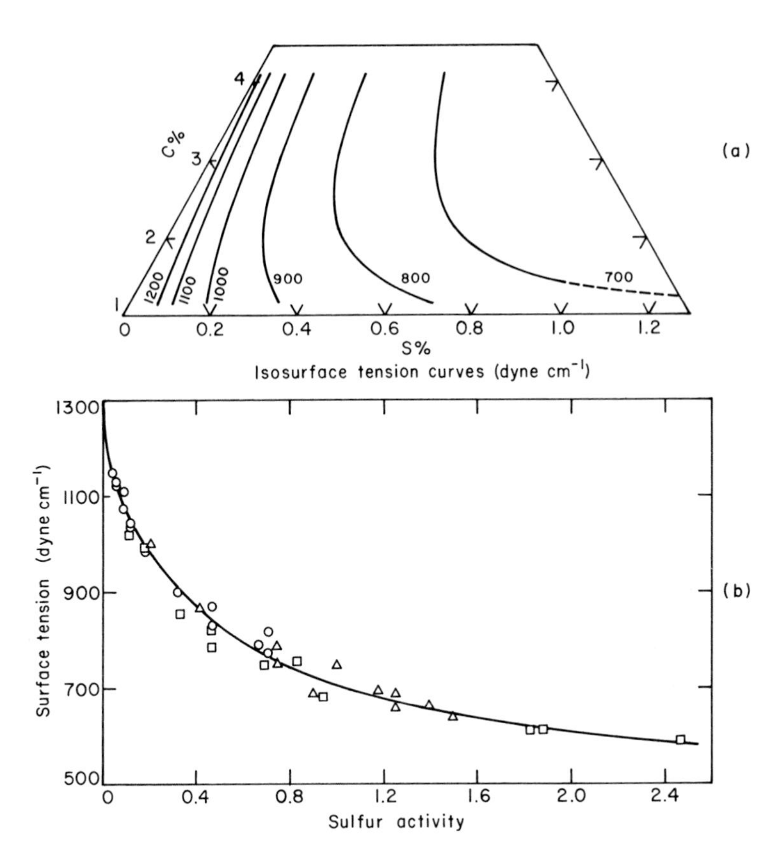

FIG. 3.27 Surface tension of liquid Fe–C–S alloys at 1450 °C; activity of sulfur defined such that $a_s \to$ wt.%S as %C $\to 0$. Legend (°C): ○, 1.25; □, 2.2; Δ, 4.0. After Kozakevitch [90a].

solute, the activity may be replaced by the solute concentration in weight or atom percent.

The surface concentration of solute is usually determined by drawing tangents to the curve in the plot of σ against $\ln a_i$. At high solute activities, the curve approaches linearity asymptotically; this limiting slope gives the surface concentration at saturation, designated by Γ_i°. The fraction of sites covered by adsorbed species at constant temperature and solute activity is given by the ratio

$$\theta_i = (\Gamma_i/\Gamma_i^\circ)_{T,a_i}. \tag{3.95}$$

Based on the kinetic theory of gases and the reaction kinetics of adsorption and desorption, Langmuir [95] derived the following adsorption isotherm for binary solutions with a fixed number of surface adsorption sites

$$a_i = \varphi_i \theta_i/(1 - \theta_i), \tag{3.96}$$

where φ_i is the proportionality factor. Fowler [96] derived the same equation through statistical mechanics on the assumption that each site can accommodate only one adsorbed species; the energy of adsorption is the same at any site and independent of the extent of surface coverage. For such an ideal monolayer, the proportionality factor φ_i, which may be considered as an activity coefficientlike term for a two-dimensional solution, is a function of temperature only.

The heat of adsorption ΔH_a for a given coverage θ_i is obtained from the temperature dependence of φ_i.

$$d \ln \varphi_i / d(1/T) = \Delta H_a / R. \tag{3.97}$$

As in most chemical reactions, chemisorption is an exothermic reaction (ΔH_a has a negative value); hence, for a given activity, the fraction of surface sites occupied by the absorbed species decreases with increasing temperature.

The assumption of an ideal monolayer does not hold for chemisorption on most solid surfaces as indicated by variations of φ_i and ΔH_a with the coverage θ_i. This led to various modifications of Eq. (3.95) by introducing empirical relations for variations of φ_i and ΔH_a with coverage [97, 98].

If chemisorption obeys the Langmuir adsorption equation for an ideal monolayer, as shown by Belton [99], Eqs. (3.95) and (3.97) may be combined with the Gibbs adsorption Eq. (3.94) and integrated to give for constant temperature

$$\sigma_\circ - \sigma = RT\Gamma_i^\circ \ln[1 + (a_i/\varphi_i)]_T, \tag{3.98}$$

where $\sigma_\circ$ is the surface tension of the pure liquid. It is perhaps of historical interest to note that in studies of surface tensions of fatty acids, Szyszkowski [100] found an empirical relation that is identical to Eq. (3.98), with activity a_i being replaced by the concentration of the solute.

The effect of oxygen on surface tension of liquid iron is shown in more detail in Fig. 3.28a, using the data of Halden and Kingery [101] and of Kozakevitch and Urbain [90]. The limiting slope drawn at 0.1% O gives $\Gamma_O^\circ = 10.5 \times 10^{14}$ O atoms cm^{-2}, which corresponds to 9.54 Å^2 for the cross-sectional area of adsorbed oxygen. This is to be compared with 6 Å^2 calculated for the ionic radius of 1.40 Å for O^{2-}, and 8 Å^2 in the plane of maximum packing in FeO. The fraction of surface coverage calculated [102] from the tangents drawn to the curve in Fig. 3.28a is given by curve I in Fig. 3.28b. According to this estimate of the adsorption isotherm, the coefficient φ_O decreases from 0.014% as $\theta_O \to 0$ to 0.0013% as $\theta_O \to 1$.

Belton [99] calculated adsorption isotherms for various metal–oxygen and sulfur systems from the available surface-tension data by assuming the formation of an ideal monolayer for which Eq. (3.98) would apply. The curve II in Fig. 3.28b calculated by Belton by assuming an ideal monolayer gives $\varphi_O = 0.0045\%$.

In the iron–sulfur alloys, σ is a linear function of ln[% S] over the range of compositions investigated (0.008–0.65% S) [90, 101, 103], indicating almost

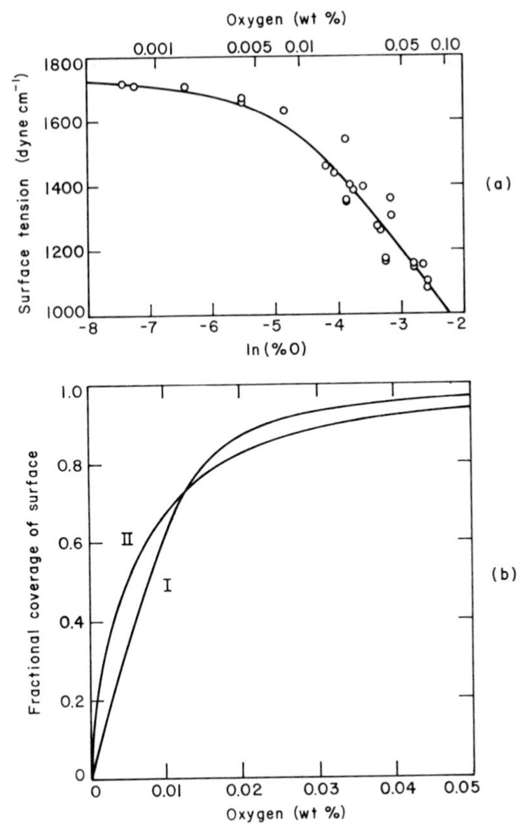

FIG. 3.28 (a) Effect of oxygen on surface tension of liquid iron at 1550 °C: Kozakevitch and Urbain [90], Halden and Kingery [101]. (b) Adsorption isotherm for oxygen on liquid iron at 1550 °C: curve I by tangent method [102]; curve II for assumed ideal monolayer [99].

complete surface coverage even at 0.008% S. Yet, as shown by Belton [99], the same data fit Eq. (3.98) reasonably well for an ideal monolayer.

Bernard and Lupis [104] suggested an adsorption model which takes into account interaction forces between adsorbed atoms. It is interesting to note that the surface-tension data analyzed by Belton are also in agreement with the model considered by Bernard and Lupis for nonideal behavior. It appears, therefore, that surface tensions are not too sensitive to the choice of the adsorption function.

Although the surface tension of a pure liquid decreases with increasing temperature, in the presence of a strongly surface active solute a positive temperature coefficient is often observed. This is due to the fact that for solutes interacting strongly with the solvent, the chemical potential term in Eq. (3.69) overcomes the entropy term, resulting in a positive temperature coefficient of the surface tension, at least in the vicinity of the liquidus temperatures.

3.3.5 Surface Tensions of Solid Metals

The surface tension plays an important role in nucleation, recrystallization, grain growth, and fracture of metals and alloys. Because of the experimental difficulties, data on the surface tension of solid metals are meager.

On the basis of a bond mechanism, Skapski [105] derived the following relation (simplified by neglecting the small entropy term) between surface tensions of solid and liquid metals at their melting points,

$$\sigma_{sv} = [(Z - Z_s)/Z](\Delta H_m/A_s) + (\rho_s/\rho_l)^{2/3}\sigma_{lv}, \tag{3.99}$$

where σ_{sv} and σ_{lv} are surface tensions of solid and liquid metal in equilibrium with its vapor; Z is the coordination number in the bulk metal and Z_s that on the surface; ΔH_m is the heat of fusion; ρ is the solid (s) and liquid (l) density; and A_s is the molar area as given by Eq. (3.72). As pointed out by Allen [66], the ratio σ_{sv}/σ_{lv} estimated from Eq. (3.99) is about 1.1, which is lower than 1.25–1.3 obtained from direct measurements of σ_{sv}.

When a polycrystalline solid is heat treated, the surface-tension effect brings about thermal grooving along the grain boundaries intersecting the surface. For the simple case of a grain boundary normal to the surface, and neglecting small variations in surface energy and grain boundary energy with anisotropy and lattice mismatch, the balance of interfacial tension vectors is given the relation

$$\sigma_{ss} = 2\sigma_{sv}\cos(\theta/2), \tag{3.100}$$

where σ_{ss} is the grain boundary energy and θ is the dihedral angle of the groove. If the grain boundary is in contact with a liquid phase, σ_{sv} in Eq. (3.100) is replaced by the solid–liquid interfacial tension σ_{sl}. The dihedral angle θ of the thermal groove for the solid–vapor equilibrium is, of course, different from that for the solid–liquid equilibrium.

The solid–liquid interfacial tension is evaluated from surface tensions σ_{sv} and σ_{lv}, and the liquid–solid contact angle ψ using the equilibrium relation

$$\sigma_{sl} = \sigma_{sv} - \sigma_{lv}\cos\psi. \tag{3.101}$$

From the theory of homogeneous nucleation and experimentally determined frequencies of formation of small droplets, Turnbull [106] found that for many pure metals σ_{sl} is related to the heat of fusion ΔH_m and the molar area.

$$\sigma_{sl} = 0.45\,\Delta H_m/A_s. \tag{3.102}$$

The role of interfacial tension on the kinetics of heterogeneous reactions, metal–slag (matte) emulsification and growth of dispersed particles are discussed in other chapters.

For small variations of surface energy with crystallographic orientation, a particle of homogeneous composition will acquire a spherical shape to minimize the surface energy, when heated for a long period of time in an uncontaminated

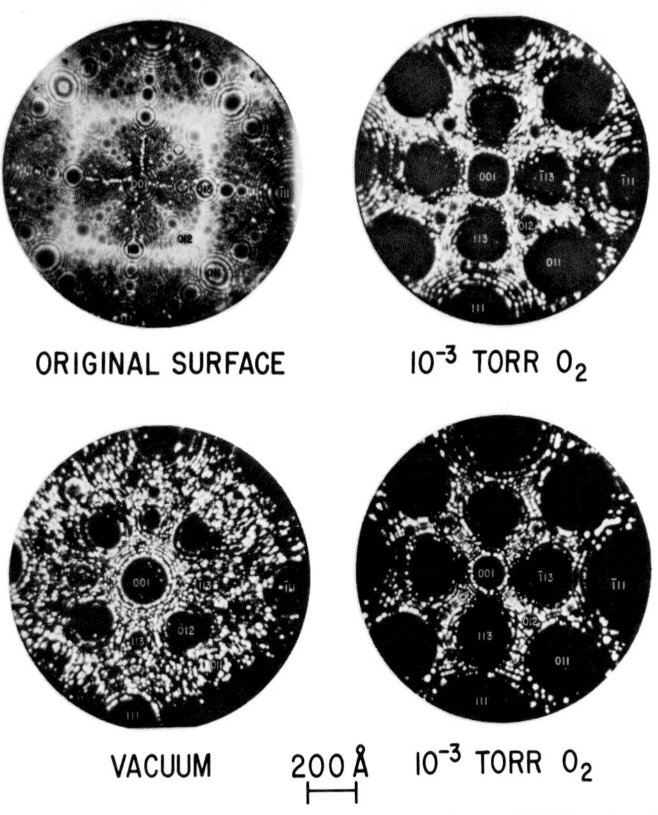

FIG. 3.29 Reversible thermal rearrangement of iridium surface at 700 °C as viewed by field-ion microscopy. After Brenner [108].

environment. If the surface energy varies with orientation because of preferential adsorption of impurities, the particle will acquire a faceted shape with preferred orientation having the lowest surface energy. For example, as shown by Sundquist [107], the faceting of nickel and copper particles when annealed in dry hydrogen becomes more pronounced in the presence of a trace amount of oxygen as in "vacuum" annealing. Another example of faceting induced by adsorption is that found by Brenner [108] using field-ion microscopy as shown in Fig. 3.29. The original field-evaporated surface of iridium when exposed to a low pressure of oxygen at 700 °C exhibits pronounced facets which shrink rapidly in high vacuum but which appear again on reexposure to oxygen.

There have been many observations of the development of striations on flat metal surfaces by so-called thermal etching, i.e., by holding at an elevated temperature in controlled atmosphere for an extended period of time [109]. An example of this is given in Fig. 3.30, which is a photomicrograph in oblique light showing striations formed after annealing a polished specimen of zone-refined iron in wet hydrogen ($p_{H_2O}/p_{H_2} = 0.035$) for 24 hr at 1000 °C. The

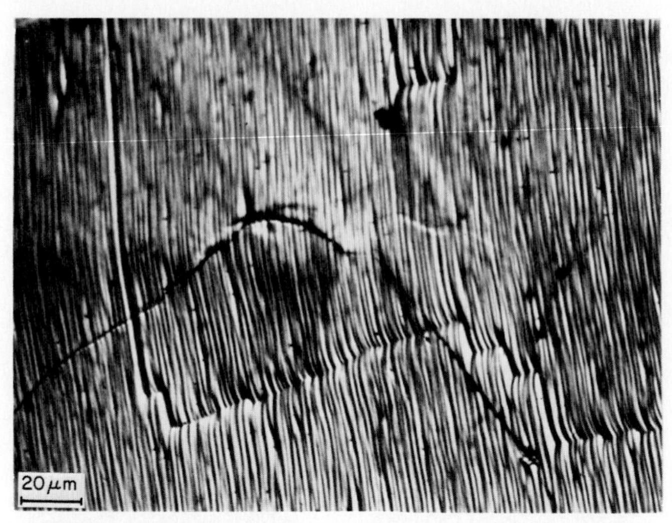

FIG. 3.30 Thermal faceting on zone-refined high-purity iron after annealing in wet hydrogen ($H_2O/H_2 = 0.035$) for 24 hr at 1000 °C.

surface of the facets are usually found to be low-index planes, especially {110} and {111}.

Direct measurement of adsorption isotherms at elevated temperatures presents many difficulties on account of the solubility of the adsorbed species in the matrix and the problem of separating surface adsorption from solution in the bulk phase. Some of these difficulties were overcome by using a radiotracer technique as in the work of Cabané-Brouty [110] to determine chemisorption of sulfur on silver. An example of her findings is shown in Fig. 3.31. The limiting surface occupancy corresponds to about 2×10^{14} S atoms cm^{-2}, roughly a monolayer. The surface saturation is reached when the activity of sulfur is about 1% of that corresponding to the Ag–Ag_2S equilibrium.

The surface energies of solid metals and alloys at elevated temperatures have been measured satisfactorily by the zero-creep technique. The principal feature of the technique is the measurement of the critical stress applied to the specimen at an elevated temperature to balance the shrinkage due to the surface-tension effect so that the strain rate is zero [111]. Although the surface energy is a function of crystallographic orientation, variations are usually within about 5%. The zero-creep technique has been used satisfactorily to determine the effect of, for example, phosphorus and nitrogen on the surface energy of γ and δ iron [112] and of oxygen on nickel and copper [113, 114].

The measurements of the surface energy by the zero-creep technique and the dihedral angle of the thermally etched grain-boundary grooves at the surface give the grain-boundary free energy σ_{ss}. The solute segregation to grain boundaries has been evaluated for many binary metal alloys by applying the Gibbs adsorption theorem to the composition dependence of the grain-boundary

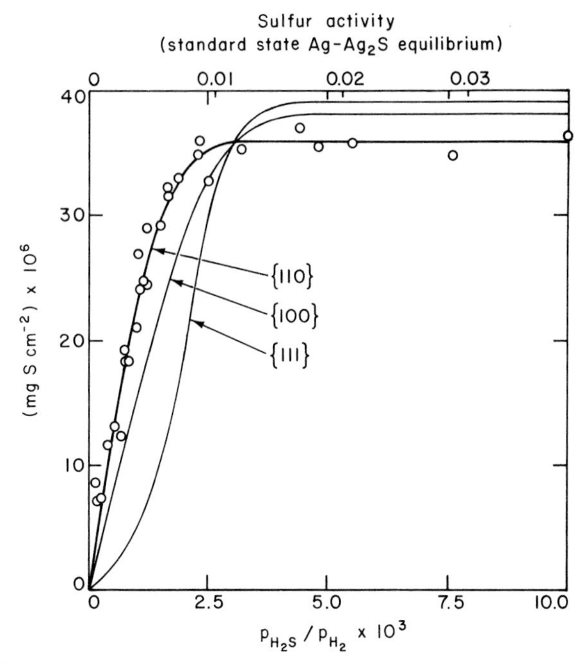

FIG. 3.31 Chemisorption of sulfur on silver {111}, {100}, and {110} planes at 400 °C. After Cabané-Brouty [110].

free energy [115]. As shown by Seah and Hondros [116], the grain-boundary enrichment ratio is inversely proportional to the solid solubility of the solute.

Based on physical processes, techniques and instruments have been developed in recent years to analyze surfaces for constituent elements such as field-ion microscopy with the atom probe, x-ray photoelectron emission, secondary-ion mass spectroscopy, ion-beam scattering, and Auger electron emission. References to the development work on these techniques of surface analysis and segregation to interfaces are given in a review paper by Hondros and Seah [117]. Also, advances are being made at present to develop high temperature Auger and x-ray photoelectron spectroscopy that will be of much value in future studies of chemisorption. For example, Grabke *et al.* [118] have shown that the Auger spectroscopy can be used to measure the chemisorption of carbon on the surface of iron at temperatures of 500–750 °C. Reference may be made to a book by Tompkins [98] for further reading on the status of modern theoretical and experimental studies of chemisorption on metal surfaces.

3.4 Viscosities of Liquid Metals

The viscosity is a measure of resistance of the fluid to flow when subjected to an external force. As proposed originally by Newton, the shear stress ε, i.e., force per unit area, causing a relative motion of two adjacent layers in a liquid is proportional to the velocity gradient du/dz normal to the direction of the applied force,

$$\varepsilon = \eta \, du/dz, \tag{3.103}$$

where the proportionality factor η is the coefficient of viscosity, or simply, viscosity of the fluid. Although not tested rigorously, most liquid metals are believed to obey Eq. (3.103), i.e., Newtonian liquids.

In honor of Poiseuille, who pioneered the earlier studies of viscosity, the unit of viscosity is called the poise (P), which is, in cgs units, 1 dyn sec cm^{-2} ($\equiv$1 g cm^{-1} sec^{-1}). For aqueous solutions and liquid metals, the unit of centipoise (cP = 0.01 P) is often used. The kinematic viscosity, defined by the ratio viscosity/density, in units of cm^2 sec^{-1} is called stokes (S); 0.01 S is centistokes (cS).

On the basis of a quasi-crystalline model of liquids, Andrade [119] derived the following semiempirical relation for viscosities of liquid metals at their melting temperature:

$$\eta = B(MT_{\mathrm{m}})^{1/2} V^{-2/3}, \tag{3.104}$$

where B is a constant to be determined from viscosity data. For all the liquid metals investigated, the variation of viscosity with temperature obeys the relation proposed by Andrade, even for the undercooled liquid metals,

$$\eta V^{1/3} = A \exp(C/VT) \quad \text{or} \quad \eta V^{1/3} = A \exp(E_{\mathrm{v}}/RT), \tag{3.105}$$

TABLE 3.8

Viscosities of Liquid Metals at Their Melting Points[a]

Element	η (cP)	E_{v} (kcal)	Element	η (cP)	E_{v} (kcal)
Ag	3.88	5.3	Li	0.55	1.33
Al	1.39	3.95	Mg	1.32	7.3
Au	5.38	5.1	Na	0.68	1.25
Bi	1.85	1.75	Ni	4.60	9.85
Ca	1.22	6.5	Pb	2.61	2.35
Cd	2.28	2.25	Pr	2.80	—
Ce	2.88	—	Pu	5.5	3.07
Co	4.49	10.6	Rb	0.60	1.23
Cs	0.60	1.15	Sb	1.48	4.05
Cu	4.10	7.3	Sn	2.00	1.3
Fe	4.95	9.9	Ti	5.2	—
Ga	1.70	1.0	Tl	2.64	—
Hg	1.88	0.6	U	6.50	7.26
In	1.89	1.15	Zn	2.82	3.03
K	0.51	1.2	Zr	8.0	—

[a] References to sources of data are given in review papers by Wilson [120] and Wittenberg [121].

where A and C are constants characteristic of the metal and $E_v = R(C/V)$ is the energy of activation for viscous flow.

Viscosities of liquid metals at their melting temperatures and energies of activation for viscous flow are listed in Table 3.8; references to sources of data are given in review articles by Wilson [120] and Wittenberg [121].

As is seen from the plot in Fig. 3.32, with the exception of Zr, Ti, Pu, U, and Sb the experimental data are in substantial agreement with Andrade's Eq. (3.104). For η is poise and V in $\mathrm{cm^3\,g\,atom^{-1}}$,

$$\eta = 5.4 \times 10^{-4}\,(MT_m)^{1/2}V^{-2/3}. \tag{3.106}$$

The variation of energy of activation with the melting temperature shown in Fig. 3.33 may be represented by

$$\log E_v = 1.36 \log T_m - 3.418, \tag{3.107}$$

where E_v is in kilocalories.

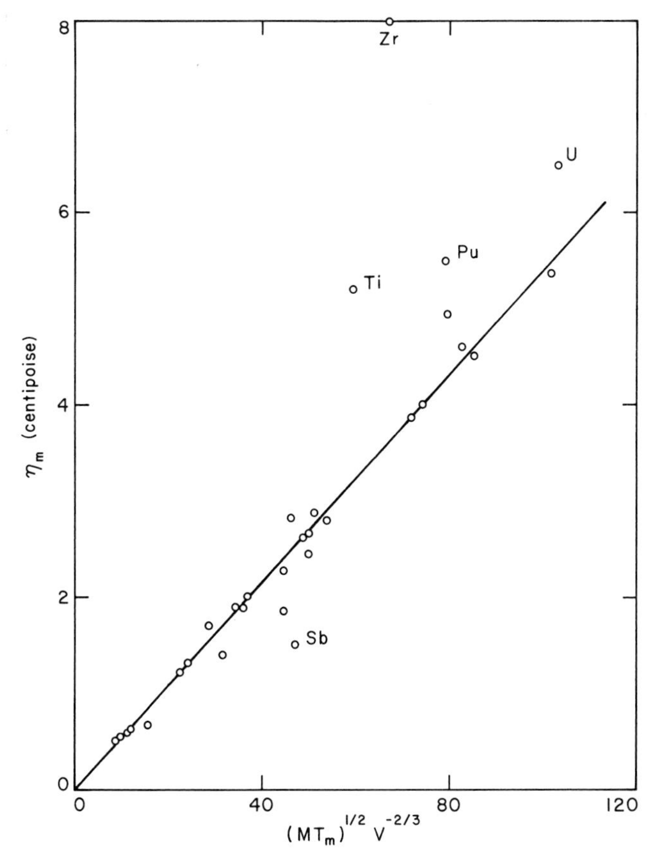

FIG. 3.32 Viscosities of liquid metals at their melting points related to atomic weight, melting temperature, and atomic volume as in Andrade's equation (3.104).

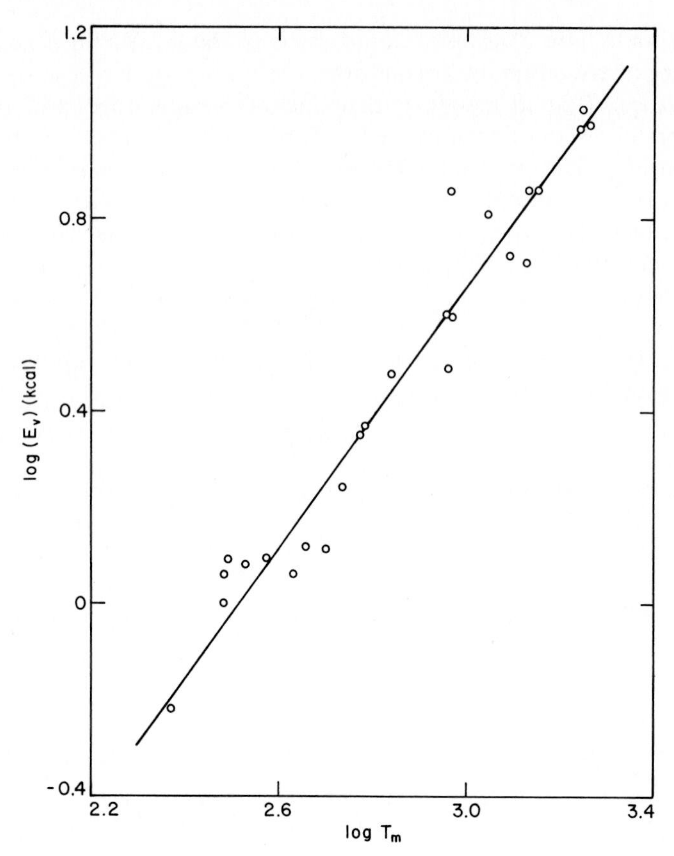

FIG. 3.33 Energy of activation for viscous flow related to melting temperature of metals.

Many advances have been made in the development of the theory of simple liquids [122] since the earlier work of Born and Green [123]. The generally accepted view is that the motion of an atom or a molecule in a liquid is similar to the motion of Brownian particles, and the dynamical properties of molecules are described in terms of a molecular friction constant ζ, a pair distribution function $g(\delta)$, and the intermolecular potential energy function $\varphi(r)$. Although theories developed give unified descriptions of the properties of liquids in terms of intermolecular potential and statistical molecular geometry, numerical evaluation of properties is not possible without developing various approximations. Despite the extent of work done on viscosities of liquid metals, no serious attempt has been made to test the applicability of equations derived from the kinetic theory of liquids.

The simplest theory of liquids is that based on the nonattracting rigid-sphere model. On the assumption that the pair distribution function of relative position is independent of the rate of strain or the temperature gradient, and that the velocity distribution is locally Maxwellian, Longuet-Higgins and Pople [124] derived the following expression for bulk viscosity of the liquid:

$$\eta = \frac{2\delta}{3}\left(\frac{mkT}{\pi}\right)^{1/2}\left(\frac{p}{kT} - \frac{N}{V}\right), \tag{3.108}$$

where m is the mass of the molecule (hard sphere) and p is the pressure of the rigid sphere of diameter δ. In terms of molecular weight M and the gas constant R,

$$\eta = \frac{2\delta}{3V}\left(\frac{MRT}{\pi}\right)^{1/2}\left(\frac{pV}{RT} - 1\right). \tag{3.109}$$

The molecular shear viscosity ζ is proportional to the bulk viscosity η.

$$\zeta = \tfrac{3}{5}\eta. \tag{3.110}$$

The effective diameter of the molecule and the compressibility factor pV/RT have been estimated with reasonable accuracy through computer calculations by using the method of *molecular dynamics* or the technique of *Monte Carlo* calculations. Such computer experiments on the behavior of molecules for a given pair distribution function gives the compressibility factor as a function of packing fraction χ, which is the fraction of the total volume that is occupied by the rigid spheres,

$$\chi = N\pi\delta^3/6V. \tag{3.111}$$

The total volume is that for random close-packed spheres, which is 14% greater than a truly close-packed arrangements.

Depending on the interpretation of the results from computer experiments [125, 126], the packing fraction of the rigid-sphere liquid at the melting point is in the range 0.46–0.49 and for the solid χ is from 0.48 to 0.54. These values of χ for solid and liquid give about 5% for change in volume on melting, in general accord with actual measurements. Therefore, for the rigid-sphere model with $\chi = 0.475 \pm 0.015$, the effective atomic diameter is given by

$$\delta = (1.146 \pm 0.012) \times 10^{-8} V^{1/3} \quad \text{cm}. \tag{3.112}$$

It is interesting to note in Fig. 3.34 that Eq. (3.112) for rigid-sphere liquid does not differ much from the linear relation between the Goldschmidt atomic diameter at 0 °K (from crystallographic data [127]) and the molar volume of liquid metals at their melting points.

The *Monte Carlo* calculations using Lennard-Jones potential give for the compressibility factor at the solid–liquid equilibrium $pV/RT = 11.2 \pm 1.2$. This value of the compressibility factor is not much altered when long-range interaction is included in the potential energy function. The near neighbor interaction as given by the Lennard-Jones potential is adequate to describe the molecular dynamics. Inserting this value of the compressibility factor in Eq. (3.109) and

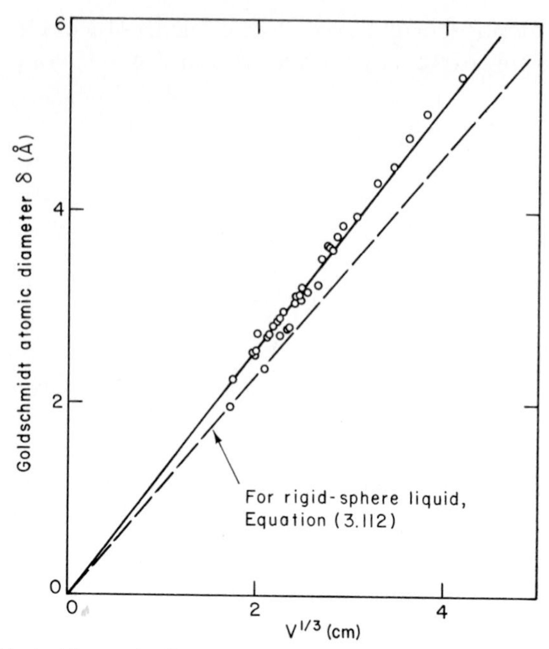

FIG. 3.34 Goldschmidt atomic diameter as a function of cube root of molar volume of liquid metals at their melting points compared with the relation for the rigid-sphere liquid.

combining with (3.112) gives for viscosity at the melting point

$$\eta = (4.0 \pm 0.5) \times 10^{-4}\,(MT)^{1/2}V^{-2/3}. \tag{3.113}$$

In a study of autocorrelation for hard spheres, Alder and Wainwright [128] have shown that the computer calculations of the pair potential between the molecules have to be corrected for what is called the *backscatter effect*. They found that the self-diffusivity calculated for rigid-sphere liquid at its melting point as derived by Longuet-Higgins and Pople [126] (discussed in Sec. 3.5) should be corrected for the backscatter effect by multiplying the calculated self-diffusivity with 0.72. Since viscosity is inversely proportional to self-diffusivity, the correction of viscosity for the backscatter effect should be made by dividing the right side of Eq. (3.113) by 0.72. With this correction, the "theoretical" equation for viscosity at the melting point for the rigid-sphere liquid becomes

$$\eta = (5.5 \pm 0.7) \times 10^{-4}\,(MT_{\mathrm{m}})^{1/2}V^{-2/3}, \tag{3.114}$$

which agrees well with Andrade's empirical equation (3.106).

The principle of corresponding states has been applied to the viscosity of liquid metals by Chapman [129]. The reduced viscosity η^* is considered to be a function of reduced volume V^* and reduced temperature T^*,

$$\eta^* = f(V^*, T^*), \tag{3.115}$$

where $\eta^* = \eta\delta^2 N/(MRT)^{1/2}$ and V^* and T^* are as given by Eqs. (3.84) and (3.85). On the assumption that all liquid metals obey the same interatomic potential energy function, and using the experimentally determined values of ϵ/k, δ, and η for sodium and potassium [130], Chapman evaluated a corresponding state plot of the function $\eta^*(V^*)^2$ versus T^* as shown in Fig. 3.35a. By using this corresponding state correlation, and assuming that the effective atomic diameter δ is the same as that of Goldschmidt's atomic diameter at 0 °K, Chapman estimated the force constants ϵ/k for several metals from the viscosity data. As is seen in Fig. 3.35b, ϵ/k is a linear function of the melting temperature of metals.

$$\epsilon/k = 5.20 T_m \quad °\text{K}. \tag{3.116}$$

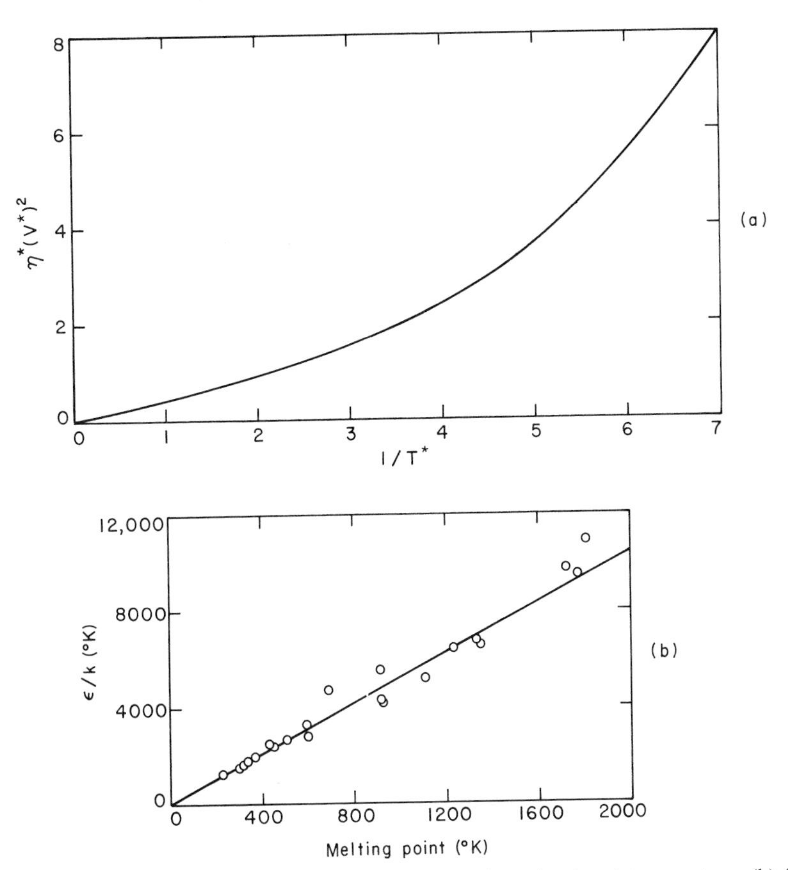

FIG. 3.35 (a) Reduced viscosity of liquids as a function of reduced temperature: (b) force constant ϵ/k as a function of melting point of metals. After Chapman [129].

By estimating ϵ/k from the melting temperature of a metal or an alloy, and δ from the molar volume, Fig. 3.34, the viscosity at any temperature may be computed from Fig. 3.35a. At the melting temperature, $1/T^*$ is 5.20 for which $\eta^*(V^*)^2$ is equal to 3.95; as would be expected, these values give an equation for viscosity the same as Eq. (3.106).

The surface tension is a thermodynamic property while the viscosity is a kinetic property; therefore, a simple connection between these properties is not readily anticipated. However, the simple theory of liquids leading to Eqs. (3.86), (3.109), and (3.110) suggests that the ratio of surface tension to viscosity may be of some significance. From these essentially theoretical equations based on the rigid-sphere (cell) models, the following ratio is evaluated:

$$\sigma/\zeta = 1.485(kT/m)^{1/2} \quad \text{cm sec}^{-1}. \tag{3.117}$$

This may be rewritten in terms of molecular velocities: for most probable molecular velocity,

$$\sigma/\zeta = 1.05(2kT/m)^{1/2}; \tag{3.118a}$$

for average molecular velocity,

$$\sigma/\zeta = 0.93(8kT/\pi m)^{1/2}. \tag{3.118b}$$

In the rigid-sphere model, the whole volume is considered to be divided up into similar cells forming Voronoi polyhedra, and the center of each particle is restricted to stay within its own cell. A molecule oscillating about its equilibrium position in a liquid at rest may be imagined to move a short distance $\Delta\delta$ within the cell with a velocity of $(2kT/m)^{1/2}$, which may be set equal to ε/ζ, where ε is the shear stress that the molecule experiences and ζ is the molecular shear viscosity. The relation in Eq. (3.118) implies that the shear stress that the molecule experiences over a distance $\Delta\delta$ is equal to the surface tension of the rigid-sphere liquid in equilibrium with its vapor.

Another observation of possible significance concerns the pressure p of the sphere exerted on the cell walls. Let us use Eq. (3.77) on the pretense that the pressure p is that arising from the radius of curvature and surface tension of the sphere, $p = 4\sigma/\delta$. With the compressibility factor of 11.2 for the rigid-sphere liquid at its melting point, p is equal to $11.2RT/V$. Substituting this in the above equation, and inserting the equivalence of V from Eq. (3.112), gives

$$\sigma\delta^2 = 2.80kT_{\mathrm{m}}, \tag{3.119}$$

which, surprisingly, is almost identical to Eq. (3.86), which was derived from a theoretical formulation of surface tension of nonpolar liquids by Lennard-Jones and Corner.

The physical significance, if any, of the correlations noted above between surface tension, viscosity, and compressibility factor waits to be resolved.

3.5 Diffusion in Metals and Alloys

The diffusive flux J is defined as the amount of a diffusing species crossing a surface of unit area, normal to the direction of flow, in unit time,

$$J = -D\,\partial c/\partial z, \tag{3.120}$$

where $\partial c/\partial z$ is the concentration gradient and D is the coefficient of diffusion, or simply diffusivity. Equation (3.120) is a statement of Fick's first law [131] for steady-state diffusion.

In non-steady-state diffusion, the flux changes with the diffusion distance z and time t:

$$\partial J/\partial z = -\partial(D\,\partial c/\partial z)/\partial z. \tag{3.121}$$

This difference in flux is equal to $-\partial c/\partial t$, a negative rate of concentration change; therefore

$$\partial c/\partial t = \partial(D\,dc/dz)/\partial z. \tag{3.122}$$

If the diffusivity D is independent of concentration (or substantially so under conditions of the experiments), Fick's second law may be written

$$\partial c/\partial t = D\,\partial^2 c/\partial z^2. \tag{3.123}$$

The solution of this equation depends on the geometry and the boundary conditions; for methods of solution of Eq. (3.123) reference may be made to, for example, Crank [132] and Jost [133].

Based on the concept of a random walk process, Einstein [134] showed that an atom or a molecule free to move in three dimensions because of thermal vibrations, will experience a mean square displacement $\bar{\Delta} r^2$ after a time t; this concept leads to the definition of self-diffusivity by the fundamental relation

$$D = \bar{\Delta} r^2/6t. \tag{3.124}$$

3.5.1 Self-Diffusivities of Pure Liquid Metals

Based on Eq. (3.124), Longuet-Higgins and Pople [124] derived the following equation for self-diffusivity of liquid metals described in terms of a rigid-sphere model:

$$D = \tfrac{1}{4}\delta(\pi kT/m)^{1/2}[(pV/RT) - 1]^{-1}. \tag{3.125}$$

As stated earlier, Alder and Wainwright [128] showed that D calculated from Eq. (3.125) for liquid metals at their melting points should be multiplied by 0.72 to correct for the backscatter effect. With this correction and taking for the compressibility factor $pV/RT = 11.2$, substituting Eq. (3.112) for δ gives for the self-diffusivity at the melting temperature

$$D = 3.27 \times 10^{-6}(T_{\mathrm{m}}/M)^{1/2}V^{1/3}. \tag{3.126}$$

As is seen from the plot in Fig. 3.36, measured self-diffusivities [135, 136] are in close agreement with those calculated from Eq. (3.126).

The form of the temperature dependence of diffusion in liquid metals still remains to be resolved. As shown for example by Nachtrieb [135], an Arrhenius-type plot, i.e., $\ln D$ versus $1/T$, gives a slightly concave curve for some metals, while a plot of D versus T is approximately a straight line. Nachtrieb pointed out that the proportionality of D to T may be justified theoretically from Eq. (3.124) on the assumption that the overall diffusion process may come from small displacement of atoms.

The apparent energy of activation E_d, taken from compiled data [135, 136], is seen in Fig. 3.37 to be essentially a linear function of the melting temperature, thus

$$E_{\mathrm{d}} = 6.6T_{\mathrm{m}} \quad \text{cal.} \tag{3.127}$$

In terms of energy per molecule ϵ,

$$\epsilon = 3.3kT_{\mathrm{m}}, \tag{3.128}$$

which is close to the average energy of vibration. It should be noted that the value of $\varepsilon/kT_{\mathrm{m}} = 5.2$ obtained by Chapman [129] from viscosity data is higher

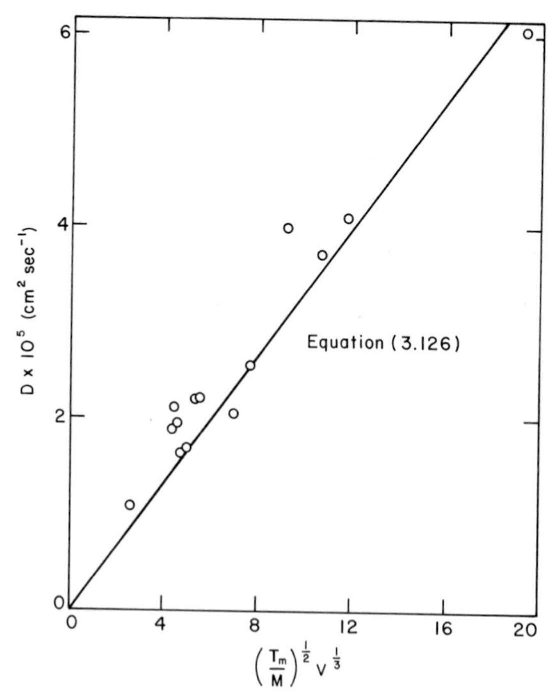

FIG. 3.36 Self-diffusivity as a function of atomic volume, atomic weight, and temperature at the melting point of metals.

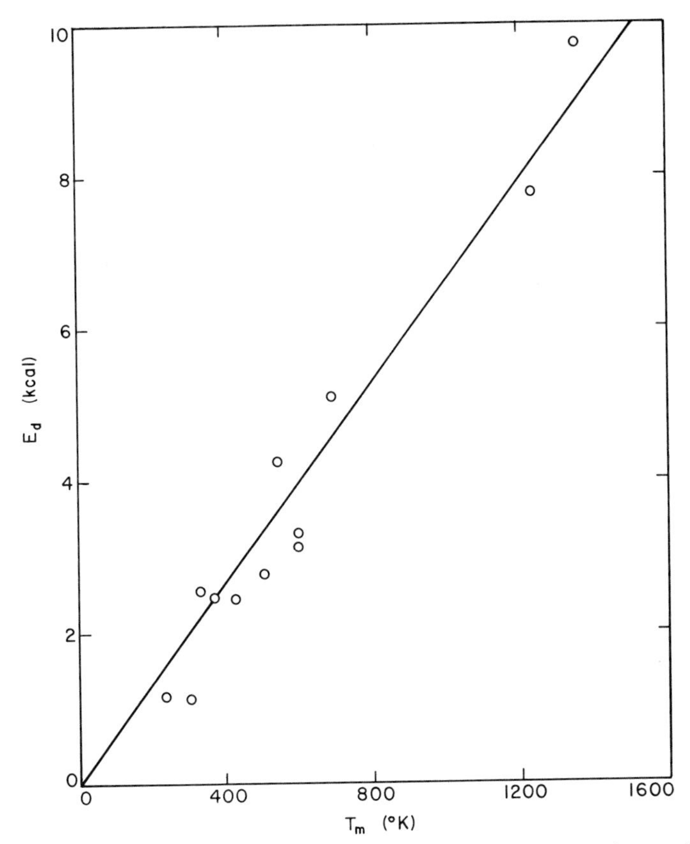

FIG. 3.37 Apparent energy of activation for self-diffusion in liquid metals as a function of melting temperature of metals.

than the value obtained above from self-diffusivity data.

Combination of Eqs. (3.110), (3.114), and (3.126) gives

$$D\eta = 18.2 \times 10^{-10} T_m V^{-1/3}. \tag{3.129}$$

The Stokes–Einstein relation for Brownian motion given in terms of molecular shear viscosity

$$D\zeta = kT/3\pi\delta \tag{3.130}$$

may be rewritten in terms of bulk viscosity, Eq. (3.110). Substituting Eq. (3.112) for δ, we obtain

$$D\eta = 21.30 \times 10^{-10} T_m V^{-1/3}. \tag{3.131}$$

This is in close agreement with Eq. (3.129) for the rigid-sphere liquid.

Combining equations for surface tension, viscosity, and self-diffusivity at melting points of liquid metals results in the following dimensionless parameter:

$$(\sigma\delta/D\eta)_{T_m} = 34.21 \qquad \text{or} \qquad (\sigma V^{1/3}/D\eta)_{T_m} = 29.85. \tag{3.132}$$

3.5.2 Self-Diffusivities of Pure Solid Metals

Because of variations in the concentration of lattice vacancies and defects, and the presence of grain boundaries and dislocations in solids, the interpretation of diffusion in solid metals in terms of atomistic models is more complex than that of diffusion in liquid metals. Excluding special cases of localized diffusion along grain boundaries, dislocation pipes, and on surfaces, we are concerned here primarily with volume diffusion, which is the process of atomic migration in the bulk solid phase involving vacant lattice sites and solutes in interstitial lattice sites. The self-diffusion is volume diffusion of solvent atoms in a pure solvent, measured by tracing the rate of migration of a radioactive isotope of the solvent element. The self-diffusion includes also the migration of solute atoms present in trace quantities in a solvent. The chemical diffusion is another form of volume diffusion that occurs under a concentration gradient in a binary or a multicomponent system.

Reference may be made to review papers for methods of measurement of diffusion in metals and alloys [136–138]. For data on diffusion in metals and alloys, reference may be made to updated compilations in "Diffusion Data" [139] and "Diffusion and Defect Data" [140].

Based on Eq. (3.124) for mean square displacement in the random walk process, Chandrasekhar [141] derived the following relation for self-diffusivity in pure solid metals on the well justified assumption that all isotopes of the same metal are equivalent:

$$D = a_\circ{}^2 x_v w, \tag{3.133}$$

where $a_\circ$ is the lattice parameter, x_v is the atom fraction of vacant sites, and w is the frequency of jump into an adjacent vacant site.

For low concentrations of vacancies

$$x_v = \exp(-\Delta F_v/RT), \tag{3.134}$$

where $\Delta F_v = \Delta H_v - T\Delta S_v$ is the molar free energy of formation of vacancies.

The jump of an atom to a neighboring vacant site must be accompanied by the movement of two adjacent restraining atoms apart. For this activated process of atomic migration, the frequency of jump is given by the relation [142]

$$w = \nu \exp(-\Delta F^*/RT), \tag{3.135}$$

where $\Delta F^* = \Delta H^* - T\Delta S^*$ is the free energy of formation of the activated complex, ΔH^* and ΔS^* being the enthalpy and entropy of activation, and ν is the Debye frequency

$$\nu = (k/h)\Theta_D, \tag{3.136}$$

where Θ_D is the characteristic Debye temperature. Inserting the values of the Boltzmann constant k and Planck constant h gives for the Debye frequency per second

$$\nu = 20.84 \times 10^9 \Theta_D. \tag{3.137}$$

Substituting Eqs. (3.134) and (3.135) in Eq. (3.133) gives for the self-diffusivity

$$D = D_\circ \exp(-E_d/RT), \tag{3.138}$$

where

$$D_\circ = a_\circ^2 \nu \exp[(\Delta S_v + \Delta S^*)/R], \tag{3.139}$$

$$E_d = \Delta H_v + \Delta H^*. \tag{3.140}$$

The application of Zener's theory [142] to self-diffusion by the vacancy mechanism leads to the following approximation:

$$T_m(\Delta S_v + \Delta S^*) \simeq (\lambda\beta)E_d, \tag{3.141}$$

where T_m is the melting point, in degrees kelvin; the value of the constant λ is within 0.5 (for fcc) to 0.9 (for bcc); β is the coefficient of the temperature dependence of the elastic moduli and has values in the range 0.25–0.45.

The entropy sum $(\Delta S_v + \Delta S^*)$ calculated with Eq. (3.139) from known values of $a_\circ$, ν, and the measured values of $D_\circ$ is plotted in Fig. 3.38 against the energy of activation. Considering that Eq. (3.141) is approximate only, the relation in Fig. 3.38 derived from experimental values of $D_\circ$ and E_d for all structures, bcc, fcc, hcp, and diamond, is in general accord with the Zener's theory. The lattice parameter $a_\circ$ was used in the calculations for hcp structures. The scatter of the points in Fig. 3.38 is due to variations in the value of $(\lambda\beta)$ from one metal to another and is random relative to the structural variations. For Hf, U, and Zr negative, values of λ are obtained.

Variations of the energy of activation with the enthalpy of fusion and the melting temperature are shown in Fig. 3.39. With the exception of some metals, these linear relations may be approximated by

$$E_d/\Delta H_m \simeq 16 \quad \text{and} \quad E_d/T_m \simeq 36 \quad \text{cal deg}^{-1}. \tag{3.142}$$

Theoretical calculations and experimental determinations of the thermodynamics of vacancies in metals will not be discussed here. For the present purpose it is sufficient to state that the experimental data cited by Peterson [143] indicate that the ratio $\Delta H/E_d$ is essentially constant.

$$\Delta H_v/E_d = 0.55 \pm 0.06 \quad \text{and} \quad \Delta H^*/E_d = 0.45 \pm 0.06. \tag{3.143}$$

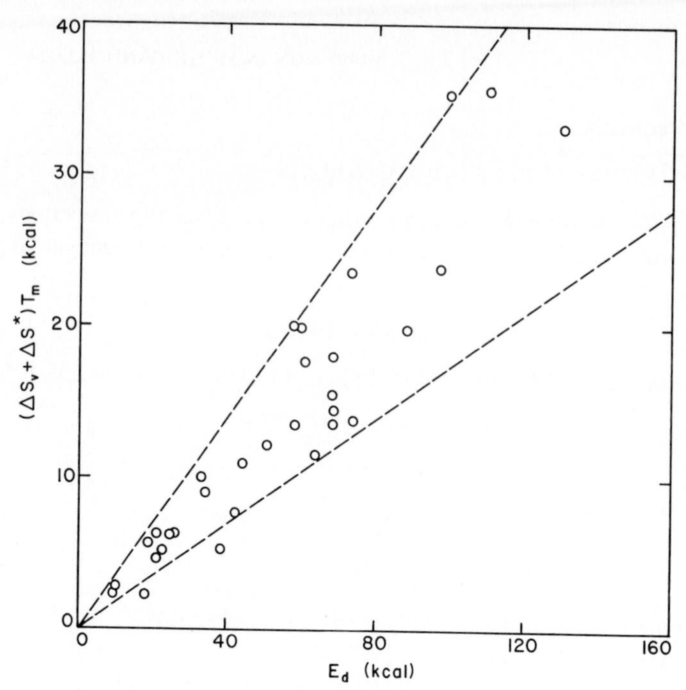

FIG. 3.38 Product of the melting temperature and the sum of the entropy of formation of vacancy and activated complex as a function of the energy of activation for self-diffusion in pure solid metals. Legend: ---, range predicted by theory.

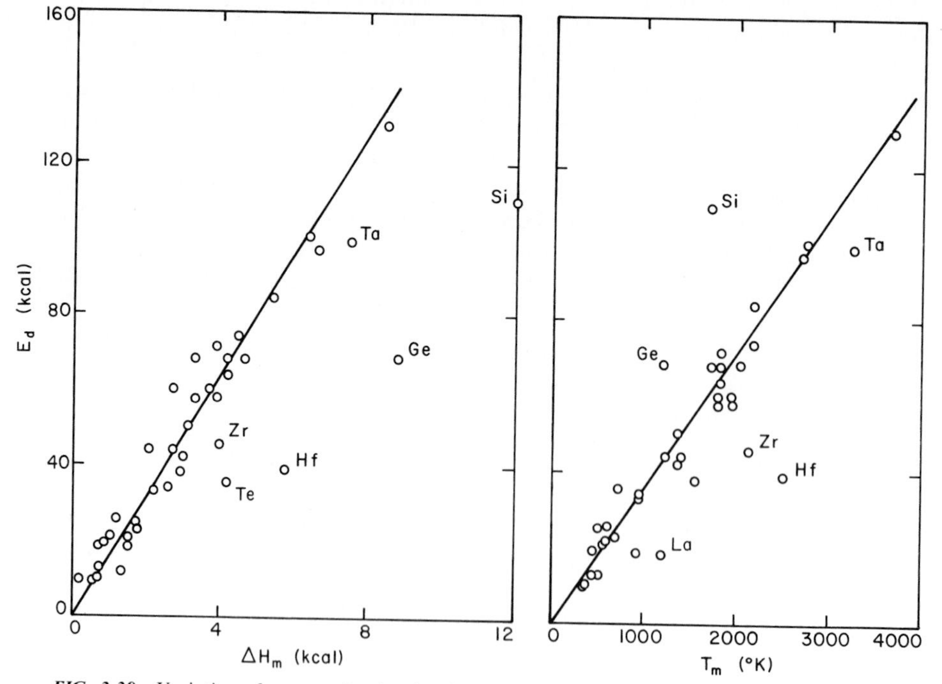

FIG. 3.39 Variation of energy of activation for self-diffusion in pure solid metals with enthalpy of fusion and temperature of the melting point.

The energy of activation for self-diffusion of solute A at low concentrations in solvent B may be considered equal to the sum of the enthalpy of formation of vacancy in the matrix of solvent B and the enthalpy of formation of the activated complex involving solute A.

$$E_d(\mathrm{A}) = \Delta H_v(\mathrm{B}) + \Delta H^*(\mathrm{A}). \tag{3.144}$$

As shown by Srikrishnan and Ficalora [144], values of $E_d(\mathrm{A})$ calculated for self-diffusion in dilute binary solutions of transition metals from Eqs. (3.143) and (3.144) and known values of E_d for transition metals are in close agreement with measured values of $E_d(\mathrm{A})$.

3.5.3 Interdiffusion in Binary Metallic Solutions

In a study of diffusion in alpha brass, Smigelskas and Kirkendall [145] found that zinc moves much faster than copper as evidenced by the relative motion of inert molybdenum wires placed on the sample as markers. This observation of the difference in the drift velocities of components of a solid solution, known as the *Kirkendall effect*, had led to numerous definitions of the coefficient of diffusion in binary metallic solutions, e.g., mutual diffusivity, volume diffusivity, chemical diffusivity, and intrinsic diffusivity. Subsequently, Stark [146] showed that the coefficient of mutual diffusion, or simply interdiffusivity, is an invariant of binary diffusive motion and, therefore, independent of the frame of reference of the diffusion process. The interdiffusivity is defined by the following expression in terms of fluxes J_1 and J_2 of components 1 and 2, concentration gradients $\partial c_1/\partial z$ and $\partial c_2/\partial z$, and the partial molar volumes $\bar{V}_1$ and $\bar{V}_2$:

$$D = -c_2\bar{V}_2 J_1/(\partial c_1/\partial z) - c_1\bar{V}_1 J_2/(\partial c_2/\partial z), \tag{3.145}$$

where the subscripts 1 and 2 refer to solvent and solute, respectively, and c is the molar concentration per unit volume.

For dilute solutions, i.e., $c_2 \to 0$ and $c_1\bar{V}_1 \to 1$, Eq. (3.145) reduces to the the form of Fick's first law,

$$J_2 = -D\,\partial c_2/\partial z. \tag{3.146}$$

By regarding the virtual driving force in diffusion as the negative gradient of the chemical potential, Darken [147] derived the following phenomenological relation between diffusivity D_i, mobility B_i, and activity a_i of the diffusing species i:

$$D_i = kTB_i(\partial \ln a_i/\partial \ln c_i). \tag{3.147}$$

For an ideal solution, this reduces to the Einstein relation $D_i = kTB_i$. From Eq. (3.147) and another form of Eq. (3.145), Darken [147] and Hartley and Crank [148] independently derived the phenomenological equation for diffusion in a binary system 1–2, showing the thermodynamic effect on the interdiffusivity:

$$D = (c_1 \bar{V}_1 D_2^* + c_2 \bar{V}_2 D_1^*)(\partial \ln a_2/\partial \ln c_2), \quad (3.148)^*$$

where D_1^* and D_2^* are self-diffusivities of components 1 and 2. It should be noted that the above equation is for constant temperature and composition. Simplification of Eq. (3.148) for dilute solution gives $D = D_2^*$ as the concentration of solute 2 approaches zero.

In liquid metals, diffusivities of solutes are in the range 10^{-5}–10^{-4} cm^2 sec^{-1} with heats of activation 8–12 kcal. The diffusivity of hydrogen in liquid metals is in the range 6×10^{-4}–2×10^{-3} cm^2 sec^{-1} with heats of activation 1–5 kcal.

An atom occupying an interstitial site in a metallic solution diffuses by jumping to a neighboring interstitial site without displacing any of the matrix solvent atoms. The free energy of activation for interstitial diffusion is that due only to the free energy of formation of the activated complex in local straining of the matrix atoms to allow the interstitial solute to jump to the adjacent interstitial site. Since no vacancy formation is involved, E_d for interstitial diffusion is smaller than the vacancy diffusion. For this reason, diffusivities of interstitial solutes are several orders of magnitude greater than those of substitutional solutes. Also, because of the difference in interatomic spacing, diffusivities of interstitials in bcc structures are 1 to 2 orders of magnitude greater and heats of activation ($\sim$20 kcal) are smaller than in fcc structures with E_d about 40 kcal.

From an atomistic model of interstitial diffusion, Wert and Zener [149] derived the following equation for $D_\circ$:

$$D_\circ = n\alpha a_\circ^2 \nu \exp(\Delta S^*/R), \quad (3.149)$$

where n is the number of nearest neighbor interstitial positions (4 for bcc and 12 for fcc) and α is the numerical coefficient whose value depends on the location of the interstitial positions ($\frac{1}{24}$ for bcc and $\frac{1}{12}$ for fcc).

The frequency of vibration of a solute in an interstitial position is given by

$$\nu = (\Delta H^*/2M\lambda^2)^{1/2}, \quad (3.150)$$

where M is the atomic weight and λ is the distance between the interstitial positions; $\lambda = a/2$ for bcc and $a/\sqrt{2}$ for fcc.

The third equation given by Wert and Zener is for the entropy of activation

$$\Delta S^* \simeq \beta \, \Delta H^*/T_m. \quad (3.151)$$

* The equation given by Darken originally is for the special case of constant and equal molar volumes for which

$$D = (x_1 D_2^* + x_2 D_1^*)(\partial \ln a_2/\partial \ln x_2), \quad (3.148a)$$

or in terms of activity coefficient γ_2,

$$D = (x_1 D_2^* + x_2 D_1^*)[1 + (\partial \ln \gamma_2/\partial \ln x_2)]. \quad (3.148b)$$

Values of D_0 for interstitial diffusion calculated from these equations of Wert and Zener have been found to be in close agreement with those determined by linear extrapolation of $\ln D$ to $1/T \to 0$.

3.6 Thermal and Electrical Conductivity of Metals

Electrons and phonons (lattice waves) are the principal carriers of heat in metals; thus, the overall thermal conductivity is given by the sum

$$\kappa = \kappa_e + \kappa_p, \tag{3.152}$$

where κ_e is the electronic component and κ_p the lattice component. In metals the electronic component predominates at and above the characteristic Debye temperature. At lower temperatures the heat conduction by phonons becomes significant. The variation of thermal conductivity with temperature below the Debye temperature is represented by the empirical relation

$$\kappa = (\alpha T^n + \beta/T)^{-1}, \tag{3.153}$$

where α, β, and n are constants for a given metal. For all elements the thermal conductivity passes through a maximum at low temperatures.

Complexity of the temperature dependence of the thermal conductivity is illustrated by examples in Fig. 3.40 for solid and liquid iron, sodium, and aluminum. With the exception of few metals, e.g., Fe, Bi, Sb, and Te, thermal

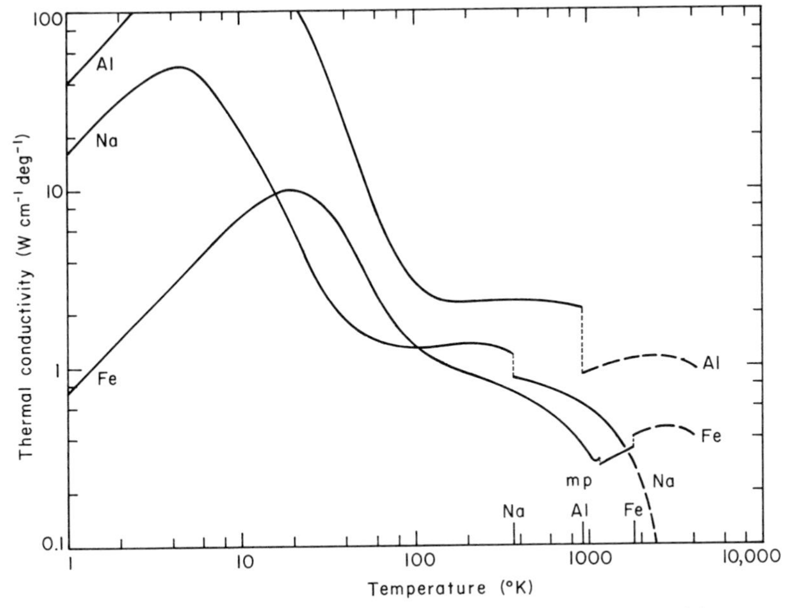

FIG. 3.40 Variation of thermal conductivity of aluminum, sodium, and iron with temperature.

conductivity of most metals decreases upon melting. Reference may be made to the work of Ho *et al.* [150] for a comprehensive review and compilation of data on the thermal conductivity of solid and liquid elements.

The electrical resistivity ρ, reciprocal of electrical conductivity σ, consists of two terms as stated in Matthiessen's rule [151]

$$1/\sigma = \rho = \rho_\circ + \rho_i, \tag{3.154}$$

where $\rho_\circ$ is the temperature-independent residual resistivity due to the presence of impurities and lattice defects, and ρ_i is the temperature-dependent intrinsic resistivity. The residual resistivity, usually measured at 4 °K is only a small fraction of ρ_i measured at room temperature. It should be mentioned in passing that another contribution to electrical resistivity is due to electron scattering from disordered spin arrangements in ferromagnetic and antiferromagnetic metals below their magnetic transition temperatures.

At ordinary pressures and in the absence of a magnetic field, the temperature dependence of electrical resistivity above the Debye temperature may be represented by the formalism,

$$\rho = a + bT - cT^{-2}, \tag{3.155}$$

where the constants *a*, *b*, and *c* are characteristics for each element. In most metals the electrical resistivity increases abruptly upon melting and continues to rise with increasing temperature.

Of particular interest to high temperature technology is the theoretical relation between electronic thermal and electrical conductivities, known as the Lorenz function [152],

$$\kappa/\sigma T = \tfrac{1}{3}\pi^2(k/e)^2 = 2.45 \times 10^{-8} \quad \mathrm{W\,\Omega\,deg^{-1}}, \tag{3.156}$$

where k is the Boltzmann constant and e is the electronic charge. This is the limiting relation at elevated temperatures where the conduction of thermal and electrical energy is by electrons only.

Although there are several compilations of transport properties of elements, no comprehensive survey has been made of the electrical resistivity of metals and alloys. References to data on resistivity are given in *Metals Handbook* [153] and in a text book by Meaden [154].

The thermal conductivity is plotted in Fig. 3.41 against the product σT_m at the melting points of solid and liquid metals. The line drawn with the slope 2.45×10^{-8} of the Lorenz function represents the data well.

An early, simple theory of liquid metals proposed by Mott [155] leads to the following relation of the ratio of liquid/solid resistivity ρ_l/ρ_s at the melting point to the entropy of fusion:

$$\rho_l/\rho_\mathrm{s} = \exp(0.33\,\Delta S_\mathrm{m}), \tag{3.157}$$

where ΔS_m is in units of $\mathrm{cal\,mol^{-1}\,deg^{-1}}$. With the exception of few metals, e.g., Bi, Ga, Hg, Sb, and Sn, most metals obey this relation within about 30%.

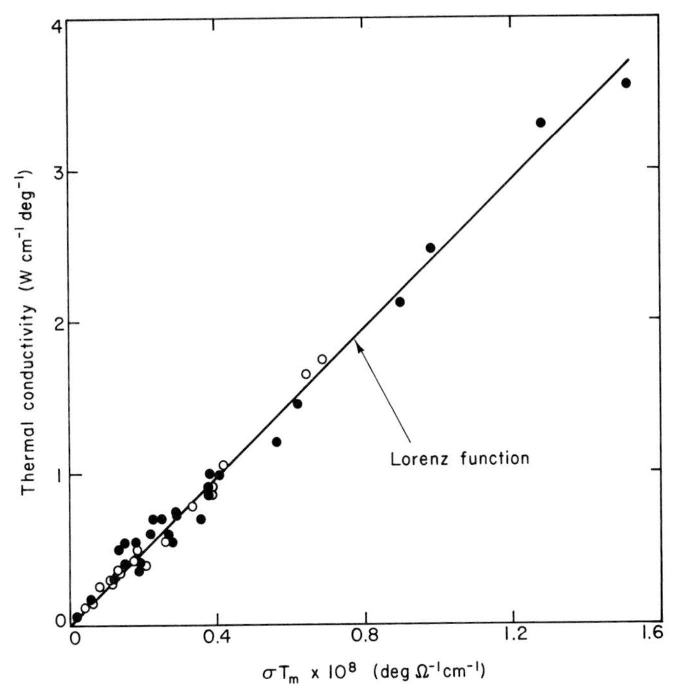

FIG. 3.41 Variation of thermal conductivity with the product of melting temperature and electrical conductivity of solid (●) and liquid (○) metals.

It should also be noted that the thermal conductivity calculated from the rigid-sphere model of Longuet-Higgins and Pople [124] discussed earlier does not apply to liquid metals because the thermal conductivity derived from the model describes the ionic contribution to the thermal conductivity, which is only a very small fraction of the electronic contribution.

References

1. R. Hultgren, P. D. Desai, D. T. Hawkins, M. Gleiser, K. K. Kelley, and D. D. Wagman, "Selected Values of the Thermodynamic Properties of the Elements." Am. Soc. Met., Metals Park, Ohio, 1973.
2. R. Hultgren, P. D. Desai, D. T. Hawkins, M. Gleiser, and K. K. Kelley, "Selected Values of the Thermodynamic Properties of Binary Alloys." Am. Soc. Met., Metals Park, Ohio, 1973.
3. M. Hansen and K. Anderko, "Constitution of Binary Alloys." McGraw-Hill, New York, 1958.
4. F. A. Shunk, "Constitution of Binary Alloys, Second Supplement." McGraw-Hill, New York, 1969.
5. M. Margules, *Sitzungsber. Akad. Wiss. Wien* **104**, 1243 (1895).
6. A. W. Porter, *Trans. Faraday Soc.* **16**, 336 (1921).
7. J. H. Hildebrand, *Proc. Natl. Acad. Sci. U.S.A.* **13**, 267 (1927).
8. K. F. Herzfeld and W. Heitler, *Z. Elektrochem.* **31**, 536 (1925).
9. G. Scatchard, *Chem. Rev.* **8**, 321 (1931).

10. J. H. Hildebrand and R. L. Scott, "Regular Solutions." Prentice-Hall, Englewood Cliffs, New Jersey, 1962.
11. E. A. Guggenheim, *Trans. Faraday Soc.* **33**, 151 (1937).
12. L. S. Darken, *Trans. Metall. Soc. AIME* **239**, 80 (1967).
13. E. T. Turkdogan and L. S. Darken, *Trans. Metall. Soc. AIME* **242**, 1997 (1968).
14. A. L. Vierk and K. Hauffe, *Z. Elektrochem.* **54**, 382 (1950).
15. A. G. Morachevskii, *J. Appl. Chem. USSR* **30**, 1307 (1957).
16. A. Roeder and W. Morawietz, *Z. Elektrochem.* **60**, 431 (1956).
17. J. J. Egan, *Acta Metall.* **7**, 560 (1959).
18. O. J. Kleppa, *J. Phys. Chem.* **58**, 354 (1955).
19. O. Redlich and A. T. Kister, *Ind. Eng. Chem.* **40**, 345 (1948).
20. G. R. Belton and R. J. Fruehan, *J. Phys. Chem.* **71**, 1403 (1967).
21. G. M. Wilson, *J. Am. Chem. Soc.* **86**, 127 (1964).
22. R. V. Orye and J. M. Prausnitz, *Ind. Eng Chem.* **57**, 18 (1965).
23. G. Scatchard and F. G. Satkiewicz, *J. Am. Chem. Soc.* **86**, 130 (1964).
24. G. Scatchard and G. M. Wilson, *J. Am. Chem. Soc.* **86**, 133 (1964).
25. S. K. Tarby and F. P. Stein, *Metall. Trans.* **1**, 2354, (1970).
26. A. Krupkowski, *Bull. Int. Acad. Pol. Sci. Lett., Cl. Sci. Math. Nat., Ser. A* **1**, 15 (1950).
27. Z. Moser and J. F. Smith, *Metall. Trans. B* **6**, 457 (1975).
28. Z. Moser, *Metall. Trans. B* **6**, 659 (1975).
29. K. Wohl, *Trans. Am. Inst. Chem. Eng.* **42**, 215 (1946).
30. G. Scatchard, *Chem. Rev.* **44**, 7 (1949).
31. R. O. Williams, *Trans. Metall. Soc. AIME* **245**, 2565 (1969).
32. C. W. Bale and A. D. Pelton, *Metall. Trans.* **5**, 2323, (1974).
33. E. A. Guggenheim, "Mixtures." Oxford Univ. Press (Claredon), London and New York, 1952.
34. J. C. Mathieu, F. Durand, and E. Bonnier, *J. Chim. Phys.* **62**, 1289, 1297 (1965).
35. C. H. P. Lupis and J. F. Elliott, *Acta Metall.* **15**, 265 (1967).
36. C. H. P. Lupis and H. Gaye, *in* "Metallurgical Chemistry" (O. Kubaschewski, ed.), p. 469. HM Stationery Off., London, 1972.
37. "Phase Stability in Metals and Alloys" (P. S. Rudman, J. Stringer, and R. I. Jaffee, eds.). McGraw-Hill, New York, 1967.
38. "Metallurgical Chemistry" (O. Kubaschewski, ed.). HM Stationery Off., London, 1972.
39. O. Kubaschewski, *in* "Phase Stability in Metals and Alloys" (P. S. Rudman, J. Stringer, and R. I. Jaffee, eds.), p. 63. McGraw-Hill, New York, 1967.
40. E. T. Turkdogan, R. J. Fruehan, and L. S. Darken, *Trans. Metall. Soc. AIME* **245**, 1003 (1969).
41. T. Azakami and A. Yazawa, *Can. Metall. Q.* **15**, 111 (1976).
42. R. A. Buckley and W. Hume-Rothery, *J. Iron Steel Inst., London* **201**, 227 (1963).
43. L. S. Darken, *Trans. Metall. Soc. AIME* **239**, 90 (1967).
44. J. Chipman, R. W. Alfred, L. W. Gott, R. B. Small, D. M. Wilson, C. N. Thomson, D. L. Guernsey, and J. C. Fulton, *Trans. Am. Soc. Met.* **44**, 1215 (1952).
45. J. Chipman and T. P. Floridis, *Acta Metall.* **3**, 456 (1955).
46. H. Schenck and G. Perbix, *Arch. Eisenhuettenwes.* **32**, 123 (1961).
47. E. T. Turkdogan, R. A. Hancock, S. I. Herlitz, and J. Denton, *J. Iron Steel Inst., London* **183**, 69 (1956).
48. E. T. Turkdogan and L. E. Leake, *J. Iron Steel Inst., London* **180**, 269 (1955).
49. E. T. Turkdogan and R. A. Hancock, *J. Iron Steel Inst., London* **179**, 155 (1955).
50. J. Chipman, *in* "Basic Open Hearth Steelmaking." p. 531, 3rd Ed. Am. Inst. Min. Metall. Eng., New York, 1964.
51. C. Wagner, "Thermodynamics of Alloys." Addison-Wesley, Reading, Massachusetts, 1952.
52. G. K. Sigworth and J. F. Elliott, *Met. Sci.* **8**, 298 (1974).
53. G. K. Sigworth and J. F. Elliott, *Can. Metall. Q.* **15**, 123 (1976).

54. G. K. Sigworth and J. F. Elliott, *Can. Metall. Q.* **13**, 455 (1974).
55. C. H. P. Lupis and J. F. Elliott, *Trans. Metall. Soc. AIME* **236**, 130 (1966).
56. J. Chipman and D. A. Corrigan, *in* "Applications of Fundamental Thermodynamics to Metallurgical Processes" (G. R. Fitterer, ed.), p. 23. Gordon & Breach, New York, 1967.
57. E. T. Jacob and J. H. E. Jeffes, *Inst. Min. Metall., Trans., Sect. C* **80**, 181 (1971).
58. C. H. P. Lupis and J. F. Elliott, *Acta Metall.* **14**, 529, 1019 (1966).
59. C. B. Alcock and F. D. Richardson, *Acta Metall.* **6**, 385 (1958).
60. C. B. Alcock and F. D. Richardson, *Acta Metall.* **8**, 882 (1960).
61. C. Wagner, *Acta Metall.* **21**, 1297 (1973).
62. T. Chiang and Y. A. Chang, *Metall. Trans. B* **7**, 453 (1976).
63. E. T. Jacob and J. H. E. Jeffes, *Inst. Min. Metall., Trans.* **80**, 32 (1971).
64. U. Block and H. Stuwe, *Z. Metallkd.* **60**, 709 (1969).
65. L. D. Lucas, *in* "Physicochemical Measurements in Metals Research" (R. A. Rapp, ed.), p. 219. Wiley (Interscience), New York, 1970.
66. B. C. Allen, *in* "Liquid Metals—Chemistry and Physics" (S. Z. Beer, ed.), p. 161. Dekker, New York, 1972.
67. J. G. Dillon, P. A. Nelson, and B. S. Swanson, *J. Chem. Phys.* **44**, 4229 (1966).
68. F. Hensel and E. U. Franck, *Ber. Bunsenges. Phys. Chem.* **70**, 1154 (1966).
69. I. K. Kikoin and A. P. Senchenkov, *Phys. Met. Metallogr. (USSR)* **24**, 74 (1967).
70. R. G. Ross and D. A. Greenwood, *Prog. Mater. Sci.* **14**, 173 (1969).
71. A. V. Grosse, *J. Inorg. Chem.* **22**, 23 (1961).
72. "The Collected Works of J. Willard Gibbs," Vol. I. Longmans, Green, New York, 1928.
73. E. A. Guggenheim, "Thermodynamics," 5th Ed. North-Holland Publ., Amsterdam, 1967.
74. C. Wagner, "Phenomenal and Thermodynamic Equations of Adsorption." *Nachr. Akad. Wiss. Goettingen, Math.—Phys. Kl., II* No. 3 (1973).
75. W. Thomson (Lord Kelvin), *Philos. Mag.* **42**, 448 (1871).
76. P. S. de Laplace, "Sur l'Action Capillaire," Suppl. to Book 10, Vol. 4, "Traité de Mecanique Céléste." Courcier, Paris, 1805.
77. A. V. Grosse, *J. Inorg. Chem.* **26**, 1349 (1964).
78. R. Eötvös, *Wied. Ann.* **27**, 456 (1886).
79. J. D. van der Waals, *Z. Phys. Chem., Stoechiom. Verwandschaftsl.* **13**, 716 (1894).
80. A. Furgeson, *Trans. Faraday Soc.* **19**, 408 (1923).
81. J. Lennard-Jones and J. Corner, *Trans. Faraday Soc.* **36**, 1156 (1940).
82. J. Corner, *Trans. Faraday Soc.* **44**, 1036 (1948).
83. J. Lennard-Jones and A. F. Devonshire, *Proc. R. Soc. London, Ser. A* **165**, 1 (1938).
84. S. W. Strauss, *Nucl. Sci. Eng.* **8**, 362 (1960).
85. V. V. Fesenko, V. N. Eremenko, and M. L. Vasiliu, *Zh. Fiz. Khim.* **35**, 1750 (1961).
86. D. W. G. White, *Metall. Trans.* **2**, 3067 (1971).
87. N. K. Dzhenilev, S. I. Popel, and B. V. Taarevskii, *Fiz. Met. Metalloved.* **18**, 77 (1964).
88. A. M. Korol'kov and A. A. Igumnova, *Izv. Akad. Nauk SSSR, Otd. Tekh. Nauk, Metall. Topl.* **95**, 480 (1961).
89. J. G. Eberhart, *J. Phys. Chem.* **70**, 1183 (1966).
90. P. Kozakevitch and G. Urbain, *Rev. Metall. (Paris)* **58**, 401, 517, 931 (1961).
90a. P. Kozakevitch, *in* "Surface Phenomena of Metals," p. 223. Soc. Chem. Ind., London, 1968.
91. K. Monma and H. Suto, *Trans. Jpn. Inst. Met.* **2**, 148 (1961).
92. A. M. Korol'kov, "Casting Properties of Metals and Alloys." Consultants Bureau, New York, 1960.
93. T. J. Whalen, S. M. Kaufman, and M. Humenik, *Trans. Am. Soc. Met.* **55**, 778 (1962).
94. G. R. Belton, *Metall. Trans.* **3**, 1465 (1972).
95. I. Langmuir, *J. Am. Chem. Soc.* **38**, 2221 (1916); **40**, 1361 (1918).
96. R. H. Fowler, *Proc. Cambridge Philos. Soc.* **31**, 260 (1935).
97. V. K. Semenchenko, "Surface Phenomena in Metals and Alloys," Engl. transl. by N. G. Anderson. Pergamon, New York, 1961.

98. F. C. Tompkins, "Chemisorption of Gases on Metals." Academic Press, New York, 1978.
99. G. R. Belton, *Metall. Trans. B* **7**, 35 (1976).
100. B. V. Szyszkowski, *Z. Phys. Chem., Stoechiom. Verwandschaftsl.* **64**, 385 (1908).
101. F. Halden and W. D. Kingery, *J. Phys. Chem.* **57**, 557 (1955).
102. J. H. Swisher and E. T. Turkdogan, *Trans. Metall. Soc. AIME* **239**, 602 (1967).
103. B. F. Dyson, *Trans. Metall. Soc. AIME* **227**, 1098 (1963).
104. G. Bernard and C. H. P. Lupis, *Surf. Sci.* **42**, 61 (1974).
105. A. S. Skapski, *Acta Metall.* **4**, 576 (1956).
106. D. Turnbull, *J. Appl. Phys.* **21**, 1022 (1950); *J. Chem. Phys.* **18**, 768, 769 (1950).
107. B. E. Sundquist, *Acta Metall.* **12**, 67, 595 (1964).
108. S. S. Brenner, *Surf. Sci.* **2**, 496 (1964).
109. A. J. W. Moore, *in* "Metal Surfaces: Structure, Energetics and Kinetics," p. 155. Am. Soc. Met., Metals Park, Ohio, 1963.
110. F. Cabané-Brouty, *J. Chim. Phys.* **62**, 1056 (1965).
111. M. C. Imman, D. McLean, and H. R. Tipler, *Proc. R. Soc. London, Ser. A* **273**, 538 (1963).
112. E. D. Hondros, *Proc. R. Soc. London, Ser. A* **286**, 479 (1965).
113. E. D. Hondros and M. McLean, *C.N.R.S. Conf.* **187**, 219 (1969).
114. C. E. Bauer, R. Speiser, and J. P. Hirth, *Metall. Trans. A* **7**, 75 (1976).
115. D. McLean, "Grain Boundaries in Metals." Oxford Univ. Press (Clarendon), London and New York, 1957.
116. M. P. Seah and E. D. Hondros, *Proc. R. Soc. London, Ser. A* **335**, 191 (1973).
117. E. D. Hondros and M. P. Seah, *Int. Met. Rev.* Dec., p. 262 (1977).
118. H. J. Grabke, G. Tauber, and H. Viefhaus, *Scr. Metall.* **9**, 1181 (1975).
119. E. N. da C. Andrade, *Philos. Mag.* **17**, 497, 698 (1934).
120. J. R. Wilson, *Metall. Rev.* **10**, 381 (1965).
121. L. J. Wittenberg, *in* "Physicochemical Measurements in Metals Research" (R. A. Rapp, ed.), Vol. IV, Part 2, p. 193. Wiley (Interscience), New York, 1970.
122. S. A. Rice and P. Gray, "The Statistical Mechanics of Simple Liquids." Wiley (Interscience), New York, 1965.
123. M. Born and H. S. Green, "A General Kinetic Theory of Liquids." Cambridge Univ. Press, London, 1949.
124. H. C. Longuet-Higgins and J. A. Pople, *J. Chem. Phys.* **25**, 884 (1956).
125. H. C. Longuet-Higgins and B. Widom, *Mol. Phys.* **8**, 549 (1964).
126. W. G. Hoover and F. H. Ree, *J. Chem. Phys.* **49**, 3609 (1968).
127. C. J. Smithells, "Metals Reference Book," Vol. 3, p. 136. Butterworth, London, 1962.
128. B. J. Alder and T. E. Wainwright, *Phys. Rev. Lett.* **18**, 988 (1967).
129. T. W. Chapman, *AIChE J.* **12**, 395 (1966).
130. R. C. Ling, *J. Chem. Phys.* **25**, 609 (1956).
131. A. Fick, *Ann. Phys.* (*Leipzig*) **170**, 59 (1855).
132. J. Crank, "The Mathematics of Diffusion." Oxford Univ. Press, London and New York, 1956.
133. W. Jost, "Diffusion in Solids, Liquids, Gases." Academic Press, New York, 1960.
134. A. Einstein, *Z. Elektrochem.* **14**, 235 (1908).
135. N. H. Nachtrieb, *in* "Properties of Liquid Metals" (P. D. Adams, H. A. Davies, and S. G. Epstein, eds.), p. 309. Taylor & Francis, London, 1967.
136. H. A. Walls, *in* "Physicochemical Measurements in Metals Research" (R. A. Rapp, ed.), p. 459. Wiley (Interscience), New York, 1970.
137. T. S. Lundy, *in* "Physicochemical Measurements in Metals Research" (R. A. Rapp, ed.), p. 379. Wiley (Interscience), New York, 1970.
138. N. A. Gjostein, *in* "Physicochemical Measurements in Metals Research" (R. A. Rapp, ed.), p. 305. Wiley (Interscience), New York, 1970.
139. "Diffusion Data" (F. H. Wöhlbier, ed.), Vol. 1, 1967 to Vol. 7, 1973. Diffusion Inf. Cent., Columbus, Ohio.

140. "Diffusion and Defect Data" (F. H. Wöhlbier ed.), Vol. 8, 1974 to Vol. 15, 1977. Transl. Tech. Publ., Bay Village, Ohio.
141. B. S. Chandrasekhar, *Rev. Mod. Phys.* **15**, 1 (1943).
142. C. Zener, *in* "Imperfections in Nearly Perfect Crystals" (W. Shockley, ed.), p. 289. Wiley, New York, 1952.
143. N. L. Peterson, *Solid State Phys.* **22**, 409 (1968).
144. V. Srikrishnan and P. J. Ficalora, *Metall. Trans. A* **6**, 2095 (1975).
145. A. C. Smigelskas and E. O. Kirkendall, *Trans. AIME* **171**, 130 (1947).
146. J. P. Stark, *Acta Metall.* **14**, 228 (1966).
147. L. S. Darken, *Trans. AIME* **174**, 184 (1948).
148. G. S. Hartley and J. Crank, *Trans. Faraday Soc.* **45**, 801 (1949).
149. C. Wert and C. Zener, *Phys. Rev.* **76**, 1169 (1949).
150. C. Y. Ho, R. W. Powell, and P. E. Liley, *J. Phys. Chem. Ref. Data* **3**, 1 (1974).
151. A. Matthiessen, *Ann. Phys. Chem.* (*Pogg Folge*) **110**, 190 (1860).
152. L. Lorenz, *Ann. Phys. Chem.* **147**, 429 (1872).
153. C. J. S. Smithells and E. A. Brandes, "Metals Handbook." Butterworth, London, 1976.
154. G. T. Meaden, "Electrical Resistance of Metals." Plenum, New York, 1965.
155. N. F. Mott, *Proc. R. Soc. London* **146**, 465 (1934).

CHAPTER 4

PHYSICOCHEMICAL PROPERTIES OF MOLTEN GLASSES, SLAGS, AND MATTES

4.1 Thermodynamic Properties

Glasses and slags are mixtures of silicates, aluminates, borates, and phosphates, classified as polymeric melts. Mattes consist of mixtures of metal sulfides and oxides. Because of the structural differences between polymeric melts and mattes, their thermodynamic properties will be discussed separately.

4.1.1 Activities of Oxides in Polymeric Melts

Since silicates are the major constituents of glasses and slags, the physicochemical properties of silicates have been investigated more extensively than the other polymeric melts. The activities of oxides in selected binary metal silicates are given in Figs. 4.1 and 4.2. The FeO–SiO_2 system is that in equilibrium with liquid iron and the activity of FeO is relative to pure liquid iron oxide saturated with iron. The total iron in the melt is represented as FeO; variations in the state of oxidation of iron in slags are discussed later. Activities in all other silicates given in Figs. 4.1 and 4.2 are relative to pure solid metal oxides and β-cristobalite; for the alkali silicates the reference state for silica is tridymite. The activities of oxides in molten calcium borates and aluminates are given in Fig. 4.3; pure liquid oxide is the standard state for B_2O_3.

Activities of FeO, MnO, and PbO in ternary oxide melts FeO–CaO–SiO_2 [13], MnO–CaO–SiO_2 [3], PbO–CaO–SiO_2 [14], and PbO–Na_2O–SiO_2 [15] are shown in Figs. 4.4 and 4.5. For molar ratios CaO/SiO_2 less than 2, the activity of the oxide MO in these systems increases with increasing concentration of the calcium oxide.

The activity of SiO_2 (relative to β-cristobalite) in the system CaO–Al_2O_3–SiO_2, measured by Rein and Chipman [16], is shown in Fig. 4.6a, where the composition is given in terms of mole fractions of CaO, SiO_2, and $AlO_{1.5}$.

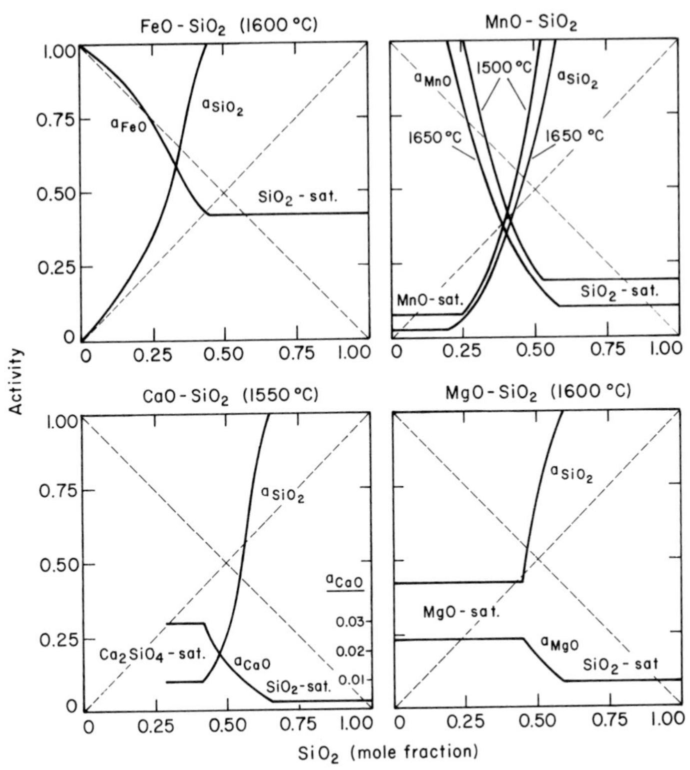

FIG. 4.1 Activities in binary metal silicates: FeO–SiO_2 [1,2], MnO–SiO_2 [3], CaO–SiO_2 [4, 5], and MgO–SiO_2 [6].

The activities of CaO and $AlO_{1.5}$ in Fig. 4.6b are obtained by the Gibbs–Duhem integration (Table 1.1).

The present knowledge of the structure of polymeric melts is based on the concept of the structure of silicate glasses originally developed by Zachariasen [17] and Warren [18] from x-ray diffraction experiments. The structure was conceived as a random network in which each silicon atom is tetrahedrally surrounded by four oxygen atoms and each oxygen atom is bonded to two silicon atoms. Subsequent studies of silicate, phosphate, and borate glasses with improved x-ray and neutron diffraction techniques have shown that the polyionic systems are made of networks with the basic building units of SiO_4 and PO_4 tetrahedra and BO_3 triangles. These units share corners with one another, forming linear or branching chains or rings. With the addition of a basic oxide such as CaO or Na_2O, covalent oxygen bonds between tetrahedra or triangles become broken and the cations are distributed in the vicinity of negatively charged polyionic units.

The structures of molten phosphates anticipated from the x-ray diffraction studies of phosphate glasses at room temperature have been confirmed by

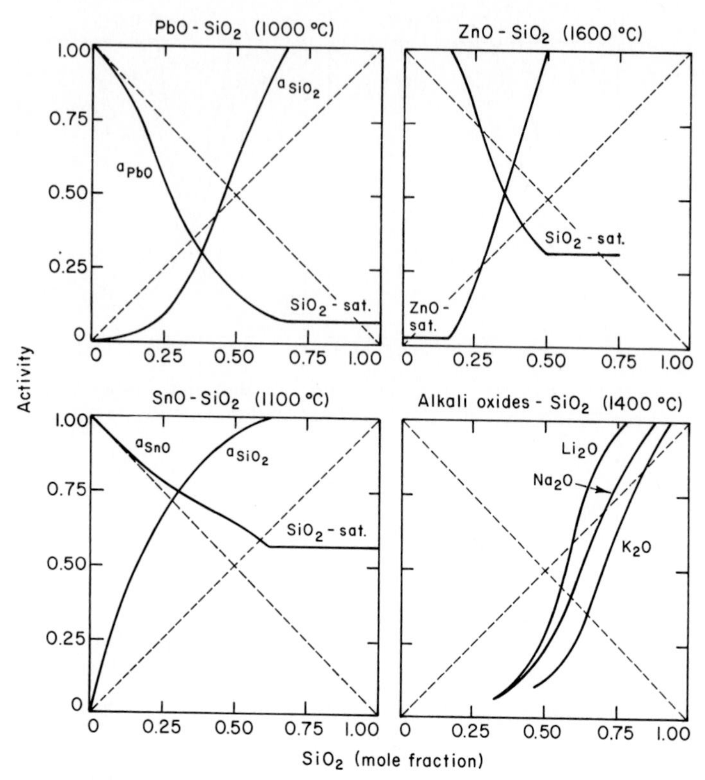

FIG. 4.2 Activities in binary metal silicates: $PbO-SiO_2$ [7,8], $ZnO-SiO_2$ [11], $SnO-SiO_2$ [9], and alkali oxides–SiO_2 [10].

chromatographic separation of anion units from aqueous solutions of quenched phosphate melts [19, 20]. This method is unique for phosphates which do not hydrolyze upon dissolution in aqueous solutions. Meadowcroft and Richardson [20] have shown that the relative proportions of different polymeric anion groups in phosphate glasses are related by the reaction equilibrium

$$2(P_nO_{3n+1})^{-n-2} = (P_{n-1}O_{3n-2})^{-n-1} + (P_{n+1}O_{3n+4})^{-n-3}. \tag{4.1}$$

For example, the equilibrium between ortho, pyro, and triphosphate anions is represented by

$$2P_2O_7^{4-} = PO_4^{3-} + P_3O_{10}^{5-}. \tag{4.2}$$

In terms of mole fractions x of polymers the equilibrium constant is given by

$$K_n = (x_{n-1})(x_{n+1})/x_n^{\,2}, \tag{4.3}$$

where n is the number of phosphorus atoms per chain, the so-called chain length. The mean number of phosphorus atoms per chain, i.e., average chain

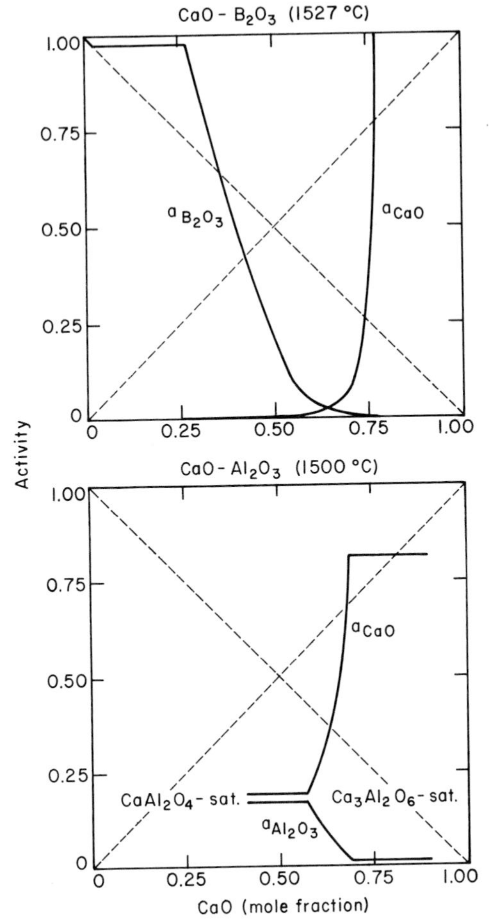

FIG. 4.3 Activities in CaO–B_2O_3, 1527 °C [11], and CaO–Al_2O_3, 1500 °C [12].

length, $\bar{n}$ is determined by the molar ratio MO/P_2O_5, where MO is the metal oxide,

$$\bar{n} = 2/[(MO/P_2O_5) - 1]. \tag{4.4}$$

For the Flory or random distribution [21], the mole fraction of the polymer is given by

$$x(P_nO_{3n+1})^{-n-2} = (1/\bar{n})[(\bar{n} - 1)/\bar{n}]^{n-1}. \tag{4.5}$$

Substituting Eq. (4.5) in Eq. (4.3) gives K_n equal to unity for all values of n and $\bar{n}$.

As is seen from the experimental data of Meadowcroft and Richardson in Fig. 4.7 for sodium and zinc phosphate glasses, the distribution of phosphorus in chains containing n phosphorus atoms differs from the Flory distribution,

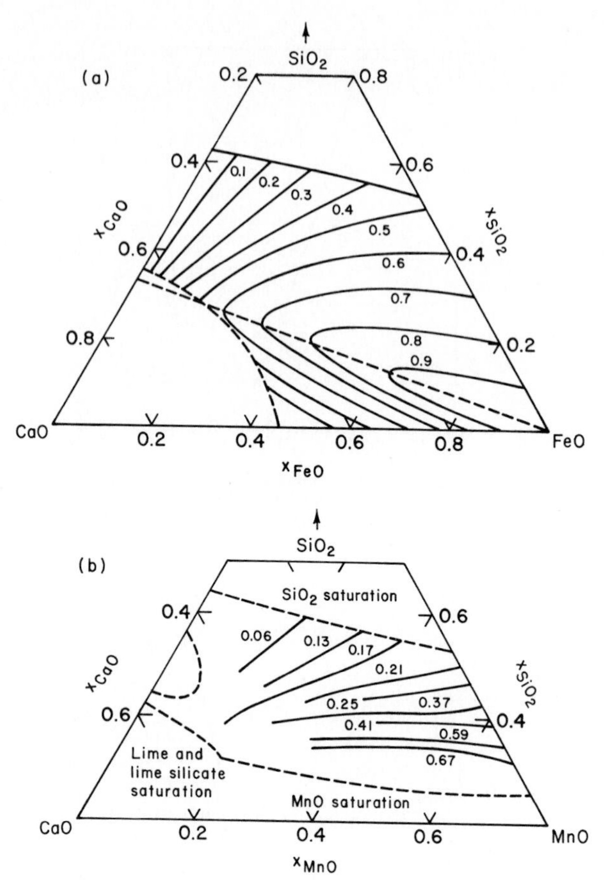

FIG. 4.4 Isoactivity curves for FeO(l) and MnO(s) in (a) $FeO-CaO-SiO_2$ melts at 1600 °C [13] and (b) $MnO-CaO-SiO_2$ melts at 1500 °C [3].

particularly in sodium phosphate glasses. The difference between the actual and Flory distribution becomes smaller with increasing n; K_n approaches unity in most phosphate glasses and melts when the average chain length exceeds 4 [11].

X-ray and neutron diffraction studies of Waseda *et al.* [22, 23] have shown that the structures of molten alkali and alkaline earth silicates are similar to those in the glassy state. That is, molten silicates consist mainly of SiO_4 tetrahedral units and Si—O—Si bonds are broken with the addition of metal oxides. This process may be represented by the reaction equilibrium

$$(\equiv\text{Si—O—Si}\equiv) + \text{MO} = 2(\equiv\text{Si—O}^-) + \text{M}^{2+}. \tag{4.6}$$

Similarly, we may consider equilibria of the type

$$2\text{O}^- = \text{O}^\circ + \text{O}^{2-}, \tag{4.7}$$

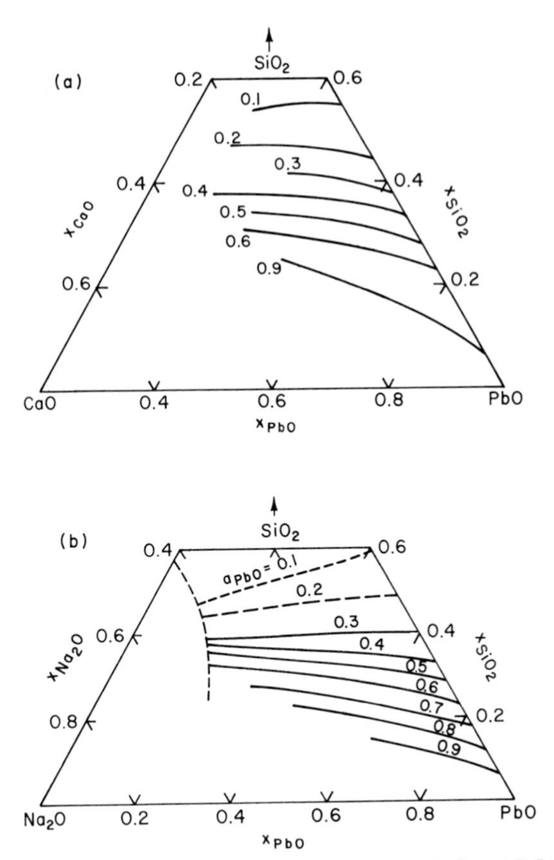

FIG. 4.5 Isoactivity curves for PbO(l) in (a) PbO–CaO–SiO_2 [14] and [b] PbO–Na_2O–SiO_2 [15] melts.

where O^- represents oxygen on the unshared corners, O° the negatively charged oxygen atom shared by two silicon atoms, and O^{2-} the free oxygen ion. The coordination number for oxygen around the cations increases with increasing radius of the cation, e.g., in alkali silicates the coordination number is 4 for Li, 6 for Na, and 7 for K; in alkaline earth silicates 5 for Mg and 7 for Ca.

Relative proportions of polymers cannot be evaluated from the distribution function obtained by x-ray and neutron diffraction. Also, the distribution of polymers in silicate glasses or quenched melts cannot as yet be determined quantitatively by chromatographic techniques because of hydrolysis upon dissolution in aqueous media. However, some advances have been made in the development of simple theories of polymers that are assumed to consist of branching chains or linear chains made up of SiO_4 units. Of the various polymer models proposed [24, 25] for basic silicates, i.e., $MO/SiO_2 > 1$, that developed by Masson [26] is perhaps more appealing.

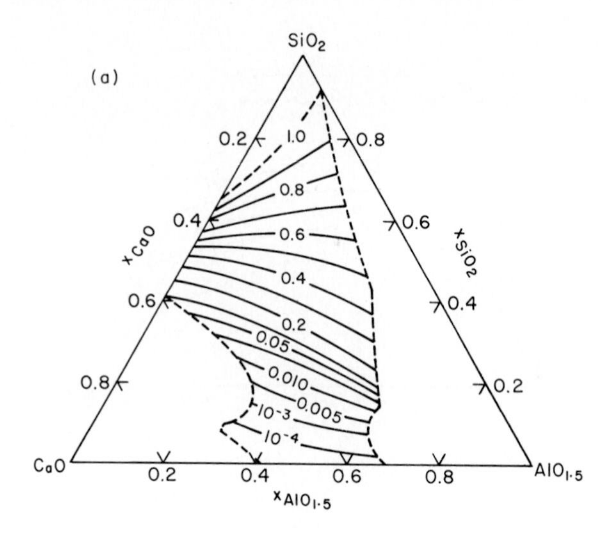

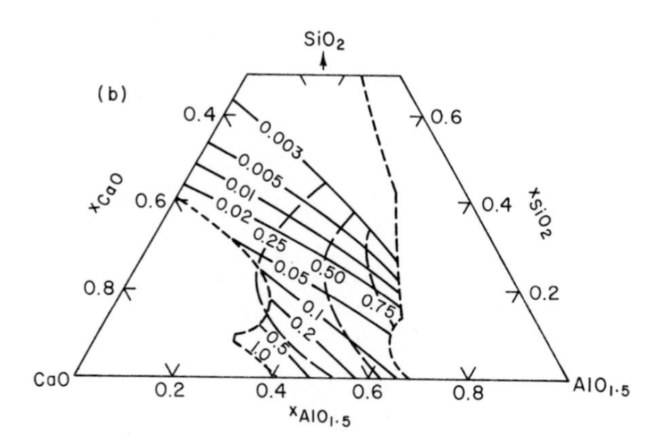

FIG. 4.6 Activities in CaO–Al_2O_3–SiO_2 melts at 1600 °C; (a) SiO_2(s) activities and (b) CaO(s) activities full lines, $AlO_{1.5}$(s) activities broken lines. After Rein and Chipman [16].

Similar to phosphate melts, a series of polymerization equilibria may be considered for silicate melts:

$$(SiO_4^{4-}) + (Si_nO_{3n+1})^{-2n-2} = (Si_{n+1}O_{3n+4})^{-2n-4} + O^{2-}. \tag{4.8}$$

With the mole fractions x, x_n, and x_{n+1} for the polymers and x_O for free oxygen ions, the equilibrium constant is represented by the equation

$$K_n = (x_{n+1})x_O/x(x_n). \tag{4.9}$$

In the model considered by Masson, it is assumed that K_1 for n equal to unity is independent of composition and all values of K_n are equal to unity when n exceeds unity. The third assumption is that the activity of the metal oxide is

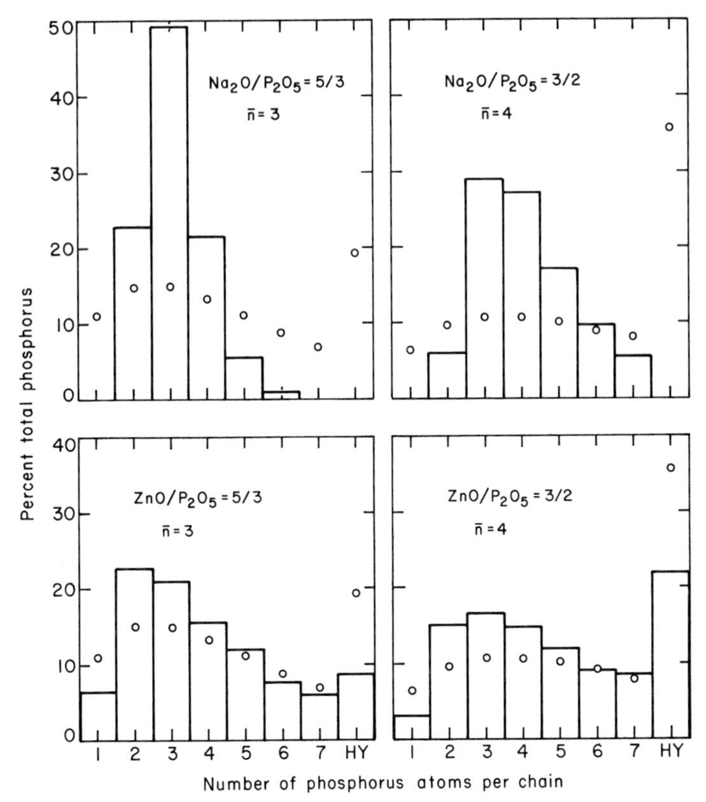

FIG. 4.7 Distribution of phosphorus [20] in chains containing n phosphorus atoms in sodium and zinc phosphate glasses; points are for Flory distribution and HY for hypoly gives phosphorus in all chains containing more than 7P atoms.

related to calculated ion fractions by the Temkin equation [27]

$$a_{MO} = (x_{M^{2+}})(x_{O^{2-}}). \tag{4.10}$$

With these assumptions and equations, Masson [26] derived the following expressions for the activity of MO as a function of composition in binary melts $MO–SiO_2$: For linear chains

$$(x_{SiO_2})^{-1} = 2 + (1 - a_{MO})^{-1} - [1 - a_{MO} + (a_{MO}/K_1)]^{-1} \tag{4.11}$$

and for branching chains

$$(x_{SiO_2})^{-1} = 2 + (1 - a_{MO})^{-1} - 3[1 - a_{MO} + (3a_{MO}/K_1)]^{-1}. \tag{4.12}$$

It should be noted that Eq. (4.11) for linear chains gives zero values of a_{MO} at the metasilicate composition $x_{SiO_2} = 0.5$; therefore, this equation is not considered satisfactory. Equation (4.12) for branching chains gives finite values of a_{MO}.

The variation of activity with composition calculated from Eq. (4.12) by Masson for various assumed values of K_1 are shown in Fig. 4.8. Calculated curves for activities of SnO, FeO, PbO, and CaO are in close agreement with the measured values given in Figs. 4.1 and 4.2.

For all the binary silicate melts considered by Masson for values of the equilibrium constant K_1 varying over a wide range, his model predicts SiO_4^{4-} as the dominant anionic species. For a given concentration of silica, ion fractions decrease in the order $SiO_4^{4-} > Si_2O_7^{6-} > Si_3O_{10}^{8-} >$, etc., for all values of K_1. Ion fractions of SiO_4^{4-}, $Si_2O_7^{6-}$, $Si_3O_{10}^{8-}$, etc., as functions of composition exhibit maxima at compositions M_2SiO_4, $M_3Si_2O_7$, $M_4Si_3O_{10}$, etc. Masson's model for linear and branching chains gives for the average chain length

$$1/\bar{n} = (1 - a_{MO})[(1/x_{SiO_2}) - 2]. \tag{4.13}$$

Relative concentrations of PO_4^{3-} ions in phosphate melts determined experimentally (Fig. 4.7) are much lower than those of SiO_4^{4-} ions calculated by Masson for silicate melts, suggesting that the free energies of phosphate melts are lower than those of the silicate melts. Such a prediction is, in fact, substantiated by the available thermodynamic data [11]. With increasing temperature and increasing MO/P_2O_5 ratio, the PO_4^{3-} ion will become the dominant species in phosphate melts.

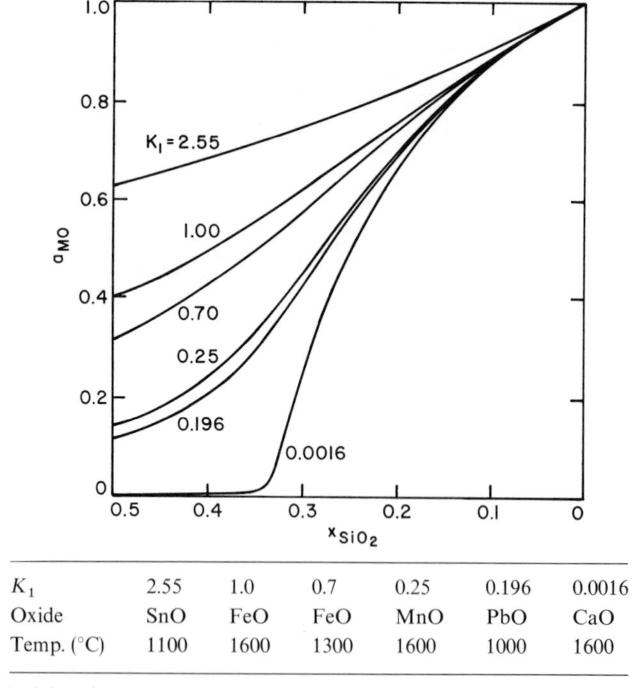

K_1	2.55	1.0	0.7	0.25	0.196	0.0016
Oxide	SnO	FeO	FeO	MnO	PbO	CaO
Temp. (°C)	1100	1600	1300	1600	1000	1600

FIG. 4.8 Activity of MO in binary silicates calculated from Eq. (4.12) for indicated values of K_1. After Masson [26].

Several attempts have been made with varying degrees of success to calculate the composition dependence of the activity of metal oxides in silicate melts containing two metal oxides. Kapoor and Frohberg [15] suggested that if the second oxide is of a more electropositive metal, ideal mixing may be assumed across the pseudobinary section between the orthosilicate and the second metal oxide such as Na_4SiO_4–PbO or Ca_2SiO_4–FeO. On this assumption, the activity is given by the ratio

$$a_{MO} = x_{MO}/(x_{MO} + x_{SiO_2}), \tag{4.14}$$

where MO is PbO, FeO, or MnO. However, they noted that measured oxide activity along the MO–Na_4SiO_4 or MO–Ca_2SiO_4 section is higher than calculated from Eq. (4.14) for ideal mixing. This departure from ideal mixing was attributed by Kapoor and Frohberg to interaction among unlike cation species. The greater the difference in the ionic charge and radius of cations, the greater is the departure from ideal mixing along the MO-orthosilicate section.

General validity of this argument is substantiated by the effect of calcium oxide on the free energy of mixing of sodium–potassium silicates measured by by Choudary *et al.* [28]. They found that the replacement of about 1 in 12 of the alkali ions by Ca^{2+} lowers the maximum excess free energy of mixing of the sodium–potassium metasilicates from about 455 to 263 cal/mole of alkali cations. A radius ratio of Ca^{2+} to O^{2-} ions (1.18/1.76 = 0.67) is between those for sodium and potassium cations. Therefore, small additions of Ca^{2+} are expected to reduce the structural differences between the sodium and potassium silicates, and, consequently, lower the excess free energy of mixing, as observed experimentally.

In an attempt to estimate the activities of oxides in ternary silicate melts, Richardson [29] considered the limiting case of ideal mixing of two binary silicates of equimolar silica concentration. That is, the heat of mixing of two binary silicates $y\mathrm{AO}\cdot\mathrm{SiO_2}$ and $y\mathrm{BO}\cdot\mathrm{SiO_2}$ is zero, and the free energy of solution is given by the configurational entropy of mixing of two types of cations A^{2+} and B^{2+}. For ideal mixing of silicates along the pseudobinary section $y\mathrm{AO}\cdot\mathrm{SiO_2}$–$y\mathrm{BO}\cdot\mathrm{SiO_2}$, activities of silicates are given by the equations

$$\begin{aligned} a_{y\mathrm{AO}\cdot\mathrm{SiO_2}} &\equiv \frac{(a^y_{\mathrm{AO}})_\mathrm{t}(a_{\mathrm{SiO_2}})_\mathrm{t}}{(a^y_{\mathrm{AO}})_\mathrm{b}(a_{\mathrm{SiO_2}})_\mathrm{b}} = x^y_{y\mathrm{AO}\cdot\mathrm{SiO_2}} \equiv \left[\frac{(x_{\mathrm{AO}})_\mathrm{t}(1+y)}{y}\right]^y, \\ a_{y\mathrm{BO}\cdot\mathrm{SiO_2}} &\equiv \frac{(a^y_{\mathrm{BO}})_\mathrm{t}(a_{\mathrm{SiO_2}})_\mathrm{t}}{(a^y_{\mathrm{BO}})_\mathrm{b}(a_{\mathrm{SiO_2}})_\mathrm{b}} = x^y_{y\mathrm{BO}\cdot\mathrm{SiO_2}} \equiv \left[\frac{(x_{\mathrm{BO}})_\mathrm{t}(1+y)}{y}\right]^y, \end{aligned} \tag{4.15}$$

where the subscripts b and t indicate the binary and ternary system.

The thermodynamic proof of Eq. (4.15) was given by Darken and Schwerdtfeger [30] by deriving the analog of Raoult's law for solutions of the type A_uX_v–B_uX_v. The lattice of the considered solution A_uX_v–B_uX_v is assumed to have two types of sites, the X sites for the X atoms and the A–B sites for the A and B atoms. All the A and B atoms are interchangeable on the A–B sites, but no X atom can occupy an A–B site and no A or B atom an X site. The partition function of the ideal assemblage of A and B atoms, the Helmholtz free energy of formation of the solution from the two terminal compounds combined with the Gibbs–Duhem equation gives

$$a_A a_X^{v/u} = I x_{A_u X_v} \quad \text{and} \quad a_B a_X^{v/u} = I x_{B_u X_v}, \tag{4.16}$$

where I is an integration constant depending on the arbitrary choice of standard state. Noting that the product of activities may also be written as

$$a_A a_X^{v/u} = I a_{AX_{v/u}} \quad \text{and} \quad a_B a_X^{v/u} = I a_{BX_{v/u}}, \tag{4.17}$$

it follows from Eq. (4.16) that activities $a_{AX_{v/u}}$ and $a_{BX_{v/u}}$ vary linearly with the mole fraction in an ideal solution $A_uX_v - B_uX_v$. Also, it follows from Eq. (4.16) that an activity $a_{A_uX_v}$ obtained from the equation

$$a_{A_uX_v} = I a_A{}^u a_X{}^v \tag{4.18}$$

is a parabolic (of the power u) function of composition

$$a_{A_uX_v} = I x^u_{A_uX_v}. \tag{4.19}$$

Therefore, in a plot of activity $a_{A_uX_v}$ against mole fractions $x_{A_uX_v}$, the curve has the following limiting slopes: da/dx is zero as x approaches zero and da/dx is equal to u as x approaches unity.

Equation (4.17) is well substantiated by the activity data for the FeO–MnO–SiO_2 ternary system in equilibrium with iron, indicating ideal mixing of the ortho- and metasilicates in the solid [31] and liquid [32, 33] states.

In many silicate melts, immiscibility occurs; the silica-rich liquid contains a few percent of the metal oxide, while the second liquid contains 50–80 mole % silica. In general, the greater the ratio of ion charge to ion radius (Z/r) for the cation, the wider is the miscibility gap. When the ratio Z/r is less than 1.5 as for alkali metals, there is complete liquid miscibility. Evidently, clustering of cations with large ratio Z/r around unshared oxygens of the polymers results in segregation of cations in the melt, which ultimately separates into two liquid phases. Partial liquid miscibility occurs also in some borates and phosphates.

4.1.2 Fluoride-Containing Polymeric Melts

Addition of fluorides to polymeric or simple oxide melts brings about structural changes as evidenced by marked decrease in viscosity of melts and the formation of liquid miscibility gaps. Small amounts of calcium fluoride added to FeO–CaO and MnO–CaO melts increase the activity of FeO and MnO appreciably [33], and in silicates the effect of fluoride on the activity of the metal oxide MO changes with concentration of silica. For example [34], in the PbO–SiO_2–PbF_2 melts containing more than 20 mole % SiO_2, the activity of PbO increases as SiO_2 is replaced by PbF_2. A reverse effect is observed at lower concentrations of silica. Similar behavior is observed in alkali and alkaline earth silicates containing alkali fluorides [35].

Generally accepted view is that fluoride ions break the covalent bonds in the Si–O—Si chains liberating free oxygen ions, thus

$$(\equiv Si—O—Si\equiv) + 2F^- = 2(\equiv Si—F) + O^{2-}. \quad (4.20)$$

The free oxygen ions may bring about further deploymerization:

$$(\equiv Si—O—Si\equiv) + O^{2-} = 2(\equiv Si—O^-). \quad (4.21)$$

In silica-rich melts, reaction (4.20) evidently predominates, as indicated by an increase in the activity of MO. At low concentrations of silica when the melt is highly depolymerized, fluoride ions may cluster around the cations, thus lowering the concentration of free oxygen ions, and hence the activity of MO. The results of the infrared absorption studies on silicate glasses [36] also substantiate the view that the structural changes occuring with the addition of calcium fluoride to high silica melts is different from that in low silica melts.

4.1.3 Solubility of Metals in Polymeric Melts

Richardson *et al.* [37, 38] have found that some metals dissolve in the elemental form in molten silicates; upon cooling, the glass contains finely dispersed metal particles. By equilibrating calcium aluminosilicate melts with vapors of copper, gold, silver, and lead, it was found that for a given vapor pressure, e.g., 1 mm Hg, the solubility of the metal decreases with increasing atomic radii of the metal. In subsequent studies by others, it was found that nickel [39], bismuth, antimony, and arsenic [40] also dissolve in the elemental form in molten iron silicates. In these latter studies, molten copper containing small amounts of Bi, Sb, and As were equilibrated with iron silicate melts at known oxygen activities. The solubilities obtained do not fit the simple relation that was found in the earlier studies with the noble metals [37, 38]. The available data on the solubility of metals are summarized in Table 4.1 in a normalized form, atom percent in slag/atom percent in metal.

TABLE 4.1

Solubility of Elemental Metal in Molten Silicates

Solute	Temperature (°C)	$\frac{\text{Atom \% in slag}}{\text{Atom \% in metal}}$	Reference
Cu	1530	5.5×10^{-4}	37
Ag	1530	1.6×10^{-3}	37
Au	1530	1.0×10^{-5}	38
Pb	1530	1.0×10^{-3}	38
Ni	1300	2.5×10^{-3}	39
Bi	1300	3.3×10^{-2}	40
Sb	1300	3.3×10^{-2}	40
As	1300	3.3×10^{-3}	40

4.1.4 Reaction of Gases with Polymeric Melts

Reaction of gases with polymeric melts is of major importance in high temperature technology. The reactions that are frequently encountered are those involving oxygen, sulfur, hydrogen, nitrogen, and carbon.

4.1.4a Oxidation–reduction reactions. The state of oxidation of iron in silicate, phosphate, and ferrite melts has been studied in detail. The oxidation–reduction reaction may be represented by the equation

$$2(Fe^{2+}) + \tfrac{1}{2}O_2 = 2(Fe^{3+}) + (O^{2-}). \tag{4.22}$$

For oxygen partial pressures below about 10^{-4} atm, gas mixtures such as H_2–CO_2, H_2–H_2O, or CO–CO_2 are used to control the oxygen potential of the gas. For the isothermal reaction

$$2(Fe^{2+}) + CO_2 = 2(Fe^{3+}) + (O^{2-}) + CO, \tag{4.23}$$

the equilibrium constant may be represented by the equation

$$K = (a_O)(\gamma_3/\gamma_2)^2(x_3/x_2)^2\, p_{CO}/p_{CO_2}, \tag{4.24}$$

where a_O is the activity of oxide ion and γ is the activity coefficients of ferric (3) and ferrous (2) ions. The concept of "free" ions is a useful way to describe the ability of slags to donate electrons in electrochemical gas–slag or metal–slag reactions. However, because the activities or activity coefficients of individual ions cannot be operationally defined, the state of equilibrium must be repre-

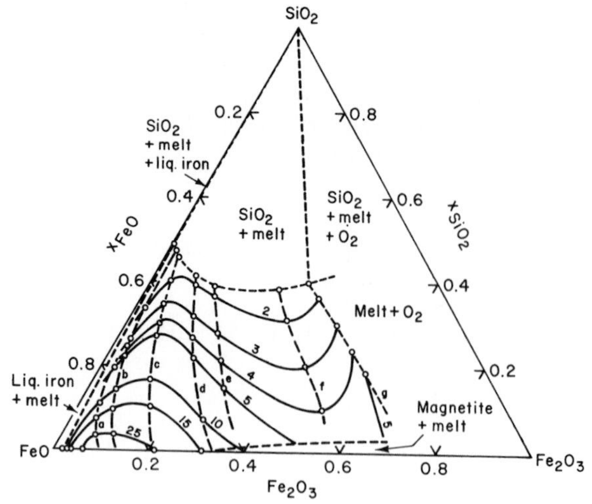

FIG. 4.9 Iso-$K_{Fe} \times 10^3$ curves in FeO–Fe_2O_3–SiO_2 system at 1550 °C. Straight line, iso-$K_{Fe} \times 10^3$ curves; dashed line, oxygen activity p_{CO_2}/p_{CO}: $a = 1.0$, $b = 2.5$, $c = 11.4$, $d = 75$, $e = 192$, $f = 1687$, $g = 3767$; dotted line, phase boundaries. After Turkdogan and Bills [41].

sented by the measured parameters

$$K(\gamma_2/\gamma_3)^2 a_O = K_{Fe} = (x_3/x_2)^2 p_{CO}/p_{CO_2}, \tag{4.25}$$

Variation of K_{Fe} with composition is shown in Figs. 4.9 and 4.10 for the ternary systems Fe–Si–O and Fe–Ca–O [41] at 1550 °C. The shape of the iso-K_{Fe} curves in iron phosphate melts is similar to that in iron silicate melts; values of K_{Fe} decrease with increasing concentration of silicate or phosphate ions. If the phosphorus concentration is given in terms of $PO_{2.5}$, the position of the curves in the composition diagram is similar to that in the silicate melts. The shape of the iso-K_{Fe} curves with maxima and minima in the composition diagram of the Fe–Ca–O system (Fig. 4.10) is reverse of that in iron silicate and phosphate melts; values of K_{Fe} increase with increasing concentration of CaO.

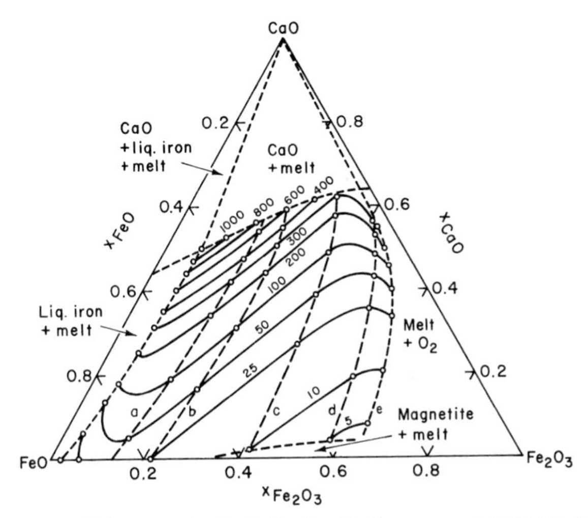

FIG. 4.10 Iso-$K_{Fe} \times 10^3$ curves in FeO–Fe_2O_3–CaO system at 1550 °C. Straight line, iso-$K_{Fe} \times 10^3$ curves: dashed line, oxygen activity p_{CO_2}/p_{CO}: $a = 2.5$, $b = 11.4$, $c = 192$, $d = 1687$, $e = 3767$; dotted line, phase boundaries. After Turkdogan and Bills [41].

Since the activity of oxygen is known over wide ranges of composition in the ternary systems Fe–Si–O and Fe–Ca–O, the activities of the oxides can be calculated by using the Gibbs–Duhem equation. Isoactivity curves thus derived [42] are shown in Figs. 4.11 and 4.12. Maxima on the iso-FeO activity curves in the FeO–CaO–SiO_2 melts (Fig. 4.4a) are more pronounced than those in the FeO–CaO–Fe_2O_3 melts, suggesting that polymerization in silicate melts is more extensive than in ferrite melts. Increase in the Fe^{3+}/Fe^{2+} ratio with the concentration of CaO is more pronounced than with MnO, indicating that the ferric ions polymerize forming ferrite anions, e.g., $Fe_2O_4^{2-}$, more extensively in the Fe–Ca–O melts than in the Fe–Mn–O melts. This is expected from the

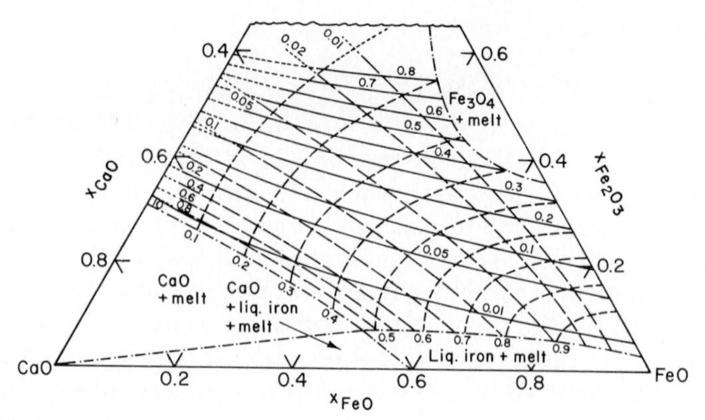

FIG. 4.11 Isoactivity curves in CaO–FeO–Fe_2O_3 melts at 1550 °C; long dashed line a_{CaO}, dashed line a_{FeO}, straight line $a_{Fe_2O_3}$. After Turkdogan [42].

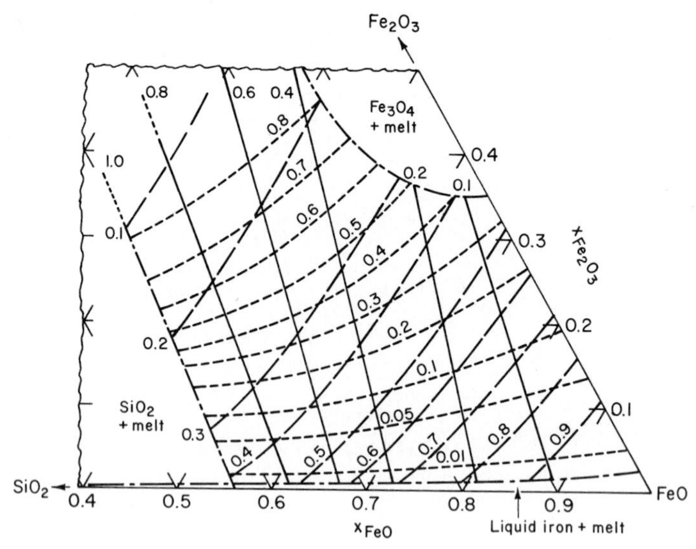

FIG. 4.12 Isoactivity curves in SiO_2–FeO–Fe_2O_3 melts at 1550 °C; straight line a_{SiO_2}, long dashed line a_{FeO}, dashed line $a_{Fe_2O_3}$. After Turkdogan [42].

thermodynamics of these systems in solid state: In the Fe–Ca–O system ferrites are formed, while there are series of solid solutions in the Fe–Mn–O system.

The chromium may exist in divalent, trivalent, or hexavalent state in glasses or polymeric melts, depending on temperature, oxygen potential, and composition. The following reaction equilibria may be considered:

$$2(Cr^{2+}) + \tfrac{1}{2}O_2 = 2(Cr^{3+}) + (O^{2-}). \qquad (4.26a)$$

$$(Cr^{3+}) + \tfrac{5}{2}(O^{2-}) + \tfrac{3}{4}O_2 = (CrO_4^{2-}). \qquad (4.26b)$$

Under oxidizing conditions the chromate anions are stable in basic glazes and enamels but decompose to trivalent state in acid glasses. Irmann [43] has shown that the equilibrium "constant" $(x_6/x_3)p_{O_2}^{-3/4}$ increases with increasing concentration of the basic oxide, i.e., increasing activity of oxygen ion, in borate melts; this is consistent with reaction (4.26b).

Recent studies* have shown that under reducing conditions the trivalent chromium is reduced to the divalent state with decreasing basicity of silicate melts. Equation (4.26a) is not consistent with these findings. It would appear that under reducing conditions, trivalent chromium may be in an anionic form; therefore, the following reaction may be more appropriate than reaction (4.26a):

$$(Cr^{2+}) + 2(O^{2-}) + \tfrac{1}{2}O_2 = (CrO_3^{2-}). \tag{4.26c}$$

The oxidation–reduction reaction is of particular importance in glass technology, because the simultaneous presence of transition-metal ions and rare-earth ions with different valence states and coordination is responsible in variations in absorption spectra and color of the glass. Johnston [44] determined oxidation–reduction equilibria of several metal oxides in solution in molten sodium disilicate and found that the variation of the ratio $M^{(x+n)+}/M^{x+}$ is as would be expected from the stoichiometry of the reaction

$$4/n(M^{(x+n)+}) + 2(O^{-2}) = 4/n(M^{x+}) + O_2. \tag{4.27}$$

The stability of the higher valency at atmospheric pressure of air or oxygen decreasing in the order Sn, Fe, Sb, Ce, Mn, Co, and Ni agrees with the sequence of stability determined from studying glass colors [45].

At 1190 °C and 1 atm air, the ratio Sb^{3+}/Sb^{5+} is unity, indicating that at higher temperatures antimony is primarily in the trivalent state and at lower temperatures is in the pentavalent state. Therefore, when Sb_2O_5 is added to molten glass, oxygen is liberated, and this facilitates degassing the melt. Upon cooling, the residual oxygen is absorbed by the oxidation of antimony to the pentavalent state.

Manganese oxides are often added to glass for decolorizing [45]. At melting temperatures in air, most of the iron in glass is in the trivalent state; however, about 1% of iron as ferrous ion is sufficient to give a greenish-blue color. If manganese oxide is present in small amounts, the following reaction occurs during cooling,

$$(Fe^{2+}) + (Mn^{3+}) \rightarrow (Fe^{3+}) + (Mn^{2+}), \tag{4.28}$$

and the glass is decolorized. Under the influence of ultroviolet as from sunlight, a glass containing a relatively small amount of manganese will develop, near the exposed surface, a purple color due to the oxidation of Mn^{2+} to Mn^{3+}. However, the violet color developed by the so-called *solarization* is bleached upon reduction of the manganese to the divalent state by annealing. The role of cerium and

* Private communication with Professor A. Muan of The Pennsylvania State University.

arsenic as decolorizing additives can be explained also in terms of *redox* equilibria.

Use is made of the redox reaction involving antimony and cerium cations in the manufacture of photosensitive glasses containing finely dispersed copper, silver, or gold. During cooling the following reaction occurs:

$$(Sb^{3+}) + 2(Ce^{4+}) \rightarrow (Sb^{5+}) + 2(Ce^{3+}). \tag{4.29}$$

The trivalent cerium is an efficient electron donor, hence absorbs energy in the ultraviolet, and the electrons emitted are trapped by neighboring copper, silver, or gold ions.

In the foregoing discussion, we considered typical examples of redox equilibria in polymeric melts. Oxidation–reduction reactions also play some role in other reactions of gases with polymeric melts discussed below.

4.1.4b Sulfidation reactions. Studies of Fincham and Richardson [46], St. Pierre and Chipman [47], and Turkdogan and Darken [48] have shown that in reaction of sulfur-bearing gases with polymeric melts, the sulfur dissolves in the melt as sulfide ions at partial pressures of oxygen below about 10^{-5} atm; above 10^{-3} atm oxygen, sulfur enters the melt as sulfate ions. This is illustrated in Fig. 4.13, showing log(%S) as a function of log(p_{O_2} atm) for a calcium ferrite and a calcium aluminosilicate melt. The limiting slopes of the lines for sulfide and sulfate reactions are as would be expected from the stoichiometry of the respective reactions.

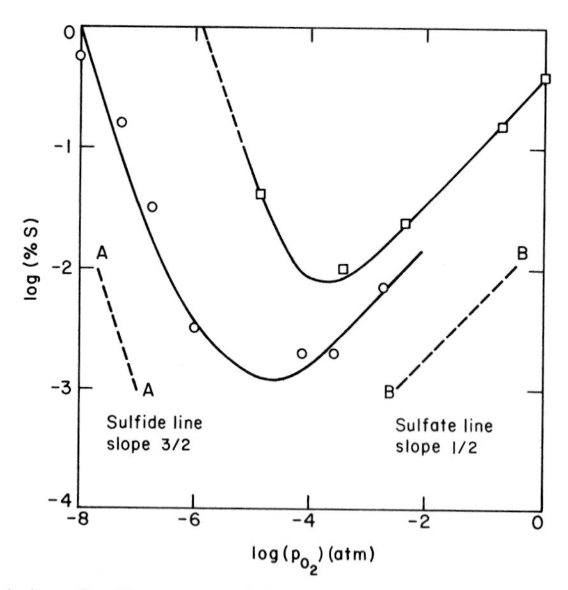

FIG. 4.13 Variation of sulfur content of ferrite and silicate melts with partial pressure of oxygen; (□) from Turkdogan and Darken [48] for calcium ferrite melts at 1620 °C and 7% SO_2, and (○) from Fincham and Richardson [46] for calcium silicate melts at 1500 °C and 2% SO_2.

$$\text{Sulfide reaction:} \quad SO_2 + (O^{2-}) = (S^{2-}) + \tfrac{3}{2}O_2. \tag{4.30}$$

$$\text{Sulfate reaction:} \quad SO_2 + \tfrac{1}{2}O_2 + (O^{2-}) = (SO_4^{2-}). \tag{4.31}$$

As shown by Turkdogan and Darken [48], there is also the pyrosulfate reaction.

$$\text{Pyrosulfate reaction:} \quad 2SO_2 + O_2 + (O^{2-}) = (S_2O_7^{2-}). \tag{4.32}$$

In gases consisting of sulfur, oxygen, hydrogen, and carbon, the sulfur exists in several gaseous forms, e.g., S, S_2, SO, SO_2, SO_3, HS, H_2S, COS, and CS. Because these species will have definite partial pressures in equilibrium with one another at the reaction temperature, any one desired form of the sulfur may be used in representing the gas–slag reaction. For simplicity, sulfide reaction may be represented by the equation

$$\tfrac{1}{2}S_2 + (O^{2-}) = \tfrac{1}{2}O_2 + (S^{2-}), \tag{4.33}$$

for which the equilibrium constant for a given temperature is represented by the equation

$$K = (\gamma_S/a_O)(p_{O_2}/p_{S_2})^{1/2} x_S. \tag{4.34}$$

Because the activity a_O of the oxygen ion and the activity coefficient γ_S of the sulfur ion cannot be evaluated, Fincham and Richardson [46] defined an equilibrium relation called the *sulfide capacity* C_S in terms of measured quantities:

$$K(a_O/\gamma_O) \equiv C_S = (\text{wt.}\,\%\,S)(p_{O_2}/p_{S_2})^{1/2}. \tag{4.35}$$

For given temperature and melt composition, the equilibrium concentration of sulfur in the melt is determined by the ratio $(p_{O_2}/p_{S_2})^{1/2}$ in the gas phase, and not by the individual partial pressures of oxygen and sulfur. It follows from Eq. (4.33) that under the restrictions that (i) electric conductance in the melt is purely ionic, (ii) no electric current is passing through the melt, and (iii) each constituent element in the melt does not change valence, the solution of an atom of sulfur in the polymeric melt must result in the evolution of an atom of oxygen, so that the electric neutrality is maintained. The validity of such a coupled reaction was demonstrated in a unique experiment by Turkdogan and Grieveson [49]. A presintered mixture of Mn–MnS–MnO and prefused mixture of Fe–S–O, contained in an iridium crucible and separated by a thin layer of soda–glass (to act as an ionic membrane for sulfur $\rightleftharpoons$ oxygen transfer), were placed in an evacuated silica capsule, as shown in Fig. 4.14. This was then kept at 1127 °C for a required time to allow reaction (4.33) to occur via the ionic membrane. The purpose of the condensed phases Mn–MnS–MnO (Mn–pellet) and Fe–S–O (Fe–melt) is to maintain the required sulfur and oxygen partial pressures on both sides of the ionic membrane. Since the phases in the Mn–pellet have fixed compositions at the temperature of reaction, the sulfur and oxygen potentials of this system, and hence the ratio p_{O_2}/p_{S_2}, remain constant at a given temperature. Therefore, the direction of sulfur and oxygen transfer

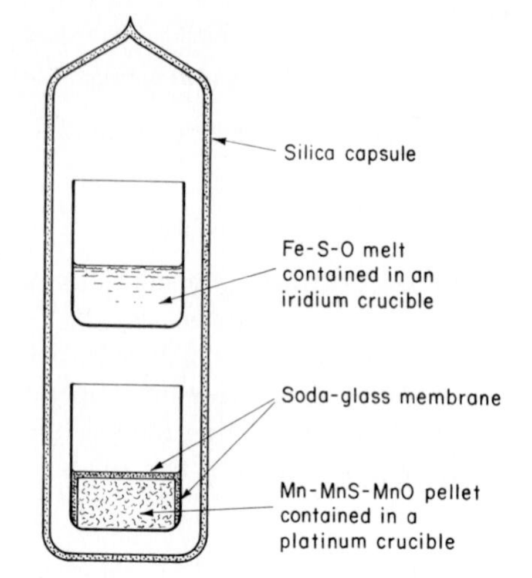

FIG. 4.14 Apparatus for study of sulfur–oxygen transfer through an ionic membrane. After Turkdogan and Grieveson [49].

from one system to the other via the gas phase and through the ionic membrane is determined by the composition of the Fe–melt.

An example of the experimental observations is given in Fig. 4.15, showing that when sulfur is transferred from the Mn–pellet to the Fe–melt through the ionic membrane, oxygen is transferred in the opposite direction and the sum of the atomic concentrations of sulfur and oxygen in both systems remains constant.

The significant feature of these observations is that although the sulfur and oxygen potentials of the system on the side of the Mn–pellet are much lower than those on the other side of the ionic membrane, as a result of coupling, sulfur can be transferred from a low to a high chemical potential through an ionic membrane when oxygen is transferred from a high to a low chemical potential, and vice versa. When the ratios $(p_{O_2}/p_{S_2})^{1/2}$ for systems on both sides of the membrane become equal, the countercurrent sulfur–oxygen transfer ceases, and the system as a whole is in a state of partial equilibrium. As shown in Section 7.2.6, the concept of partial equilibrium plays an important role in understanding the state of reactions in complex systems.

Variation of sulfide capacity C_S with composition is shown in Fig. 4.16 for numerous binary polymeric melts. Of these systems, the calcium ferrite melts have the highest capacity for sulfur. A small increase in the concentration of basic oxide in aluminate, silicate, and phosphate melts increases the sulfide capacity markedly. Based on this observation, Richardson [11] concluded that sulfur does not readily replace oxygen atoms bonded to two silicon or phosphorus atoms in the anion complexes. It is more likely that the oxygen atoms participating in the sulfide, sulfate, and pyrosulfate equilibria, Eqs. (4.30)–(4.32), are primarily free oxygen ions.

The solubility of CaS [5] in $CaO–SiO_2$ melts at temperatures of 1500–1550 °C increases from about 2.5% S at $CaO/SiO_2 = 1.5$ to 5% S at $CaO/SiO_2 = 0.5$.

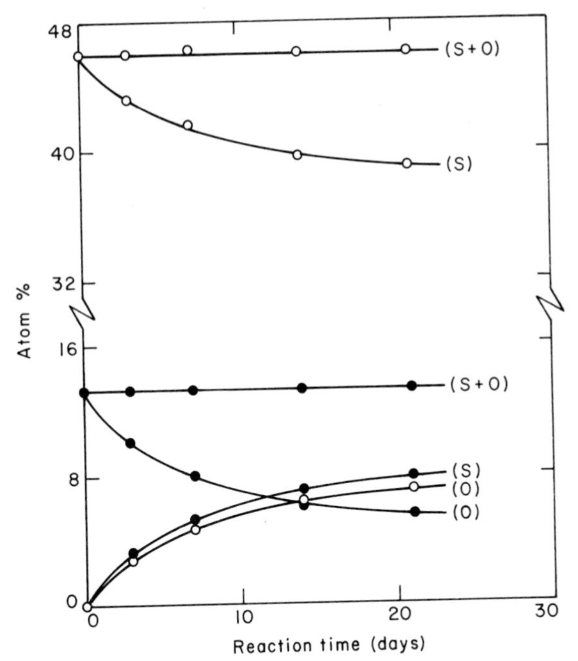

FIG. 4.15 Results showing stoichiometric displacement of sulfur and oxygen at 1127 °C to and from Fe melt (○) and Mn pellet (●) separated by an ionic membrane. After Turkdogan and Grieveson [49].

This finding suggests that when there is no oxygen transfer accompanying the dissolution of sulfur in silicate melts, as with the dissolution of CaS, reaction may occur with doubly bonded oxygen atoms, thus

$$(\equiv Si-O-Si\equiv) + CaS = (\equiv Si-O^-) + (\equiv Si-S^-) + Ca^{2+}. \quad (4.36)$$

Since silicon and sulfur interact forming volatile species SiS and crystalline form SiS_2, Si–S bonding in polymerized silicates would be expected.

Grieveson and Turkdogan [50] have found large liquid immiscibility in the calcium–sulfate–ferrite and sodium–sulfate–ferrite melts, indicating that ferrite and sulfate polymers are not structurally compatible. Large liquid immiscibility in sulfate–ferrite systems is expected from the composition dependence of the activity coefficient ratio $\gamma_{CaSO_4}/\gamma_{CaO}$ in calcium ferrite melts [48]. The ratio $\gamma_{CaSO_4}/\gamma_{CaO}$ is 10 at 72 mole % CaO and increases to 2200 in the ferrite melt containing 20 mole % CaO.

Kor and Richardson [51] have shown that sulfur does not replace fluorine to any significant extent from liquid calcium fluoride or CaF_2–Al_2O_3 melts. However, in the ternary system CaF_2–CaO–Al_2O_3, the sulfide capacity for a given lime content increases as Al_2O_3 is replaced by CaF_2. Since CaF_2 increases the solubility of CaO, the addition of CaF_2 increases the activity of the oxide ion, hence increases the sulfide capacity of the slag.

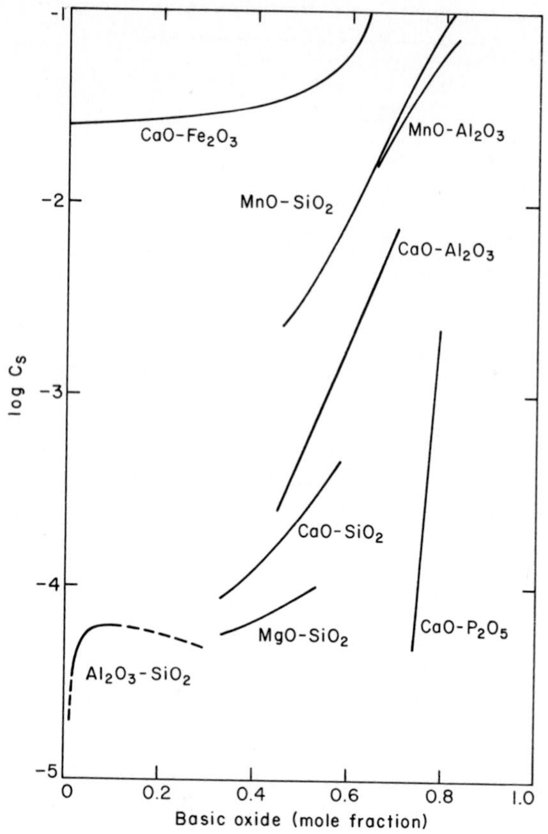

FIG. 4.16 Sulfide capacities of binary aluminate, silicate, and phosphate melts at 1650 °C [46], and calcium ferrite melts at 1620 °C [48].

4.1.4c Reaction of water vapor with molten silicates and aluminates. The equilibrium studies have shown that the solubility of H_2O in polymeric melts is proportional to the square root of the partial pressure of water vapor in the gas phase. Dry hydrogen has negligible solubility in silicates. As suggested by Tomlinson [52] and Russell [53], in polymeric melts H_2O behaves as an amphoteric oxide, i.e., a basic or an acidic oxide, depending on the extent of polymerization of the melt.

In polymerized melts, i.e., high silica contents, H_2O behaves as a basic oxide and reacts with doubly bonded oxygen as represented by the equation

$$(\equiv\text{Si—O—Si}\equiv) + H_2O = 2(\equiv\text{Si—OH}). \tag{4.37}$$

In partially depolymerized melts, H_2O reacts with singly bonded oxygen and acts much like a network former as indicated by the reaction

$$2(\equiv\text{Si—O}^-) + H_2O = (\equiv\text{Si—O—Si}\equiv) + 2(\text{OH}^-). \tag{4.38}$$

In melts of high basicity, H_2O reacts with free oxygen ions; thus

$$(O^{2-}) + H_2O = 2(OH^-). \tag{4.39}$$

The overall reaction may be written in a general form by an equation of the type

$$(O^*) + H_2O = 2(OH^*), \tag{4.40}$$

where O* represents oxygen as doubly or singly bonded, or as free ion; OH* is either bonded to silicon or is free hydroxyl ion. Since the amount of OH in solution is small, reaction (4.40) will not change the initial activity of the oxide ion in the melt; therefore, for a given temperature and melt composition, the state of the equilibrium may be represented in terms of the H_2O capacity of the melt,

$$C_H = c_{OH}/(p_{H_2O})^{1/2}, \tag{4.41}$$

where the concentration of OH, c_{OH} is usually given in terms of ppm H_2O in solution.

According to reaction (4.37), the H_2O capacity should increase with decreasing basicity of the melt; reaction (4.39) also suggests that C_H should increase with increasing basicity of the melt. That is, a minimum solubility of H_2O is anticipated in melts of intermediate basicity, the so-called neutral slags. As is seen from the solubility data in Fig. 4.17 for CaO–MgO–SiO_2 melts in equilibrium with p_{H_2O} equal to 289 mm Hg at 1550 °C, the experimental results of Iguchi and Fuwa [54] substantiate the theoretical expectation. In the lower diagram, the basicity is represented by the molar ratio $(x_{CaO} + x_{MgO})/x_{SiO_2}$; the minimum solubility occurs at the metasilicate composition. Kurkjian and Russell [55] also observed minima on H_2O solubilities in alkali silicate melts; in these melts the minimum occurs near the disilicate composition. When alkali oxides are added to calcium silicate melts [56], the solubility of H_2O increases and the minimum solubility occurs at lower values of CaO/SiO_2. The composition dependence of H_2O solubility obtained by Uys and King [57] differs noticeably from those mentioned above.

As in the polymeric melts, the solubility of H_2O in high purity vitreous silica is proportional to the square root of the partial pressure of water vapor [58]. At 1000 °C and 1 atm H_2O, the solubility is about 1250 ppm H_2O.

Much like the trend with sulfide capacities, H_2O capacities of calcium aluminate melts [59] are higher than those of calcium silicates for the same concentration of CaO. The temperature effect on the H_2O solubility is small for all the polymeric melts investigated, e.g., $\Delta H \sim 1.5$–3 kcal.

4.1.4d Solution of carbon and nitrogen in slags. The solubility of nitrogen in aluminosilicate melts is negligible under oxidizing conditions [60]. In fact, most of the solubility measurements have been made with graphite-saturated molten aluminates and silicates. Of the many studies made, those of Schwerdtfeger and Schubert [61] are probably more reliable and readily interpretable.

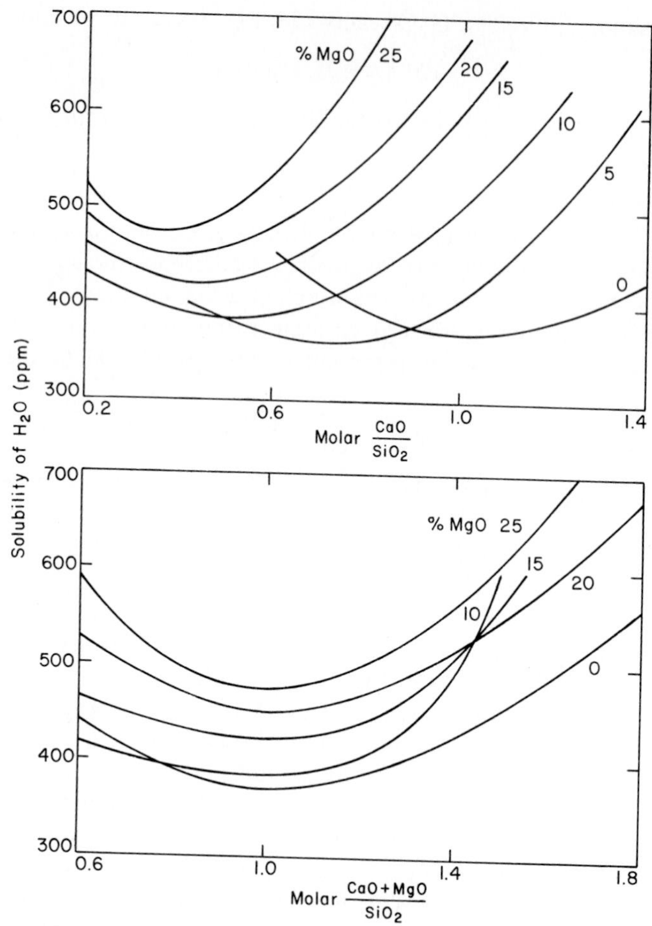

FIG. 4.17 Solubility of H_2O in CaO–MgO–SiO_2 melts at 289 mm Hg and 1550 °C. After Iguchi and Fuwa [54].

They equilibrated molten calcium aluminates with N_2–CO–Ar gas mixtures at 1600 °C and analyzed quenched melts for total nitrogen, carbon, and cyanide. The experimental results are interpreted in terms of nitride, cyanide, and carbide in solution at known partial pressures of N_2 and CO. The equilibrium relations are given in terms of the nitride, cyanide, and carbide capacities for the reactions represented by the following equations.

Nitride reaction:

$$\tfrac{3}{2}(O^{2-}) + \tfrac{3}{2}C + \tfrac{1}{2}N_2 = (N^{3-}) + \tfrac{3}{2}CO,$$

$$C_N = (\%\ N)(p_{CO}^{3/2}/p_{N_2}^{1/2}). \tag{4.42}$$

Cyanide reaction:

$$\tfrac{1}{2}(O^{2-}) + \tfrac{3}{2}C + \tfrac{1}{2}N_2 = (CN^-) + \tfrac{1}{2}CO,$$
$$C_{CN} = (\%\,CN)(p_{CO}/p_{N_2})^{1/2}. \tag{4.43}$$

Carbide reaction:

$$(O^{2-}) + 3C = (C_2^{2-}) + CO,$$
$$C_C = (\%\,C)p_{CO}. \tag{4.44}$$

Variations of nitride, cyanide, and carbide concentrations, in a melt of fixed composition, with the partial pressures of N_2 and CO as depicted by the above equations have been substantiated in the work of Schwerdtfeger and Schubert. The composition dependence of capacities C_X is shown in Fig. 4.18. As in the case of sulfide, sulfate, and hydroxyl capacities, the cyanide and carbide capacities increase with increasing basicity, suggesting that cyanide and carbide

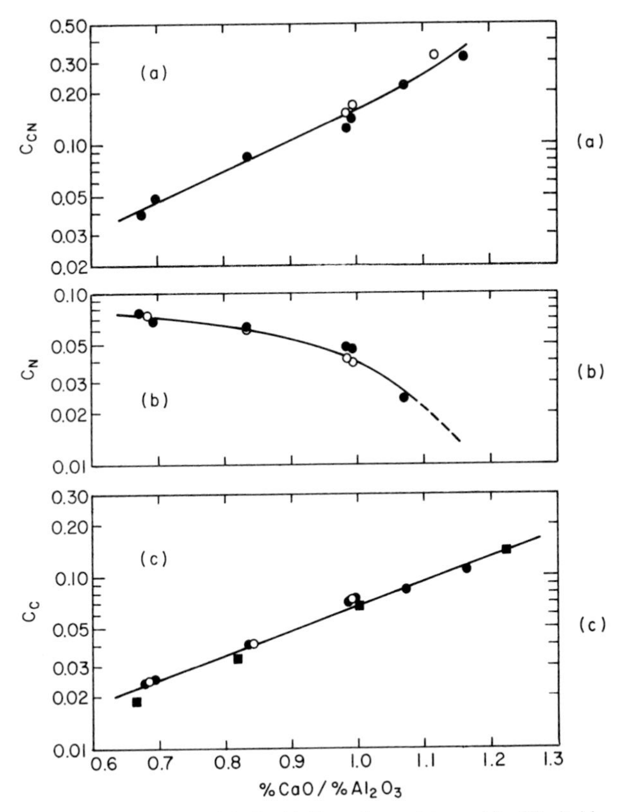

FIG. 4.18 Effect of composition of CaO–Al_2O_3 melts on (a) cyanide, (b) nitride, and (c) carbide capacities at 1600 °C. After Schwerdtfeger and Schubert [61].

species react with free oxygen ions as in Eqs. (4.43) and (4.44). Increase in the nitride capacity of calcium aluminate melts with decreasing basicity would suggest reaction of nitrogen with oxygen atoms bonded to aluminum atoms in the aluminate network.

In a similar study made by Davies and Meherali [62], they found that the total nitrogen content of graphite-saturated calcium aluminosilicate melts is proportional to $p_{N_2}^{1/2}/p_{CO}$. On the other hand, the sum of Eqs. (4.42) and (4.43) gives for the total nitrogen

$$(\%\ N)_{tot}(p_{CO}/p_{N_2}) = (C_N/p_{CO}) + \tfrac{14}{26}C_{CN}. \tag{4.45}$$

Although their data are limited, the results of Davies and Meherali can be interpreted in accord with Eq. (4.45). The values of C_N and C_{CN} derived show composition dependence similar to those in Fig. 4.18.

In the studies of Dancy and Janssen [63], graphite-saturated aluminosilicate melts were equilibrated with nitrogen (no CO was added). The total nitrogen was found to be proportional to the square root of the partial pressure of nitrogen. In order to interpret their data, p_{CO} in the system has to be estimated by assuming equilibrium between graphite and the melt. Of several equilibrium constants evaluated by Dancy and Jassen, those similar to the nitride capacity in Eq. (4.42) were found to decrease with increasing concentration of CaO. This trend is in general accord with other data [61, 62].

Schwerdtfeger and Schubert [64] investigated also the effect of CaF_2 addition to $CaO–Al_2O_3$ melts on reactions with N_2 and CO. The values of C_{CN} and C_C in the melt with 35% CaF_2 and % CaO/% $Al_2O_3 = 1$ are about twice those in the fluoride free melts, and C_N decreases with the addition of CaF_2. These findings are consistent with the effect of CaF_2 on the activity of oxygen ions, hence the corresponding effect on the cyanide, carbide, and nitride reactions.

4.1.4e Solubility of carbon dioxide in polymeric melts. The solution of carbon dioxide in polymeric melts may be represented by the reaction equilibrium

$$(O^{2-}) + CO_2 = (CO_3^{2-}), \tag{4.46}$$

for which the carbonate capacity C_{car} is represented by the equation

$$C_{ear} = (\%\ CO_2)/p_{CO_2}. \tag{4.47}$$

Pearce [65] measured the solubility of CO_2 in molten sodium silicates and borates at temperatures of 1000–1200 °C. In both systems the carbonate capacity increases with increasing concentration of sodium oxide, indicating that carbon dioxide reacts with free oxygen ion to form free carbonate ion as stated in Eq. (4.46). At 1 atm pressure of CO_2 and 1000 °C, the concentration of CO_2 is 0.005% in the silicate containing 25 mole % Na_2O and increases to about 3% CO_2 at the metasilicate composition. The solubility at 1200 °C is about one-tenth that at 1000 °C. Pearce found that the solubility of CO_2 in sodium borate melts is 2–3 times less than in silicate melts. This is in accord with free energies of formation of borates being more negative than those of the silicates.

4.1.5 Activities in Mattes

The solubility of sulfur in liquid metals of common metallurgical interest is shown in Fig. 4.19. With the exception of the Cu–S and Fe–S systems, Henry's law approximation may be made for dilute solutions of sulfur in several liquid metals. In lead and tin the solubility of sulfur is limited by the formation of PbS and SnS at the indicated activities of sulfur. The limit of solubility in copper at 1200 °C is that corresponding to saturation with liquid matte. There is complete liquid solution in the Fe–S and Ni–S systems; in the Co–S system there is an invariant equilibrium at about 1075 °C and 60 atom % S between two liquid sulfides and a nonstoichiometric phase $Co_{1-x}S$.

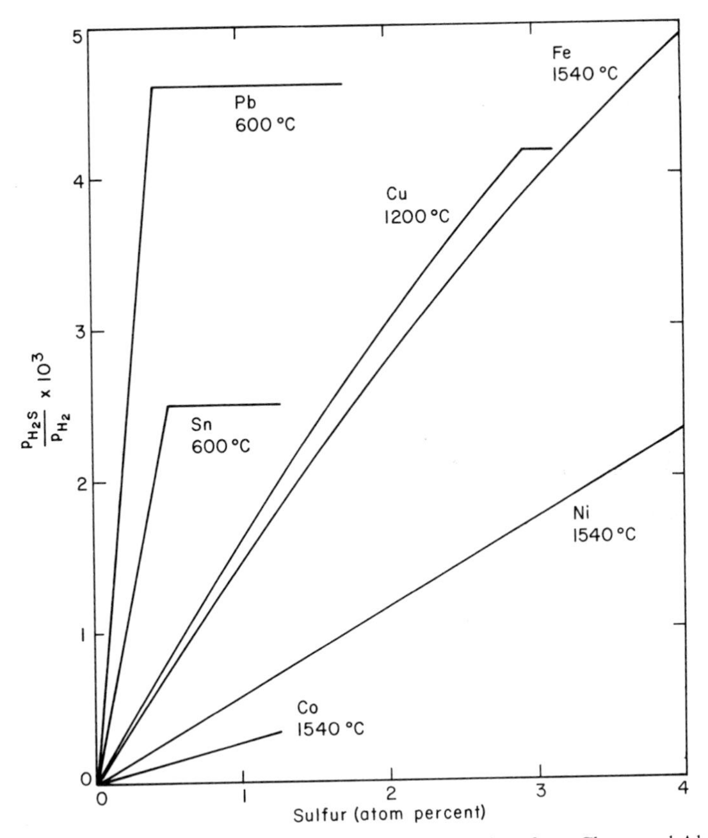

FIG. 4.19 Solubility of sulfur in some liquid metals: Pb and Sn from Cheng and Alcock [66]; Cu from Bale and Toguri [67] and Yagihashi [68]; Fe, Ni, and Co from Alcock and Cheng [69].

The effect of alloying elements on the activity coefficient of sulfur in liquid iron, cobalt, and copper was given in Tables 3.1–3.3. Activities of sulfur in copper, nickel and iron mattes at 1200 °C are shown in Fig. 4.20.

Krivsky and Schuhmann [72] measured the activity of sulfur as a function of phase compositions in three major regions of the Cu–Fe–S system at temperatures of 1150–1350 °C. The three regions studied were (a) liquid mattes, (b) the liquid metal–liquid matte miscibility gap, and (c) the γ-iron alloy–liquid metal–liquid matte region. By Gibbs–Duhem integrations, they calculated activities of Cu_2S, FeS, copper, and iron from the experimental data relating the activity of sulfur to composition. The isothermal phase diagram evaluated by Krivsky and Schuhmann for the Cu–Fe–S system at 1350 °C is shown in Fig. 4.21, where the composition is given in terms of equivalent ratios

$$E_{\mathrm{Fe}} = n_{\mathrm{Fe}}/(n_{\mathrm{Fe}} + \tfrac{1}{2}n_{\mathrm{Cu}}), \quad E_{\mathrm{S}} = n_{\mathrm{S}}/(n_{\mathrm{Fe}} + \tfrac{1}{2}n_{\mathrm{Cu}}).$$

For the stoichiometric FeS, E_{S}, and E_{Fe} are equal to unity; for pseudobinary mixtures of Cu_2S and FeS, E_{S} is equal to unity and E_{Fe} is equal to the mole fraction of FeS.

Activities of Cu_2S and FeS along the pseudobinary section read off from Fig. 4.21 along the line joining Cu_2S and FeS show small negative departures from Raoult's law for ideal mixing. In the matte-saturated liquid–metal solutions

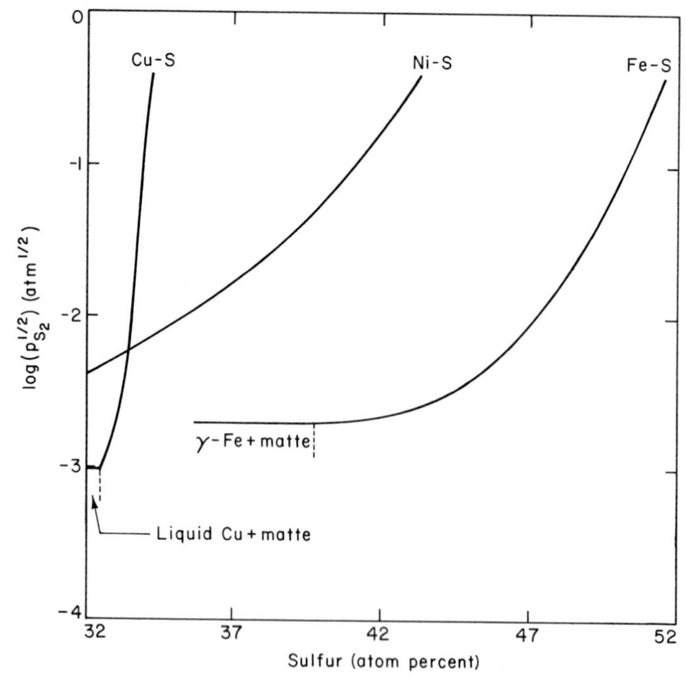

FIG. 4.20 Activity of sulfur in some molten binary mattes at 1200 °C: Cu–S from Bale and Toguri [67]; Ni–S from Meyer *et al.* [70]; Fe–S from Bale and Toguri [67] and Burgmann *et al.* [71].

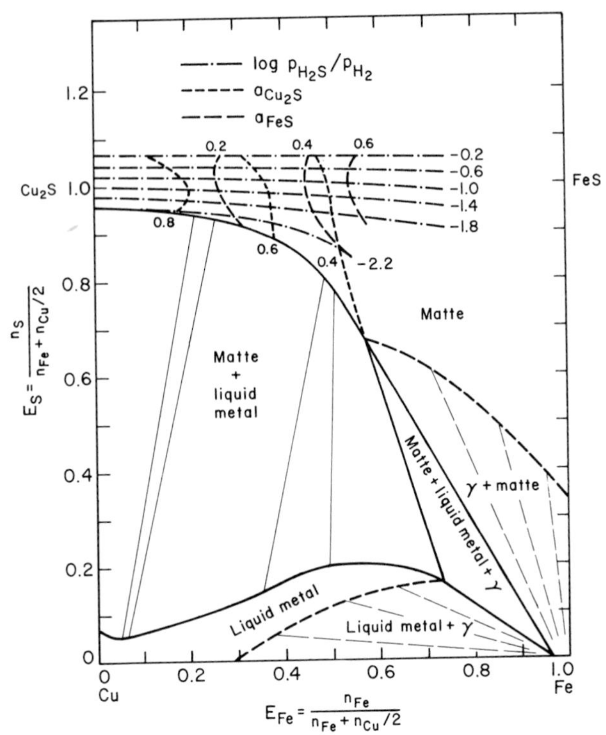

FIG. 4.21 Isothermal phase diagram for Cu–Fe–S system at 1350 °C. After Krivsky and Schuhmann [72].

the activities of Cu and Fe show strong positive deviations from ideality, indicating the tendency toward immiscibility at 1350 °C.

Matousek and Samis [73] measured the activity of sulfur in the Cu–Ni–S system at 1200 °C by equilibrating the liquid matte with gas of known sulfur potential as determined by the ratio p_{H_2S}/p_{H_2} in the gas phase. Activities of Cu_2S anb Ni_3S_2 were calculated by Gibbs–Duhem integrations from measured sulfur activities with the simplifying assumption of ideal mixing across the pseudobinary section Cu_2S–Ni_3S_2. Their results are shown in Fig. 4.22. Within the region of the miscibility gap, the activity of Cu_2S is found to range from 0.958 to 0.87.

Thermochemical modeling of molten sulfides has been studied by Kellogg [74] and co-workers. Their empirical formulation is based on the assumption of hypothetical species of metal and metal sulfides in the melt. For example, the species assumed for the Cu–Ni–S system are Cu, Cu_2S, CuS, Ni, Ni_3S_2, and NiS. Activity coefficients are assigned to all the species with empirical interaction parameters to fit Margules equations for the multicomponent system (see Section 3.1). According to this model, at constant temperature each pair of species is characterized by two parameters, k_{ij} and k_{ji}, so that a solution of three species

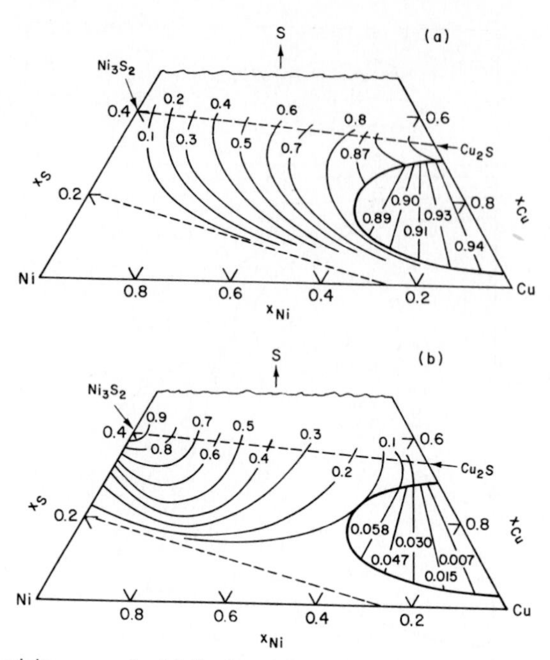

FIG. 4.22 Isoactivity curves for (a) Cu_2S and (b) Ni_3S_2 in Cu–Ni–S system at 1200 °C. After Matousek and Samis [73].

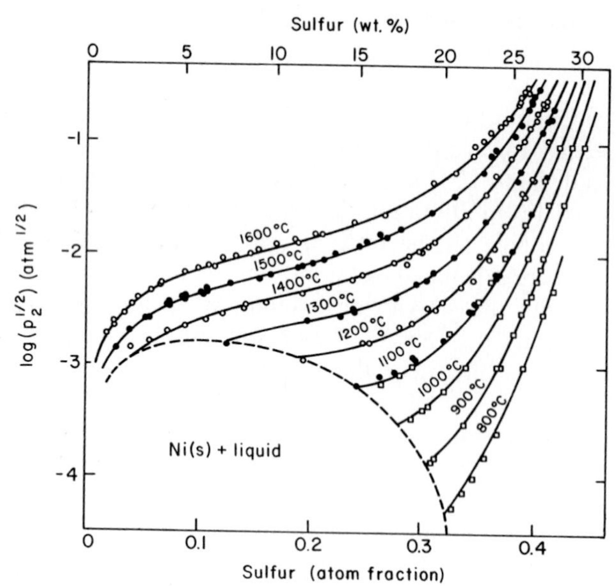

FIG. 4.23 Activities of sulfur in Ni–S system measured by (○, ●) Meyer *et al.* [70] and (□) Nagamori and Ingraham [75]; lines are calculated from a thermochemical model. After Kellogg [74].

would require six parameters for complete description. Isotherms drawn in Fig. 4.23 for activities of sulfur in Ni–S melts for temperatures of 800–1600 °C are calculated by Kellogg [14] and co-workers using their empirical model.

From a study of reaction equilibrium between SO_2 and Cu–S–O melts at temperatures of 1100–1250 °C, Schmiedl *et al.* [76] have found that within the miscibility gap region the concentration of oxygen in the liquid copper and liquid matte is proportional to the square root of the partial pressure of SO_2. Isobars of SO_2 for 0.099 and 0.97 atm are shown for 1100 and 1250 °C in Fig. 4.24a. Below 1 atm pressure of SO_2, the concentration of oxygen in liquid copper and liquid matte decreases with decreasing pressure of SO_2, accompanied by a corresponding increase in the concentration of sulfur. Isoactivity curves of Cu_2O are shown in Fig. 4.24b.

The extent of the miscibility gap and the liquidus isotherms in the Fe–S–O system, determined by Hilty and Crafts [77] and by Bog and Rosenqvist [78], are shown in Fig. 4.25. The dotted curve gives the composition of the liquid oxysulfide in equilibrium with either wustite or pyrrhotite. Isothermal sections

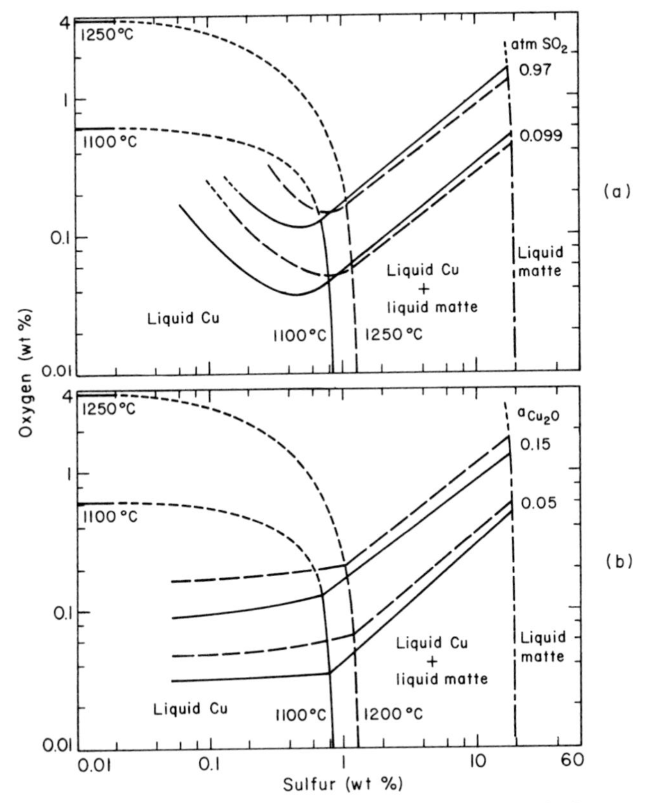

FIG. 4.24 (a) Sulfur dioxide isobars and (b) isoactivity of Cu_2O in Cu–S–O system at 1100 and 1250 °C. After Schmiedl *et al.* [76].

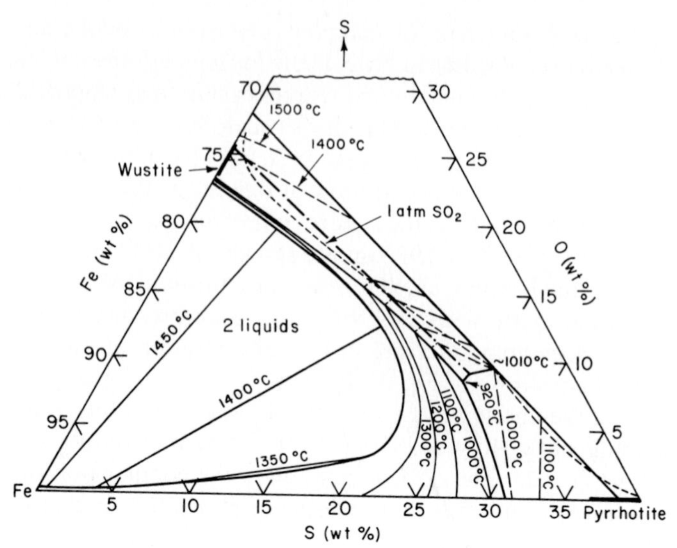

FIG. 4.25 Liquidus isotherms and miscibility gap in Fe–S–O system. After Hilty and Crafts [77] and Bog and Rosenqvist [78].

showing the phase equilibria in this system were given in Fig. 2.6 in Chapter 2. The SO_2 chemical potentials and phase equilibria in the Pb–S–O system were given in Figs. 2.7 and 2.8.

As would be expected from the phase relations in the Cu–S–O and Fe–S–O systems, the solubility of iron oxide in liquid mattes decreases with increasing ratio Cu_2S/FeS.

4.2 Surface Tensions of Molten Glasses, Slags, and Mattes

Surface tensions of liquid oxides and their mixtures are in the range 200–600 dyn cm^{-1}. As shown by King [79], surface tensions of silicate melts decrease with increasing silica content, indicating positive chemisorption of silica.

Surface tensions of $FeO–MnO–SiO_2$ and $FeO–CaO–SiO_2$ systems at 1400 °C measured by Kozakevitch [80] are shown in Fig. 4.26. The surface tension is essentially independent of the ratio FeO/MnO; in the $FeO–CaO–SiO_2$ melts, however, the surface tension at a given silica content decreases with increasing ratio FeO/CaO, indicating that the adsorption of silica on the surface increases with increasing iron oxide content of the melt. As is seen from the data of Kozakevitch in Fig. 4.27, P_2O_5 and Na_2O have large effects on the surface tension of liquid iron oxide at 1400 °C, and are more surface active than SiO_2 and TiO_2. Studies of Cooper and Kitchener [81] have shown that P_2O_5 is surface active also in calcium silicate melts, but to a lesser extent than in liquid iron oxide.

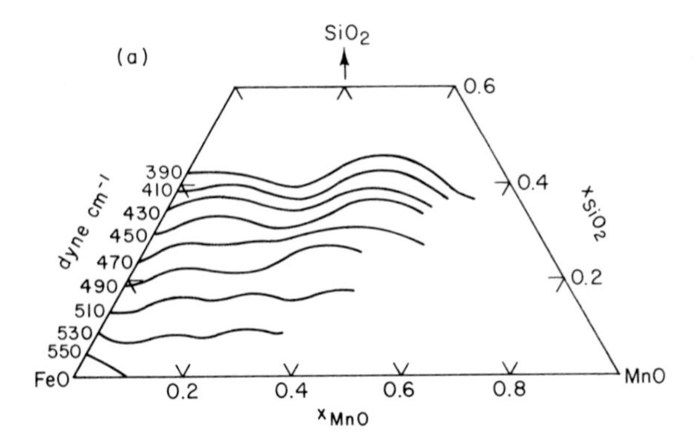

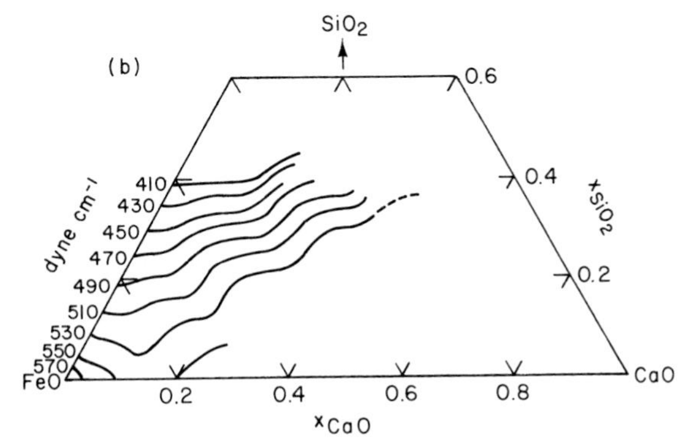

FIG. 4.26 Isosurface tension curves in (a) $FeO–MnO–SiO_2$ and (b) $FeO–CaO–SiO_2$ system at 1400 °C. After Kozakevitch [80].

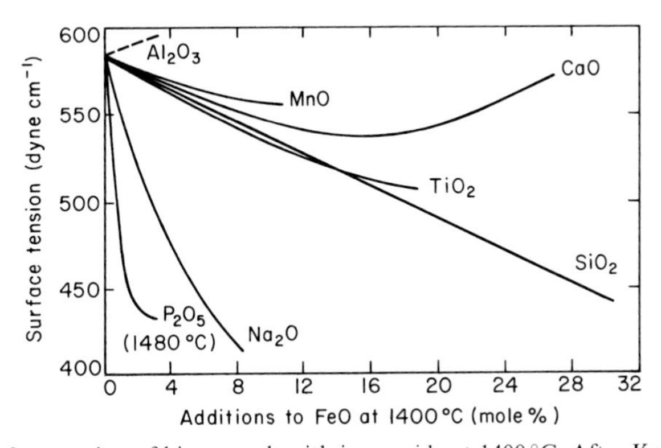

FIG. 4.27 Surface tension of binary melt with iron oxide at 1400 °C. After Kozakevitch [80].

Boni and Derge [82] measured the effect of sulfur on the surface tension of molten aluminosilicates and found sulfur to be surface active; the surface activity of sulfide ion being slightly greater in acid slags than in basic slags. The fluorides also lower the surface tension of silicates [83]; for example, in an equimolar mixture of CaO and SiO_2, the surface tension at 1550 °C is lowered from 480 to about 380 dyn cm^{-1} with the addition of 10 mole % CaF_2.

In an attempt to correlate structure and properties of glasses, Dietzel [84] made use of the concept of electrostatic field strength of the ions in polymeric systems. Representing the ion–oxygen interaction or ionic potential by the ratio of valency of the cation to ionic radius, Z/r, Dietzel found systematic variations in properties of glasses with Z/r. King [79] showed that the effect of composition and temperature on the surface tension of liquid silicates may be explained by considering the ion–oxygen interaction. He showed that the temperature coefficient of surface tension, $\partial\sigma/\partial T$, becomes more positive with increasing Z/r and silica content. For example, for metasilicates $\partial\sigma/\partial T$ increases from -0.076 for $Z/r = 0.75$ (K^+) to $+0.09$ for $Z/r = 2.56$ (Mg^{2+}). King also observed that the tendency for $\partial\sigma/\partial T$ to become more positive increases with silica content to a greater extent for the cations with higher ionic potentials. As stated by King, these observations suggest that the breakdown of the silicate network would be greater with cations of higher ion–oxygen attraction. This is consistent with widening of the miscibility gap in the temperature-composition diagram of silicates as the ionic potential of the basic cation increases.

Surface tension of mattes are in the range 100–600 dyn cm^{-1}. In liquid solutions of FeS and Cu_2S, the surface tension at 1250 °C increases nonlinearly from 340 dyn cm^{-1} for 100% FeS to 400 dyn cm^{-1} for 100% Cu_2S [84a]; the FeS–Ni_3S_2 mixtures show a similar behavior with σ for Ni_3S_2 equal to 500 dyn cm^{-1}.

Limited information is available on the interfacial tension for slag–metal and slag–matte systems. For liquid iron–slag systems the interfacial tension is in the range 900–1200 dyn cm^{-1}, usually decreasing with increasing silica content [85]. In iron silicate–matte systems the interfacial tension is in the range 5–90 dyn cm^{-1} [84a].

The energy of adhesion W of two phases at the interface is given in terms of surface energies as deduced originally by Dupré (1869); for metal/slag interface $W_{m/s}$ is given by the equation

$$W_{m/s} = \sigma_{g/m} + \sigma_{g/s} - \sigma_{m/s}, \tag{4.48}$$

where the subscripts g/m, g/s, and m/s indicate gas/metal, gas/slag, and metal/slag surfaces, respectively. When a surface active solute is transferred from metal to slag, the interfacial tension decreases, and as a result, the adhesion between metal and slag increases.

Kozakevitch [86] attributed the lowering of the interfacial tension between liquid metal and slag to the formation of double layers of orientated ions such as M^{2+}–SiO_4^{4-}, M^{2+}–PO_4^{3-}, M^{2+}–S^{2-}, the cations facing the metal and anions being orientated towards the slag. Dramatic effect on the interfacial tension of transfer of surface active ion units such as $Fe^{2+}S^{2-}$ from iron to slag is well demonstrated by Kozakevitch *et al.* [87].

A calcium aluminosilicate was poured on a drop of liquid iron containing 3% C and 0.7% S, and change in the shape of the metal drop was observed from the x-ray image. At the first instant of contact between metal and slag, the metal drop flattened and the estimated interfacial tension was less than 5 dyn cm^{-1}. When the reaction approached equilibrium, with the residual sulfur in iron being about 0.01%, the drop recovered its normal shape and the interfacial tension rose to about 800 dyn cm^{-1}.

Drastic lowering of interfacial tension resulting from interfacial reactions involving surface active ion units may bring about emulsification of metal in the slag, particularly when the reaction is accompanied by gas evolution. This subject is discussed further in Section 4.3.3.

4.3 Transport Properties of Polymeric Melts

Of particular interest to high temperature technology, the transport properties considered in this section are density, viscosity, diffusion, electrical, and thermal conductivity.

4.3.1 Density and Molar Volume

Densities of glasses and polymeric melts have been measured for many systems. Reference may be made to Morey [88] for much of the early work on the density of glass and methods of empirical formulation of the composition dependence of density. The data of Morey and Merwin [89] in Fig. 4.28a are typical examples of the variation of density with composition of annealed soda glasses. As indicated by the data, for small variations in composition, the additivity rule may be used in estimating the density for a particular composition.

The densities of MO–FeO–SiO_2 melts containing 29 mole % SiO_2 and saturated with iron at 1410 °C, determined by Gaskell *et al.* [90], are shown in Fig. 4.28b. The change in density of the molten silicate by replacing FeO with CaO, MnO, NiO, or CoO does not follow a simple additivity rule. Numerous structural models and ion interactions have been proposed to account for the variation of molar volume with temperature and composition of polymeric melts. For further reading on the subject, reference may be made to papers, for example by Lee and Gaskell [91] and by Grau and Masson [92].

Of the several density determinations on the CaF_2–CaO–Al_2O_3 melts, those of Mitchell and Joshi [93], for temperatures of 1400–1750 °C, are probably more reliable. The density of liquid CaF_2, 2.52 g cm^{-3} at 1600 °C, increases nonlinearly to 2.64 g cm^{-3} with the addition of 30% Al_2O_3 and to 2.62 g cm^{-3} with the addition of 30% CaO.

The molar volumes of liquid sulfides Ag–S, Cu–S, Fe–S, and Ni–S determined by Nagamori [94] are shown in Fig. 4.29. The Ag–S system shows an ideal

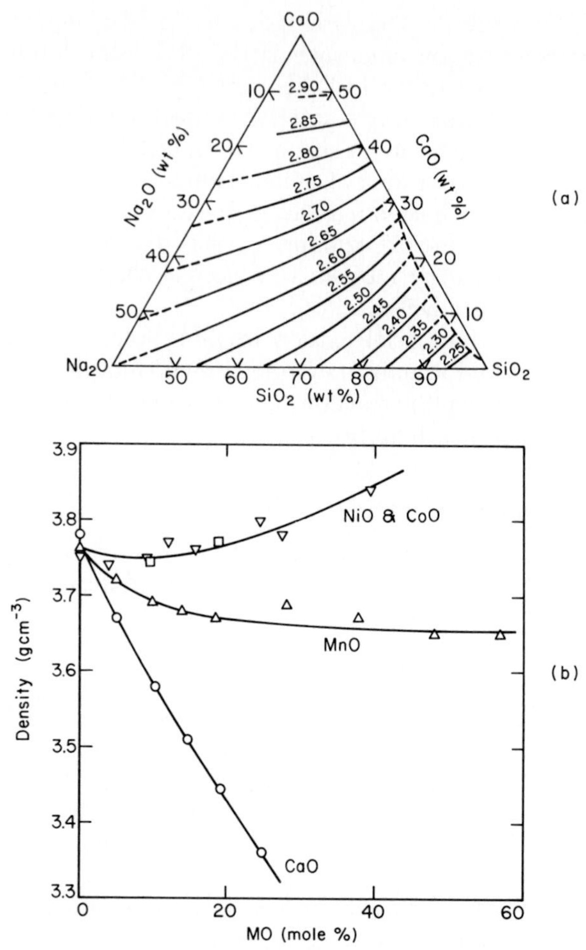

FIG. 4.28 Densities of (a) annealed soda glasses at 20 °C, after Morey and Merwin [89], and (b) MO–FeO–SiO_2 melts containing 29 mole % SiO_2 and saturated with iron at 1410 °C, after Gaskell *et al.* [90].

behavior as regards to the molar volume; however, there are small negative departures from ideality in the Fe–S and Ni–S systems.

4.3.2 Viscosity

Much work has been done in the 1950s on the viscosity of aluminosilicate melts, notably by Bockris and Kozakevitch and their co-workers. Studies of viscosity and other transport properties were of much help in the development of the present knowledge of the structure of polymeric melts and their depolymerization with the addition of metal oxides and with increasing temperature.

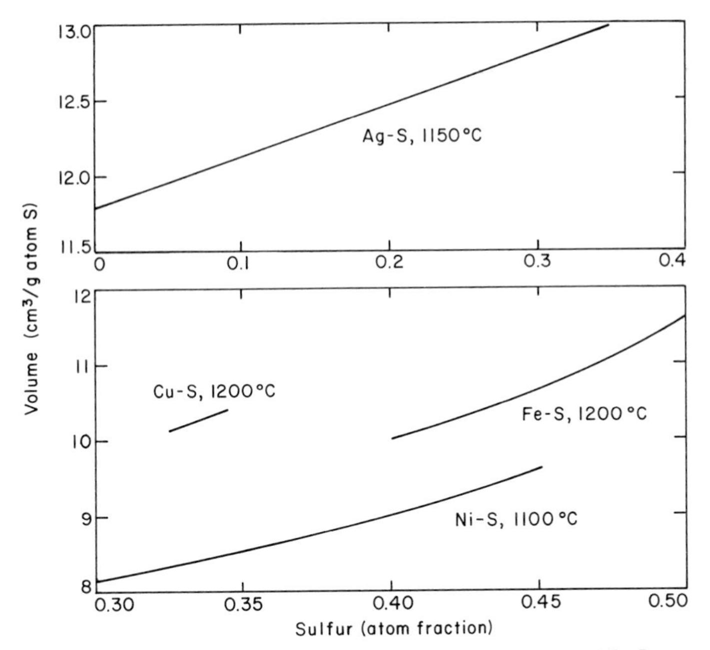

FIG. 4.29 Molar volumes of liquid mattes. After Nagamori [94].

Bockris *et al.* [95] have found that for a given molar concentration of silica, the alkaline earth oxides in silicate melts are interchangeable without affecting viscosity. Alkali oxides on molar basis also have similar lowering effect on the viscosity of alkali silicates. The apparent heat of activation for viscous flow decreases continuously with increasing concentration of basic oxides, the effect being greater with alkali oxides. For example, heat of activation for pure liquid silica is about 135 kcal and decreases to about 40 kcal at the metasilicate composition of alkaline earth silicates and to 25 kcal at the composition of alkali metasilicates. The initial breakdown of the silica network with the addition of 10–15 mole % of a basic oxide has a pronounced lowering effect on viscosity and the heat of activation. The decrease in viscosity $\partial\eta/\partial x_{MO}$ becomes less with further increase in x_{MO}.

Another useful generalization was that found by Turkdogan and Bills [96] for the silica equivalence of alumina x_a for any given temperature and viscosity:

$$x_{Al_2O_3} \equiv x_a = x_{SiO_2}\,(\text{binary}) - x_{SiO_2}\,(\text{ternary}).$$

Values of x_a have been calculated from isothermal viscosity data for the CaO–SiO_2 [97] and CaO–Al_2O_3–SiO_2 [98] system for several mole fractions of Al_2O_3 and molar ratios Al_2O_3/CaO. The relation obtained is shown in Fig. 4.30, where the mole fraction x_{MO} refers to the sum of mole fractions of divalent

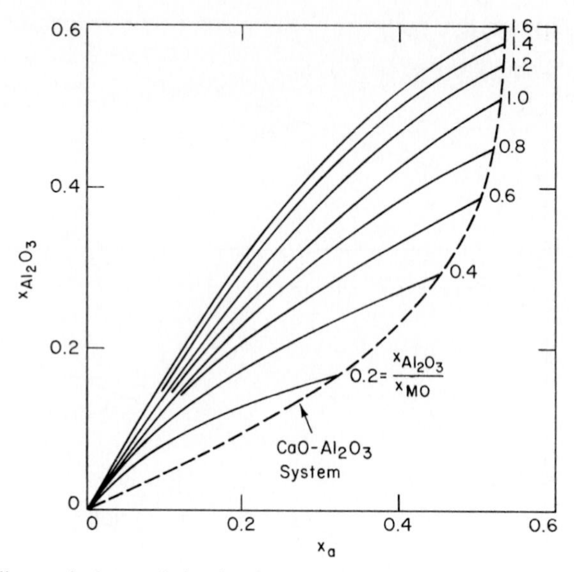

FIG. 4.30 Silica equivalence of alumina for composition dependence of viscosity; x_{MO} is the sum of mole fractions of divalent metal oxides. After Turkdogan and Bills [96].

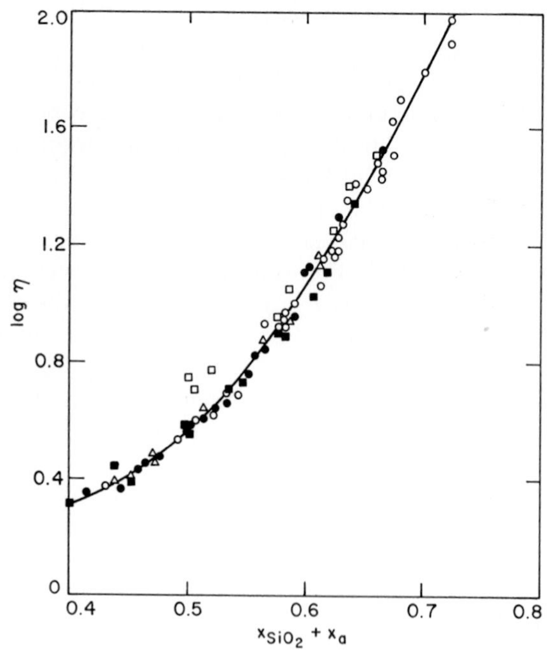

FIG. 4.31 Variation of viscosity with composition of CaO–MgO–Al_2O_3–SiO_2 melts with the sum of mole fraction of silica and the silica equivalence of alumina, using the data of Bockris and Lowe [97] for CaO–SiO_2, Kozakevitch [98] for CaO–Al_2O_3–SiO_2, and Machin [99] for CaO–MgO–Al_2O_3–SiO_2. After Turkdogan and Bills [96].

metal oxides; this relation was found to be essentially independent of temperature. As is seen from Fig. 4.31, the viscosity of CaO–MgO–Al_2O_3–SiO_2 melts at 1500 °C is a single function of the composition parameter $x_{SiO_2} + x_a$.

The temperature dependence of viscosity is shown in Fig. 4.32 for values of $x_{SiO_2} + x_a$ from 0.45 to 0.70. As indicated by the shape of the curves, for a given concentration of acid oxides, the apparent heat of activation for viscous flow decreases with increasing temperature; this change in heat of activiation with temperature becomes less pronounced with increasing basicity. This behavior is a manisfestation of the complimentary effects of basicity and temperature on the depolymerization of silicate melts. In neutral and basic melts, SiO_4^{4-} anions predominate as predicted by Masson's model, and hence there is little additional depolymerization with increasing temperature. In neutral and basic polymeric melts, therefore, the heat of activation for viscous flow will not change much with temperature. The effect of temperature on depolymerization is greater at higher silica contents because such melts are in a state of high level of polymerization. Consequently, the effect of temperature on the heat of activation for viscous flow is greater in high silica melts.

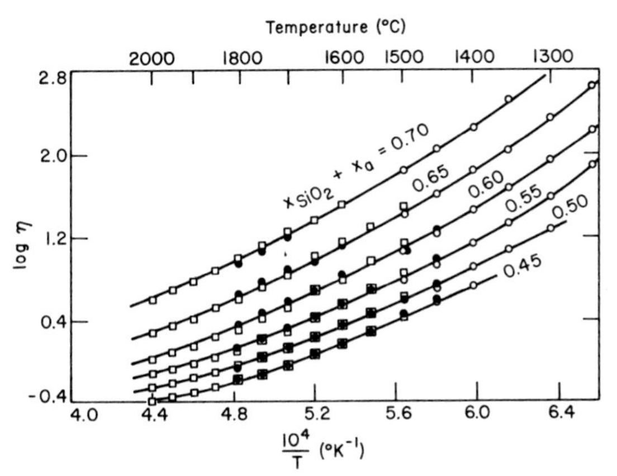

FIG. 4.32 Variation of viscosity with temperature and composition of aluminosilicate melts. After Turkdogan and Bills [96].

In viscosity measurements made by Bills [100] on complex slags, relations in Figs. 4.30 and 4.32 were found to hold well even for complex slags containing up to 10% BaO and 15% FeO. Therefore, with the relation in Fig. 4.30 for the silica equivalence of alumina and taking x_{MO} as the sum of the mole fractions of all the divalent basic oxides, the viscosity data may be represented in a condensed form as in Fig. 4.32 over wide ranges of temperature and composition.

Although earlier studies of Bockris *et al.* [95] have shown that molten silicates are Newtonian liquids, i.e., viscosity is independent of shear stress, some subsequent investigators claimed non-Newtonian behavior. To resolve this controversy, Michel and Mitchell [101] studied the rheological behavior of slags in the system CaO–SiO_2–Al_2O_3–CaF_2, using a viscometer that is capable of measuring shear stresses up to 10^4 dyn cm^{-2}, 3 orders of magnitude greater than those employed in previous studies. Over this wide range of shear stress, no effect was observed on the measured viscosity; that is, molten aluminosilicates with or without fluoride are Newtonian liquids.

Because the molecular configuration in glasses are metastable, the physical characteristics of glass such as density, refractive index, and viscosity change with the time of annealing at a given temperature. The lower the annealing temperature, the longer it takes to reach a pseudoequilibrium state. The temperature in the transformation region at which the structure is stabilized is called the *fictive temperature*. The physical characteristics of glasses should be determined at their fictive temperatures. For example, Hetherington *et al.* [102] have shown that the temperature dependence of vitreous silica, *I. R. Vitreosil*, *O. G. Vitreosil*, and *Spectrosil*, is much affected by the fictive temperature. At temperatures above and below the fictive temperature, the viscosity is higher and lower, respectively, of the equilibrium values. To obtain the true value, the viscosity measurement should be made at a temperature the same as the fictive temperature. Hetherington *et al.* have also shown that the viscosity of vitreous silica and the apparent heat of activation for viscous flow are lowered markedly with a slight increase in the hydroxyl content. For example, in I. R. Vitreosil containing 0.0003% (OH), $\eta_\circ$ is 3.1×10^{-13} P with E_v equal to 170 kcal, while in Spectrosil containing 0.12% (OH), $\eta_\circ$ is 1.9×10^{-7} P with E_v equal to 122 kcal.

Kozakevitch [103] and Bills [100] have shown that viscosities of molten aluminosilicates decrease with the addition of CaF_2. For example, at 1300 °C the viscosity of an aluminosilicate melt decreases from 39 to 16 P with an addition of 6% CaF_2; this lowering effect on viscosity decreases with increasing temperature. The breakdown of the Si–O–Si bond by the fluoride ion, manisfested by the decrease in viscosity, is consistent with the interpretation given to the effect of the fluoride ion on activities of basic oxides in polymeric melts and on their surface tension. However, the addition of CaS has no effect on the viscosity of aluminosilicate melts.

Viscosities of mattes are low and are of the same order of magnitude as those of liquid metals, indicating that there is little or no polymerization in mattes.

4.3.3 Slag Foam and Emulsion

A foam is a heterogeneous medium consisting of gas bubbles dispersed in a liquid. Similarly, an emulsion is dispersion of finely subdivided particles of a liquid in an immiscible liquid. The formation of foams and emulsions and their relative stabilities are strongly influenced by the surface tension and viscosity. Fast rate of gas injection or gas evolution, as in pneumatic metal refining processes, brings about slag foaming and dispersion of metal or matte droplets in the slag. The energy required to create foam or emulsion increases with increasing surface tension of the liquid.

The foam stability is determined by the rate of drainage of the liquid film between the bubbles; the lower the viscosity the faster the rate of drainage.

Consequently, liquids with high surface tension and low viscosity, such as metals and mattes, cannot foam. Compared to liquid metals, molten glasses and slags have relatively lower surface tensions and much higher viscosities, hence they are susceptible to foaming in processes involving extensive gas evolution. The foam is unstable; when the gas supply is terminated the foam collapses under the influence of interfacial tension (the Marangoni effect) and gravity force. The lower the interfacial tension and the higher the viscosity, the slower would be the rate of collapse of the foam.

Cooper and Kitchener [81] have studied the foaming characteristics of metallurgical slags. They found that the relative foam stability increases with decreasing temperature, decreasing CaO/SiO_2 ratio and increasing P_2O_5 content of the slag. Swisher and McCabe [104] observed similar effects of the additions of B_2O_3 and Cr_2O_3 to silicate melts. These changes in slag composition lower the surface tension and raise the viscosity; when the ratio σ/η is lowered by changing temperature and/or composition, the relative foam stability increases. It was stated in Section 4.2 that a decrease in surface tension with increasing silica content is a manifestation of adsorption of silica at the bubble surface as a monomolecular film of Si–O–Si anions which, in turn, increase the surface viscosity. Incorporation of other anion groups such as phosphates, borates, or chromates in the surface layer increases the viscoelasticity of the surface film, hence increases the foam stability.

Suspended solid matter in the liquid will also increase the relative stability of the foam by increasing the apparent viscosity of the medium and by retarding the coalescence of the bubbles on which the solid particles are attached by surface tension.

Conditions that favor foam stability also bring about emusification of droplets of a liquid of high surface tension and low viscosity in an immiscible liquid of low surface tension and high viscosity. Based on interfacial energy effects, Minto and Davenport [105] proposed three types of behavior when gas bubbles cross the interface separating two liquids such as liquid metal, or matte and slag.

(a) The bubble carries an intact film of metal to the slag layer when

$$\sigma_{s/g} - \sigma_{m/g} - \sigma_{m/s} > 0. \tag{4.49}$$

(b) The metal film around the bubble shutters, but some metal droplets become attached to bubbles on which they are floated when

$$\sigma_{s/g} - \sigma_{m/g} + \sigma_{m/s} > 0. \tag{4.50}$$

(c) The metal film shutters but droplets are not floated when

$$\sigma_{s/g} - \sigma_{m/g} + \sigma_{m/s} < 0. \tag{4.51}$$

The transport of one liquid to another via gas bubble crossing the interface has been demonstrated by Davenport *et al.* [106] with photographic studies of bubbles of nitrogen passing from mercury to an acidified water. As is seen from

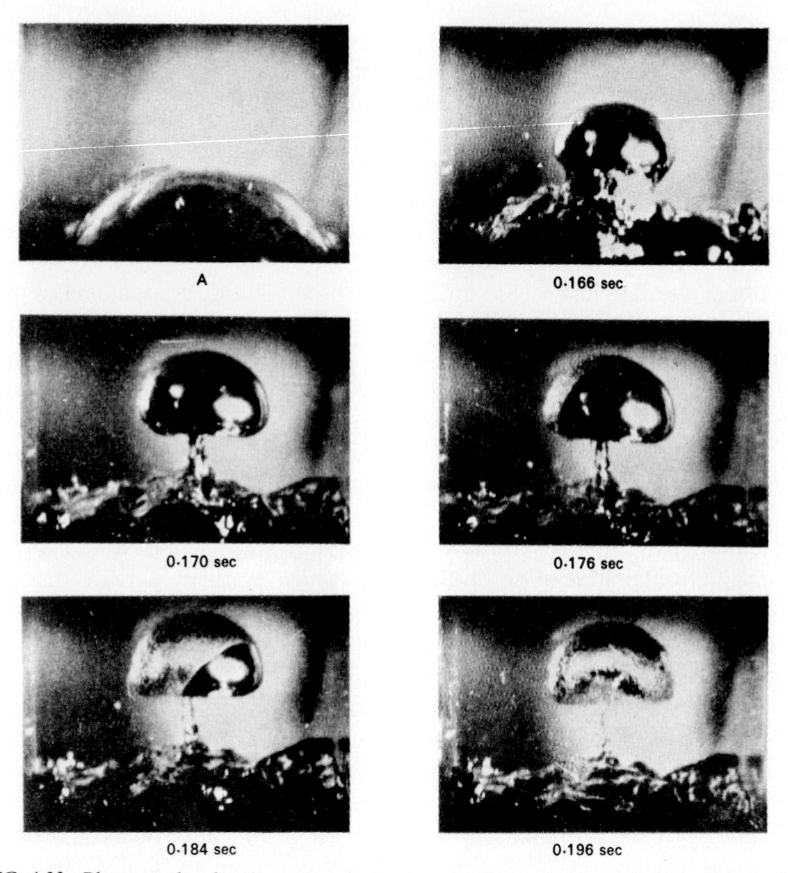

FIG. 4.33 Photographs showing stages in the transit of a sperical cap bubble 5 cm in basal diameter from mercury to acidified water. After Davenport *et al.* [106].

a series of photographs in Fig. 4.33, initially a large spherical dome forms from which a mercury coated bubble merges. Ultimately, the mercury coating ruptures and quickly drains away. The effect of viscosity on metal emulsion has been demonstrated by Poggi *et al.* [107] by bubbling argon through (a) mercury covered with a layer of water containing glycerin, (b) molten lead covered with a fused salt, and (c) molten copper covered with a slag. The amount of metal dispersed in the overlying liquid was found to increase greatly with increasing viscosity of the upper liquid phase and with increasing gas flow rate.

As demonstrated by Turkdogan [108], the impact of a gas jet on the overlying liquid may also bring about emulsification of the underlying liquid. The schematic drawings next to the photographs in Fig. 4.34 illustrate the observed liquid motion imparted by a gas jet impinging on the surface of oil overlaid on water. A negative pressure brought about by the motion of the lower liquid causes the formation of a dome at the two-liquid interface below the depression in the overlying liquid. On further increase in jet momentum, the depression and the

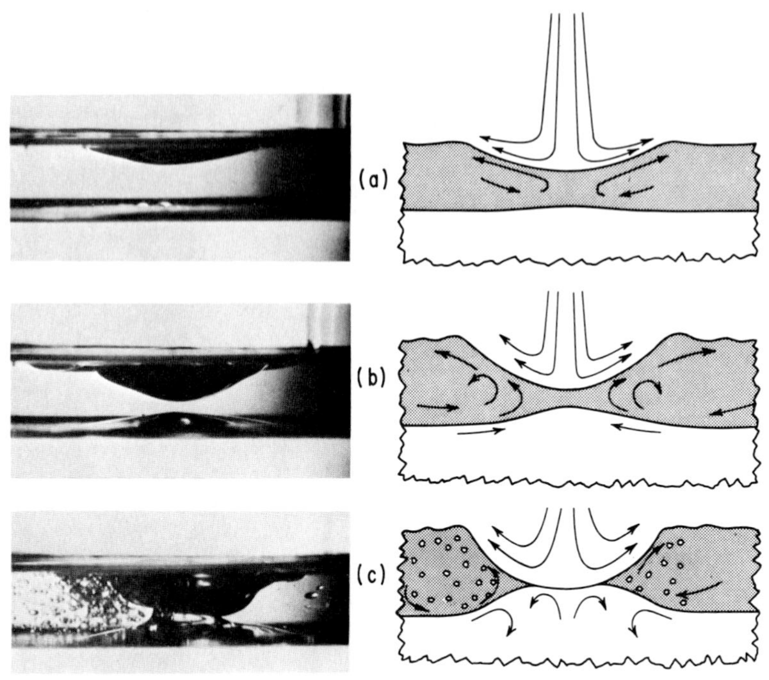

FIG. 4.34 Photographs and drawings demonstrating the dome formation at the oil–water interface beneath the depression in the oil layer caused by impinging argon jet stream; photograph (c) shows the beginning of water emulsification in oil when the water surface is exposed to the jet stream. After Turkdogan [108].

dome meet, and, subsequently, the water surface is exposed to the central part of the jet stream. This situation immediately brings about water emulsification in the oil layer. Beyond a certain critical jet momentum, splashing occurs and gas bubbles become entrapped in the oil–water emulsion. The higher the surface tension of the liquid, the higher is the impact pressure of the gas jet for the onset of splashing [108].

From experiments of gas jets impinging on the surface of liquids, Chatterjee and Bradshaw [109] obtained the following correlation for the onset of splashing in terms of the critical depth of depression $\bar{n}_c$, density ρ, surface tension σ, and viscosity η of the liquid:

$$\bar{n}_c(g\rho/\sigma)^{1/2} = 0.53\log(g\eta^4/\rho\sigma^3) + 11.33. \tag{4.52}$$

4.3.4 Diffusion

The diffusive transport in ionic media is approximated by the Nernst–Planck equation [110],

$$J_i = -D_i{}^*\left(\frac{\partial c_i}{\partial z} + Z_i c_i \frac{\mathscr{F}}{RT}\frac{\partial \phi}{\partial z}\right), \tag{4.53}$$

where the subscript i indicates the diffusion species, D_i^* is the self-diffusivity, c_i the concentration, Z_i the valence, $\partial\phi/\partial z$ the electrostatic potential gradient, and $\mathcal{F}$ the Faraday constant. In this simplified basic equation, Onsager cross effects and small variations in activity coefficients with small changes in composition are omitted. Cations are the primary mobile species in polymeric melts. For the limiting case of D_a^* (anion) $\ll D_i^*$ (cation), Okongwu *et al.* [111] derived the following set of equations for fluxes J_1 and J_2 for an anionic system of two mobile species 1 and 2 in a fixed solvent 3:

$$J_1 = -D_{11}\frac{\partial c_1}{\partial z} - D_{12}\frac{\partial c_2}{\partial z}, \qquad J_2 = -D_{21}\frac{\partial c_1}{\partial z} - D_{22}\frac{\partial c_2}{\partial z}, \tag{4.54}$$

where

$$D_{11} = \frac{D_1^*D_2^*Z_2^2c_2}{D_1^*Z_1^2c_1 + D_2^*Z_2^2c_2}, \qquad D_{12} = -\frac{Z_2}{Z_1}D_{22},$$
$$D_{22} = \frac{D_1^*D_2^*Z_1^2c_1}{D_1^*Z_1^2c_1 + D_2^*Z_2^2c_2}, \qquad D_{21} = -\frac{Z_1}{Z_2}D_{11}. \tag{4.55}$$

For uniform concentration of essentially immobile anion, these equations are simplified to the form of Fick's first law

$$J_1 = -\tilde{D}\,\partial c_1/\partial z \qquad \text{and} \qquad J_2 = -\tilde{D}\,\partial c_2/\partial z, \tag{4.56}$$

where $\tilde{D} = D_{11} + D_{22}$. For cations of mole fractions x_1 and x_2 with the same valency, the chemical diffusivity is simplified to

$$\tilde{D} = D_1^*D_2^*/(x_1D_1^* + x_2D_2^*). \tag{4.57}$$

For low concentrations of species 1, i.e., $x_1 \ll x_2 \to 1$, the chemical diffusivity approaches the value of the self-diffusivity, $\tilde{D} \simeq D_1^*$.

These equations are well substantiated by diffusion experiments in several ternary glasses [111]. Reference may be made to a comprehensive compilation of data by Frishchat [112] on the self-diffusivity and chemical diffusivity in glasses. For example, in soda glasses the pre-exponential factor D_o is in the range 10^{-3}–10^{-2} cm^2 sec^{-1}, and the apparent heat of activation for diffusive flux E_d is 15–25 kcal.

Self-diffusivities have been measured in molten calcium aluminosilicates, containing 10–20% Al_2O_3 and $CaO/SiO_2 \simeq 1$, using isotope tracer techniques. Measured self-diffusivities at 1450 °C in units of square centimeters per second are in the following decreasing order: ^{17}O [113]: 7.0×10^{-6}, ^{45}Ca [114]: 1.5×10^{-6}, ^{35}S [115]: 8.5×10^{-7}, ^{26}Al [116]: 4.5×10^{-7}, ^{31}Si [113]: 1.5×10^{-7}. The heat of activation for self-diffusion in these melts is in the range 60–80 kcal. In contrast to these aluminosilicate melts, the self-diffusivity of oxygen in silica glasses is many orders of magnitude lower than those of the sodium and calcium; e.g., at 1200 °C, D^* for ^{17}O is about 10^{-12} cm^2 sec^{-1}, while for ^{35}Ca about 10^{-6} cm^2 sec^{-1} [112].

Nowak and Schwerdtfeger [117] have measured by a galvanostatic technique the mobilities of Fe^{2+}, Co^{2+}, Ni^{2+}, and Ca^{2+} cations in silica-saturated MO–CaO–SiO_2 melts. For Ca^{2+} ions the diffusivity is 3.5×10^{-6} cm^2 sec^{-1} at 1600 °C; for the other three cations of the transition elements the diffusivity is 2.6×10^{-5} cm^2 sec^{-1}, which is close to the self-diffusivity of iron, 1.3×10^{-5} cm^2 sec^{-1}, in liquid iron and calcium orthosilicates at 1600 °C [118].

Mori and Suzuki [119] measured the iron–oxygen interdiffusivity in liquid iron oxide and iron and calcium orthosilicates. On the assumption that $D_{Fe}^* \gg D_O^*$ the self-diffusivity of iron may be calculated from the interdiffusivity and the activity data, $\partial \ln a_{FeO}/\partial \ln x_{FeO}$, by using the Darken equation (3.148). The self-diffusivity of Fe^{2+} thus derived is in close agreement with the self-diffusivity measured directly by Agarwall and Gaskell [118].

We see from these data that the self-diffusivity of iron is essentially independent of the composition of silicate melts, suggesting that the degree of polymerization of anion groups has little or no effect on the self-diffusion of iron cations. Similarly, the self-diffusivity of sodium, 6×10^{-5} cm^2 sec^{-1} at 1242 °C in molten sodium silicophosphates [120] does not change much with composition.

Self-diffusivities of molten mattes are in the range $2–10 \times 10^{-5}$ cm^2 sec^{-1}. For example, in a 50:50 mixture by weight of FeS and Cu_2S at 1160 °C, the self-diffusivity of iron is 2.9×10^{-5} cm^2 sec^{-1} and of copper 5.5×10^{-5} cm^2 sec^{-1} [121].

4.3.5 Electrical Conductivity

In polymeric melts with cations of fixed valency the electrical conduction is by cations only; that is, the transport numbers of the cations are unity. In alkali silicates the electrical conductivity is in the range 0.7–8 Ω^{-1} cm^{-1} and increases with increasing ratio MO/SiO_2 [122]. The conductivity is lower in alkaline earth silicates, 0.04–1.0 Ω^{-1} cm^{-1}, again increases with increasing ratio MO/SiO_2. For a given ratio MO/SiO_2, the energy of activation for ionic migration increases and the electrical conductivity decreases with increasing ionic potential Z/r.

In iron and cobalt silicates, the transport number for ionic conduction decreases with increasing ratio M^{3+}/M^{2+}. For example, in iron silicate of composition $19FeO \cdot SiO_2$ at 1400 °C, the conductivity is 125 Ω^{-1} cm^{-1}, which is due primarily to electronic conduction, about 10% of the conduction being ionic [123]. The conductivity decreases to 5 Ω^{-1} cm^{-1} for the composition $2FeO \cdot SiO_2$ for which about 90% of the conduction is ionic.

As discussed in Section 3.5.1, the transport of momentum (viscous flow) and the transport of matter (diffusion) in liquid metals involves the same structural entities (atoms), and hence the interrelation between the self-diffusivity and viscosity is described well by the Stokes–Einstein equation (3.130). In polymeric melts the viscous flow involves anion units, while the cations are the mobile species in diffusion; therefore, the self-diffusivity of free ion is not related to viscosity. In ionic melts involving similar mechanisms of the transport process of diffusion and conduction, the relation between ionic transport by chemical potential gradient and that due to electrical potential gradient is given by the Nernst–Einstein equation

$$\Lambda = \frac{\mathcal{F}^2}{RT} \sum_i Z_i D_i^*, \tag{4.58}$$

where Λ is the equivalent electrical conductance of the ionic melt, in units of $\Omega^{-1}\,\text{cm}^2\,\text{gew}^{-1}$.

The Nernst–Einstein equation is valid only when the same charged particles are involved in diffusion and conduction. However, experience with salts has shown that the self-diffusivity calculated from the measured electrical conductivity, by using Eq. (4.58), is lower than the measured diffusivity. This discrepancy was explained by Borucka *et al.* [124] by assuming that different entities are involved in these two transport processes. Conductive transport depends only on the charged particles, while in the diffusive flux charged species as well as united neutral species ($M^{n+}X^{n-}$) may participate. Contrary to the case with salts, in silicate melts the self-diffusivity of cations calculated from the conductivity is much higher than those measured by the tracer technique [114, 118]. This discrepancy cannot be explained, of course, by the above argument involving united neutral species. It is an exception rather than a rule that the Nernst–Einstein equation may hold for polymeric melts.

The electrical conductivity of glasses varies over a wide range 10^{-12}–10^{-2} $\Omega^{-1}\,\text{cm}^{-1}$, depending on composition and temperature. As in the case of polymeric melts, the higher the Z/r ratio of cations in the glass, the lower is the conductivity and the higher the energy of activation for ionic migration.

Much like metals, mattes are electronic conductors. For example, in Cu_2S–FeS mattes the specific conductivity at 1300 °C increases from 100 $\Omega^{-1}\,\text{cm}^{-1}$ in 100% Cu_2S to 930 $\Omega^{-1}\,\text{cm}^{-1}$ in a mixture of 25% Cu_2S and 75% FeS [125].

4.3.6 Thermal Conductivity

Because most glasses and polymeric melts are partially transparent to infrared radiation, there is a radiative component (photon) of thermal conductivity, in addition to the lattice (phonon) and electronic conductivity. Relative contributions of these heat conduction processes depend on temperature and composition of the medium. At moderate temperatures, phonon contribution to heat conduction predominates. Photon contribution becomes significant at elevated temperatures; hence the thermal conductivity increases with increasing temperature.

If a material is completely transparent to infrared radiation, as for a thin plate of glass, the total heat flux through the plate is the sum of that transported by phonons and electrons plus that radiated through without interaction with the material. For a thick block of glass, the transmitted intensity of the radiation is negligible. However, absorption of radiation in layers close to the high temperature source raises the local temperature, and consequently, increases the intensity of radiation from these layers. As this process of absorption and reradiation progresses through the successive layers, there develops an additional flux of heat through the material.

The overall, or effective, thermal conductivity is the sum of thermal conductivities due to electronic κ_e, phonon κ_p, and photon κ_f, heat conduction:

$$\kappa = \kappa_e + \kappa_p + \kappa_f. \tag{4.59}$$

Unidirectional heat flux in an isotropic medium under steady state conditions is represented by the Fourier equation,

$$J_{\mathrm{H}} = \kappa\, \partial T/\partial z. \tag{4.60}$$

When the thermal properties are independent of temperature, a simplifying assumption for small increments of temperature, the time dependence of temperature is given by the second differential

$$\partial T/\partial t = (\kappa/\rho C_p)\, d^2 T/dz^2, \tag{4.61}$$

where ρ is the density and C_p the specific heat at constant pressure. Thomson (Lord Kelvin) introduced the term thermal diffusivity, defined by the ratio

$$\alpha = \kappa/\rho C_p. \tag{4.62}$$

Since the thermal diffusivity has the dimensions $(\text{length})^2 \times (\text{time})^{-1}$, Eq. (4.61) is an analog of Fick's second law of mass diffusion. It should be noted, however, that the diffusion equation (4.61) for radiation heat transfer is an acceptable approximation only for optically thick heat transfer media. The medium is said to be optically thick when the mean free path of a photon within the medium is small compared to the dimensions of the medium. This depends on the absorption spectrum of the additives in the glass or polymeric melts. For example, in metallurgical slags, FeO is the strongest absorber [45]; the higher the concentration of iron oxide the higher the absorption coefficient and the lower the radiation conductivity.

Thermal diffusivities of solid and molten igneous rocks [126–128], coal ash [129], and metallurgical slags [130, 131] at temperature of 1000–1400 °C are in the range $3\text{–}8 \times 10^{-3}\ \text{cm}^2\ \text{sec}^{-1}$. Because of the complexity of heat transfer processes in glasses and polymeric melts, and the complexity of their structure, it is not possible to derive even an empirical relation for thermal conductivity or thermal diffusivity as a function of temperature or composition or as a function of other transport properties.

References

1. R. Schuhmann and P. J. Ensio, *Trans. AIME* **191**, 401 (1951).
2. E. T. Turkdogan, *Trans. Metall. Soc. AIME* **224**, 294 (1962).
3. K. P. Abraham, M. W. Davies, and F. D. Richardson, *J. Iron Steel Inst., London* **196**, 82 (1960).

4. R. A. Sharma and F. D. Richardson, *J. Iron Steel Inst., London* **200**, 373 (1962).
5. D. A. R. Kay and J. Taylor, *Trans. Faraday Soc.* **56**, 1372 (1960).
6. S. R. Mehta and F. D. Richardson, *J. Iron Steel Inst., London* **203**, 524 (1965).
7. Z. Kozuka and C. S. Samis, *Metall. Trans.* **1**, 871 (1970).
8. M. L. Kapoor and M. G. Frohberg, *Arch. Eisenhuettenwes.* **42**, 5 (1971).
9. Z. Kozuka, O. P. Siahaan, and J. Moriyama, *Trans. Jpn. Inst. Met.* **9**, 200 (1968).
10. J. F. Elliott, M. Gleiser, and V. Ramakrishna, "Thermochemistry for Steelmaking," Vol. 2. Addison-Wesley, Reading, Massachusetts, 1963.
11. F. D. Richardson, "Physical Chemistry of Melts in Metallurgy," Vol. 1. Academic Press, New York, 1974.
12. R. A. Sharma and F. D. Richardson, *J. Iron Steel Inst., London* **198**, 386 (1961).
13. C. R. Taylor and J. Chipman, *Trans. AIME* **154**, 228 (1943).
14. T.C. Pillay and F.D. Richardson, *Inst. Min. Metall., Trans.* **66**, 309 (1956–1957).
15. M. L. Kapoor and M. G. Frohberg, *Metall. Trans. B* **8**, 15 (1977).
16. R. H. Rein and J. Chipman, *Trans. Metall. Soc. AIME* **233**, 415 (1965).
17. W. H. Zachariasen, *J. Am. Chem. Soc.* **54**, 3841 (1932).
18. B. E. Warren, *J. Appl. Phys.* **8**, 645 (1937).
19. A. E. R. Westman and P. A. Gartaganis, *J. Am. Ceram. Soc.* **40**, 293 (1957).
20. T. R. Meadowcroft and F. D. Richardson, *Trans. Faraday Soc.* **61**, 54 (1965).
21. P. J. Flory, *J. Am. Chem. Soc.* **58**, 1877 (1936); **64**, 2205 (1942).
22. Y. Waseda and H. Suito, *Trans. Iron Steel Inst. Jpn.* **17**, 82 (1977).
23. Y. Waseda and J. M. Toguri, *Metall. Trans. B* **8**, 563 (1977).
24. G. W. Toop and C. S. Samis, *Trans. Metall. Soc. AIME* **224**, 878 (1962).
25. M. L. Kapoor and M. G. Frohberg, *Arch. Eisenhuettenwes.* **41**, 1035 (1970); *in* "Chemical Metallurgy of Iron and Steel," p. 17. Iron Steel Inst., London, 1973.
26. C. R. Masson, *Proc. R. Soc. London, Ser. A* **287**, 201 (1965); *in* "Chemical Metallurgy of Iron and Steel," p. 3. Iron Steel Inst., London, 1973.
27. M. Temkin, *Zh. Fiz. Khim.* **20**, 105 (1946).
28. U. V. Choudary, D. R. Gaskell, and G. R. Belton, *Metall. Trans. B* **8**, 67 (1977).
29. F. D. Richardson, *Trans. Faraday Soc.* **52**, 1312 (1956).
30. L. S. Darken and K. Schwerdtfeger, *Trans. Metall. Soc. AIME* **236**, 208 (1966).
31. H. B. Bell, *J. Iron Steel Inst., London* **201**, 116 (1963).
32. K. S. Song and D. R. Gaskell, *Metall. Trans. B* **10**, 15 (1979).
33. M. W. Davies, *in* "Chemical Metallurgy of Iron and Steel," p. 43. Iron Steel Inst., London, 1973.
34. A. E. Grau, W. F. Caley, and C. R. Masson, *Can. Metall. Q.* **15**, 267 (1976).
35. H. Suito and D. R. Gaskell, *Metall. Trans. B* **7**, 559, 567 (1976).
36. D. Kumar, R. G. Ward, and D. J. Williams, *Trans. Faraday Soc.* **61**, 1850 (1965).
37. F. D. Richardson and J. C. Billington, *Inst. Min. Metall., Trans.* **65**, 273 (1955–1956).
38. H. W. Meyer and F. D. Richardson, *Inst. Min. Metall., Trans.* **71**, 201 (1961–1962).
39. E. J. Grimsey and A. K. Biswas, *Inst. Min. Metall., Trans., Sect. C* **85**, 200 (1976); **86**, 1 (1977).
40. M. Nagamori, P. J. Mackey, and P. Tarassoff, *Metall. Trans. B* **6**, 295 (1975).
41. E. T. Turkdogan and P. M. Bills, *in* "Physical Chemistry of Process Metallurgy" (G. R. St. Pierre, ed.), Vol. 1 p. 207. Wiley (Interscience), New York, 1961.
42. E. T. Turkdogan, *Trans. Metall. Soc. AIME* **221**, 1090 (1961); **224**, 294 (1962).
43. F. Irmann, *J. Am. Chem. Soc.* **74**, 4767 (1952).
44. W. D. Johnston, *J. Am. Ceram. Soc.* **48**, 184 (1965).
45. W. A. Weyl, "Coloured Glasses." Dawsons of Pall Mall, London, 1959.
46. C. J. B. Fincham and F. D. Richardson, *Proc. R. Soc. London, Ser. A* **223**, 40 (1954).
47. G. R. St. Pierre and J. Chipman, *Trans. AIME* **206**, 1474 (1956).
48. E. T. Turkdogan and L. S. Darken, *Trans. Metall. Soc. AIME* **221**, 464 (1961).
49. E. T. Turkdogan and P. Grieveson, *Trans. Metall. Soc. AIME* **224**, 316 (1962).
50. P. Grieveson and E. T. Turkdogan, *Trans. Metall. Soc. AIME* **224**, 1086 (1962).

51. G. J. W. Kor and F. D. Richardson, *Trans. Metall. Soc. AIME* **245**, 319 (1969).
52. J. W. Tomlinson, *J. Soc. Glass Technol.* **40**, 25T (1956).
53. L. E. Russell, *J. Soc. Glass Technol.* **41**, 304T (1957).
54. Y. Iguchi and T. Fuwa, *Trans. Iron Steel Inst. Jpn.* **10**, 29 (1970).
55. C. R. Kurkjian and L. E. Russell, *J. Soc. Glass Technol.* **42**, 130T (1958).
56. Y. Iguchi, S. Ban-ya, and T. Fuwa, *in* "Chemical Metallurgy of Iron and Steel," p. 28. Iron Steel Inst., London, 1973.
57. J. M. Uys and T. B. King, *Trans. Metall. Soc. AIME* **227**, 492 (1963).
58. G. Hetherington and K. H. Jack, *Phys. Chem. Glasses* **3**, 129 (1962).
59. K. Schwerdtfeger and H. G. Schubert, *Metall. Trans. B* **9**, 143 (1978).
60. H.-O. Mulfinger and H. Meyer, *Glastech. Ber.* **36**, 481 (1963).
61. K. Schwerdtfeger and G. Schubert, *Metall. Trans. B* **8**, 535 (1977).
62. M. W. Davies and S. G. Meherali, *Metall. Trans.* **2**, 2729 (1971).
63. E. A. Dancy and D. Janssen, *Can. Metall. Q.* **15**, 103 (1976).
64. K. Schwerdtfeger and G. Schubert, *Metall. Trans. B* **8**, 689 (1977).
65. M. L. Pearce, *J. Am. Ceram. Soc.* **47**, 342 (1964); **48**, 175 (1965).
66. L. L. Cheng and C. B. Alcock, *Trans. Metall. Soc. AIME* **221**, 295 (1961).
67. C. W. Bale and J. M. Toguri, *Can. Metall. Q.* **15**, 305 (1976).
68. T. Yagihashi, *Nippon Kinzoku Gakkaishi* **17**, 483 (1953).
69. C. B. Alcock and L. L. Cheng, *J. Iron Steel Inst., London* **195**, 169 (1960).
70. G. A. Meyer, J. S. Warner, Y. K. Rao, and H. H. Kellogg, *Metall. Trans. B* **6**, 229 (1975).
71. W. Burgmann, G. Urbain, and M. G. Frohberg, *Rev. Metall. (Paris)* 65, 567 (1968).
72. W. A. Krivsky and R. Schuhmann, Jr., *Trans. AIME* **209**, 981 (1957).
73. J. W. Matousek and C. S. Samis, *Trans. Metall. Soc. AIME* **227**, 980 (1963).
74. H. H. Kellogg, *in* "Physical Chemistry in Metallurgy—Darken Conference" (R. M. Fisher, R. A. Oriani, and E. T. Turkdogan, eds.), p. 49. U.S. Steel Corp., Monroeville, Pennsylvania, 1976.
75. M. Nagamori and T. R. Ingraham, *Metall. Trans.* **1**, 1821 (1970).
76. J. Schmiedl, V. Repcak, and S. Cempa, *Inst. Min. Metall., Trans., Sect. C* **86**, 88 (1977).
77. D. C. Hilty and W. Crafts, *Trans. AIME* **194**, 1307 (1952).
78. S. Bog and T. Rosenqvist, *Proc. Natl. Phys. Lab. Symp.* No. 9, p. 6B (1958).
79. T. B. King, *J. R. Tech. Coll. Glasgow* **5**, 217 (1950); *J. Soc. Glass Technol.* **35**, 241 (1951).
80. P. Kozakevitch, *Rev. Metall. (Paris)* **46**, 505, 572 (1949).
81. C. F. Cooper and J. A. Kitchener, *J. Iron Steel Inst., London* **193**, 48 (1959).
82. R. E. Boni and G. Derge, *Trans. AIME* **206**, 59 (1956).
83. A. Ejima and M. Shimoji, *Trans. Faraday Soc.* **66**, 99 (1970).
84. A. Dietzel, *Z. Elekrochem.* **48**, 9 (1942).
84a. V. Y. Zaytsev, A. V. Vanyukov, and V. S. Kolosova, *Izv. Akad. Nauk SSSR Met., Engl. Trans.* **5**, 29 (1968).
85. A. Adachi, K. Ogino, and T. Suetaki, *Technol. Rep. Osaka. Univ.* **14**, 713 (1964).
86. P. Kozakevitch, *in* "Liquids: Structure, Properties, Solid Interactions," p. 243. Elsevier Amsterdam, 1965.
87. P. Kozakevitch, G. Urbain, and M. Sage, *Rev. Metall.* (Paris) **52**, 161 (1955)..
88. G. W. Morey, "The Properties of Glass," 2nd Ed. Reinhold, New York, 1954.
89. G. W. Morey and H. E. Merwin, *J. Opt. Soc. Am.* **22**, 632 (1932).
90. D. R. Gaskell, A. McLean, and R. G. Ward, *Trans. Faraday Soc.* **65** 1498 (1969).
91. Y. E. Lee and D. R. Gaskell, *Metall, Trans.* **5**, 853 (1974).
92. A. E. Grau and C. R. Masson, *Can. Metall. Q.* **15**, 367 (1976).
93. A. Mitchell and S. Joshi, *Metall. Trans.* **3**, 2306 (1972).
94. M. Nagamori, *Trans. Metall. Soc. AIME* **245**, 1897 (1969).
95. J. O'M. Bockris, J. D. Mackenzie, and J. A. Kitchener, *Trans. Faraday Soc.* **51**, 1734 (1955).
96. E. T. Turkdogan and P. M. Bills, *Am. Ceram. Soc., Bull.* **39**, 682 (1960).
97. J. O'M. Bockris and D. C. Lowe, *Proc. R. Soc. London, Ser. A* **226**, 423 (1954).

98. P. Kozakevitch, *in* "Physical Chemistry of Process Metallurgy" (G. R. St. Pierre, ed.), p. 97. Wiley (Interscience), New York, 1961.
99. J. S. Machin, T. B. Yee, and D. L. Hanna, *J. Am. Ceram. Soc.* **31**, 200 (1948); **35**, 322 (1952).
100. P. M. Bills, *J. Iron Steel Inst., London* **201**, 133 (1963).
101. J. R. Michel and A. Mitchell, *Can. Metall. Q.* **14**, 153 (1975).
102. G. Hetherington, K. H. Jack, and J. C. Kennedy, *Phys. Chem. Glasses* **5**, 130 (1964).
103. P. Kozakevitch, *Rev. Metall. (Paris)* **51**, 569 (1954).
104. J. H. Swisher and C. L. McCabe, *Trans. Metall. Soc. AIME* **230**, 1669 (1964).
105. R. Minto and W. G. Davenport, *Inst. Min. Metall., Trans., Sect. C* **81**, 36 (1972).
106. W. G. Davenport, A. V. Bradshaw, and F. D. Richardson, *J. Iron Steel Inst., London* **205**, 1034 (1967).
107. D. Poggi, R. Minto, and W. G. Davenport, *J. Met.* **21**, 40 (1969).
108. E. T. Turkdogan, *Chem. Eng. Sci.* **21**, 1133 (1966).
109. A. Chatterjee and A. V. Bradshaw, *J. Iron Steel Inst., London* **210**, 179 (1972).
110. I. Prigogine, S. R. DeGroot, and P. Mazur, *J. Chem. Phys.* **50**, 146 (1953).
111. D. A. Okongwu, W.-K. Lu, A. E. Hamielec, and J. S. Kirkaldy, *J. Chem. Phys.* **58**, 777 (1973).
112. G. H. Frischat, "Ionic Diffusion in Oxide Glasses," Trans. Tech. Publ. Aedermannsdorf, No. 3/4 (1975).
113. T. B. King and P. Koros, *Trans. Metall. Soc. AIME* **224**, 299 (1962).
114. H. Towers and J. Chipman, *Trans. AIME* **209**, 769 (1957).
115. T. Saito and Y. Kawai, *Sci. Rep. Res. Inst., Tohuku Univ., Ser. A* **5**, 460 (1953).
116. J. Henderson, L. Yang, and G. Derge, *Trans. Metall. Soc. AIME* **221**, 56 (1961).
117. N. Nowak and K. Schwerdtfeger, *in* "Metal–Slag–Gas Reactions and Processes" (Z. A. Foroulis and W. W. Smeltzer, eds.), p. 98. Electrochem. Soc., Princeton, New Jersey, 1975.
118. D. P. Agarwall and D. R. Gaskell, *Metall. Trans. B* **6**, 263 (1975).
119. K. Mori and K. Suzuki, *Trans. Iron Steel Inst. Jpn.* **9**, 409 (1969).
120. P. O. Perron and H. B. Bell, *Trans. Br. Ceram. Soc.* **66**, 347 (1967).
121. L. Yang, S. Kado, and G. Derge, *in* "Physical Chemistry of Process Metallurgy" (G. R. St. Pierre, ed.), p. 535. Wiley (Interscience), New York, 1961.
122. J. O'M. Bockris, J. A. Kitchener, S. Ignatowicz, and J. W. Tomlinson, *Trans. Faraday Soc.* **48**, 75 (1952).
123. J. D. Mackenzie, *Inorg. Chem. Radiochem.* **4**, 293 (1962).
124. A. Z. Borucka, J. O'M. Bockris, and J. A. Kitchener, *J. Chem. Phys.* **54**, 1282 (1956); *Proc. R. Soc. London, Ser. A* **241**, 554 (1957).
125. G. M. Pound, G. Derge, and G. Osuch, *Trans. AIME* **203**, 481 (1955).
126. T. Murase and A. McBirney, *Science* **170**, 165 (1970).
127. R. L. Gibby and J. L. Bates, *Therm. Conductiv. Conf., 10th* p. IV 7. Dynatech R/D Co., Cambridge, Massachusetts (1970).
128. K. Kawada, *Bull. Earthquake Res. Inst., Tokyo Univ.* **44**, 1071 (1966).
129. J. L. Bates, *Int. Conf. Magnetohydrodyn. Electr. Power Generat.* (J. E. Klepels, ed.), Vol. II, p. 163. Natl. Tech. Inf. Serv., Springfield, Virginia (1975).
130. J. Nauman, G. Foo, and J. F. Elliott, *Int. Symp. Copper Extract. Refin.* p. 237. AIME, New York (1976).
131. H. A. Fine, T. Engh, and J. F. Elliott, *Metall. Trans. B* **7**, 277 (1976).

CHAPTER 5

PHYSICOCHEMICAL PROPERTIES OF MOLTEN SALTS

5.1 Structure and Thermodynamic Properties

Molten salts are used in many areas of high temperature technology, for example, as electrolytes in electrochemical processes, heat transfer media in nuclear reactors, and molten solvents for fluorides of fissile materials (UF_4, PuF_3) for breeder reactors. Extensive knowledge accumulated on the physical chemistry of molten salts arising from the research of the past three or four decades cannot be covered adequately in this chapter. The present discussion is therefore confined to a brief presentation of certain properties of molten salts with a list of references to text books and review papers for further reading on the subject.

5.1.1 Nature of Molten Salts

Molten alkali halides are the simplest form of dissociated ionic melts where cations are surrounded by anions and vice versa, as evidenced from the radial distribution functions obtained by x-ray and neutron diffraction studies of Levy *et al.* [1]. They found that the interionic distance M^+–X^- is about 0.2 Å shorter than that in the solid at the melting temperature with a coordination number of 4–5, as compared to 6 in the solid halide. On the other hand, the average distance between the like ions, M^+–M^+, X^-–X^-, is about 0.2 Å longer than that in the solid with a coordination number between 7 and 12, as compared with 12 in the solid. Although there is little change in the interionic distances, most salts expand upon melting, usually 10%–30% volume expansion. These findings indicate that molten salts have open structures with holes, vacancies, or free volume.

One of the early fundamental theoretical developments was the concept of lattice energy of ionic crystals derived from simple theoretical considerations of the potential energy of interaction between ions. The lattice energy is a free energy change accompanying the formation of one mole of the lattice from free gaseous ions at infinite distances apart. In the theory developed by Born, Haber, Landé, Madelung, and others (references to early work are given in a review paper by Sherman [2]), the major contributions to the lattice energy are considered to arise from the Coulomb force of interaction, the potential of which is inversely proportional to the interionic distance, and the repulsive force of overlapping electron clouds. In subsequent developments of the theory of lattice energy, other ion interactions have been included; reference may be made to a review paper by Stillinger [3] for detailed discussion of potential energy interactions.

Experience has shown that the theory in its simplified form gives an adequate description of the lattice energy E_L represented by the equation

$$E_L = -AZ_cZ_a(Ne^2/\delta)(1 - 1/n), \tag{5.1}$$

where A is Madelung's constant for a given structure, Z is the valency of cation (c) and anion (a), N is Avogadro's number, e is the electronic charge ($Ne^2 =$ 329.7 kcal cm mole^{-1}), δ is the interionic distance M^+–X^- when the potential energy is a minimum, and n is the Born exponent, which is close to 9 for all crystals.

Accepted values of Madelung's constant are given for various lattice structures:

Lattice	Madelung's constant	Lattice	Madelung's constant
NaCl	1.748	CaF_2	5.038
CsCl	1.763	Cu_2O	4.432
ZnS	6.522		

The so-called *Born–Haber cycle* gives the lattice energy of the opposite sign, that is the free energy of decomposition of the halide to gaseous ionic species at 0 °K, denoted by U_0

$$MX(s) = M^+(g) + X^-(g), \tag{5.2}$$

$$U_0 = -E_L = Q + S + I + \tfrac{1}{2}D - E, \tag{5.3}$$

where Q is the heat of formation of MX (s) from elements, S the heat of vaporization of metal, I the ionization energy of metal vapor, D the heat of dissociation of halogen molecule, and E the electron affinity of halogen atom. The direct measurements of U_0 have been shown to be consistent with those calculated from Eq. (5.1) [4].

For salts with divalent or trivalent cations or complex anions such as nitrates, carbonates, and sulfates, Eq. (5.1) does not hold because of covalent bonding and the formation of associated groups (complex ions, molecules) [4].

There is little difference, if any, in interionic distances determined by x-ray diffraction and those calculated from the Goldschmidt ionic radii of crystals for sixfold coordination. The interionic distance, taken as the sum of the Goldschmidt ionic radii of cation and anion, is shown in Fig. 5.1a as a function of the cube root of the molar volume of molten salts at their melting temperatures. The line drawn passes through the origin of the coordinates. There is a slight departure from direct proportionality at high values of δ; approximate direct proportionality may be represented by the equation

$$\delta = 0.85 \times 10^{-8} \; V^{1/3} \quad \text{cm.} \tag{5.4}$$

The dotted line is for the rigid-sphere model, Eq. (3.112), which was shown in Fig. 3.34 to hold well for liquid metals. Departure from the rigid-sphere model for salts is indicative of the presence of holes, vacancies, or free volume in the structure of molten salts, as previously noted from the large volume expansion upon melting.

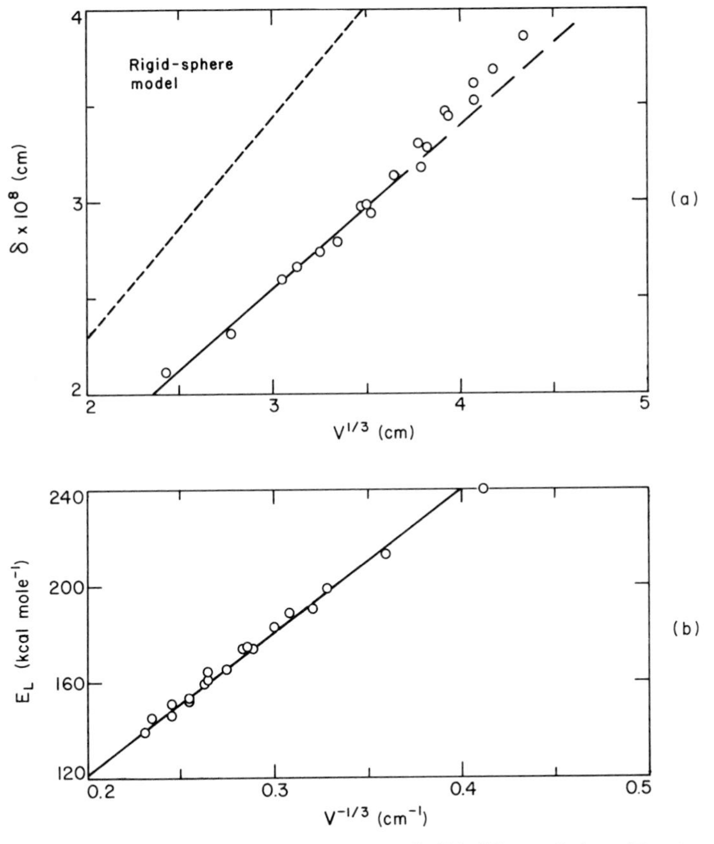

FIG. 5.1 Cube root of molar volumes of molten alkali halides at their melting temperatures related to (a) interionic distance and (b) lattice energy at 298 °K.

If a rigid-sphere model, or a cell structure, is assumed for dissociated molten salts, we must concede that some of the cells are unoccupied. As pointed out in Section 3.4, computer calculations using the method of *molecular dynamics* for the rigid-sphere model yield $\chi = 0.475$ for the fraction of the total volume of the liquid occupied by the rigid spheres. If this model is assumed for dissociated molten salts, we must consider two types of empty space in the liquid: (1) Each cell containing an ion (hard sphere) has a free volume fraction $1 - \chi = 0.525$ and (2) there is random distribution of unoccupied cells which may be considered as holes in the liquid. On the basis of this assumption, we would estimate from Eq. (3.112) and (5.2) an upper limit of about 59% unoccupied cells in dissociated molten salts. Various views on the concept of holes in molten salts have been discussed in several review papers [3, 6]. By considering the surface tension

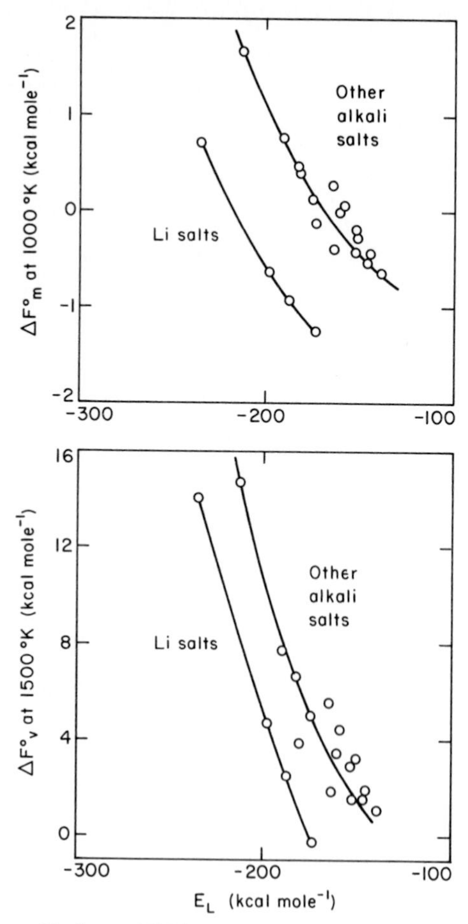

FIG. 5.2 Free energy of fusion at 1000 °K and vaporization at 1500 °K as a function of lattice energy at 298 °K.

effect, Fürth [7] estimated the averaged size of the hole to be of the same order of magnitude as the average size of ions.

As would be expected from the relations in Eqs. (5.1) and (5.4), the lattice energy is inversely proportional to the cube root of the molar volume of liquid halides at their melting temperatures; this is shown by the plot in Fig. 5.1b.

Variations of free energies of fusion of alkali halides at 1000 °K and vaporization at 1500 °K with the lattice energy are shown in Fig. 5.2. In both cases, the relation for lithium halides differs from the other alkali halides.

5.1.2 Halide Mixtures

The ideal mixing of monovalent molten salt solutions, as defined by Temkin [8], is based on the assumption that different types of cations A^+, B^+ are randomly distributed among the cations, and anions X^-, Y^- among the anions with zero enthalpy change, and no mixing of cations and anions. For this random mixing, the molar free energy of mixing of pure salts AX and BY is given by the equation

$$\begin{aligned}\Delta F^M &= -T\Delta S^M \\ &= -RT(x_A \ln N_A + x_B \ln N_B + x_X \ln N_X + x_Y \ln N_Y),\end{aligned} \tag{5.5}$$

where the x's are atom fractions of ionic species and the N's are the ion fractions,

$$N_A = \frac{x_A}{x_A + x_B}, \qquad N_B = \frac{x_B}{x_A + x_B}, \qquad N_X = \frac{x_X}{x_X + x_Y}, \qquad N_Y = \frac{x_Y}{x_X + x_Y}. \tag{5.6}$$

The relative partial molar free energy of solution of, for example, component AX (under isobaric and isothermal conditions) is given by the equation

$$\Delta \bar{F}_{AX} = \left(\frac{\partial\, \Delta F^M}{\partial x_{AX}}\right)_{x_B, x_Y} = RT(\ln N_A + \ln N_X) \tag{5.7}$$

and the activity is given by

$$a_{AX} = N_A N_X. \tag{5.8}$$

For nonideal solutions the activity coefficient is defined by the ratio

$$\gamma_{AX} = a_{AX}/N_A N_X. \tag{5.9}$$

The extension of Temkin rule to a component AX_v gives for the activity in ideal solutions, with the assumption of no vacancy in the lattice,

$$a_{AX_v} = N_A N_X{}^v. \tag{5.10}$$

Not many molten salt solutions are ideal. It has been found that the thermodynamics of these melts are better represented in terms of regular solution behavior or departures therefrom. For binary salt mixtures with common cations or anions, the activity coefficients γ_{AX}, γ_{BX} and the molar heat of solution ΔH^M for AX–BX mixtures obeying the regular solution law are

$$\begin{aligned} \ln \gamma_{AX} &= \beta N_B^2, \qquad \ln \gamma_{BX} = \beta N_A^2, \\ \Delta H^M &= RT\beta N_A N_B, \end{aligned} \tag{5.11}$$

where β is the interaction parameter and $RT\beta$ does not vary much with temperature. In mixtures with a common ion, the ion fractions N_A, N_B are the same as the mole fractions x_{AX}, x_{BX}.

We see from a detailed presentation by Lumsden [9] of the thermodynamics of molten salt solutions that most mixtures of monovalent halides with common cation or anion exhibit near regular solution behavior; that is, $\ln \gamma_{AX}$ is essentially a linear function of x_{BX}^2. At 1000 °K, the interaction parameter β for salt solutions with a common anion has values in the range from -2 to 0. Kleppa *et al.* [10, 11] measured the heats of solution of a number of salt mixtures with common cations and anions, and found that in many systems the ratio $\Delta H^M/x_{AX}x_{BX}$ does not change much with composition, indicating little or only modest departures from regular solution behavior.

The negative departure from ideal mixing, i.e., decrease in energy, is attributed by Forland [12] to the difference between the sizes of the cations. Mixing of cations of different sizes causes a decrease of coulombic energy; an additional decrease of energy is due to the polarization of the anion surrounded by cations of different sizes. The cations are less polarizable, and hence salt mixtures with common cations are closer to ideal solutions.

In mixtures with two different cations and two different anions, the so-called *reciprocal salt mixtures*, we may consider a reaction of the type

$$AX + BY = AY + BX, \tag{5.12}$$

indicating different types of cation–anion interactions: A^+–X^-, B^+–Y^-, A^+–Y^-, and B^+–X^-. Flood *et al.* [13] extended the formulation for regular solutions to reciprocal salt mixtures by introducing four ternary interaction parameters for next-nearest neighbor interaction.

$$\begin{aligned} &\beta_{XBY} \text{ for BX–BY}, \qquad \beta_{XAY} \text{ for AX–AY}, \\ &\beta_{AXB} \text{ for AX–BX}, \qquad \beta_{AYB} \text{ for AY–BY}. \end{aligned} \tag{5.13}$$

With the assumption of ideal configurational entropy of mixing of cations among cations and anions among anions, Flood *et al.* derived equations for the excess partial molar free energies of the components. For the component AX, the activity coefficient γ_{AX} is given by the equation

$$\ln \gamma_{AX} = N_{BY}^2[\Delta F^\circ/RT + \beta_{XBY} + \beta_{AYB} + 2N_{AX}(\beta_{YAX} + \beta_{AXB} - \beta_{XBY} - \beta_{AYB})], \tag{5.14}$$

where ΔF° is the standard free-energy change accompanying reaction (5.12).

Lumsden [9] calculated the values of $RT \ln \gamma$ for LiF and NaF from phase diagrams of the systems LiF–KCl and NaF–KCl and compared the results with those calculated from Eq. (5.14). As is seen from Lumsden's calculations in Fig. 5.3, agreement with Eq. (5.14) is good for the NaF–KCl system. The agreement is poor for the LiF–KCl system, indicating lack of random mixing of cations; that is, there is apparently preferential clustering of Li^+ around F^- and K^+ around Cl^-. In fact, extensive miscibility gaps exist in LiF–CsCl and LiF–CsBr systems.

Equations similar to (5.14) have been derived for mixtures of ions with different valencies and for mixtures of complex anions such as nitrates, carbonates,

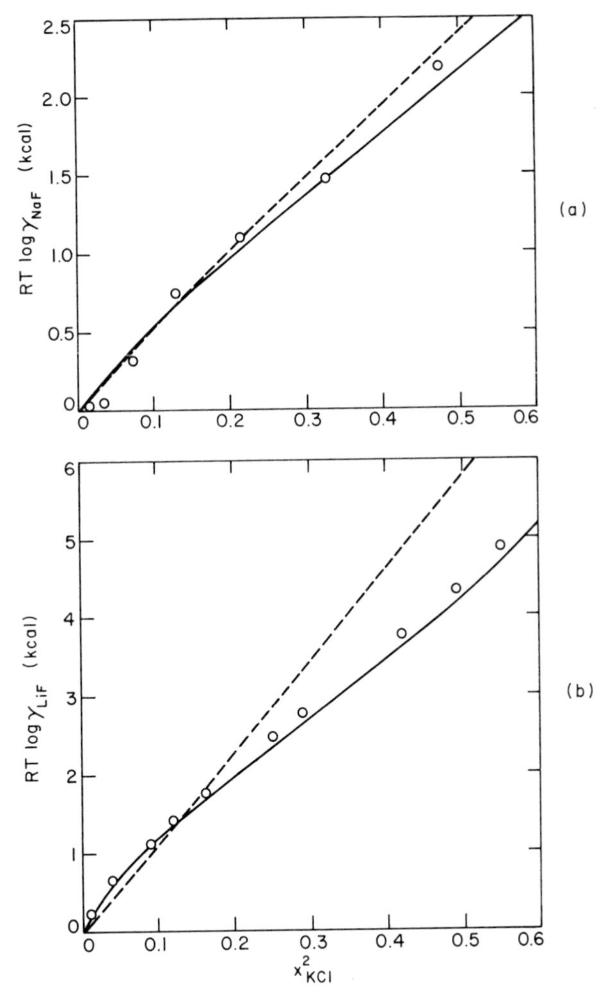

FIG. 5.3 Test of regular solution behavior in (a) NaF–KCl melts, (b) LiF–KCl melts. (○) phase diagram; dashed line, from Eq. (5.14). After Lumsden [9].

and sulfates [14]. However, the thermodynamic equations derived on the basis of random mixing turn out to have limited applicability in satisfactory representation of the experimental data. For further reading on the subject, reference may be made to Lumsden's book [13] and to a review paper by Forland [14].

5.1.3 Cryolite

As an electrolyte and solvent for alumina, the cryolite is of particular importance for the manufacture of aluminum by cathodic reduction. At temperatures above the melting point of cryolite (1012 °C at the stoichiometric composition Na_3AlF_6), there is complete liquid miscibility in the NaF–AlF_3 system. Because of its practical importance as an electrolyte, several attempts have been made to identify the ionic entities in liquid cryolite. Much of the controversy on the nature of the ion complexes in liquid cryolite seems to have been resolved by Dewing [15] in a study of the thermodynamics of NaF–AlF_3 melts.

Dewing measured the activity of NaF in liquid mixtures NaF–AlF_3 at 1020 and 1080 °C and along the cryolite liquidus. Representing the composition in terms of mole fractions of NaF and $NaAlF_4$, Dewing obtained the relation in Fig. 5.4 between $RT \ln \gamma_{NaF}$ and $(1 - x_{NaF})^2$. At temperatures and compositions of the cryolite liquidus, there is regular solution behavior with $\beta = -12.2 \times 10^3/RT$; at higher temperatures, the curves approach the line for regular solu-

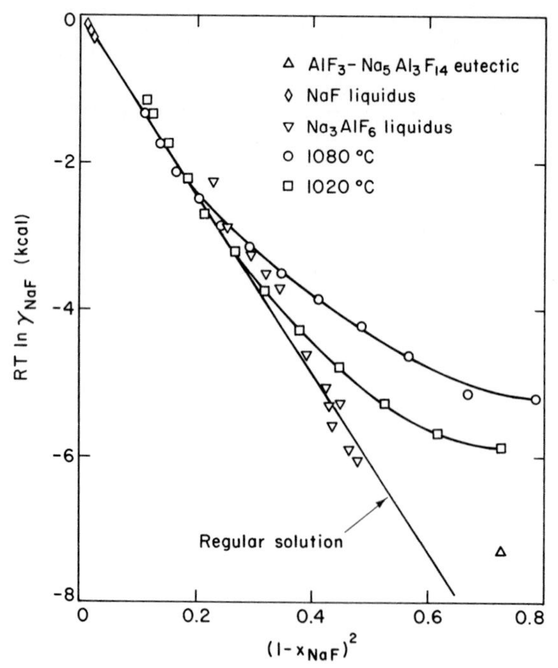

FIG. 5.4 Test of regular solution behavior in cryolite melts assumed to contain F^- and AlF_4^- ions. After Dewing [15].

tion at high concentrations of NaF. Dewing concluded from these findings that anions F^- and AlF_4^- predominate in melts with decreasing temperature and increasing concentration of NaF. Departure from regular solution behavior with increasing temperature and decreasing concentration of NaF is attributed to the dissociation of the anion complex AlF_4^-.

The mode of disproportionation of the AlF_4^- ion appears to be

$$2AlF_4^- = Al_2F_7^- + F^-. \tag{5.15}$$

The ion $Al_2F_7^-$ is analogous to $Al_2Cl_7^-$, which is shown to be present in KCl–$AlCl_3$ melts from studies of Raman spectra on these salt mixtures [16].

In commercial aluminum reduction cells, cryolite contains 5%–10% CaF_2, because of the buildup of lime impurity in the alumina charged to the cell (~0.05% CaO). From available data on the solubility of alumina in molten cryolite, Dewing [17] derived the pseudobinary liquidus diagrams, shown in Fig. 5.5, for melts containing 5% and 10% CaF_2. Addition of AlF_3, i.e., lowering the ratio NaF/AlF_3, lowers the temperature of the cryolite liquidus, but has little effect on the alumina liquidus. Addition of CaF_2 lowers the cryolite liquidus and reduces the alumina solubility.

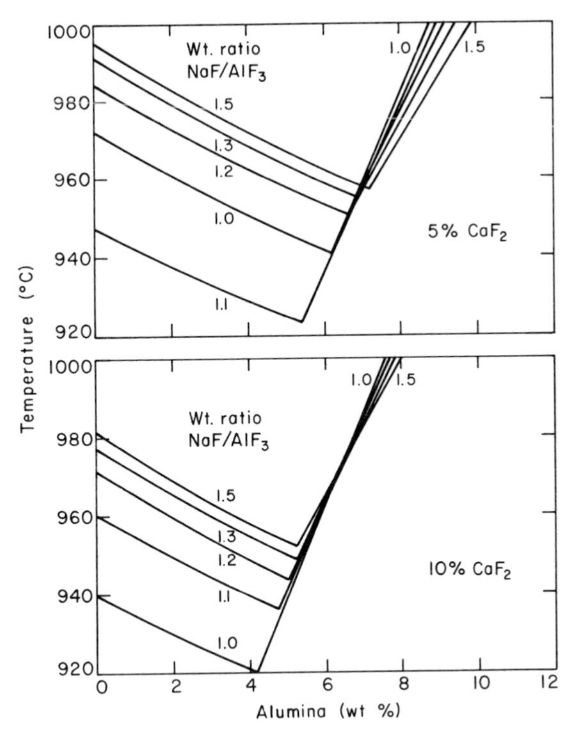

FIG. 5.5 Effect of CaF_2 on liquidus temperature and solubility of alumina in liquid cryolite. After Dewing [17].

Alumina is considered to dissolve in molten cryolite as oxyfluoride complexes $AlOF_n^{(n-1)}$; with $n = 2$, the following reaction may be considered [17]:

$$3NaF + AlF_3 + Al_2O_3 = 3NaAlOF_2. \tag{5.16}$$

Vetyukov *et al.* [18] and Rolin [19] measured the activity of alumina in liquid cryolite of stoichiometric composition Na_3AlF_6. Their data for 1300 °K may be represented by the equation

$$a_{Al_2O_3} = (\text{wt.}\%Al_2O_3)^{2.77}, \tag{5.17}$$

where $a_{Al_2O_3}$ is the activity of alumina relative to pure solid alumina. This activity relation is approximately in accord with the stoichiometry of reaction (5.16).

Another thermodynamic property of practical importance is the solubility of Al_4C_3 in cryolite-base melts. The solubility measured by Dewing [20] is represented by the equation

$$\log(\text{wt.}\%\ Al_4C_3) = 1.298 - 1.174(\text{wt. ratio NaF}/AlF_3). \tag{5.18}$$

This is independent of temperature in the range 985–1080 °C. Additions of Al_2O_3, NaCl, CaF_2, or Lif up to 10% do not affect the solubility, provided that LiF and CaF_2 are replaced by molar equivalent NaF in calculating the ratio NaF/AlF_3. Through thermodynamic calculations, Dewing showed that the solubility of Al_4C_3 in these melts is approximately proportional to $(a_{AlF_3})^{2/3}$, and that the dissolution may be represented by the reaction

$$2AlF_3 + Al_4C_3 = 3Al_2CF_2. \tag{5.19}$$

His analysis indicates that no complex carbide ions are formed; however, traces of C_2^{2-} ions are present in melts with the ratio NaF/AlF_3 above 1.5.

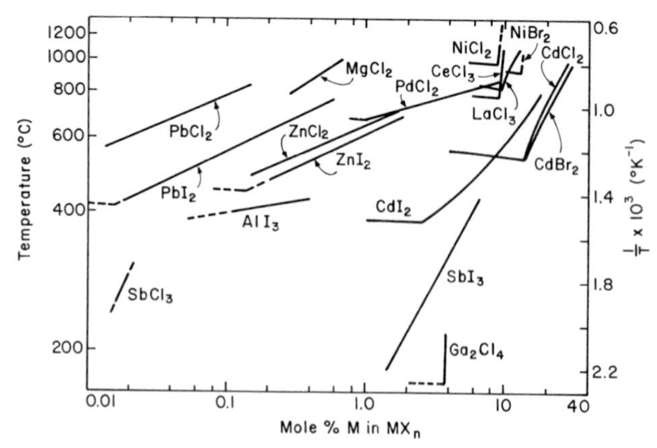

FIG. 5.6 Solubilities of metals in their halides. After Corbett [21], from "Fused Salts," edited by B. R. Sundheim, copyright 1964 and used with permission of the McGraw-Hill Book Company.

5.1.4 Solubilities in Metal–Metal Halide Systems

There are large miscibility gaps in most metal-metal salt systems. Depending on the system, the miscibility gaps close at temperatures of 1000–1300 °C above which there is complete liquid miscibility.

As is seen from the solubility data in Fig. 5.6 [21], the extent of metal solubility in its halide varies over a wide range from one system to another. For example, at 600 °C, the solubility of lead in $PbCl_2$ is about 0.02 mole %, while for $CdCl_2$, the solubility is about 20 mole %.

Selected data are given in Fig. 5.7 for solubilities of chlorides in their metals [22]. No particular trend could be found between solubilities and other physicochemical properties of metals and their salts.

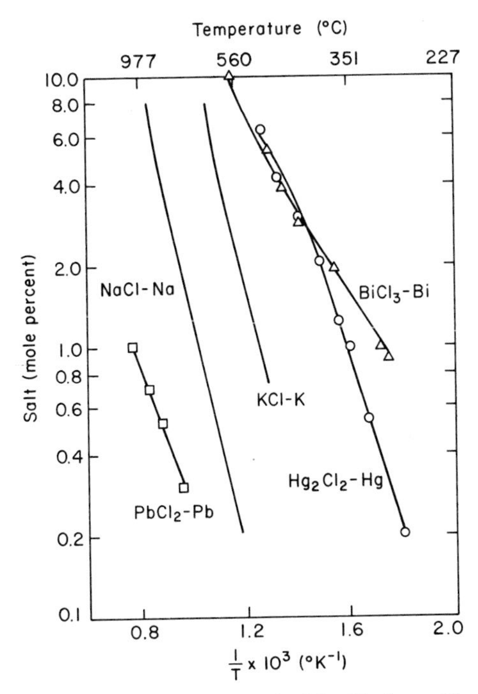

FIG. 5.7 Solubilities of chlorides in their metals. After Yosim and Luchsinger [22].

Metals also dissolve in liquid salts of other metals and vice versa. This is considered as an oxidation–reduction reaction such as

$$2Bi + 3PbCl_2 = 2BiCl_3 + 3Pb. \qquad (5.20)$$

The extent of dissolution of foreign metals in salts depends on the free energy change for the reaction and also on activities in metal (1)–metal (2) and salt (1)–salt (2) solutions. The data of Yosim and Luchsinger [22] on the solubilities of molten chlorides in molten bismuth are compared in Fig. 5.8 with the solubility of $BiCl_3$ in bismuth. It is seen that the solubilities of metal salts in bismuth are considerably less than that of $BiCl_3$; in fact, in no case does the solubility in a metal exceed that of its own salt. On the other hand, a salt can dissolve to a greater extent in another metal than in its own metal. For example, the solubility of $PbCl_2$ in Bi is about 2 mole% at 950 °C, and in Pb about 1 mole% at 1000 °C.

Several different views have been postulated on the formation of various neutral and charged entities when metals dissolve in salts or vice versa. Such views are somewhat speculative, therefore will not be dwelled upon here.

Molten salts are beginning to gain importance in the development of nuclear breeder reactors in the production of nuclear power. Rosenthal *et al.* [23], for example, considered the use of fluorides of fissile materials (UF_4, PuF_3) dissolved in a mixture of fluorides of lithium, beryllium, and thorium as a fuel. The fused salt also acts as a heat transfer medium in the fused salt circuit with anticipated minimum and maximum operating temperatures of about 500 and 700 °C. With these objectives in mind, the solubility of fluorides of fissile mate-

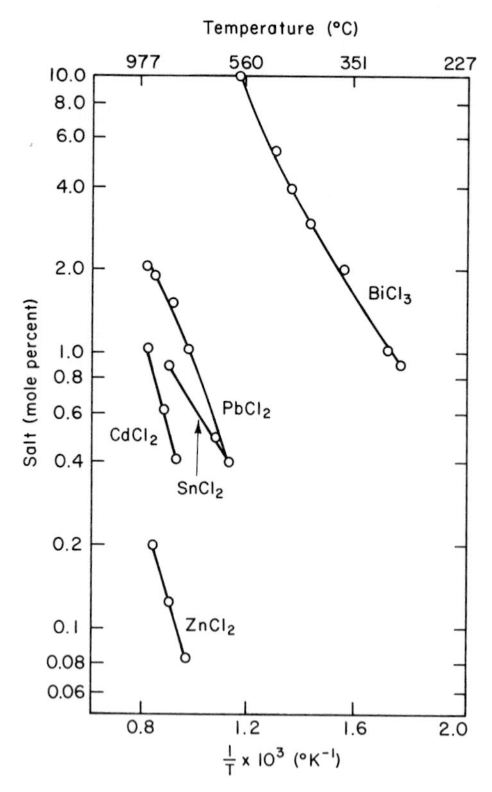

FIG. 5.8 Solubilities of chlorides in bismuth. After Yosim and Luchsinger [22].

rials in molten salts have been measured [24–26]. As an example, the data of Sood and Iyer [27] are shown in Fig. 5.9 for the solubility of solid PuF_3 (mp 1426 °C) in $LiF–BeF_2$ and $LiF–ThF_4$ melts at 600, 700, and 800 °C. The decrease in solubility with increasing concentration of BeF_2 or ThF_4 is attributed to a decrease in the concentration of F^- ions because of the formation of complexes BeF_4^{2-}, ThF_5^-, and ThF_7^{3-} as BeF_2 and ThF_4 are added to LiF. The effect of BeF_2 on the solubility of PuF_3 in LiF is much greater than that of ThF_4. This difference is anticipated since the charge/size ratio (Z/r) of 3.0 for Pu^{3+} is closer to $Z/r = 4.0$ for Th^{4+} than to 6.7 for Be^{2+}. This is similar to the solubility behavior of CeF_3 in fluoride mixtures [28].

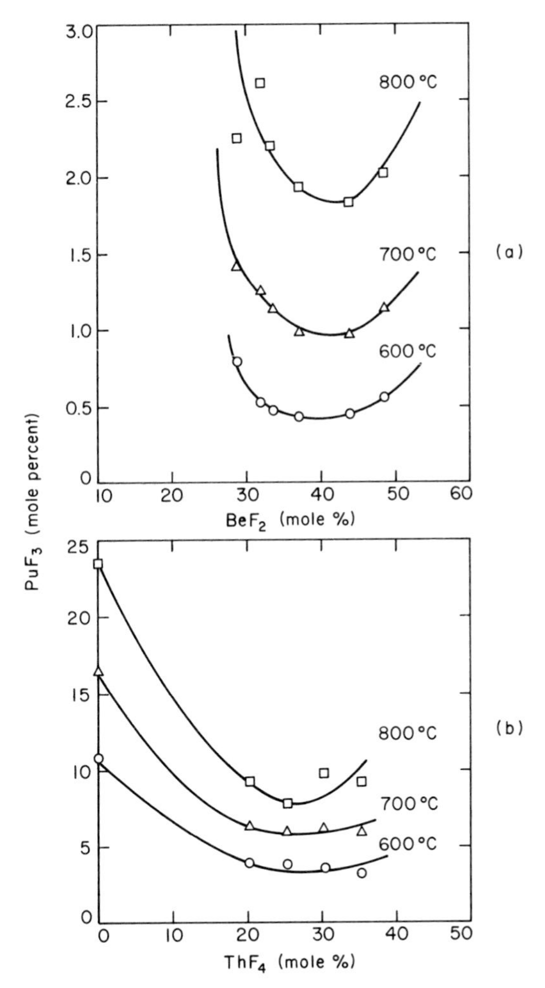

FIG. 5.9 Solubility of solid PuF_3 in (a) $LiF–BeF_2$ melts and (b) $LiF–ThF_4$ melts at indicated temperatures. After Sood and Iyer [27].

5.1.5 Volatility of Halides

Much use is made of the high volatility of halides in metallurgical halogenation processes such as the Kroll process for the production of titanium, zirconium, and tantalum, the segregation roasting of copper ores, and the purification of metals which have high affinity for oxygen. Some of these are discussed in more detail later.

Attention is drawn here to specific cases, as examples, of enhanced vaporization that occurs in mixtures of halides. For example, the volatility of NaCl is enhanced in the presence of volatile trivalent chlorides, e.g., $AlCl_3$, $FeCl_3$, [29, 30] or by the addition of cuprous chloride [31], due to the formation of volatile double chlorides.

Dewing [32] investigated the effect of $AlCl_3$ and $FeCl_3$ vapors on the volatility of divalent metal chlorides, and found that both monomer and dimer complex vapor species are formed

$$\begin{aligned} MCl_2(s) + 2AlCl_3(g) &= MAl_2Cl_8(g), \\ MCl_2(s) + 3AlCl_3(g) &= MAl_3Cl_{11}(g). \end{aligned} \tag{5.21}$$

As is shown schematically in Fig. 5.10, MCl_2 is the dominant vapor species at elevated temperatures. Because reaction (5.21) is exothermic, at lower temperatures the vapor pressure of the complex species exceeds that of MCl_2. At much lower temperatures, $AlCl_3$ is dimerized, and consequently, the partial pressure of MAl_2Cl_8 ultimately decreases with decreasing temperature. However, even with Al_2Cl_6, the vaporization of MCl_2 is greatly increased at relatively low temperatures.

As another example, let us consider the vaporization of cryolite. Kuxmann and Tillessen [33] found that the vapor pressure of liquid cryolite increases with increasing ratio AlF_3/NaF as a result of the reaction

$$2AlF_3 + Na_3AlF_6 = 3NaAlF_4(g). \tag{5.22}$$

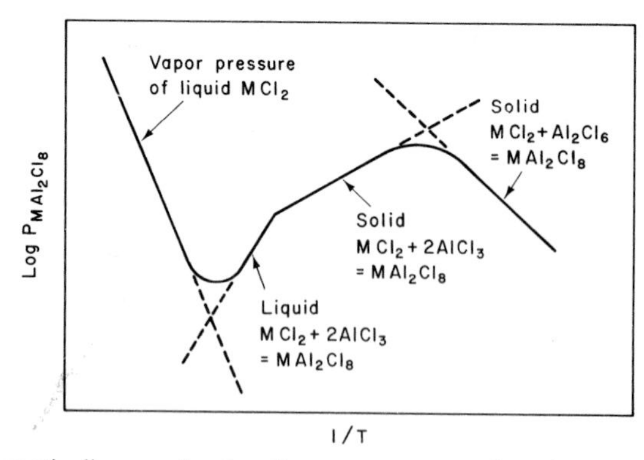

FIG. 5.10 Schematic diagram showing the vapor pressure of a salt MCl_2 as affected by temperature in the presence of $AlCl_3$ vapor. After Dewing [32].

For example, at 1000 °C the total vapor pressure is 0.58 mm Hg for a melt with the molar ratio AlF_3/NaF equal to $\frac{1}{8}$; the pressure increases to 32.8 mm Hg when AlF_3/NaF increases to 1.0. In the presence of aluminum, there is a complete change in the mode of vaporization of cryolite because of the reaction

$$2Na_3AlF_6 + Al = 3NaAlF_4(g) + 3Na(g) \tag{5.23}$$

and the sodium monomer becomes the dominant vapor species.

We see from the foregoing few examples that the vaporization of halides can be greatly affected by mixing with other halides or with suitable metals.

5.2 Surface Tension

The available data on the density and surface tension of several molten salts have been compiled by Janz and Dijkhuis [34] as functions of temperature with references to the source of data. Molar volumes and surface tensions of selected molten salts at their melting temperatures are given in Table 5.1. The values of surface tensions are in the range 20–350 dyn cm^{-1}; for most salts the range is within 80–160 dyn cm^{-1}. In general, the surface tension increases with decreasing lattice energy. The relation in Eq. (3.87) based on the intermolecular potential function and the concept of free volume for nonpolar liquids does not hold for molten salts. For molten alkali halides, however, the surface tension is approximately proportional to $T_m V^{-2/3}$ at the melting point. The proportionality factor is about 3.5 times less than that given by Eq. (3.87), which holds reasonably will for liquid metals (Fig. 3.23). The surface tension is approximately proportional to $\Delta H_v V^{-2/3}$, but the proportionality factor is about a third that for liquid metals, for which Eq. (3.88) is a good approximation.

The surface tension of binary molten salt mixtures vary nonlinearly with composition. The data of Bertozzi and Sternheim [35] in Fig. 5.11 for mixtures of alkali nitrates are typical examples of the compositional dependence of the surface tension. On the assumption of regular solution behavior, Guggenheim [36] derived a theoretical equation for the variation of surface tension with the composition of binary salt mixtures; the equation was further developed subsequently by Defay *et al.* [37]. Nissen and Van Domelen [38] have shown that the surface tension of salt mixtures calculated from the surface tensions of the components and known interaction parameter β for regular solution behavior are within 2–4% of the measured values.

Dewing and Desclaux [39] measured the interfacial tension between liquid aluminum and cryolite saturated with alumina. They found that the interfacial tension decreases with increasing ratio NaF/AlF_3. For example, at 1000 °C, σ is 650 dyn cm^{-1} for weight ratio NaF/AlF_3 equal to 0.6 and decreases to 400 dyn cm^{-1} when NaF/AlF_3 is 2.0. Since the activity of sodium in the melt increases with increasing ratio NaF/AlF_3, as anticipated from the reaction

$$3NaF + Al = AlF_3 + 3Na, \tag{5.24}$$

TABLE 5.1

Molar Volumes and Surface Tensions of Molten Salts at Their Melting Temperatures[a]

Salt	mp (K)	V ($cm^3\ mole^{-1}$)	σ ($dyn\ cm^{-1}$)	Salt	mp (K)	V ($cm^3\ mole^{-1}$)	σ ($dyn\ cm^{-1}$)
Fluorides				Bromides			
LiF	1121	13.36	236	$CaBr_2$	1015	61.54	120
NaF	1269	19.84	186	$SrBr_2$	916	63.27	150
KF	1131	27.84	167	$BaBr_2$	1127	69.98	153
RbF	1048	—	127	$ZnBr_2$	675	60.42	51
CsF	954	38.89	107	$CdBr_2$	841	62.32	111
ThF_4	1383	49.15	238	$HgBr_2$	514	60.15	65
UF_4	1309	46.48	196	$BiBr_3$	492	82.79	99
Chlorides				Iodides			
LiCl	883	26.16	126	LiI	472	43.03	—
NaCl	1074	34.30	114	NaI	933	49.95	101
KCl	1044	44.22	101	KI	954	61.35	79
RbCl	988	48.56	98	RbI	913	65.99	74
CsCl	918	54.62	106	CsI	894	74.10	87
CuCl	703	25.39	92	CaI_2	1052	80.54	87
AgCl	728	28.05	179	SrI_2	788	78.54	107
				BaI_2	1013	86.94	135
$MgCl_2$	987	54.10	67				
$CaCl_2$	1045	50.45	148	Carbonates			
$SrCl_2$	1146	55.04	168	Li_2CO_3	993	38.20	244
$BaCl_2$	1235	61.98	166	Na_2CO_3	1123	50.52	212
				K_2CO_3	1174	68.64	119
$ZnCl_2$	556	51.31	54				
$CdCl_2$	841	50.75	100	Nitrates			
$SnCl_2$	518	51.12	104	$LiNO_3$	527	35.74	116
$HgCl_2$	550	52.76	56	$NaNO_3$	583	40.51	120
$PbCl_2$	771	51.83	138	KNO_3	610	48.86	111
$GaCl_3$	351	67.15	27	$RbNO_3$	589	53.78	108
$BiCl_3$	505	69.46	65	$CsNO_3$	687	62.11	91
				$AgNO_3$	483	40.06	150
Bromides							
NaBr	1020	40.12	104	Sulfates			
KBr	1007	50.58	89	Li_2SO_4	1132	51.97	225
RbBr	953	54.93	89	Na_2SO_4	1157	64.52	195
CsBr	909	61.38	84	K_2SO_4	1342	87.44	143
AgBr	703	32.03	154	Rb_2SO_4	1347	97.85	355
LiBr	823	34.32	—	Cs_2SO_4	1292	113.07	294

[a] Calculated from data compiled by Janz and Dijkhuis [34].

accompanying decrease in the interfacial tension is a manisfestation of chemisorption of sodium. By applying the Gibbs adsorption equation (3.94), Dewing and Desclaux concluded that complete surface coverage with sodium occurs when the molar ratio NaF/AlF_3 is about 2.8, for which the activity of sodium at 1000 °C is about 0.02. The addition of MgF_2, or the substitution of Na_3AlF_6 with Li_3AlF_6, raises the interfacial tension because of the lowering of the sodium

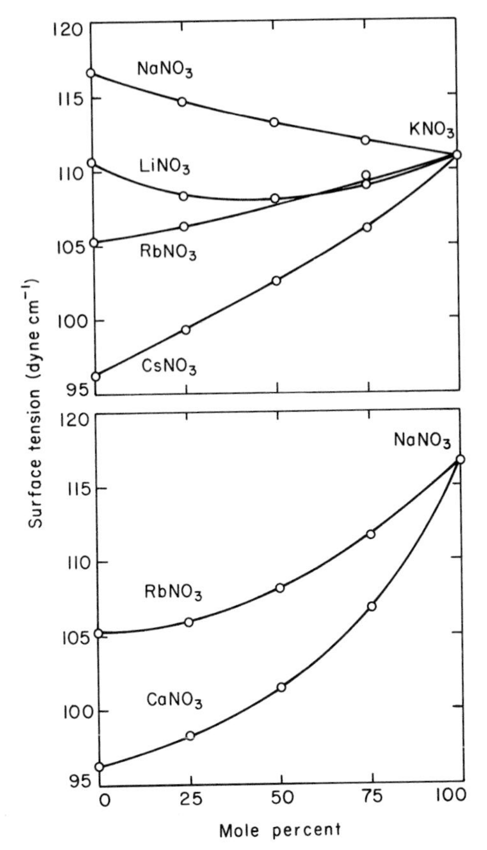

FIG. 5.11 Variation of surface tension with composition in binary alkali nitrate melts at 350 °C. After Bertozzi and Sternheim [35].

activity, hence the lowering of the sodium adsorption. The chemisorption of sodium at the interface is expected to have a dominant effect on the electrode kinetics for the deposition of aluminum in the cathodic reduction of alumina.

5.3 Transport Properties

The tracer (self) diffusivity of cations and anions in molten chlorides and nitrates at their melting points, computed from the available data [40–43], are listed in Table 5.2. Also included in this table are the viscosity and equivalent electrical conductance at the melting temperatures, computed from the data compiled by Janz *et al.* [44].

Self-diffusivities of ions in molten salts are in the range 1–8 $\times 10^{-5}$ cm^2 sec^{-1}. In most cases, self-diffusivities of cations and anions are similar, and vary with temperature in accord with the exponential equation (3.138). Viscosities are in the 1–10 cP range and decrease exponentially with decreasing reciprocal of the

TABLE 5.2

Self-Diffusivities of Cations and Anions, Viscosity, and Electrical Equivalent Conductance of Molten Salts at Their Melting Temperatures[a]

Salt	$D^b \times 10^5$, ($cm^2 sec^{-1}$) Cation	Anion	Reference	η (cP)	Λ ($\Omega^{-1} cm^2 gew^{-1}$)
LiCl	—	—		1.79	161
NaCl	7.38	5.82	40	1.46	134
KCl	6.51	5.77	40	1.19	106
RbCl	4.37	3.92	41	1.35	80
CsCl	3.13	3.38	41	1.36	87
$CaCl_2$	1.98	2.66	40	3.51	54
$SrCl_2$	1.98	3.75	40	3.83	58
$BaCl_2$	1.66	4.20	40	5.76	67
$CdCl_2$	1.84	1.88	40	2.50	50
$PbCl_2$	0.93	1.67	42	4.64	41
$LiNO_3$	1.31	0.46	43	7.41	31
$NaNO_3$	1.77	1.12	43	2.98	44
KNO_3	1.38	1.23	43	2.91	34
$RbNO_3$	—	—		3.87	26
$CsNO_3$	1.85	1.79	43	2.26	37
$AgNO_3$	1.01	0.57	43	5.04	29

[a] Values of η and Λ are calculated from data compiled by Janz *et al.* [44].
[b] In terms of specific conductivity σ, Ω^{-1} cm^{-1}, molar volume V, and valence Z, equivalent conductance is $\Lambda = V/Z$.

absolute temperature. The apparent energies of activation for self-diffusion and viscous flow are similar and in the range 5–10 kcal.

The Stokes–Einstein relation (3.130) holds well for molten salts as demonstrated in Fig. 5.12 for melting temperatures. The lower line is that calculated from Eq. (3.130)

$$D\zeta = kT/3\pi\delta = 1.46 \times 10^{-17}(T/\delta) \quad \text{erg cm}^{-1}, \tag{5.25}$$

where ζ is the molecular shear viscosity and δ is the interionic distance. In the rigid-sphere model of liquid metals, the molecular shear viscosity is $\frac{3}{5}$ times the bulk viscosity, Eq. (3.110). The upper line in Fig. 5.12 is that calculated from Eqs. (3.110) and (5.25) for the bulk viscosity. The data points are seen to be within these two limiting lines.

It should be pointed out that the self-diffusivity used for the plot in Fig. 5.12 is the lower of the values for cation and anion in the salt.

Bockris and Hooper [41] have shown that of the various structural models the hole model proposed by Fürth [7] gives a better description of the transport

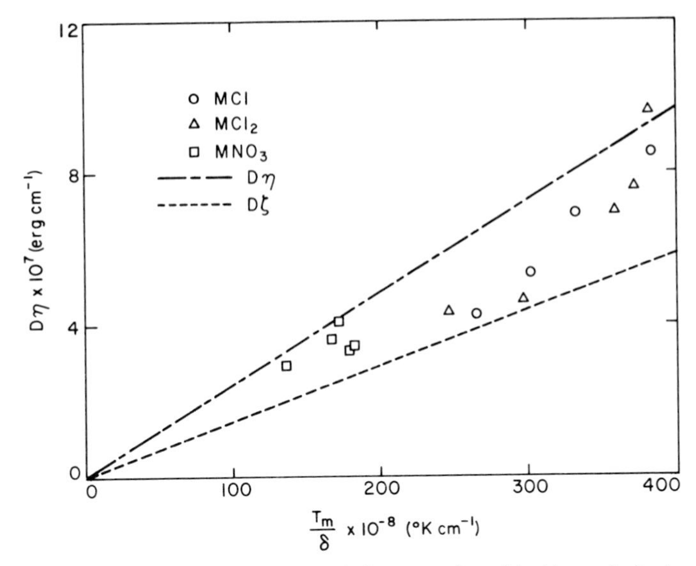

FIG. 5.12 Test of Stokes–Einstein relation on molten chlorides and nitrates.

properties of molten salts. Fürth's model is often called the *density fluctuation hole model*, because the sizes of holes are thermally distributed. The work of hole formation per mole of holes and their most probable volume are given by the equations

$$W_{\mathrm{h}} = \delta_{\mathrm{h}}{}^2 \sigma N, \tag{5.26}$$

$$V_{\mathrm{h}} = 0.68N(kT/\sigma)^{3/2}, \tag{5.27}$$

where δ_{h} is the average diameter of the hole. For simple salts, V_{h} is approximately one-half of the gram-ionic volume of the melt.

Bockris and Hooper showed on the basis of the density fluctuation hole model that the enthalpy of formation of holes ΔH_{h} at the melting point T_{m} is equal to $3.55RT_{\mathrm{m}}$. This is close to the proportionality factor for the apparent energy of activation for diffusion E_{d} and viscous flow E_{v}; experimental data give

$$E_d \simeq E_v \simeq 3.74RT_{\mathrm{m}} = 7.43T_{\mathrm{m}} \text{ cal.} \tag{5.28}$$

This is similar to the relation in Eq. (3.127) for liquid metals. On the basis of these observations, Bockris and Hooper concluded that the energy of activation for the jump process for diffusion or viscous flow is only a small fraction of the enthalpy of formation of holes. These predictions are further substantiated from a study of diffusion in molten salts at constant volume by Nagarajan and Bockris [45]. They deduced from their diffusion measurements at pressures up to 1200 atm that the energy of activation for the jump process in simple salts is about 20% of the total experimental enthalpy of activation for transport at constant pressure.

The diffusivities of foreign ions in molten salts are similar to the self-diffusivities of the ions. However, the diffusivity of the foreign ion usually decreases with increasing interionic size of the salt. For example, diffusivities of U^{4+} [46] and Zr^{4+} [47] in molten NaCl at 800 °C are about 4.6×10^{-5} cm^2 sec^{-1} and decrease to about 1.7×10^{-5} cm^2 sec^{-1} in CsCl; the values in KCl and RbCl lie in the intermediate range.

As pointed out in Section 4.3.5, the equivalent electrical conductance predicted from the self-diffusivity by using the Nernst–Einstein equation (4.58) is 10%–50% greater than the measured value, even in simple salts. This deviation from the theoretical Nernst–Einstein relation is attributed to the formation of neutral M^+X^- species which contribute to diffusion, but not to electrical conduction.

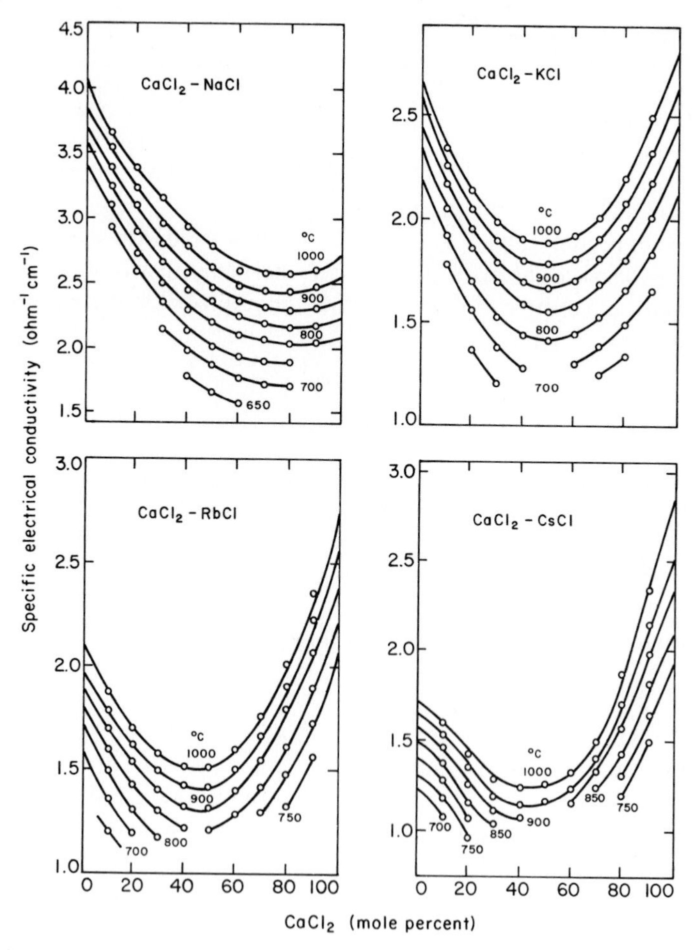

FIG. 5.13 Specific electrical conductivity of $CaCl_2$–alkali chloride melts. After Emons *et al.* [48].

The electrical conductivity increases with increasing temperature, and for most salts the temperature dependence of the equivalent conductance may be represented by the exponential function

$$\Lambda = \Lambda_0 \exp(-E_c/RT). \tag{5.29}$$

The apparent energy of activation E_c for conduction is almost always lower than that for diffusion or viscous flow. In some salts there are departures from the relation in Eq. (5.29); that is, E_c increases with decreasing temperature, indicating polymerization at low temperatures.

The transport numbers of ions, i.e., the fraction of the total current carried by the ion, usually decrease with increasing ionic size. For example, the transport number for Li^+ (0.68 Å) in molten LiCl is 0.75, while for K^+ (1.33 Å) in KCl it is 0.62. By invoking the Stokes–Einstein relation, the following equation is obtained for the transport number in terms of the ionic radii r_c,r_a of cation and anion and their valences Z_c,Z_a:

$$t_c = Z_c r_a/(Z_c r_a + Z_a r_c).$$

The specific electrical conductivity of salt mixtures varies nonlinearly with composition, as seen from the data in Fig. 5.13 for binary mixtures of $CaCl_2$ with alkali chlorides [48]. Minima on the conductivity curves are indicative of the formation of neutral species between cations and anions.

In metal–metal halide solutions, the specific conductivity increases with increasing concentration of the metal in solution. For example, in K–KCl melts, the specific conductivity increases by about an order of magnitude to 22 Ω^{-1} cm^{-1} with the addition of about 10 mole % K. The electronic contribution to conduction is attributed to the release of electrons by reaction of the type $M \rightarrow M^{2+} + 2\varepsilon$ [49]. Reference may be made to a review paper by Corbett [21] for other examples of conductivity in metal–metal halide mixtures.

References

1. H. A. Levy, P. A. Agron, M. A. Bredig, and M. D. Danford, *Ann. N.Y. Acad. Sci.* **79**, 762 (1960).
2. J. Sherman, *Chem. Rev.* **11**, 93 (1932).
3. F. H. Stillinger, *in* "Molten Salt Chemistry" (M. Blander, ed.), p. 1. Wiley (Interscience), New York, 1964.
4. J. R. Partington, "An Advanced Treatise on Physical Chemistry." Longmans, Green, London, 1952.
5. I. S. Yaffe and E. R. van Artsdalen, *J. Phys. Chem.* **60**, 1125 (1956).
6. H. Bloom and J. O'M. Bockris, *in* "Fused Salts" (B. R. Sundheim, ed.), p. 1. McGraw-Hill, New York, 1964.
7. R. Fürth, *Proc. Cambridge Philos. Soc.* **37**, 281 (1941).
8. N. Temkin, *Acta Physicochim. URSS* **20**, 411 (1945).
9. J. Lumsden, "Thermodynamics of Molten Salt Mixtures." Academic Press, London, 1966.
10. O. J. Kleppa, L. S. Hersh, and J. M. Toguri, *Acta Chem. Scand.* **17**, 2681 (1963).
11. L. S. Hersh and O. J. Kleppa, *J. Chem. Phys.* **42**, 1309 (1965).
12. T. Forland, *Discuss. Faraday Soc.* **32**, 126 (1961).

13. H. Flood, T. Forland, and K. Grjotheim, *Z. Anorg. Allg. Chem.* **276**, 289 (1954).
14. T. Forland, *in* "Fused Salts" (B. R. Sundheim, ed.), p. 63. McGraw-Hill, New York, 1964.
15. E. W. Dewing, *Metall. Trans.* **3**, 495 (1972).
16. C. J. Cyvin, P. Klaboe, E. Rytter, and H. A. Oye, *J. Chem. Phys.* **52**, 2776 (1970).
17. E. W. Dewing, *Can. Metall. Q.* **13**, 607 (1974).
18. M. M. Vetyukov and N. van Ban, *Tsvetn. Met.*, **44**, 13 (1971).
19. M. Rolin, *Rev. Int. Hautes Temp. Refract.* **9**, 333 (1972).
20. E. W. Dewing, *Trans. Metall. Soc. AIME* **245**, 2181 (1969).
21. J. D. Corbett, *in* "Fused Salts" (B. R. Sundheim, ed.), p. 341. McGraw-Hill, New York, 1964.
22. S. J. Yosim and E. B. Luchsinger, *Ann. N.Y. Acad. Sci.* **79**, 1079 (1960).
23. M. W. Rosenthal, P. N. Haubenreich, and R. B. Briggs, U.S. Atomic Energy Comm. Rep. CRNL-4812 (1972).
24. C. J. Barton, *J. Phys. Chem.* **64**, 306 (1960).
25. J. C. Mailen, F. J. Smith, and L. M. Ferris, *J. Chem. Eng. Data* **16**, 68 (1971).
26. C. E. Bamberger, R. G. Ross, and C. F. Baes, U.S. Atomic Energy Comm. Rep. CRNL-4622 (1971).
27. D. D. Sood and P. N. Iyer, *in* "Thermodynamics of Nuclear Materials, 1974," Vol. II, p. 489. IAEA, Vienna, 1975.
28. C. J. Barton, M. A. Bredig, L. O. Gilpatrick, and J. A, Fredericksen, *Inorg. Chem.* **9**, 307 (1970).
29. C. M. Cook and W. E. Dunn, *J. Phys. Chem.* **65**, 1505 (1961).
30. R. R. Richards and N. W. Gregory, *J. Phys. Chem.* **68**, 3089 (1964).
31. P. Gross, *Can. Metall. Q.* **12**, 359 (1973).
32. E. W. Dewing, *Metall. Trans.* **1**, 2169 (1970).
33. U. Kuxmann and U. Tillessen, *Z. Erzbergbau Metallhuettenwes.* **20**, 147 (1967).
34. G. J. Janz and C. G. M. Dijkhuis, "Molten Salts," Natl. Stand. Ref. Data Ser., Vol. 2. Natl. Bur. Stand., Washington, D.C., 1969.
35. G. Bertozzi and G. Sternheim, *J. Phys. Chem.* **68**, 2908 (1964).
36. E. A. Guggenheim, "Mixtures." Oxford Univ. Press, London, 1952.
37. R. Defay, I. Prigogine, A. Bellemans, and D. H. Everett, "Surface Tension and Adsorption." Wiley, New York, 1966.
38. D. A. Nissen and B. H. Van Domelen, *J. Phys. Chem.* **79**, 2003 (1975).
39. E. W. Dewing and P. Desclaux, *Metall. Trans. B* **8**, 555 (1977).
40. J. O'M. Bockris and A. K. N. Reddy, "Modern Electrochemistry," Vol. 1. Plenum, New York, 1973.
41. J. O'M. Bockris and G. W. Hooper, *Discuss. Faraday Soc.* **32**, 218 (1961).
42. G. Perkins, R. B. Escue, J. F. Lamb, and J. W. Wimberley, *J. Phys. Chem.* **64**, 1792 (1960).
43. A. S. Dworkin, R. B. Escue, and E. R. Van Artsdalen, *J. Phys. Chem.* **64**, 872 (1960).
44. G. J. Janz, F. W. Dampier, G. R. Lakshminarayanan, P. K. Lorenz, and R. P. T. Tomkins, "Molten Salts," Natl. Stand. Ref. Data Ser., Vol. 1. Natl. Bur. Stand., Washington, D.C., 1968.
45. M. K. Nagarajan and J. O'M. Bockris, *J. Phys. Chem.* **70**, 1854 (1966).
46. M. V. Smirnov, V. E. Komarov, N. P. Borodina, and Yu. N. Kramov, *Elektrokhimiya* **10**, 770 (1970).
47. V. E. Komarov, M. V. Smirnov, and N. P. Borodina, *Tr. Inst. Elektrokhim., Ural. Nauchn. Tsentr., Akad. SSSR* **17**, 49 (1971).
48. H. H. Emons, G. Braeutigam, and H. Vogt, *Z. Anorg. Allg. Chem.* **394**, 263 (1972).
49. A. S. Dworkin, H. R. Bronstein, and M. A. Bredig, *J. Phys. Chem.* **66**, 1201 (1962).

CHAPTER 6

TRANSPORT PROPERTIES OF GASES

6.1 Dilute Gases

The molecular transfer of mass, momentum, and energy are interrelated transport processes of diffusion under a concentration gradient, viscous flow in a velocity gradient, and heat conduction in a thermal gradient. The earlier derivation of the transport properties from an oversimplified kinetic theory of gases were based on the assumption that (i) the molecules are nonreacting rigid spheres of diameter δ, (ii) all the molecules travel with the same speed, and (iii) equal number of molecules travel parallel to each of the three coordinate axes (one-sixth in $+x$ direction, one-sixth in $-x$ direction, and similarly for the y and z directions). From this simple model the following equations were derived for the transport properties in terms of mean free path l and the arithmetic mean speed of molecules $\bar{v}$:

$$D = \tfrac{1}{3}\bar{v}l \qquad \text{(diffusivity)}, \tag{6.1}$$

$$\eta = \tfrac{1}{3}nm\bar{v}l \qquad \text{(viscosity)}, \tag{6.2}$$

$$\kappa = \tfrac{1}{3}nc_v\bar{v}l \qquad \text{(thermal conductivity)}, \tag{6.3}$$

where $\bar{v} = (8kT/\pi m)^{1/2}$, $l = (\sqrt{2}\,n\pi\delta^2)^{-1}$, $nm = pm/kT$, the density of the gas, and c_v is the specific heat at constant volume.

These equations, though obsolete, give approximate estimates of the transport properties and are useful in describing some transport phenomena in simple terms.

The subsequent development of the rigorous kinetic theory of gases, known as the *Chapman–Enskog theory*, is based on the knowledge of the molecular distribution function $f_i(r, v_i, t)$. This function represents the number of molecules of ith species at time t in a unit volume element about the point r which has a velocity of v_i. The derivation of the transport properties from the rigorous kinetic theory of gases, described in depth by Chapman and Cowling [1] and

Hirschfelder *et al.* [2], is based on the evaluation of the intermolecular energy of attraction ϵ, the collision diameter δ, and the collision integral $\Omega^{(l,s)}$ involving explicitly the dynamics of a molecular encounter and hence the intermolecular force law.

Of the various intermolecular potential energy functions, the Lennard-Jones (6–12) potential energy function $\varphi(r)$ has been found to be more realistic. For spherical and nonpolar molecules

$$\varphi(r) = 4\epsilon[(\delta/r)^{12} - (\delta/r)^6], \tag{6.4}$$

where ϵ is the depth of the potential well (the maximum energy of attraction) and δ is the collision diameter for low energy collisions, i.e., the value of r at $\varphi(r) = 0$. The inverse sixth power represents molecular attraction; the repulsive contribution to the potential function is represented by the inverse twelfth-power term.

For collisions between unlike molecules of species 1 and 2, the maximum energy of attraction is approximated as the geometrical mean of the pure components,

$$\epsilon_{12} = (\epsilon_1 \epsilon_2)^{1/2}, \tag{6.5}$$

and the collision diameter is taken as the arithmetical mean of the diameters of the components,

$$\delta_{12} = \tfrac{1}{2}(\delta_1 + \delta_2). \tag{6.6}$$

The transport properties of dilute gases, i.e., low density gases at ordinary pressures, derived from the Chapman–Enskog theory are given by the following equations:

self-diffusivity ($\mathrm{cm^2\,sec^{-1}}$),

$$D = 1.8583 \times 10^{-3}\,(T^{3/2}/P\delta^2\Omega^{(1,1)*})(2/M)^{1/2}; \tag{6.7}$$

binary interdiffusivity ($\mathrm{cm^2\,sec^{-1}}$),

$$D_{12} = 1.8583 \times 10^{-3}(T^{3/2}/P\delta_{12}^2\Omega^{(1,1)*})(1/M_1 + 1/M_2)^{1/2}, \tag{6.8*}$$

where P is the pressure, M_1 and M_2 are molecular weights, and $\Omega^{(1,1)*}$ is the collision integral described below;

viscosity of a pure gas [$\mathrm{dyn\,sec\,cm^{-2} \equiv g\,cm^{-1}\,sec^{-1}}$ (P)],

$$\eta = 2.6693 \times 10^{-5}\,(MT)^{1/2}/\delta^2\Omega^{(2,2)*}; \tag{6.9}$$

viscosity of binary gas mixtures (P),

$$\eta_{12} = \frac{x_1\eta_1}{x_1\phi_{11} + x_2\phi_{12}} + \frac{x_2\eta_2}{x_1\phi_{21} + x_2\phi_{22}}, \tag{6.10}$$

* The Chapman–Enskog first approximation gives D_{12} independent of composition; the value of $(D_{12})_2$ for the second approximation, though more accurate, differs from D_{12} by less than 3%.

where x_1 and x_2 are mole fractions and

$$\phi_{ij} = 8^{-1/2}(1 + M_i/M_j)^{-1/2}[1 + (\eta_i/\eta_j)^{1/2}(M_j/M_i)^{1/4}]^2, \qquad \phi_{ii} = \phi_{jj} = 1; \tag{6.11}$$

thermal conductivity of a pure gas (cal cm^{-1} sec^{-1} deg^{-1}), for monatomic gases

$$\kappa = (15R/4M)\eta, \tag{6.12}$$

and for polyatomic gases

$$\kappa = [c_p + (9R/4M)]\eta, \tag{6.13}$$

where c_p is the specific heat capacity at constant pressure;
thermal conductivity of binary gas mixtures (cal cm^{-1} sec^{-1} deg^{-1}),

$$\kappa_{12} = \frac{x_1\kappa_1}{x_1\phi_{11} + x_2\phi_{12}} + \frac{x_2\kappa_2}{x_1\phi_{21} + x_2\phi_{22}}, \tag{6.14}$$

where the coefficients ϕ are the same as those given in Eq. (6.11);
thermal diffusivity (cm^2 sec^{-1}),

$$D^{\mathrm{T}} = \kappa/\rho c_p, \tag{6.15}$$

where ρ is the density.

The kinetic theory from which the above equations have been derived is for nonreacting mixtures of simple gas molecules with no internal degrees of freedom, i.e., monatomic gases, and for gas molecules which are nonpolar, spherically symmetric, and have densities sufficiently low so that the effect of collisions involving more than two molecules is negligible. Although inelastic collision occurs between molecules with internal degrees of freedom (polyatomic gases), because of the conservation of mass and momentum in collision, the expressions for coefficients of diffusion and viscosity derived from the Chapman–Enskog theory for monatomic gases are applicable to polyatomic gases with internal degrees of freedom, provided that the molecules are nonpolar and not completely nonspherical. In the case of the thermal conductivity of polyatomic gases, however, a term is added to Eq. (6.12) to account approximately for the transfer of energy between translational and internal degrees of freedom in the molecules. Inserting $c_p = 3R/2M$ for monatomic gases simplifies Eq. (6.13) to (6.12).

The dynamics of binary collision between the nonpolar spherical molecules is accounted for by the collision integral $\Omega^{(l,s)}$, the derivation of which is usually based on the Lennard-Jones (6–12) potential. This collision integral is a complex function of (i) the reduced mass of the colliding species, (ii) the reduced initial relative speed of the colliding molecules, (iii) the angle by which the molecules are deflected in the center of gravity of the coordinate system, and (iv)

TABLE 6.1

Omega Integrals for Transport Properties of Gases at Low Densities[a]

kT/ϵ or kT/ϵ_{12}	$\Omega^{(2,2)*}$ (for viscosity and thermal conductivity)	$\Omega^{(1,1)*}$ (for mass diffusivity)	kT/ϵ or kT/ϵ_{12}	$\Omega^{(2,2)*}$ (for viscosity and thermal conductivity)	$\Omega^{(1,1)*}$ (for mass diffusivity)
0.30	2.785	2.662	2.60	1.081	0.9878
0.35	2.628	2.476	2.70	1.069	0.9770
0.40	2.492	2.318	2.80	1.058	0.9672
0.45	2.368	2.184	2.90	1.048	0.9576
0.50	2.257	2.066	3.00	1.039	0.9490
0.55	2.156	1.966	3.10	1.030	0.9406
0.60	2.065	1.877	3.20	1.022	0.9328
0.65	1.982	1.798	3.30	1.014	0.9256
0.70	1.908	1.729	3.40	1.007	0.9186
0.75	1.841	1.667	3.50	0.9999	0.9120
0.80	1.780	1.612	3.60	0.9932	0.9058
0.85	1.725	1.562	3.70	0.9870	0.8998
0.90	1.675	1.517	3.80	0.9811	0.8942
0.95	1.629	1.476	3.90	0.9755	0.8888
1.00	1.587	1.439	4.00	0.9700	0.8836
1.05	1.549	1.406	4.10	0.9649	0.8788
1.10	1.514	1.375	4.20	0.9600	0.8740
1.15	1.482	1.346	4.30	0.9553	0.8694
1.20	1.452	1.320	4.40	0.9507	0.8652
1.25	1.424	1.296	4.50	0.9464	0.8610
1.30	1.399	1.273	4.60	0.9422	0.8568
1.35	1.375	1.253	4.70	0.9382	0.8530
1.40	1.353	1.233	4.80	0.9343	0.8492
1.45	1.333	1.215	4.90	0.9305	0.8456
1.50	1.314	1.198	5.0	0.9269	0.8422
1.55	1.296	1.182	6.0	0.8963	0.8124
1.60	1.279	1.167	7.0	0.8727	0.7896
1.65	1.264	1.153	8.0	0.8538	0.7712
1.70	1.248	1.140	9.0	0.8379	0.7556
1.75	1.235	1.128	10.0	0.8242	0.7424
1.80	1.221	1.116	20.0	0.7432	0.6640
1.85	1.209	1.105	30.0	0.7005	0.6232
1.90	1.197	1.094	40.0	0.6718	0.5960
1.95	1.186	1.084	50.0	0.6504	0.5756
2.00	1.175	1.075	60.0	0.6335	0.5596
2.10	1.156	1.057	70.0	0.6194	0.5464
2.20	1.138	1.041	80.0	0.6076	0.5352
2.30	1.122	1.026	90.0	0.5973	0.5256
2.40	1.107	1.012	100.0	0.5882	0.5170
2.50	1.093	0.9996			

[a] From Hirschfelder *et al.* [2].

the impact parameter. For convenience, reduced quantities of the omega integral are used in the derivation of transport properties of gases. The reduced quantity denoted by $\Omega^{(l,s)*}$ is the ratio of the omega integral for a given potential energy function to that for a rigid spherical molecule, so that for all rigid spherical molecules $\Omega^{(l,s)*}$ becomes unity. In the derivation of the diffusion equation to the first approximation, only one term in the Sonine polynomial expansion is necessary, and the corresponding omega integral is designated by $\Omega^{(1,1)*}$. In the derivation of equations for viscosity and thermal conductivity to a first approximation, two terms are necessary in the Sonine polynomial expansion, and the collision integral is denoted by $\Omega^{(2,2)*}$.

The values of $\Omega^{(l,s)*}$ as functions of reduced temperature $T^* = kT/\epsilon$ have been tabulated by Hirschfelder et al. [2] for the Lennard-Jones (6–12) potential. For convenience, these values are reproduced in Table 6.1.

The force constants ϵ/k and δ derived from the viscosity data for the Lennard-Jones (6–12) potential for common gases and vapors are given in Table 6.2. Hirschfelder *et al.* [2] have shown that the transport properties calculated from Eqs. (6.7)–(6.15) are in close agreement with measured values over a wide range of temperatures.

TABLE 6.2

Force Constants ϵ/k and δ from Viscosity Data for the Lennard-Jones (6–12) Potential and b_0 from Second Virial Coefficients[a]

Gas	ϵ/k (°K)	δ (Å)	b_0 ($cm^3\ mole^{-1}$)	Gas	ϵ/k (°K)	δ (Å)	b_0 ($cm^3\ mole^{-1}$)
He	10.22	2.576	21.07	SO_2[b]	252.0	4.290	—
Ne	35.7	2.789	27.10	F_2	112.0	3.653	—
Ar	124.0	3.418	49.58	Cl_2	357.0	4.115	—
H_2	33.3	2.968	31.67	Br_2	520.0	4.268	—
O_2	113.0	3.433	52.26	I_2	550.0	4.982	—
N_2	91.5	3.681	63.78	H_2S[b]	221.0	3.733	—
Air	97.0	3.617	60.34	H_2O[b]	380.0	2.65	—
CO[b]	110.0	3.590	67.22	NH_3[b]	320.0	2.60	—
CO_2	190.0	3.996	85.05	HCl[b]	360.0	3.305	—
CH_4	137.0	3.882	70.16	Hg	851.0	2.898	—

[a] From Hirschfelder *et al.* [2].

[b] These are polar molecules for which the force constants for the Lennard-Jones (6–12) potential are approximate; data from Krieger [3] and Rowlinson [4].

As examples, the transport properties calculated for helium and argon are shown in Fig. 6.1. The dotted curve marked v is the kinematic viscosity η/ρ.

With the force constants for the Lennard-Jones (6–12) potential, the principle of the corresponding states gives essentially constant values for the reduced pressure P^*, volume V^*, and temperature T^* for nonpolar and almost spherical

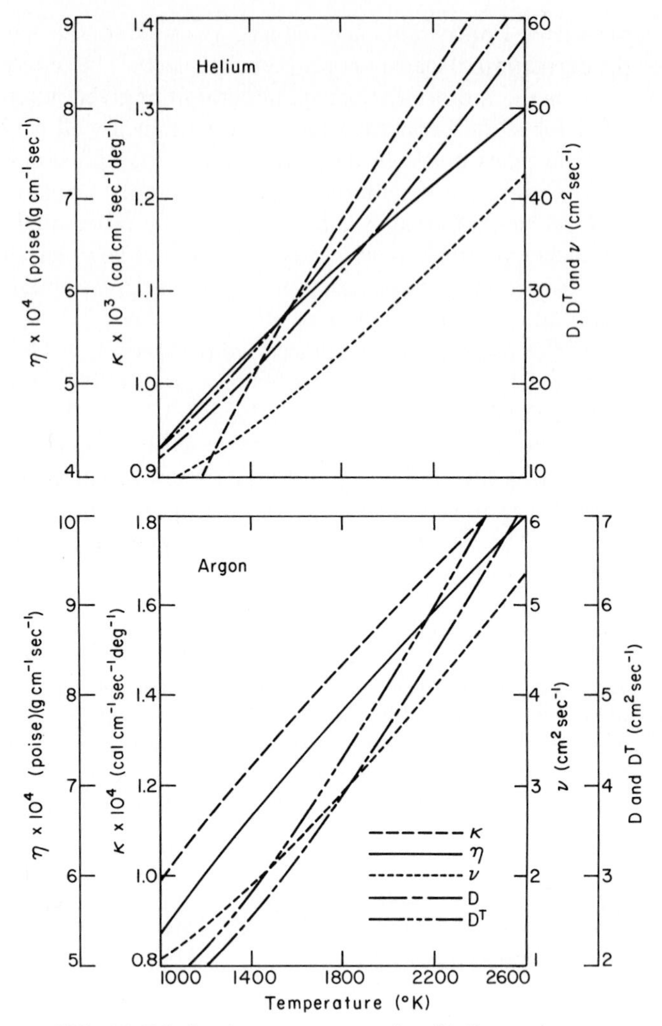

FIG. 6.1 Calculated transport properties of helium and argon.

molecules. With the exception of light gases, hydrogen and helium, the following approximations apply [2]:

$$P^* = P_c\delta^3/\epsilon = 0.126 \pm 0.015, \tag{6.16}$$

$$V^* = V_c/N\delta^3 = 3.00 \pm 0.32, \tag{6.17}$$

$$T^* = kT_c/\epsilon = 1.29 \pm 0.03, \tag{6.18}$$

$$P_cV_c/RT_c = 0.292 \pm 0.002, \tag{6.19}$$

where N is the Avogadro's number and the subscript c indicates P, V, and T at the critical point.

By using these empirical relations, the force constants may be estimated from the following relations involving critical, boiling, and melting temperature and molar volume:

$$\epsilon/k = 0.77T_c, \quad \epsilon/k = 1.15T_b, \quad \epsilon/k = 1.92T_m \qquad (^\circ\mathrm{K}); \tag{6.20}$$

$$\delta = 0.84V_c^{1/3}, \quad \delta = 1.17V_b^{1/3}, \quad \delta = 1.22V_m^{1/3} \qquad (\text{Å}). \tag{6.21}$$

It should be noted that the coefficient 1.22 for $V_m^{1/3}$ is close to that given by Eq. (3.112) for the rigid-sphere model of liquid metals, or that obtained from the Goldschmidt atomic diameter in Fig. 3.34.

As is shown in Fig. 6.2a, the collision diameters calculated [5] from the molar volumes are in close agreement with those derived from the viscosity data. The

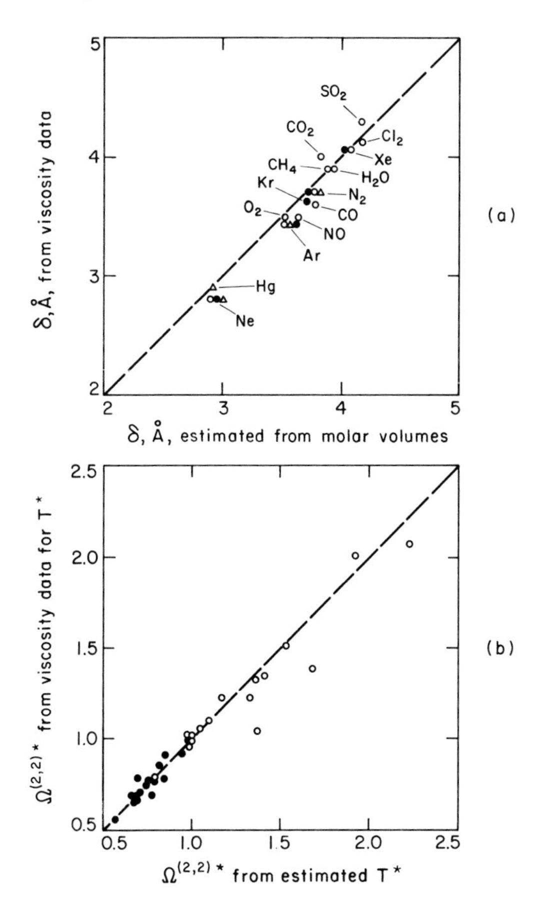

FIG. 6.2 Comparison of collision diameter and collision integral derived from viscosity data with those estimated from empirical relations in Eq. (6.20) and (6.21). In (a), δ is calculated from (●) V_m, (△) V_b, (○) V_c. In (b), T^* is at (○) 273 °K, (●) 1873 °K. After Turkdogan [5], reprinted with permission of the MIT Press from "Steelmaking: The Chipman Conference," edited by J. L. Elliott and T. R. Meadowcroft. Copyright © 1958, by Massachusetts Institute of Technology.

collision integrals $\Omega^{(2,2)*}$ for reduced temperature T^* at 273 and 1873 °K derived from the viscosity data are compared in Fig. 6.2b with those for T^* (at 273 and 1873 °K) estimated from the melting, boiling, and critical temperatures. The interdiffusivities in binary argon–metal vapors measured by Grieveson and Turkdogan [6] are shown in Fig. 6.3 to be in good agreement with those calculated from Eq. (6.8) with the force constants estimated from Eqs. (6.20) and (6.21).

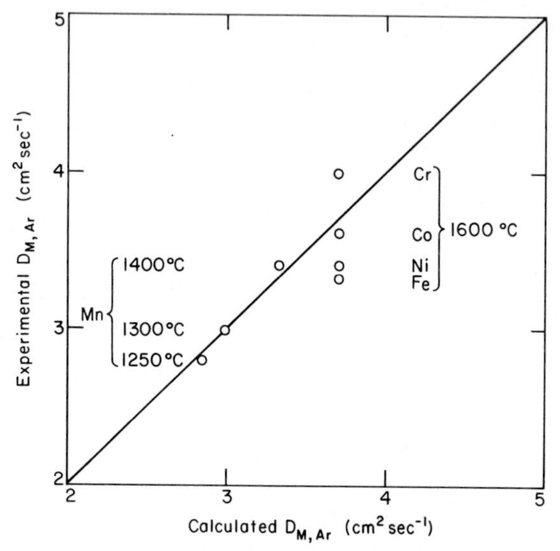

FIG. 6.3 Comparison of experimental interdiffusivities of metal vapor–argon mixtures at elevated temperatures with those calculated from Eq. (6.8) using force constants estimated from empirical relations (6.20) and (6.21). After Grieveson and Turkdogan [6].

In a system with both a concentration gradient and a temperature gradient, four types of fluxes are envisaged: (i) mass flux due to the concentration gradient (ordinary diffusion), (ii) energy flux due to the temperature gradient (thermal conductivity), (iii) mass flux due to the temperature gradient (thermal diffusion known as the *Soret effect*), and (iv) energy flux due to the concentration gradient (known as the *Dufour effect*). The thermodynamics of irreversible processes, the fundamentals of which were founded by Onsager [7], will not be discussed here. For the present purpose, only a brief reference is made below to the Soret effect in binary gas mixtures.

Let us consider a binary gas mixture contained in two large vessels connected to each other with a narrow tube. If the vessels are at different temperatures, there will be a mass flux through the tube due to the temperature gradient

$$J_1^{\mathrm{T}} = -cD_1^{\mathrm{T}}(d\ln T/dz), \tag{6.22}$$

where c is the gas concentration, D_1^T the thermal diffusivity of component 1, and z the diffusion distance. The resulting concentration gradient is opposed by the ordinary diffusion which tends to equalize composition. The opposed mass flux due to the concentration gradient is

$$J_1^D = -cD_{12}\, dx_1/dz. \tag{6.23}$$

When a steady state is reached, there will be no net flux; hence

$$J_1^T + J_1^D = -cD_{12}\left(\frac{D_1^T}{D_{12}}\frac{d\ln T}{dz} + \frac{dx_1}{dz}\right) = 0. \tag{6.24}$$

The ratio D_1^T/D_{12} is known as the *thermal diffusion ratio* k_T; inserting this in Eq. (6.24) gives for the thermal diffusion ratio

$$k_T = -(dx_1/dz)/(d\ln T/dz). \tag{6.25}$$

Since the diffusivity D_{12} is equal to D_{21}, the thermal diffusivities of components 1 and 2 are numerically the same but of opposite sign:

$$D_1^T = -D_2^T. \tag{6.26}$$

The generally adopted convention is that when k_T is positive the heavy component 1 migrates to the cold region; when k_T is negative the heavy component 1 moves to the hot region.

Although equations have been derived to calculate k_T by invoking the Chapman–Enskog theory [1], the computational procedure is cumbersome and the results are no better than $\pm 30\%$ of experimental values. Typical examples of thermal diffusion ratios are given in Table 6.3 for some binary 1:1 gas mixtures.

TABLE 6.3

Examples of Thermal Diffusion Ratios for 1:1 Gas Mixtures

Gas mixture	$\bar{T}$ (°K)	$k_T(\bar{T})$	Reference	Gas mixture	$\bar{T}$ (°K)	$k_T(\bar{T})$	Reference
He–H_2	330	0.0481	8	CH_4–H_2	445	0.0730	10
CO–H_2	142	0.0583	8	CO_2–CO	323	0.0142	11
H_2O–H_2	368	0.0588	9	CO_2–N_2	323	0.0142	11
CH_4–H_2	236	0.0530	10	CO_2–O_2	323	0.0136	11

Integration of Eq. (6.25) for a mean temperature of $\bar{T}$ gives the thermal segregation for a binary gas mixture in a closed system with a temperature difference $T'' > T'$.

$$\Delta x_1 = -k_T(\bar{T})\ln(T''/T'). \tag{6.27}$$

Since k_T is a complex function of temperature, measurements made over small temperature differences are used to obtain the value of k_T at a mean temperature of $\bar{T}$. Usually the mean temperature is that calculated from the equation suggested by Brown [12],

$$\bar{T} = [T''T'/(T'' - T')] \ln(T''/T'). \tag{6.28}$$

6.2 Dense Gases

In the rigorous kinetic theory of gases, it is assumed that the molecular diameter is small compared with the average distance between the molecules and that there are two-body collisions only. These assumptions are valid at low pressures and high temperatures for which the ideal gas law applies. In dense gases, where the volume of the molecule (rigid sphere) is comparable to the volume of the gas, the transfer of molecular properties is no longer due solely to the free motions of molecules between collisions, as was assumed to be the case for dilute gases. The kinetic theory developed by Enskog [13] for dense gases is for gases assumed to be made up of rigid spherical molecules. There are two basic features of the theory. (1) *Collisional transfer of momentum and energy*: in a two-body collision there is an instantaneous transfer of momentum and energy from the center of one molecule to the center of the other. (2) *Change in collision frequency*: because the diameter of the molecule δ (rigid sphere) is comparable to the mean free path, the collision frequency increases; on the other hand, since the molecules are close enough to shield one another from oncoming molecules, the frequency of collision decreases. The net result is that the frequency of collision in a dense gas made up of rigid spherical molecules differs by a factor of Y from that in a dilute gas.

The function Y is closely related to the *equation of state*, which is generally represented by

$$\frac{pV}{RT} = 1 + \frac{B(T)}{V} + \frac{C(T)}{V^2} + \frac{D(T)}{V^3} + \cdots, \tag{6.29}$$

where the temperature dependent functions $B(T)$, $C(T)$, ... are called the second, third, ... *virial coefficients*. The virial coefficients are positive and independent of temperature for the rigid-sphere model, for which the virial coefficients have been calculated in terms of the volume of the rigid spherical molecules [14, 15].

$$\frac{pV}{RT} = 1 + \frac{b_0}{V} + \frac{5}{8}\left(\frac{b_0}{V}\right)^2 + 0.2869\left(\frac{b_0}{V}\right)^3 + 0.1103\left(\frac{b_0}{V}\right)^4 + \cdots \tag{6.30}$$

and

$$Y = 1 + 0.6250\left(\frac{b_0}{V}\right) + 0.2869\left(\frac{b_0}{V}\right)^2 + 0.1103\left(\frac{b_0}{V}\right)^3 + \cdots, \tag{6.31}$$

where

$$b_0 = \tfrac{2}{3}\pi N\delta^3. \tag{6.32}$$

This, known as the *covolume* of the molecules, is four times the volume of the rigid molecules. The values of b_0 determined from virial coefficients are given in Table 6.3; when virial coefficients are not known, b_0 may be estimated from the collision diameters.

The transport properties of dense fluids derived by Enskog [13] by applying the rigid-sphere model to the Boltzmann distribution function, are summarized below in terms of dimensionless quantities relative to the properties in dilute gases, indicated by $^\circ$, and the reduced volume y, defined by

$$y = (b_0/V)Y. \tag{6.33}$$

shear viscosity:
$$\frac{\zeta}{\zeta^\circ}\frac{V}{b_0} = y^{-1} + 0.8 + 0.761y; \tag{6.34}$$

bulk viscosity:
$$\frac{\eta}{\eta^\circ}\frac{V}{b_0} = 1.00y; \tag{6.35}$$

thermal conductivity:
$$\frac{\kappa}{\kappa^\circ}\frac{V}{b_0} = y^{-1} + 1.20 + 0.755y; \tag{6.36}$$

self-diffusivity*:
$$\frac{D}{D^\circ}\frac{V}{b_0} = y^{-1}. \tag{6.37}$$

It should be noted that for rigid spherical molecules, y is given by the compressibility factor minus 1:

$$y = (pV/RT) - 1. \tag{6.38}$$

Enskog suggested that, in order to determine y function directly from the experimental p–V–T data, the pressure p in the above equation should be replaced by the *thermal pressure*, $T(\partial p/\partial T)_V$, thus

$$y = (V/RT)[T(\partial p/\partial T)_V] - 1. \tag{6.39}$$

For rigid-sphere molecules, this equation is identical to Eq. (6.38).

For high density fluids (when $b_0/V > 1.6$), the values of y obtained from the radial distribution function by Kirkwood *et al.* [16] are somewhat lower than those calculated from Eqs. (6.31) and (6.33); this is shown in Fig. 6.4.

The dimensionless transport properties calculated from the Enskog equations (6.34)–(6.37) are shown in Fig. 6.5 as functions of y; the values of y are those calculated by Kirkwood *et al.* for high density fluids. It should be noted that in dilute gases (ordinary pressures), the value of y approaches b_0/V and Y approaches unity. Therefore, as b_0/V approaches zero, the ratios ζ/ζ°, η/η°, κ/κ°, and D/D° approach unity.

*D° is the self-diffusivity at 1 atm pressure.

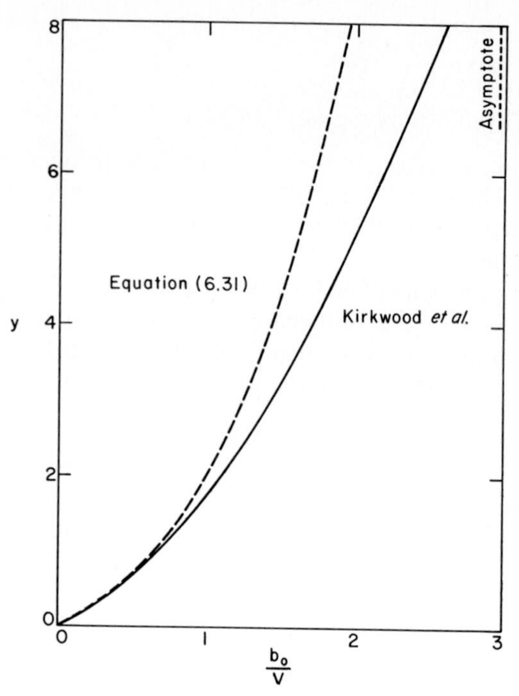

FIG. 6.4 Variation of y function with dimensionless volume b_0/V as calculated from Eq. (6.31) and that by Kirkwood *et al.* [16].

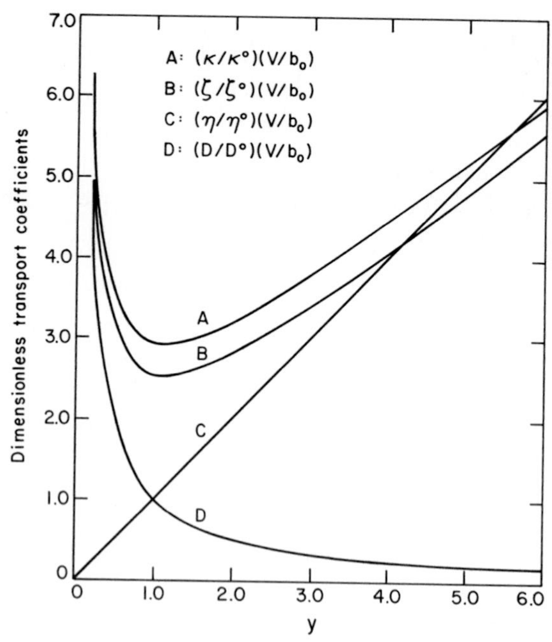

FIG. 6.5 Transport coefficients of a gas composed of rigid spherical molecules according to the Enskog theory.

As was shown by Chapman and Cowling [1], the experimental data on the viscosity of nitrogen at pressures up to 1000 atm are in close agreement with those calculated from Eq. (6.35). The minima on the curves for reduced shear viscosity and thermal conductivity times the reduced volume occur at y equal to 1.146 and 1.151, respectively.

The theoretical equations derived by Longuet-Higgins and Pople [17] for the transport properties of a fluid of rigid spheres are based on two assumptions: (1) the spatial pair distribution function depends only on the temperature and density, and not on the temperature gradient or rate of strain, and (2) the velocity distribution function of a single particle is Maxwellian with a mean equal to the local hydrodynamic velocity and a spread is determined by the local temperature. The equations they derived for bulk viscosity and self-diffusivity are identical to Enskog's expressions. It was shown in Section 3.4 that experimentally determined viscosities and self-diffusivities of liquid metals at their melting temperatures are in close agreement with those predicted from the theoretical equations for fluids of rigid spheres.

6.3 Gas Diffusion in Porous Media

Many scientists and engineers have contributed to the development of our present knowledge of the phenomenon of gas diffusion in porous media, with contributions by Knudsen, Clausing, Wicke, Present, Carman, and Hoogschagen—just to name a few—among those of the earlier workers.

6.3.1 Theory of Gas Diffusion in Pores

Since there are many comprehensive review papers on the development of the theory of gas diffusion in pores, further chronological review of the subject would be redundant. For the present purpose, it is adequate to give the basic equations describing the present state of the theory, with only a minimum of commentary on the development of the concepts and the derivation of the equations.

The total flux J_t in a porous medium consists of diffusive flux J and forced flow F:

$$J_t = J + F. \tag{6.40}$$

The pressure effect on diffusive flux is different from that for the forced flow. First we shall discuss the diffusive flux.

In the Chapman–Enskog derivation of the equation for mass diffusion in gas mixtures, the term for the pressure diffusion results from the momentum transfer in the collision of gas molecules. In a subsequent study, Zhadanov *et al.* [18] have evaluated the coefficient of the pressure diffusion for a binary system by considering the effect of viscous momentum transfer on diffusion. As shown by Mason *et al* [19], however, when the equation for molecular diffusion in binary gas mixtures is applied to the dusty gas model of Evans *et al.*

[20], the pressure and viscous-diffusion terms drop out of the equation, thus leading to the following expression for the unidirectional isothermal diffusive flux in a porous medium:

$$J_1 = -D_{e1}\, dc_1/dz + \delta_1 x_1(J_1 + J_2), \tag{6.41a}$$

$$J_2 = -D_{e2}\, dc_2/dz + \delta_2 x_2(J_1 + J_2), \tag{6.41b}$$

where dc_i/dz is the concentration gradient ($i = 1$ or 2), x_i is the mole fraction of components, D_{ei} is the effective diffusivity of component i, $\delta_i = D_{ei}/D_{12e}$, and D_{12e} is the effective molecular interdiffusivity.

These equations are for binary diffusion in large pores where the diffusion is via the collision between the gas molecules. For a free gas stream (no wall restriction) δ_i is unity, and Eq. (6.41) reduces to

$$J_1 = -D_{12}\, dc_1/dz + x_1(J_1 + J_2). \tag{6.42}$$

This is known as the *Maxwell–Stephan equation*, which is readily derived for binary gas mixtures at constant pressure from the general equation (3.145) for any binary solution for which the diffusivity is an invariant property of the system for a given temperature, pressure, and composition.

In discussing capillary or pore diffusion, it is more convenient to adopt the simpler (approximate) form of the kinetic theory of gases, for which the self-diffusivity is given by Eq. (6.1). When the pores are small enough such that the mean free path l of the molecules is comparable to the dimension of the pore, diffusion occurs via the collision of molecules with the pore walls. Knudsen [21] showed that the flux involving collision of molecules with and reflection from the surface of the capillary wall is represented by the equation

$$J = -\tfrac{2}{3}\bar{v}r\, dc/dz, \tag{6.43}$$

where $\bar{v}$ is the mean thermal molecular velocity, equal to $(8kT/\pi m)^{1/2}$, r is the radius of capillary tube, and dc/dz is the concentration gradient along the capillary. The coefficient is known as the Knudsen diffusivity; for r in units of centimeters and D_K in square centimeters per second,

$$D_K = \tfrac{2}{3}\bar{v}r = 9.7 \times 10^3 r(T/M)^{1/2}, \tag{6.44}$$

where M is the molecular weight of the diffusing species.

For molecular diffusion, the diffusivity is inversely proportional to pressure, while Knudsen diffusivity is independent of pressure. Therefore, in any porous medium, Knudsen diffusion predominates at low pressures where the mean free path is larger than the dimension of the pore.

Pollard and Present [22] derived a theoretical equation for the intermediate pressure range where Knudsen diffusion and molecular diffusion have to be considered. Figure 6.6 is reproduced from their paper and shows the dimensionless parameter $D/\bar{v}r$ as a function of the dimensionless pore size r/l. Curve IIIa, calculated by Pollard and Present, applies to diffusion in an infinitely long capillary at constant pressure. This curve is below curve II, which is for

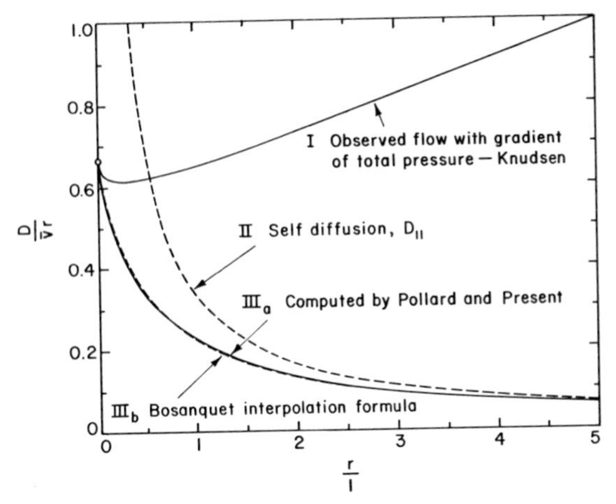

FIG. 6.6 Dimensionless diffusion parameter $D/\bar{v}r$ as a function of the ratio of tube radius to mean free path r/l. After Pollard and Present [22].

molecular self-diffusion, Eq. (6.1), but approaches curve III at large values of r/l. When r/l approaches zero, the diffusivity is that given by Eq. (6.44).

For the mixed region of molecular and Knudsen diffusion in capillaries, Bosanquet [23] derived the following expression based on a random walk model in which the successive movements of molecules are terminated by collision either with the capillary wall or with other molecules:

$$(D)^{-1} = (D_{Ki})^{-1} + (D_{ii})^{-1} = 2\bar{v}r/3(1 + 2r/l), \tag{6.45}$$

where D_{ii} is the molecular self-diffusivity. As is seen in Fig. 6.6, curve IIIb, calculated from Eq. (6.45) agrees well with that computed by Pollard and Present.

For a porous medium of uniform pore structure with pores of equal size, the Bosanquet interpolation formula gives for the effective diffusivity of component i,

$$D_{ei} = (\varepsilon/\tau)D_{12}D_{Ki}/(D_{12} + D_{Ki}), \tag{6.46}$$

where $\varepsilon =$ is the volume fraction of connected pores, equal to the connected pore volume V_p times the bulk density ρ_b, τ the tortuosity factor, D_{12} the molecular interdiffusivity for a binary mixture 1–2, and D_{Ki} the Knudsen diffusivity of component i for a given uniform pore radius r.

From the consideration of momentum transfer to the pore walls, Hoogschagen [24], Wicke and Hugo [25], Evans *et al.* [20], and Scott and Dullien [26] have shown that regardless of whether the mechanism of diffusion is Knudsen, molecular, or transitional, in all open systems involving isothermal pore diffusion of a binary gas mixture at uniform pressure the ratio of fluxes is inversely proportional to the square root of the ratio of molecular weights.

$$-J_2/J_1 = (m_1/m_2)^{1/2} = 1 - \alpha. \tag{6.47}$$

This theoretical relation has been verified experimentally over wide ranges of pore sizes, up to $r \sim 500\ \mu$m, and over a thousandfold pressure range [24–28]. Inserting this in Eq. (6.41) gives for the diffusive flux of component i

$$J_i = -[D_{ei}/(1 - \alpha\delta_i x_i)]\, dc_i/dz. \tag{6.48}$$

Next, let us consider the forced flow, which consists of two types of flow: viscous flow and slip flow. The slip flow is important only at low pressures, when the Knudsen diffusion predominates. For most cases of practical interest, the slip flow term can be omitted, and the forced flow under a pressure gradient across the porous medium is that due to viscous (Poiseuille) flow, which in terms of permeability is represented by

$$F = -(B_0 P/\eta RT)\, dP/dz. \tag{6.49}$$

From Eqs. (6.40), (6.41), (6.47), and (6.49), the total flux of the component i under isothermal conditions is given by

$$J_{it} = -\frac{D_{ei}}{1 - \alpha\delta_i x_i}\frac{dc_i}{dz} - \delta_i x_i \frac{B_0 P}{\eta RT}\frac{dP}{dz}, \tag{6.50}$$

where B_0 is the permeability in square centimeters, P the total pressure, and η the viscosity of the gas mixture. It should be noted that in accurate use of this equation due account should be taken of the pressure dependence of the diffusivity and the composition dependence of the viscosity.

Taking sum of the equations for components 1 and 2, as in Eq. (6.50), and substituting the corresponding values of D_{e1} and D_{e2} from Eq. (6.46) gives

$$\begin{aligned}(1-\alpha)J_{1t} + J_{2t} = &-\frac{\varepsilon}{\tau}\frac{D_{12}}{RT}\frac{1-\alpha}{1 - \alpha x_1 + (D_{12}/D_{K1})}\frac{dP}{dz} \\ &-\left[\frac{(1-\alpha)x_i}{1 + (D_{12}/D_{K1})} + \frac{1 - x_1}{1 + (D_{12}/D_{K2})}\right]\frac{B_0 P}{\eta RT}\frac{dP}{dz}.\end{aligned} \tag{6.51}$$

By rearrangement and change of notations, Eq. (6.51) can be expressed in different equivalent forms, as found in various publications.

The integration of Eq. (6.51) is complex. However, Mason *et al.* [19] and Gunn and King [29] obtained at least a partial integration in a closed form which gives mole fraction as a function of total pressure for the linear steady state case. They also demonstrated the application of these equations to experimental data on countercurrent diffusion of binary gas mixtures in porous materials under nonisobaric conditions.

When the Knudsen diffusion predominates ($D_{K1}, D_{K2} \ll D_{12}$, and therefore $\delta_1, \delta_2 \to 0$), the viscous flow term drops out, and Eq. (6.51) is simplified to

$$(1-\alpha)J_1 + J_2 = -\frac{\varepsilon}{\tau}\frac{D_{K1}}{RT}(1-\alpha)\frac{dP}{dz}. \tag{6.52}$$

For this limiting case, the slip flow term should, of course, be included.* When dP/dz approaches 0, the required relation is obtained, $-J_2/J_1 = 1 - \alpha$.

When molecular diffusion predominates ($D_{K1}, D_{K2} \gg D_{12}$, and therefore $\delta_1, \delta_2 \to 1$), Eq. (6.51) is simplified to

$$(1 - \alpha)J_{1t} + J_{2t} = -\frac{\varepsilon}{\tau}\frac{D_{12}}{RT}\frac{1 - \alpha}{1 - \alpha x_1}\frac{dP}{dz} - (1 - \alpha x_1)\frac{B_0 P}{\eta RT}\frac{dP}{dz}. \tag{6.55}$$

The following special cases of molecular diffusion are of particular experimental interest.

(1) Equimolar countercurrent diffusion $J_1 = -J_2$, for which Eq. (6.55) is transformed to

$$J_{1t} = \frac{\varepsilon}{\tau}\frac{D_{12}}{RT}\frac{1 - \alpha}{\alpha(1 - \alpha x_1)}\frac{dP}{dz} + \frac{1 - \alpha x_1}{\alpha}\frac{B_0 P}{\eta RT}\frac{dP}{dz}. \tag{6.56}$$

For equal molecular weights, $\alpha = 0$, the total pressure gradient becomes zero.

(2) Stagnant diffusion $J_2 = 0$, for which Eq. (6.55) is transformed to

$$J_{1t} = -\frac{\varepsilon}{\tau}\frac{D_{12}}{RT}\frac{1}{1 - \alpha x_1}\frac{dP}{dz} - \frac{1 - \alpha x_1}{1 - \alpha}\frac{B_0 P}{\eta RT}\frac{dP}{dz}. \tag{6.57}$$

For equal molecular weights, $\alpha = 0$, this equation is further reduced to

$$J_{1t} = -\frac{\varepsilon}{\tau}\frac{D_{12}}{RT}(1 + HP)\frac{dP}{dz}, \tag{6.58}$$

where $H = \tau B_0/\varepsilon\eta D_{12}$. For this simple limiting case ($J_2 = 0$, $\alpha = 0$), the integration gives the following relation between p_1 and P at any position in the porous plug [32]:

$$p_1 = P - (P^\circ - p_1^\circ)\exp[-H(P - P^\circ)], \tag{6.59}$$

where the superscript $^\circ$ indicates pressure at the outer surface of the plug exposed to the free gas stream.

6.3.2 Permeability

When the contribution of the diffusive flux and slip flow to the total flow under a pressure gradient is negligible, the permeability is derived from the

* The contribution to flow due to slip at the wall becomes important at low pressures and in small pores. The forced flow term due to slip at the wall, as derived by Scott and Dullien [30] for capillaries is represented (with modification for tortuosity in the porous medium) by

$$F = -(\varepsilon/\tau)(4r/3m\bar{v})\,dp/dz. \tag{6.53}$$

Wakao *et al.* [31] have suggested that for a binary gas mixture the average momentum may be expressed by

$$m\bar{v} = m_1\bar{v}_1 x_1 + m_2\bar{v}_2 x_2. \tag{6.54}$$

measured steady state flux by using the Poiseuille or Darcy equation (6.49). The permeability of a well-compacted powder mixture consisting of spheroidal particles of uniform size can be estimated from the Kozeny–Carman equation derived from pore-structure considerations [33],

$$B_0 = (1/bS_0{}^2)[\varepsilon^3/(1-\varepsilon)^2], \tag{6.60}$$

where b is a constant, ε is the fractional volume of connected interparticle pores, and S_0 is the surface area per unit volume of the particle. Carman [33] has shown that for most packed beds $b = 5$ is a good approximation. It should also be remembered that for spheroidal particles of any uniform size $\varepsilon = 0.476$ for bcc packing and $\varepsilon = 0.26$ for fcc packing. For well-compacted powder mixtures, $\varepsilon = 0.26$ is more appropriate. For a mixture of nonuniform particle size, ε can be less than 0.26.

Equation (6.60) should apply to packed beds made up of porous grains; in this case the values of S_0 and ε are those for the interparticle pores, exclusive of those for the micropores of the grains. This may be demonstrated by the following experimental observation.

In connection with a study of reactions in packed powder mixtures of porous Mn_3O_4 and C, permeability and other related measurements were made by Tien and Turkdogan [32]. As is seen from their data in Fig. 6.7, the measured permeabilities for reacted and unreacted beds are similar, $(2.3\text{–}3.2) \times 10^{-9}$ cm^2. An average value of B_0 equal to 3×10^{-9} cm^2 may be taken for this packed powder mixture. The sized powder Mn_3O_4 had the following properties of interest: particle diameter $d = 59 \pm 15$ μm, total pore area $S_p = 0.05$ m^{2}g^{-1}, $\rho_b = 2.89$ g cm^{-3}, ρ_b' (for stoichiometric mixture with 2.55% coconut charcoal) $= 2.76$ g cm^{-3} and ρ_b'' (for packed powder mixture) $= 2.2$ g cm^{-3}. For spheroidal par-

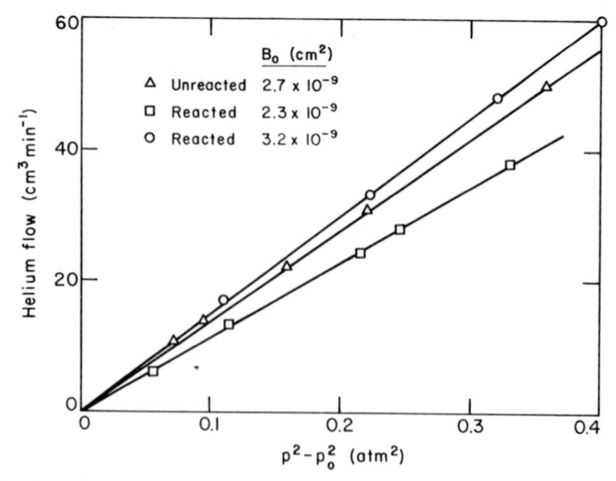

FIG. 6.7 Results of permeability experiments at 25°C and $P_0 = 0.96$ atm; sample cross-sectional area 1.77 cm^2. After Tien and Turkdogan [32].

ticles, the calculated average interparticle pore area is

$$S = 6/\rho_b' d = (0.039 \pm 0.01) \quad m^2\, g^{-1}$$

which is, as it should be, smaller than the total pore surface area including micropores in the grains. The interparticle void fraction is

$$\varepsilon = [1 - (\rho_b''/\rho_b')] = 0.20$$

which is close to $\varepsilon = 0.26$ for close packing of spherical particles of uniform size.

Inserting these values in Eq. (6.60) gives for the calculated permeability $(3.03 \pm 0.87) \times 10^{-9}\ cm^2$, which fortuitously is in excellent agreement with the experimental results in Fig. 6.7. After the Mn_3O_4 was reduced by carbon to MnO, the total pore surface area increased to $0.27\ m^2\, g^{-1}$, but the permeability of the reacted bed did not change; evidently the values of S and ε for the interparticle pores are not altered much upon reduction of the oxide particles in the bed. This example illustrates the usefulness and reliability of Eq. (6.60) in the calculation of the permeability from structural considerations of packed spherical particles.

6.3.3 Departures from Ideal Pore Structure

The ideal pore structure is defined such that the porous material appears to have circular pores of uniform size all of which are interconnected and intersect each other at an angle of 45°. The average pore radius $\bar{r}$ for an ideal pore structure is therefore given by the equation

$$\bar{r} = 2V_p/S_p, \tag{6.61}$$

where V_p is the pore volume per unit mass and S_p is the pore surface area per unit mass. The average pore length L_p is given by

$$L_p = \sqrt{2}V_b/S_b, \tag{6.62}$$

where V_b and S_b, respectively, are bulk volume and bulk (external) surface area of the porous sample.

It follows from the above definition of parameters for an ideal pore structure that the unidirectional flux relative to the superficial area is equal to $(\varepsilon/\sqrt{2})J'$ and the superficial distance is $(1/\sqrt{2})z'$, for which the tortuosity factor $\tau = 2$.

As stated by the Hagen–Poiseuille law, the viscous flow per unit area of a capillary of radius r is given by

$$F' = -\frac{r^2}{8\eta}\frac{P}{RT}\frac{dP}{dz'}. \tag{6.63}$$

Combining this with Eq. (6.49) and invoking the concept of the ideal pore structure gives the following relationship between the permeability and the average effective pore radius:

$$B_0 = \varepsilon\bar{r}^2/16. \tag{6.64}$$

In terms of the tortuosity factor τ, a more general form of this relation is

$$B_0 = \varepsilon\bar{r}^2/8\tau. \tag{6.65}$$

These equations are often used in estimating the flow characteristics of porous media from structural considerations. However, some caution should be exercised in the use of these and other similar equations because of varying departures from the ideal pore structure, as demonstrated below.

6.3.3a Diffusion in graphite. Evans *et al.* [34] measured the effective diffusivity (He–Ar) and permeability in a high purity graphite (AGOT–CS grade) along the extrusion direction. There was little or no Knudsen diffusion, the diffusivity ratio was found to be $D_e/D_{12} = 0.0086$, and with $\varepsilon = 0.22$, the tortuosity factor τ was estimated to be 25.7. From the measurement of pore size distribution, the average size of macropores $\bar{r}$ was found to be 2.5 μm. Inserting these values in Eq. (6.65) gives 0.67×10^{-10} cm^2 for the permeability B_0; this is about a third of the permeability measured experimentally. This comparison illustrates the expected order of accuracy of estimating the permeability in consolidated materials by using equations such as (6.65) from structural considerations.

In a study of the pore characteristics of carbons, Turkdogan *et al.* [35] found that at and above atmospheric pressure the molecular diffusion mechanism prevails in countercurrent diffusion of H_2–H_2O and CO–CO_2 mixtures in AUC electrode graphite. The temperature dependence of effective diffusivities at atmospheric pressure is shown as a log–log plot in Fig. 6.8. The straight lines are drawn with a slope of 3:2, corresponding to the temperature dependence of the molecular diffusion. Additional evidence in support of this conclusion is that the ratio $D_e(H_2\text{–}H_2O)/D_e(CO\text{–}CO_2)$ of 5.5 obtained from the measured values is close to the theoretical ratio for interdiffusivities in the free gas stream, $D_{12}(H_2\text{–}H_2O)/D_{12}(CO\text{–}CO_2) = 5.9$. The measured diffusivity ratio $D_e/D_{12} = 0.009 \pm 0.002$ and the fractional porosity $\varepsilon = 0.25$ gives $\tau = 29 \pm 6$, which is similar to the value obtained by Evans *et al.* [34].

In a low permeability graphite ($B_0 = 9.6 \times 10^{-14}$ cm^2), Evans *et al.* [36] found that, at pressures below 4 atm, there was a strong Knudsen diffusion effect on the countercurrent He–Ar diffusion. The effective interdiffusivity at atmospheric pressure and 25 °C was 1.06×10^{-4} cm^2 sec^{-1}; with D_{12}(He–Ar) = 0.747 cm^2 sec^{-1} and $\varepsilon = 0.11$, the tortuosity factor as defined by Eq. (6.46) is 775. Such a high tortuosity is indicative of a high degree of heterogeneity in the pore structure where there are large variations in the size of the diffusion path with many narrow passages or bottlenecks.

It is important to point out that the equations for diffusive flux derived from the kinetic theory of gases, the dusty gas model, and the Bosanquet relation are for homogeneous porous media having essentially a uniform pore size distribution. In such media, with macropores or micropores, there are modest departures from the ideal pore structure; consequently, the tortuosity factor is in the

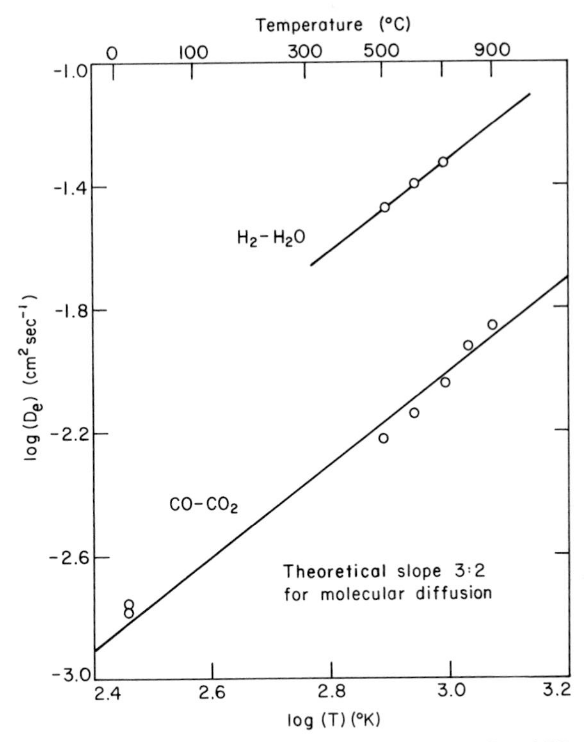

FIG. 6.8 Temperature dependence of effective diffusivity for H_2–H_2O and CO–CO_2 in electrode graphite. After Turkdogan *et al.* [35].

range from 2 to about 10. Constrictions on pores lead to higher values of τ, indicative of the heterogeneity of the medium in which the mechanism of flow cannot be described rigorously by equations intended for homogeneous media. However, these equations do provide a convenient means of difining the overall effective diffusivity.

6.3.3b Diffusion in porous iron. Studies made by Turkdogan *et al.* [37] of gas diffusion in porous iron, produced by hydrogen reduction of lump hematite ore or sintered hematite pellets, provided useful information on the pore structure of reduced iron. The diffusion of CO_2 in a stagnant atmosphere of helium was measured at 20 °C and at pressures from 0.01 to 10 atm with samples previously reduced in hydrogen at 800–1000 °C. The pressure drop across the sample was small enough to be neglected, and hence, a simplified form of the diffusion equation could be used, which in an integrated form is

$$J_{CO_2} = \frac{P(D_{12})_e}{LRT} \ln\left[\frac{1 + (D_{12})_e/(\bar{D}_K)_e}{1 - x_{CO_2} + (D_{12})_e/(\bar{D}_K)_e}\right], \tag{6.66}$$

where the subscript e indicates the effective diffusivity; $(\bar{D}_K)_e$ is the effective Knudsen diffusivity CO_2 for the average micropore radius $\bar{r}_a$, and L is the sample

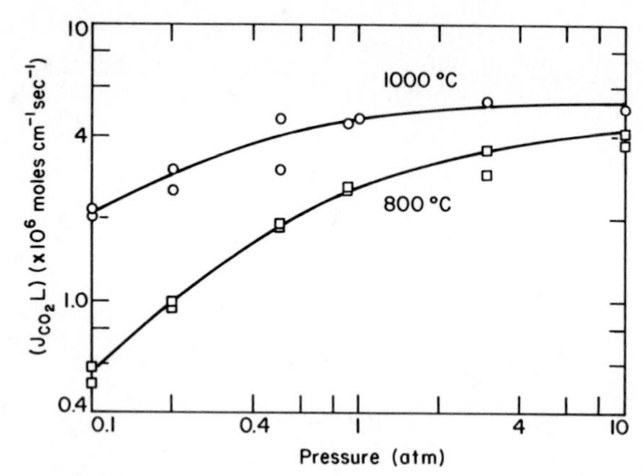

FIG. 6.9 Flux × sample length for diffusion of CO_2 in stagnant helium at 20 °C as a function of total pressure. After Turkdogan *et al.* [37].

thickness. The curves drawn through the data points in Fig. 6.9 for the experimental parameter (LJ_{CO_2}) versus P were calculated from Eq. (6.66), and the values of $(D_{12})_e$ and $(\bar{D}_K)_e$ chosen to give the best fit to the data. These are given in Table 6.4, together with other measured or calculated values.

Close agreement between the calculated and measured diffusivities indicates that the ideal pore structure ($\tau = 2$) is a good approximation, and that both molecular and Knudsen diffusion processes occur mainly via the macropores which constitute most of the pore volume. It looks as though this material has essentially uniform macropores intertwined with a network of micropores which evidently do not constrict the diffusion path. Such a medium is considered to be a *bidisperse* or *biomodal* pore system.

In this type of bimodal pore system, the average pore radius $\bar{r}$ derived from Eq. (6.61) will give a wrong estimate of the Knudsen diffusivity. For the examples in Table 6.4, S_p is equal to 4.2 and 0.48 m^2 g^{-1} for iron reduced at 800 and

TABLE 6.4

Comparison of Measured Effective Diffusivities for He–CO_2 *at* 20 °C *in Reduced Iron with Those Calculated for an Ideal Pore Structure*

Reduction temperature (°C)	Porosimeter data[a] (μm)		Measured at 20 °C (cm^2 sec^{-1})		Calculated[b] (cm^2 sec^{-1})	
	$\bar{r}_a$	$\bar{r}_i$	$P(D_{12})_e$	$(D_K)_e$	$P(D_{12})_e$	$(D_K)_e$
800	0.3	0.01	0.16	0.3	0.19	0.26
1000	1.5	0.2	0.19	1.6	0.19	1.28

[a] $\varepsilon = 0.68$, $V_p = 0.26$ cm^3 g^{-1}.
[b] PD_{12}(He–CO_2) = 0.56 cm^2 sec^{-1} at 20 °C and for $\tau = 2$.

1000 °C; with V_p equal to 0.26 cm^3 g^{-1} in both cases, the estimated values of $\bar{r}$ are 0.12 and 1.08 μm, and the corresponding Knudsen diffusivities are lower than those for $\bar{r}_a$.

6.3.4 Random Pore Model

Complexities in the understanding of pore diffusion arising from the heterogeneity of the pore structure have long been realized. Of the various models proposed, that conceived by Wakao and Smith [38] is simpler and well substantiated by the experimental data. In the model proposed by Wakao and Smith, diffusion is considered to take place simultaneously along three paths, two in parallel and one in series:

(1) through macropores of average radius $\bar{r}_a$ and void fraction ε_a,
(2) through micropores of average radius $\bar{r}_i$ and void fraction ε_i, and
(3) through macropores and micropores in series.

The sum of these three diffusive fluxes gives the total diffusive flux, which for constant pressure simplifies to (here given in a slightly different and simpler form)

$$J_1 = -\left[\varepsilon_a{}^2 D_a + \varepsilon_i{}^2 D_i + 4\varepsilon_a\varepsilon_i \frac{D_a D_i}{D_a + D_i}\right]\frac{dc_1}{dz}, \tag{6.67}$$

$$D_a = \frac{D_{12}}{(1 - \alpha x_1) + D_{12}/D_{K1a}}, \qquad D_i = \frac{D_{12}}{(1 - \alpha x_1) + D_{12}/D_{K1i}}.$$

Since the total porosity is $\varepsilon = \varepsilon_a + \varepsilon_i$, for a homogeneous pore system either ε_a or ε_i is zero, and Eq. (6.67) is reduced to

$$J_1 = -\varepsilon^2 \frac{D_{12}}{(1 - \alpha x_1) + D_{12}/D_{K1}} \frac{dc_1}{dz}. \tag{6.68}$$

This is analogous to Eq. (6.48), and inserting the equivalence of D_{ei} from Eq. (6.46), the random pore model gives for the tortuosity of a homogeneous pore system

$$\tau = 1/\varepsilon. \tag{6.69}$$

This relation would hold for an ideal pore structure only when ε is equal to 0.5.

The bidisperse pore model of Wakao and Smith was later extended to a tridisperse pore system by Cunningham and Geankoplis [39]. An even more complicated model was developed by Foster and Butt [40].

6.3.5 Gas Diffusion in Coals and Zeolites

It is now a well established fact that the pore structure of coal is essentially of bimodal type: that is, it consists of (1) macropores and micropores due to cracks, fissures, and interparticle voids, and (2) ultrafine pores in the colloidal body of the coal. This broad classification of the pore structure of coal was derived from the earlier findings of Maggs [41] and of Zwietering and van Krevelen [42]. They observed marked increases in the sorption in coal of gases such as methane, carbon dioxide, nitrogen, and hydrogen with increasing temperature. These findings were indicative of activated diffusion in the colloidal body of the coal, which has a capacity for gas absorption greater than that of the surfaces of macropores and micropores.

The subject of ultrafine pore structure of coals is well covered in review papers by Bond and Spencer [43], Marsh [44], and others. However, a comment or two should be made on what is meant by ultrafine pores. Surely, the so-called ultrafine pores, 5–10 Å diameter, are no more than cavities in the network of covalently bonded constituent atoms in the coal polymer, as in silicate glasses or zeolites. The capability of such media to absorb gas or liquid depends on the type and arrangement of the atoms in the vicinity of the network cavities in the colloidal body of the coal and on the polar interaction of the cavities with the gas or liquid that is being absorbed.

It is still customary to interpret the sorption data in terms of the monolayer coverage on the walls of the ultrafine pores. The internal surface areas thus derived are, of course, 1 or 2 orders of magnitude greater than the surface areas of macropores and micropores in the coal, as normally determined by the N_2-adsorption (BET) technique. However, as pointed out by Spencer and Bond [45], the surface areas of ultrafine pores in coals should not be reported, because of the variations in the values obtained from one sorbate to another.

It is more meaningful to speak of the gas sorption capacity of the coal without any attempt to interpret the solubility data as though they represent monolayer coverage on the hypothetical "walls" of the lattice vacancies or network cavities. Since the network cavities in polymers such as glasses, zeolites, and coals are of intermolecular dimensions, they should not be considered as pores. That is, a subunit particle having the property of a molecular sieve in relation to sorption and diffusion is not a porous particle. On the basis of this criterion, nitrogen adsorption at $-195\,^\circ\mathrm{C}$ should give a truer estimate of the pore surface area, because of negligible absorption of nitrogen in the subunit solid body of the coal particles at such low temperatures.

This line of reasoning was, in fact, evident from Barrer's earlier studies of molecular diffusion in zeolites. For example, Barrer [46] interpreted the kinetics of sorption of gases and vapors in zeolites in terms of diffusion in a single-phase solid solution. The rate equation he used was a solution of Fick's second law, which when simplified for small particles and short reaction times is given by

$$F = (2S/V)(Dt/\pi)^{1/2}, \tag{6.70}$$

where F is the fraction of gas absorbed or desorbed, S the external surface area of the subunit particle, and V the volume of the subunit particle. Since this is an activated diffusion process, the temperature dependence of the diffusivity is given by

$$D = D_0 \exp(-E/RT), \tag{6.71}$$

where E is the heat of activation required for the diffusing species to jump from one network cavity to another. Zwietering *et al.* [47], and subsequently others, interpreted the rate of sorption of gases in coal particles by using Eq. (6.70) or its equivalence.

In later studies, Walker *et al.* [48, 49] and Ruthven *et al.* [50, 51] employed a more exact transient solution of the diffusion equation for longer reaction times; the diffusion parameter derived from the data on the rate of sorption is usually given in terms of the ratio $D^{1/2}/a$, where a is the radius of the subunit coal particle free of pores. For spherical particles, a is equal to $3V/S$, and for cylindrical particles a is equal to $2V/S$; the value of a is estimated from the volume V of the particles measured by the helium displacement method and the external surface area S, measured by the (BET) N_2 adsorption at $-195\,°C$. According to the measurements of Nandi and Walker [49] for various types of high- and low-rank coals, the estimated value of a is in the range 0.03–0.09 μm, which is several orders of magnitude smaller than the granules used in the sorption experiments, 100–500 μm diameter. These authors determined the values of ${D_0}^{1/2}/a$ and E for methane in various types of coals; these parameters were also evaluated by Nelson and Walker [48] for propane, ethane, argon, and nitrogen in Linde type A synthetic zeolites.

From a plot of the available data in Fig. 6.10, we see that the heat of activation for diffusion is a simple function of $D_0^{1/2}/a$ for vastly different gases diffusing in

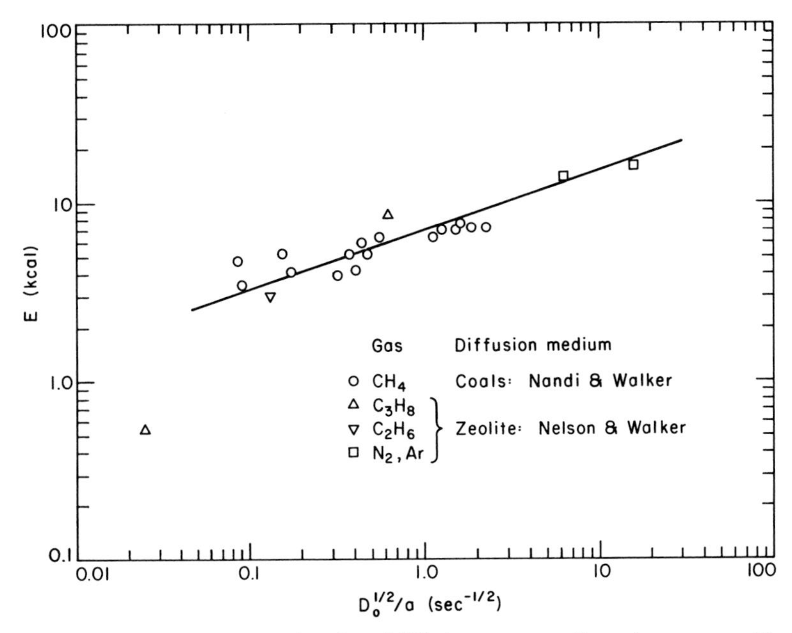

FIG. 6.10 Heat of activation as a function of diffusion parameter for various gases and in coals and zeolites.

various types of coal and zeolite. For the systems considered, this correlation may be represented by

$$E = 7(D_0^{1/2}/a)^{1/3}, \tag{6.72}$$

with E in kilocalories and $D_0^{1/2}/a$ in (seconds)$^{-1/2}$.

For methane in anthracite, the diffusivity is about 10^{-14} cm^2 sec^{-1} at 100 °C. This is to be compared with the interstitial diffusion of nitrogen and carbon in α iron at 100 °C: $D_N = 7.8 \times 10^{-14}$ and $D_C = 2.5 \times 10^{-15}$ cm^2 sec^{-1}. The diffusivity of H_2O in silica at 1000 °C is 3×10^{-10} cm^2 sec^{-1}; the extrapolation of the data for methane in anthracite to the hypothetical temperature of 1000 °C would give for the methane diffusivity 0.7×10^{-10} cm^2 sec^{-1}. Such similarities in diffusivities substantiate the view that the mechanism of diffusion of gases in coals is not much different from the interstitial or vacancy diffusion in other solids.

In large samples, the rate of sorption will also be affected by diffusion of gas in the pores surrounding the subunit solid particles of the coal. Denoting the effective diffusivity in the pores by D_p and the radius of the spherical coal sample by a_p, the critical size of the sample for the onset of the pore diffusion control is obtained from the relation

$$a_p > (D_p/D)^{1/2} a. \tag{6.73}$$

For example, if D_p is 10^{-4} cm^2 sec^{-1} and $D^{1/2}/a$ is 10^{-2} sec$^{-1/2}$ at 100 °C, the rate of sorption will be controlled primarily by diffusion in the subunit particles of the coal when the sample radius is less than about 1 cm. It follows from this simple reasoning that the rate of methane drainage in the coal seam is controlled by diffusion and viscous flow of gas in the pores and cracks in the coal bed. Some aspects of migration of methane in coal seams have been investigated in some detail by Kissell [52].

References

1. S. Chapman and T. G. Cowling, "The Mathematical Theory of Nonuniform Gases." Cambridge Univ. Press, London, 1939 (3rd Ed., 1970).
2. J. O. Hirschfelder, C. F. Curtiss, and R. B. Bird, "Molecular Theory of Gases and Liquids." Wiley, New York, 1954.
3. F. J. Krieger, "The Viscosity of Polar Gases," Proj. Rand Rep. RM-646 (July 1951).
4. J. S. Rowlinson, *Trans. Faraday Soc.* **45**, 974 (1949).
5. E. T. Turkdogan, *in* "Steelmaking: The Chipman Conference" (J. F. Elliott and T. R. Meadowcroft, eds.), p. 77. MIT Press, Cambridge, Massachusetts, 1958.
6. P. Grieveson and E. T. Turkdogan, *J. Phys. Chem.* **68**, 1547 (1964).
7. L. Onsager, *Phys. Rev.* **37**, 405 (1931); **38**, 2265 (1931).
8. E. R. S. Winter, *Trans. Faraday Soc.* **46**, 81 (1950).
9. Z. Shibata and H. Kitagawa, *J. Fac. Sci., Hokkaido Imp., Univ.* **2**, 223 (1938); recalculated by E. Whalley, *J. Chem. Phys.* **19**, 509 (1951).
10. H. G. Drickamer, S. L. Downey, and N. C. Pierce, *J. Chem. Phys.* **17**, 408 (1949).
11. T. L. Ibbs and L. Underwood, *Proc. Phys. Soc.* **39**, 227 (1926–1927).
12. H. Brown, *Phys. Rev.* **58**, 661 (1940).

13. D. Enskog, *K. Sven. Vetenskapsakad. Handl.* **63**, No. 4 (1921).
14. R. Clausius, *Mech. Warmetheorie* **3** (2nd Ed.), 57 (1879).
15. M. N. Rosenbluth and A. W. Rosenbluth, *J. Chem. Phys.* **22**, 881 (1954).
16. J. G. Kirkwood, E. K. Maun, and B. J. Alder, *J. Chem. Phys.* **18**, 1040 (1950).
17. H. C. Longuet-Higgins and J. A. Pople, *J. Chem. Phys.* **25**, 884 (1956).
18. V. Zhadanov, Y. Kagan, and A. Sazykin, *Sov. Phys.—JEPT* **15**, 596 (1962).
19. E. A. Mason, A. P. Malinauskaus, and R. B. Evans, III, *J. Chem. Phys.* **46**, 3199 (1967).
20. R. B. Evans, III, G. M. Watson, and E. A. Mason, *J. Chem. Phys.* **35**, 2076 (1961); **36**, 1894 (1962).
21. M. Knudsen, *Ann. Phys.* (*Leipzig*) **28**, 75 (1909).
22. W. G. Pollard and R. D. Present, *Phys. Rev.* **73**, 762 (1948).
23. C. H. Bosanquet, Br. TA Rep. BR-507 (September 1944).
24. J. Hoogschagen, *Ind. Eng. Chem.* **47**, 906 (1955).
25. E. Wicke and P. Hugo, *Z. Phys. Chem.* **28**, 401 (1961).
26. D. S. Scott and F. A. L. Dullien, *Chem. Eng. Sci.* **17**, 771 (1962).
27. H. A. Kramers and J. Kistemaker, *Physica* (*Utrecht*) **10**, 699 (1943).
28. J. P. Henry, R. S. Cunningham, and C. J. Geankoplis, *Chem. Eng. Sci.* **22**, 11 (1967).
29. R. D. Gunn and C. J. King, *AIChE J.* **15**, 507 (1969).
30. D. S. Scott and F. A. L. Dullien, *AIChE J.* **8**, 293 (1962).
31. N. Wakao, S. Otani, and J. M. Smith, *AIChE J.* **11**, 435 (1965).
32. R. H. Tien and E. T. Turkdogan, *Metall. Trans. B* **8**, 305 (1977).
33. P. C. Carman, "Flow of Gases Through Porous Media." Buttersworth, London, 1956.
34. R. B. Evans, III, J. Truitt, and G. M. Watson, *J. Chem. Eng. Data* **6**, 522 (1961).
35. E. T. Turkdogan, R. G. Olsson, and J. V. Vinters, *Carbon* **8**, 545 (1970).
36. R. B. Evans, III, G. M. Watson, and J. Truitt, *J. Appl. Phys.* **33**, 2682 (1962).
37. E. T. Turkdogan, R. G. Olsson, and J. V. Vinters, *Metall. Trans.* **2**, 3189 (1971).
38. N. Wakao and J. M. Smith, *Chem. Eng. Sci.* **17**, 825 (1962).
39. R. S. Cunningham and C. J. Geankoplis, *Ind. Eng. Chem., Fundam.* **7**, 535 (1968).
40. R. N. Foster and J. B. Butt, *AIChE J.* **12**, 180 (1966).
41. F. A. P. Maggs, *Res. Corres.* **6**, 13 (1953).
42. P. Zwietering and D. W. van Krevelen, *Fuel* **33**, 331 (1954).
43. R. L. Bond and D. H. T. Spencer, *in* "Industrial Carbon and Graphite," p. 231. Soc. Chem. Ind., London, 1958.
44. H. Marsh, *Fuel* **44**, 253 (1965).
45. D. H. T. Spencer and R. L. Bond, *in* "Coal Science Conference," p. 724. Pennsylvania State Univ., University Park, 1964.
46. R. M. Barrer, *Trans. Faraday Soc.* **45**, 358 (1949).
47. P. Zwietering, J. Overseem, and D. W. van Krevelen, *Fuel* **35**, 66 (1956).
48. E. T. Nelson and P. L. Walker, *J. Appl. Chem.* **11**, 358 (1961).
49. S. P. Nandi and P. L. Walker, *Fuel* **49**, 309 (1970); **54**, 81 (1975).
50. D. M. Ruthven and K. F. Loughlin, *Chem. Eng. Sci.* **26**, 577 (1971).
51. K. F. Loughlin, R. I. Derrah, and D. M. Ruthven, *Can, J. Chem. Eng*, **49**, 66 (1971).
52. F. N. Kissell, Rep. Invest. RI 7649 and RI 7667, Methane Migration Characteristics of the Pocahontas No. Coalbed. U.S. Bur. Mines, Washington, D.C. (1972).

PART II

Applications

CHAPTER 7

RATE PHENOMENA

7.1 Introduction

The theme of Part II is the application of the principles of thermodynamics and physical chemistry in high temperature technology. Particular emphasis will be given to understanding the technical workings of some industrial processes as an aid in the development of methods of optimization and control of processes. So-called high temperature technology extends over a wide range of diversified fields; in addition, what is meant by high temperature technology depends on one's point of view. However, the topics chosen for detailed discussion with reference to particular industrial processes are common to many areas of high temperature technology.

The principles of rate phenomena are a prerequisite to a discussion of specific aspects of high temperature reactions and the characteristic features of high temperature systems that are frequently encountered in industrial processes.

During the past two decades, vast advances have been made in the application of the concepts of chemical engineering technology to the study of rates of reactions of interest to high temperature technology. The advent of computer science played a particularly significant role in the advancements made in the development of mathematical methods of systems analysis for the study of industrial processes and their control. This aspect of rate phenomena is beyond the scope of Chapter 7; for chemical engineering aspects of rate phenomena, reference may be made, for example, to textbooks by Bird *et al.* [1] and by Szekely and Themelis [2].

This chapter is confined to the discussion of certain physicochemical aspects of rate phenomena, with reference to the limiting cases of rate controlling steps that are often used in the mathematical formulation of the systems analysis in the study of industrial processes. The following limiting cases of rate phenomena are discussed: kinetics of interfacial reactions, nucleation and growth, heat and mass transfer in the gas–film layer, heat transfer by radiation, heat and mass transfer in the product layer, reaction of gases with porous media, mass transfer across liquid interfaces, and gas bubbles in liquids.

7.2 Kinetics of Interfacial Reactions

In most high temperature reactions of experimental or practical interest, we are concerned primarily with heterogeneous reactions involving surfaces and two or more phases. With application to experiments in mind, only the phenomenological concepts of kinetics of interfacial reactions are discussed here, with particular reference to the role of adsorption of surface active elements on the rates of high temperature reactions. The discussion of the atomistic details of reaction kinetics is not warranted here because of the lack of experimental evidence to substantiate the mechanisms of interfacial reactions postulated from the atomistic models and the kinetic theory.

7.2.1 Theoretical Considerations

Unlike the laws of thermodynamics, reaction rate theory is far from being rigorous. However, much progress has been made in the understanding of this subject during the past five decades, and many have contributed to the development of reaction rate theory; the contributions made by Eyring are the most outstanding [3–5]. The theory is based essentially on two principal concepts: (i) the formation of an activated complex in equilibrium with the reactants adsorbed on the surface and (ii) the universal specific rate for the decomposition of the activated complex.

For the case of a fast rate of transport of reactants and products to and from the reaction site and a rapid rate of nucleation of a second phase, the rate is controlled by a chemical reaction occurring in the adsorbed layer at the interface. The reaction between adsorbed species L and M on the surface producing Q via an activated complex $(LM)^\ddagger$ is represented by

$$L + M = (LM)^\ddagger \rightarrow \text{product } Q. \tag{7.1}$$

For a given temperature the equilibrium constant $K_\ddagger$ is given by

$$K_\ddagger = a_\ddagger / a_L a_M, \tag{7.2}$$

where $a_\ddagger$ is the activity of the activated complex and a_L and a_M are the activities of the reactants in the chemisorbed layer. The equilibrium constant $K_\ddagger$ may be represented in a more general form by

$$K_\ddagger = a_\ddagger / \prod a_i^{\nu_i}, \tag{7.3}$$

where $\prod a_i^{\nu_i}$ is the product of the activities of the reactants in equilibrium with the activated complex.

Next to be considered is the specific rate of decomposition of the activated complex to the overall reaction product, represented by

$$\mathscr{R} = (kT/h)\Gamma_\ddagger = (kT/h)\Gamma_\circ \theta_\ddagger \qquad \text{(specific rate)}, \tag{7.4}$$

where kT/h is the universal rate (k, the Boltzmann constant, $= 1.380 \times 10^{-16}$ erg deg^{-1}, and h, Planck's constant $= 6.626 \times 10^{-27}$ erg sec), $\Gamma_{\ddagger}$ the concentration of the activated complex in the adsorbed layer, in molecules per square centimeter and $\Gamma_{\circ}$ the total number of adsorption sites at the surface, $\sim 10^{15}$ cm^{-2}, and hence the fractional coverage by the activated complex $\theta_{\ddagger} = \Gamma_{\ddagger}/\Gamma_{\circ}$.

For single-site occupancy by the activated complex in the adsorbed layer, the activity of the complex is represented by

$$a_{\ddagger} = \varphi_{\ddagger}\theta_{\ddagger}/(1 - \theta), \tag{7.5}^*$$

where $\theta = \sum \theta_i$ the total fractional occupancy of the sites by the adsorbed species i, and $\varphi_{\ddagger}$ is the activity coefficient of the complex in the adsorbed layer.

Assuming that the activated complex is the same for the forward and reverse reactions, i.e., microscopic reversibility of the reaction, the isothermal net reaction rate is given by

$$\mathcal{R} = (kT/h)\Gamma_{\circ}(K_{\ddagger}/\varphi_{\ddagger})(1 - \theta)\left(\prod a_i^{\nu_i} - \prod_{\text{eq}} a_i^{\nu_i}\right), \tag{7.6}$$

where $\prod \text{eq}\, a_i^{\nu_i}$ is the value that would prevail at equilibrium with the products and is found from the equilibrium constant and the activities of the reaction products.

At low site fillage, the term $1 - \theta$ is approximately unity, and since $\theta_{\ddagger}$ is very small, $\varphi_{\ddagger}$ is essentially constant, and Eq. (7.6) reduces to

$$\mathcal{R} = (kT/h)\Gamma_{\circ}(K_{\ddagger}/\varphi_{\ddagger})\left(\prod a_i^{\nu_i} - \prod_{\text{eq}} a_i^{\nu_i}\right). \tag{7.7}$$

At high site fillage, the term $1 - \theta$ approaches zero. For this limiting case of almost complete surface coverage by a single species p,

$$1 - \theta = \varphi_{\text{p}}'/a_{\text{p}}. \tag{7.8}$$

This solution is usable only when the high site fillage is dominantly by a single species. It is only in this case that each site has essentially the same surroundings in the adsorbed layer and hence the activity coefficients φ_{p}' and $\varphi_{\ddagger}'$ are constant. Therefore, for the limiting case of $1 - \theta \to 0$, the rate equation is

$$\mathcal{R} = (kT/h)\Gamma_{\circ}(K_{\ddagger}\varphi_{\text{p}}'/\varphi_{\ddagger}')\left(\prod a_i^{\nu_i} - \prod_{\text{eq}} a_i^{\nu_i}\right)/a_{\text{p}}, \tag{7.9}$$

where $\varphi_{\ddagger}'$ and φ_{p}' are the values of $\varphi_{\ddagger}$ and φ_{p} as $1 - \theta$ approaches zero; for the Langmuir-type ideal monolayer discussed in Section 3.3.4, $\varphi_{\ddagger}$ is equal to $\varphi_{\ddagger}'$ and φ_{p} is equal to φ_{p}'.

The heat of activation of the rate controlling reaction is obtained from the slope of the plot of the logarithm of the rate constant, $\ln \Phi$ versus the reciprocal of the absolute temperature.†

* When the activated complex occupies x number of sites, $a_{\ddagger} = \varphi_{\ddagger}\theta_{\ddagger}/(1 - \theta)^x$.

† Over a wide temperature range, it is more correct to plot $\ln(\Phi/T)$ against $1/T$.

The true heat of activation of the reaction is that associated with the coefficient of temperature dependence of $\ln K_{\ddagger}$. However, the temperature dependence of the rate constant determined experimentally involves three parameters, $K_{\ddagger}$, φ_p, and $\varphi_{\ddagger}$, which are all temperature dependent. Hence, the temperature dependence of the rate constant obtained experimentally gives only the apparent heat of activation, which may change with the value of $1 - \theta$, or the activity of strongly adsorbed species.

For an ideal monolayer involving a single adsorbed species p, Eq. (3.96) is transformed to

$$1 - \theta = \varphi_p/(a_p + \varphi_p). \tag{7.10}$$

Inserting this in Eq. (7.6) gives, for the net rate of reaction in an ideal monolayer,

$$\mathscr{R} = (kT/h)\Gamma_{\circ}(K_{\ddagger}\varphi_p/\varphi_{\ddagger})\left(\prod a_i^{\nu_i} - \prod_{\text{eq}} a_i^{\nu_i}\right)/(a_p + \varphi_p). \tag{7.11}$$

With the exception of a few isolated cases, the thermodynamics of chemisorption are not known for many systems of experimental interest to warrant detailed interpretation of the rates of interfacial reactions in terms of the kinetic theory. In view of this limitation of our knowledge of the thermodynamics of chemisorbed layers, it is more realistic to write the isothermal rate equation (7.6) in a general phenomenological form of the type

$$\mathscr{R} = \Phi(1 - \theta)\left(\prod a_i^{\nu_i} - \prod_{\text{eq}} a_i^{\nu_i}\right), \tag{7.12}$$

where the overall rate constant Φ is an exponential function of temperature, and $1 - \theta$ is a function of temperature and activity of strongly adsorbed species. If there is more than one surface active species in the system, the formulation of the equivalence of $1 - \theta$ and hence the interpretation of the rate data, is subject to speculation.

For a given surface coverage, the temperature dependence of the rate constant may be represented by the Arrhenius-type equation

$$\Phi = [\Phi_{\circ} \exp(-\Delta H_e/RT)]_{\theta}, \tag{7.13}$$

where $\Phi_{\circ}$ is a preexponential constant and ΔH_e is the apparent or experimental heat of activation for the reaction at a given site fillage θ. The apparent heat of activation should not be confused with the enthalpy of formation of the activated complex, denoted by $\Delta H_{\ddagger}$.

In the foregoing consideration of the rates of interfacial reactions, the chemisorbed layer is assumed to be in equilibrium with the bulk environment, i.e., the bulk gas phase and the bulk condensed phase. At elevated temperatures, rates of many interfacial reactions on uncontaminated surfaces, i.e., where $\theta \to 1$, are fast relative to the diffusional processes or rate of adsorption. The diffusional effects are discussed in the subsequent sections; now let us consider the rate of adsorption as the limiting rate controlling step.

The isothermal rate of adsorption is directly proportional to the rate of collision of gas molecules with the surface. From the kinetic theory of gases, the isothermal rate of adsorption $\dot{n}$ (molecules $cm^{-2}\ sec^{-1}$, for a given surface coverage, is given by the equation

$$\dot{n} = \tfrac{1}{4}sc\bar{v} = sc(kT/2\pi m)^{1/2}, \tag{7.14}$$

where $\bar{v}$ is the mean molecular speed, s is the sticking coefficient, and c is the number of molecules per cubic centimeter. The sticking coefficient is defined as the number of activated collisions of molecules with the surface leading to adsorption divided by the total number of collisions, i.e., the fraction of activated collisions which are successful. By introducing the appropriate exponential term for the activated complex, Eq. (7.14) may be written in a more general form in terms of the absolute reaction rate theory:

$$\dot{n} = sc(kT/h)[h/(2\pi mkT)^{1/2}]\exp(-\epsilon_a/kT), \tag{7.15}$$

where ϵ_a is the energy of activation per molecule. By replacing ckT with the pressure p of the gas and introducing molar quantities, the rate $\mathscr{R}_a$ of adsorption in terms of mole $cm^{-2}\ sec^{-1}$ is given by the equation

$$\mathscr{R}_a = s[p/(2\pi MRT)^{1/2}]\exp(-E_a/RT). \tag{7.16}$$

With s equal to unity and zero activation energy for adsorption, Eq. (7.16) is reduced to the classical Hertz–Knudsen equation for the number of moles striking a unit surface area per unit time; this also gives the maximum rate of vaporization, or so-called free vaporization, from an uncontaminated surface at low pressures,

$$\mathscr{R}_{max} = p_i/(2\pi M_i RT)^{1/2}, \tag{7.17}$$

where p_i is the vapor pressure and M_i the molecular weight of the vaporizing species. This limiting rate law applies only when (i) the composition of the vaporizing species is the same as that in the condensed phase, (ii) the surface of the condensed phase is not contaminated with impurities, and (iii) the pressure is sufficiently low that the vaporizing species do not collide and do not return back to the surface of the condensed phase. Noting that 1 atm is equivalent to 1.0132×10^6 dyn cm^{-2} and that the gas constant R is equal to 8.1344×10^7 erg $mole^{-1}\ deg^{-1}$, for vapor pressure in terms of atmosphere, the above equation reduces to

$$\mathscr{R}_{max} = 44.3p_i(M_i T)^{-1/2} \quad \text{mole cm}^{-2}\ \text{sec}^{-1}. \tag{7.18}$$

Rates of interfacial reactions measured by conventional laboratory techniques give the overall rate, often with limited or no information on the chemisorption equilibrium. Therefore, the rate parameters derived from the experimental data are subject to interpretation. Also, disparities are often seen in the results of different investigators on the same reaction, possibly because of variations in surface contaminations from one experimental technique to another. Recent advances made in the application of Auger electron spectroscopy at elevated temperatures, coupled with low energy electron diffraction (LEED) studies, will eventually provide better means of studying the kinetics of gas–solid interfacial reactions.

Problems encountered and some of the successes made in the study of the kinetics of interfacial reactions are discussed in Section 7.2.2 with reference to a few diversified specific cases as typical examples. For many other examples of our present understanding of the kinetics of interfacial reactions, reference may be made to review papers by Darken and Turkdogan [6], Wagner [7], Belton [8], and Grabke and Horz [9].

7.2.2 Nitrogen Transfer across an Iron Surface

Several investigators have demonstrated that the rate of nitrogen dissolution in, or evolution from, liquid or solid iron is markedly retarded by the presence of small amounts of surface active species in the system. Pehlke and Elliott [10] and later Mowers and Pehlke [11] studied the reaction of molecular nitrogen with liquid iron by using the Sieverts technique whereby the rate of absorption or desorption by the liquid iron was measured continuously without disturbing the system. The liquid metal, ~2 cm deep, was inductively stirred to avoid diffusional effects, and desired amounts of oxygen, sulfur, selenium, or tellurium were added to the melt to measure their effect on the rate. The rate of nitrogenation at 1 atm N_2, in units of g-atm. N cm^{-2} sec^{-1}, is shown in a log–log plot

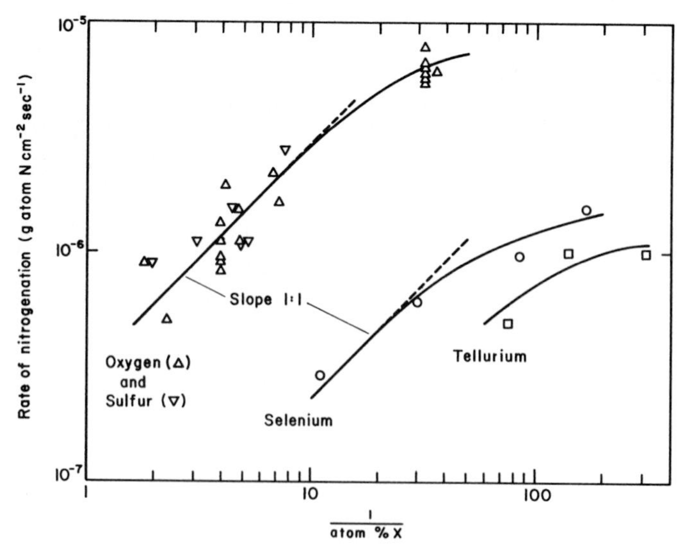

FIG. 7.1 Rate of nitrogenation of inductively stirred liquid iron at 1 atm N_2 and 1600 °C as affected by surface active solutes oxygen, sulfur [10], selenium, and tellurium [11]

in Fig. 7.1 as a function of the reciprocal of the atom percent of the surface active solute.

As indicated by the surface tension data in Fig. 3.25, the selenium, and presumably tellurium, in liquid iron is more surface active than sulfur and oxygen. This is consistent with the greater retarding effect of selenium and tellurium on the rate of nitrogenation. In terms of atom concentrations, the sulfur and oxygen have similar retarding effects on the rate of nitrogen reaction, and at solute concentrations above about 0.08 atom percent, the rate is essentially inversely proportional to the concentration of oxygen or sulfur, in accord with the limiting rate equation (7.9). When sulfur and oxygen are present as impurities in iron, the sum of their atomic concentrations may be used in estimating the value of $1 - \theta$.

These findings of Pehlke *et al.* [10, 11] are in substantial agreement with those of Schenck *et al.* [12] and other subsequent studies [13–16]. However, there are disparities in the rate data as regards to the order of the reaction relative to the concentration of nitrogen. The earlier studies [10–12] suggest a first-order kinetics, i.e., rate is proportional to the square root of the nitrogen pressure, while the later studies [13–16] indicate a second-order-type kinetic relation, i.e., rate is proportional to the partial pressure of nitrogen. With the data available, no unequivocal decision can be made on the variations of the rate with the concentration of nitrogen. Also, noting from the surface tension data in Fig. 3.25 that nitrogen is a moderately surface active solute, the rate may be proportional to $p_{N_2}^x$, with the exponent x in the range 0.5–1.0, depending on temperature and the concentration of the other surface active solutes in the metal.*

Turkdogan and Grieveson [17] showed that chemisorbed oxygen markedly retards the transfer of nitrogen across the surface of solid iron. In the nitrogenation and denitrogenation experiments with 0.05-cm-thick iron strips in N_2–H_2–H_2O mixtures at 1000 °C, the rate was found to be inversely proportional to the oxygen activity in the gas ($\equiv$ratio p_{H_2O}/p_{H_2}), and a first-order type relative to the concentration of nitrogen. In a more detailed study, Grabke [18] used thin iron foils ($\sim$10 μm) and determined the rates of forward and reverse reactions by measuring the change in the electrical resistivity of the sample during the progress of the reaction. He found that the rate is of a second-order type; that is, the dissociation of adsorbed N_2 is rate determining for the forward reaction and the recombination of two adsorbed nitrogen atoms is rate determining for the reverse reaction.

In the overall reaction

$$N_2\,(g) = 2\underline{N}\ \text{(dissolved in iron)}, \tag{7.19}$$

* Recent experimental work of R. J. Fruehan (Research Laboratory, U.S. Steel Corporation, Monroeville, Pennsylvania) has confirmed the findings of other investigators [13–16] showing that the rate of nitrogen reaction at the surface of liquid iron follows a second-order-type kinetic relation.

adsorbed N_2 is assumed to be in equilibrium with gaseous N_2 and the rate controlling reaction step is that involving the activated complex $N_2^{\ddagger}$ (ad).

$$N_2(g) = N_2(ad), \tag{7.20}$$

$$N_2(ad) = N_2^{\ddagger}(ad) \rightarrow 2\underline{N}. \tag{7.21}$$

From Eq. (7.12) the rate is represented by

$$d[\%N]/dt = (2/l)\Phi(1-\theta)(p_{N_2} - [\%N]^2/K^2), \tag{7.22}$$

where K is the equilibrium constant, defined by the ratio $[\%N]/p_{N_2}^{1/2}$, and l is the thickness of the sample. Because of relatively strong chemisorption of nitrogen on solid iron [19], Grabke suggested that $1-\theta$ should decrease with increasing nitrogen content of the metal. With the assumption of the Langmuir-

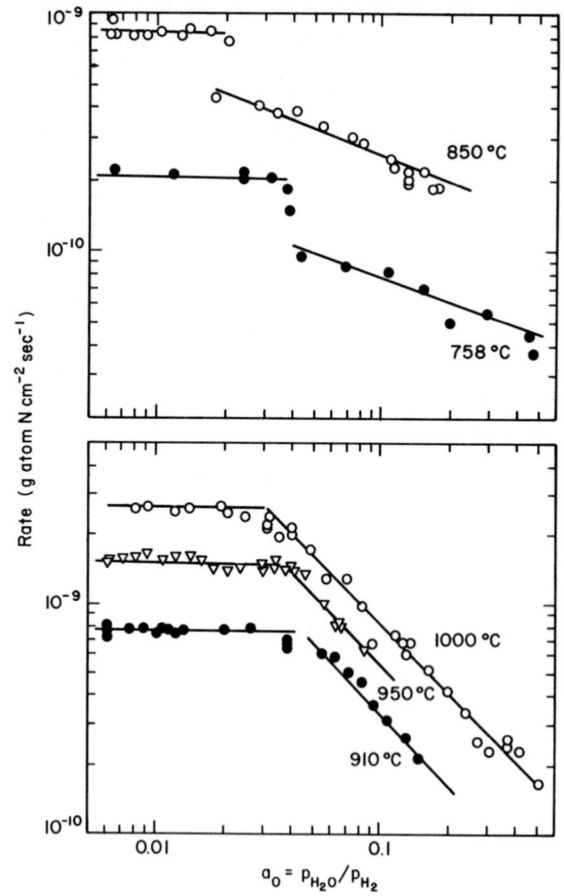

FIG. 7.2 Effect of oxygen activity on the rate of nitrogenation of solid iron foils at 1 atm N_2. After Grabke [20].

type ideal monolayer for single site occupancy, the isothermal rate expression from Eq. (7.11) is

$$\frac{d[\%\mathrm{N}]}{dt} = \frac{2}{l}\Phi \frac{1}{1 + \%\mathrm{N}/\varphi}\left(p_{\mathrm{N}_2} - \frac{[\%\mathrm{N}]^2}{K^2}\right). \tag{7.23}*$$

According to this rate equation, at low nitrogen contents the forward rate, i.e., the initial rate of nitrogenation, is directly proportional to p_{N_2} in the gas. For denitrogenation in an inert gas at high nitrogen contents, i.e., when $\% N/\varphi \gg 1$, the rate is a linear function of the nitrogen content; at low nitrogen levels the rate of denitrogenation is proportional to the square of the nitrogen concentration. These predictions are substantiated by Grabke's experimental results, and the following rate constants are obtained for α- and γ-iron with the apparent heats of activation in calories per mole.

$$\Phi_\alpha = 44 \exp(-56{,}200/RT) \ \text{mole cm}^{-2}\,\text{sec}^{-1}\,\text{atm}^{-1}, \tag{7.24}$$

$$\Phi_\gamma = 12 \exp(-52{,}500/RT) \ \text{mole cm}^{-2}\,\text{sec}^{-1}\,\text{atm}^{-1}. \tag{7.25}$$

The rate constant for the reverse reaction is given by the ratio Φ/K^2, where K is the equilibrium constant for the solution of nitrogen in iron.

$$K = [\%\mathrm{N}]/p_{\mathrm{N}_2}^{1/2}. \tag{7.26}$$

Grabke [20] showed that this mechanism of nitrogen reaction is valid also when there is chemisorption of oxygen on iron. As is seen from his data in Fig. 7.2, beyond certain oxygen activity for a given temperature when the surface becomes striated as in Fig. 3.30 indicating high surface coverage, the rate of nitrogenation decreases with increasing oxygen activity, giving the following relations for the fraction of sites not occupied by oxygen:

$$(1 - \theta_\mathrm{O})_\alpha \propto a_\mathrm{O}^{-1/3}, (1 - \theta_\mathrm{O})_\gamma \propto a_\mathrm{O}^{-1}. \tag{7.27}$$

Grabke pointed out that the proportionality of $(1 - \theta_\mathrm{O})_\alpha$ to $a_\mathrm{O}^{-1/3}$ may be interpreted by a saturation coverage on the assumption that the coverage is limited to one oxygen atom per three iron atoms, i.e., triple-site occupancy by adsorbed oxygen, which would be in accordance with the adsorption structures observed by LEED studies at lower temperatures [21].

The inverse proportionality to oxygen activity for γ-iron is the same as that found by Turkdogan and Grieveson [17] for 1000 °C. However, they observed a first-order-type kinetics for the nitrogen reaction, i.e., $\mathscr{R} \propto p_{N_2}^{1/2}$, and their rate constants are about an order of magnitude lower than those of Grabke [20]. It is difficult to account for these differences, particularly in view of the fact that in similar studies made by Grieveson and Turkdogan [22] with dry $N_2(+H_2)$

* The second term in this equation differs slightly from that of Grabke by the omission of the coefficient $1/\varphi$ for $[\% \mathrm{N}]^2$. Grabke assumed that on denitrogenation, N_2 is formed between N adsorbed on special sites and N adsorbed on normal sites. Equation (7.23) is for the same type of single site occupancy for both the forward and reverse reactions.

the measured rates were found to be controlled by diffusion in the metal. Although perhaps not relevant to the point in question, there are large differences in reaction times in these two types of experiments. In experiments with the resistivity technique using thin foils [18, 20] the total reaction time was in the range 10–150 sec; in experiments with iron strips [17] involving chemical analysis for nitrogen, the reaction time was in the range 1–74 hr. It is well to bear in mind that the kinetics of interfacial reactions may be affected by changes occurring in the surface structure, such as crystal growth, during reactions of long duration.

7.2.3 Rate of Decarburization of Iron in Wet Hydrogen

Grabke and Tauber [23] studied the kinetics of decarburization of α- and γ-iron in H_2O–H_2 mixtures using the electrical resistivity technique. With thin iron foils and sufficiently high gas flow rates, all diffusional effects were overcome and measured rates could be interpreted in terms of the kinetics of interfacial reactions. In the overall reaction

$$\underline{C} + H_2O = CO + H_2 \tag{7.28}$$

the oxygen chemisorption was assumed to have been maintained in equilibrium with the oxygen activity in the gas, i.e., the ratio p_{H_2O}/p_{H_2}, and the rate controlled by the formation of CO from chemisorbed oxygen and carbon.

$$H_2O = O(ad) + H_2, \tag{7.29}$$

$$O(ad) + \underline{C} = CO^{\ddagger} \rightarrow CO(g). \tag{7.30}$$

For this reaction mechanism, the isothermal rate of decarburization is represented by the equation

$$d[\%C]/dt = -(2/l)\Phi(1-\theta)(p_{H_2O}/p_{H_2})[\%C]. \tag{7.31}$$

In this system the oxygen is the only strongly adsorbed species; therefore, substituting the equation for $1-\theta$ in terms of p_{H_2O}/p_{H_2} with the assumption of an ideal monolayer gives

$$\frac{d[\%C]}{dt} = -\frac{2}{l}\Phi\frac{\varphi}{\varphi + p_{H_2O}/p_{H_2}}\frac{p_{H_2O}}{p_{H_2}}[\%C], \tag{7.32}$$

where φ is the constant of the chemisorption equilibrium and Φ is the rate constant in centimeters per second. Integration gives the logarithm of the fractional carbon removal as a function of time,

$$\ln\frac{[\%C]}{[\%C]_0} = -\frac{2}{l}\Phi\frac{\varphi}{\varphi + p_{H_2O}/p_{H_2}}\left(\frac{p_{H_2O}}{p_{H_2}}\right)t, \tag{7.33}$$

where $[\%C]_0$ is the initial carbon content of the metal.

In this system the chemisrobed oxygen, and hence the ratio p_{H_2O}/p_{H_2}, has a dual role: It poisons the surface and is a reactant, leading to two limiting rate expressions. At low oxygen activities, when $1 - \theta \rightarrow 1$,

$$\ln([\%\,C]/[\%\,C]_\circ) = -\frac{2}{l}\Phi(p_{H_2O}/p_{H_2})t \tag{7.34}$$

at high oxygen activities, when $1 - \theta \rightarrow 0$, the rate is independent of p_{H_2O}/p_{H_2}:

$$\ln([\%\,C]/[\%\,C]_\circ) = -(2/l)\Phi\varphi t. \tag{7.35}$$

The rate equation (7.33) is well substantiated by the data of Grabke and Tauber [23]; they obtained the following rate parameters for α- and γ-iron:

$$\Phi_\alpha = 1.8 \times 10^6 \exp(-41{,}000/RT) \quad \mathrm{cm\,sec^{-1}}, \tag{7.36}$$

$$\Phi_\gamma = 1.0 \times 10^7 \exp(-47{,}000/RT) \quad \mathrm{cm\,sec^{-1}}, \tag{7.37}$$

$$\varphi_\alpha = 3.18 \exp(-6000/RT). \tag{7.38}$$

With Eq. (7.38), the thermodynamic quantities for the reaction $H_2O = H_2 + \frac{1}{2}O_2$ give the enthalpy and entropy data for dissociative chemisorption of oxygen on α-iron.

$$\tfrac{1}{2}O_2(g) = O(ad),$$

$$\Delta H_a^\circ = -65 \quad \mathrm{kcal}, \qquad \Delta S_a^\circ = -13.3 \quad \mathrm{cal\,mole^{-1}\,deg^{-1}}. \tag{7.39}$$

7.2.4 Gasification of Tantalum in Oxygen

Rosner *et al.* [24] investigated the rate of vaporization of tantalum as volatile oxides TaO and TaO_2 in molecular and atomic oxygen at temperatures above 2400 °K. A 0.38-mm-diameter tantalum wire was heated in an oxygen stream in a discharge tube at pressures of 2.63×10^{-7}–6.56×10^{-5} atm and temperatures of 2415–2700 °K. For a given filament temperature and oxygen pressure, the concentration of oxygen in tantalum reached a constant level and TaO and TaO_2 vaporized under steady state conditions. The rate was determined by measuring the decrease in the diameter of the filament by the electrical resistivity technique. The vaporization of Ta was negligible in comparison to TaO and TaO_2.

The rate of volatilization is considered to be controlled by the formation of TaO and TaO_2 in the chemisorbed layer where the chemisorbed oxygen is taken to be in equilibrium with the oxygen in solution in tantalum; the dissociative chemisorption of molecular oxygen was assumed to be irreversible. The following elementary reaction steps are considered:

$$O_2(g) \rightarrow 2O(ad), \tag{7.40a}$$

$$O(g) \rightarrow O(ad), \tag{7.40b}$$

$$O(ad) = O\ (\text{in tantalum}), \tag{7.40c}$$

$$Ta + 2O(ad) = TaO_2^{\ddagger}(ad) \rightarrow TaO_2(g), \tag{7.40d}$$

$$Ta + O(ad) = TaO^{\ddagger}(ad) \rightarrow TaO(g). \tag{7.40e}$$

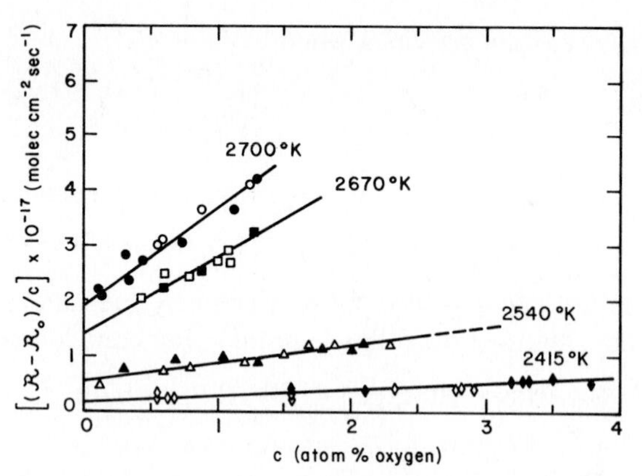

FIG. 7.3 Rate of vaporization of tantalum in molecular and atomic oxygen as volatile suboxides TaO and TaO_2 shown as a function of oxygen concentration in tantalum in accord with Eq. (7.42). After Rosner *et al.* [24].

For the limiting case of low surface coverage with adsorbed species, i.e., $1 - \theta \to 1$, the total isothermal rate of vaporization of tantalum suboxides is given by

$$\mathcal{R} - \mathcal{R}_\circ = \Phi_1 c + \Phi_2 c^2, \tag{7.41}$$

where $\mathcal{R}_\circ$ is the rate of vaporization of tantalum in vacuo and c is a known concentration of oxygen in solution in tantalum for a given temperature and oxygen pressure. Equation (7.41) may be rearranged to

$$(\mathcal{R} - \mathcal{R}_\circ)/c = \Phi_1 + \Phi_2 c. \tag{7.42}$$

As shown by Rosner *et al.*, their results given in Fig. 7.3 are in general accord with Eq. (7.42) for $1 - \theta \to 1$; the rate constants Φ_1 for TaO and Φ_2 for TaO_2 derived from the intercepts and slopes of the lines are given in Table 7.1. The

TABLE 7.1

Rate Constants for the Formation of Volatile TaO and TaO_2 on Tantalum Surface in Diatomic and Monatomic Oxygen[a]

Temperature (°K)	$\Phi_1 \times 10^5$ (mole cm^{-2} sec^{-1})	$\Phi_2 \times 10^3$ (mole cm^{-2} sec^{-1})
2415	0.23	0.15
2540	0.75	0.63
2670	2.49	2.24
2700	3.09	3.02

[a] Determined by Rosner *et al.* [24].

temperature dependence of the rate constants may be represented by the equations

$$\Phi_1 = 1.15 \times 10^5 \exp(-118{,}230/RT) \quad \text{mole cm}^{-2}\,\text{sec}^{-1}, \tag{7.43}$$

$$\Phi_2 = 1.60 \times 10^8 \exp(-132{,}590/RT) \quad \text{mole cm}^{-2}\,\text{sec}^{-1}. \tag{7.44}$$

The relatively large heat of activation for the formation of volatile suboxides of tantalum is to be compared with 178 kcal for the enthalpy of vaporization of tantalum.

Assuming that the rate of dissociative adsorption of oxygen is nonactivated, from Eq. (7.16) with $E_a = 0$, the total rate of oxygen adsorption is given by

$$\mathscr{R}_a = (2s_2 p_{O_2}/\sqrt{32} + s_1 p_O/\sqrt{16})(2\pi RT)^{-1/2}, \tag{7.45}$$

where s_2 and s_1 are sticking coefficients for O_2 and O, respectively. In the experiments of Rosner *et al.*, the efficiency of the dissociation of molecular oxygen was 28%, and setting $\mathscr{R}_a$ equal to $\mathscr{R} - \mathscr{R}_o$, they estimated s_1 to be close to unity and s_2 to be in the range 0.3–0.5, increasing with increasing temperature. However, it would seem unlikely that the sticking coefficient for dissociative adsorption should be much less than unity. If we were to assume a value of 4 kcal for the heat of activation for dissociative adsorption of molecular oxygen, then the data of Rosner *et al.* would give s_2 close to unity.

7.2.5 Vaporization of SiS from Liquid Iron Alloys

The rate of removal of sulfur from liquid iron alloys in vacuo has been studied by several investigators [25–30]. As is seen from the results of rate measurements of Fruehan and Turkdogan [29] in Fig. 7.4, addition of carbon and silicon increases the rate of vaporization of sulfur from liquid iron alloys, partly because carbon and silicon increase the activity of sulfur, and more importantly because of the formation of volatile SiS. In these experiments with inductively stirred melts at low pressures ($\sim 10^{-3}$ atm), the rate of diffusion of sulfur to the surface of the melt became sluggish only at low sulfur levels ($<0.01\%$). The rate of desulfurization was found to be about an order of magnitude lower than the rate calculated from Eq. (7.18) for free vaporization of SiS. Fruehan and Turkdogan proposed that the rate of evolution of SiS is controlled by its formation at the surface of the melt involving an activated complex $(SiS_2)^\ddagger$. The equilibrium with the activated complex is represented by

$$\underline{Si} + 2\underline{S} = (SiS_2)^\ddagger \rightarrow SiS(g) + \underline{S} \tag{7.46}$$

for which the rate equation is

$$d[\%\,S]/dt = -(\Phi/l)(1-\theta)a_{Si}a_S^{\,2}, \tag{7.47}$$

where l is the depth of melt and a's are activities of reactants in the metal.

In these experiments the sulfur activity in the metal is high enough to satisfy the limiting case of $1-\theta$ approaching zero. Therefore, substituting φ/a_S for

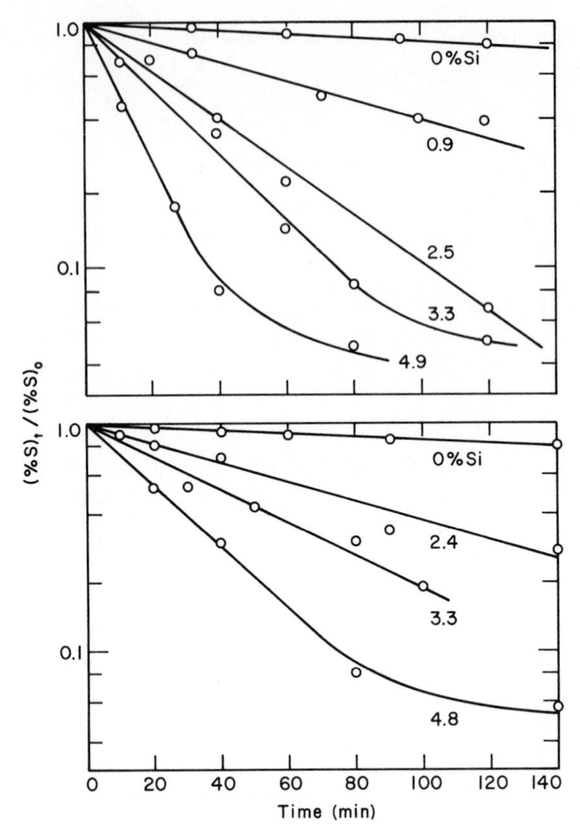

FIG. 7.4 Desulfurization of liquid iron alloys at 1600 °C and 10 μm pressure: (a) Fe–Si–S alloys, (b) graphite saturated Fe–C–Si–S alloys. The weight of the melt is 15 kg and the surface area 123 cm². After Fruehan and Turkdogan [29].

$1 - \theta$ and integrating gives

$$\ln([\%\,\mathrm{S}]/[\%\,\mathrm{S}]_\circ) = -(\Phi\varphi/l)a_{\mathrm{Si}}f_{\mathrm{S}}t, \tag{7.48}$$

where φ is the chemisorption constant and f_S is the activity coefficient of sulfur in the melt. As is seen from the results of several investigators in Fig. 7.5 for inductively stirred melts at 1600 °C, the rate of desulfurization, in units of g S $\mathrm{cm}^{-2}\,\mathrm{min}^{-1}$ divided by f_S, is a linear function of the activity of silicon in the metal, in accord with Eq. (7.48). In these inductively stirred laboratory melts, the liquid-phase mass-transfer coefficient is about 0.05 cm sec^{-1}, a value deduced from the rate measurements of Fruehan [29a]. The mass-transfer controlled rate would be approached only at high silicon activities when the rate of SiS evolution would become essentially independent of the activity of silicon.

The intercept of the line in Fig. 7.5 at a_{Si} equal to zero gives the rate constant for vaporization of S and S_2 not included in Eq. (7.47) for the sake of simplicity.

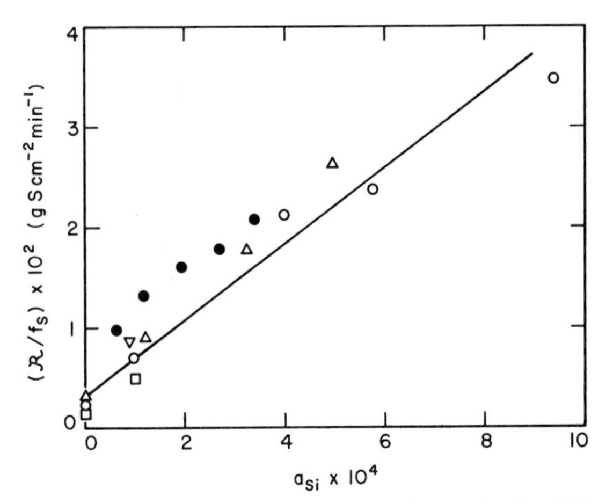

FIG. 7.5 Ratio of the rate of desulfurization/activity coefficient of sulfur $\mathcal{R}/f_S$ as a linear function of the activity of silicon (relative to pure silicon) in Fe–C–Si–S melts at 1600 °C. After Fruehan and Turkdogan [29].

In the absence of silicon, the rate of vaporization of sulfur from iron–carbon–sulfur melts in vacuo was found to be that due to free vaporization of monatomic and diatomic sulfur from the surface [30].

7.2.6 Coupled Reactions and Partial Equilibrium

Because of the ionic nature of molten slags and the nonpolar nature of metals, the transfer of an element from metal to slag is accompanied by exchange of electrons between the reacting species. This is well demonstrated by King and Ramachandran [31] and by Nilas and Frohberg [32], who investigated the transfer of sulfur from iron to slags saturated with graphite. When the sulfur is transferred from metal to slag, electrons must be provided; thus

$$\underline{S} + 2\varepsilon = S^{2-}. \tag{7.49}$$

Since electroneutrality is maintained in the slag and metal, in the absence of an electric field across the slag–metal interface, the solutes in the metal become oxidized to provide the electrons needed for reaction (7.49).

$$\begin{aligned} &\underline{Fe} = Fe^{2+} + 2\varepsilon, && \underline{Si} = Si^{4+} + 4\varepsilon, \\ &\underline{Mn} = Mn^{2+} + 2\varepsilon, && \underline{C} + O^{2-} = CO + 2\varepsilon, \\ &\underline{Al} = Al^{3+} + 3\varepsilon. && \end{aligned} \tag{7.50}$$

Therefore, in the presence of these solutes in iron, the transfer of n_S moles of sulfur from metal to slag should be accompanied by the transfer of an equivalent total number of moles of solutes to satisfy the stoichiometric requirements,

$$2n_S = 2n_{Fe} + 2n_{Mn} + 3n_{Al} + 4n_{Si} + 2n_C. \tag{7.51}$$

The sign of n is positive for metal → slag transfer and negative for slag → metal transfer. Typical examples of the results of King and Ramachandran [31] are given in Fig. 7.6. In the top diagram, silicon and sulfur contents of iron, sulfur, and iron oxide contents of slag and the amount of carbon monoxide evolved are plotted against the time of reaction. In the lower diagram, the change in iron and silicon contents and carbon monoxide evolution are given in terms of sulfur equivalence. It is seen that the amount of sulfur transferred from metal to slag calculated from n_{Fe}, n_{Si}, and n_{CO} as in Eq. (7.51) agrees well with that observed experimentally. In the example considered, as slag and metal move towards equilibrium with respect to the slag–metal sulfur partition, the silicon and iron

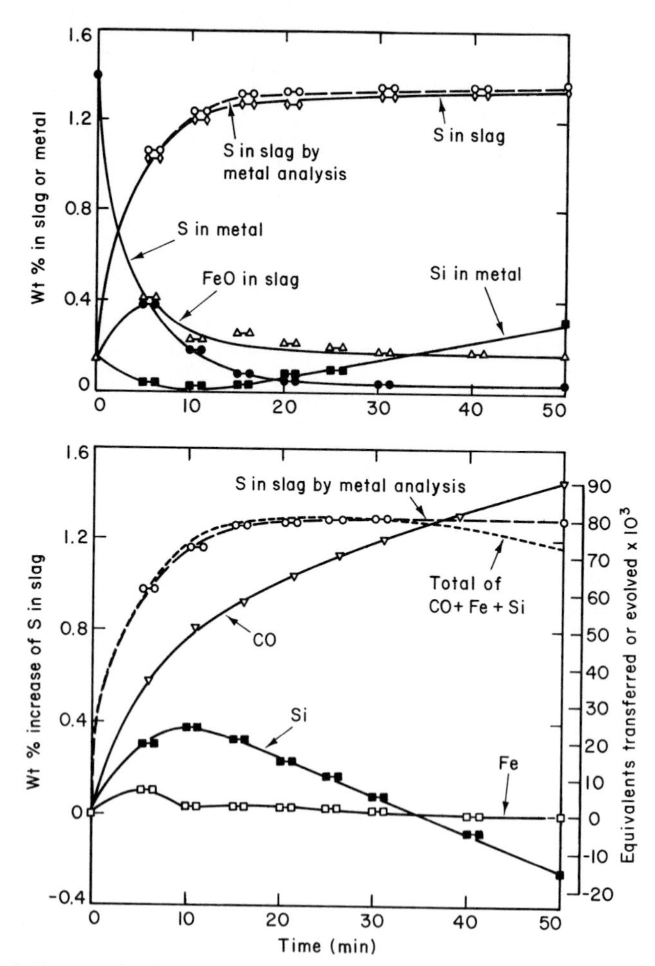

FIG. 7.6 Sulfur transfer from metal to slag accompanied with equivalent iron and silicon transfer and CO evolution from graphite-saturated iron reacting at 1550 °C with a slag (48% CaO, 21% Al_2O_3, and 31% SiO_2). After King and Ramachandran [31]. Reprinted with permission from "The Physical Chemistry of Steelmaking," edited by J. F. Elliot, copyright © 1958 by the Massachusetts Institute of Technology.

initially move away from equilibrium. These coupled reactions are considered as electrochemical. The relative rates of these electrochemical reactions are determined by their relative potential-current relations. It follows from these considerations that, if the initial concentrations of manganese, silicon, aluminum, etc., in the metal are higher than the ultimate equilibrium values for a given metal and slag system, the rate of sulfur transfer from metal to slag will increase. This conclusion is also borne out by the results of Goldman *et al.* [33].

For another example of an electrochemical-coupled reaction, reference may be made to a study of Turkdogan and Bills [34]. They investigated the reaction between iron silicate melts with gases containing CO, CO_2, and SO_2. Two reactions occur simultaneously:

$$\tfrac{1}{2}S_2 + (O^{2-}) = (S^{2-}) + \tfrac{1}{2}O_2, \tag{7.52}$$

$$\tfrac{1}{2}O_2 + 2(Fe^{2+}) = (O^{2-}) + 2(Fe^{3+}). \tag{7.53}$$

As suggested by the results in Fig. 7.7, reaction (7.52) is faster than reaction (7.53), with the tesult that the sulfur content of the melt reaches a maximum at some stage of the reaction when the following equilibrium is approached.

$$\tfrac{1}{2}S_2 + 2(Fe^{2+}) = (S^{2-}) + 2(Fe^{3+}). \tag{7.54}$$

However, the system as a whole is not at equilibrium. As the oxidation of iron in the melt continues, the concentration ratio Fe^{3+}/Fe^{2+} increases, and since the sulfur potential in the gas is maintained constant, the sulfur content of the

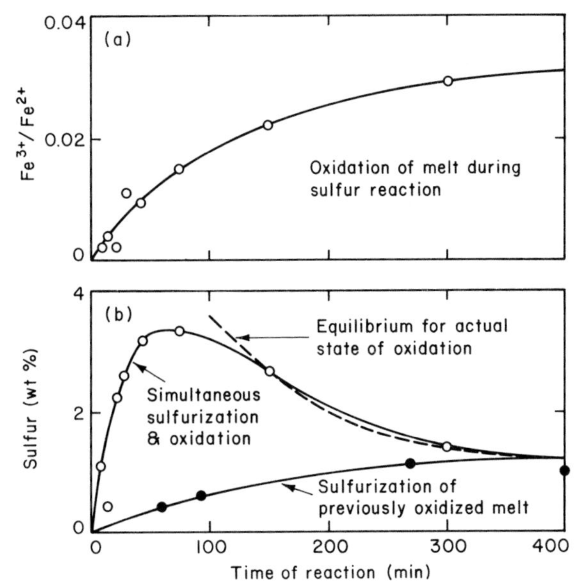

FIG. 7.7 Oxidation and sulfurization of an iron silicate melt (40% SiO_2) in a CO–CO_2–SO_2 gas mixture at 1550 °C with ingoing gas composition $p_{CO}/p_{CO_2} = 3.0$ and $p_{SO_2} = 0.028$ atm. After Turkdogan and Bills [34].

melt must decrease as dictated by reaction (7.54). As the whole system is approaching equilibrium, reaction (7.54) remains at a state of pseudoequilibrium; that is, the system is at a state of partial equilibrium. When the ferrous silicate melt is first equilibrated with the prevailing oxygen partial pressure of the gas, the sulfurization proceeds in a normal manner, as shown by the lower curve in Fig. 7.7.

7.3 Nucleation and Growth

Bubble formation in liquids, condensation of vapors, and formation of a reaction product as a second phase are of technological importance and of experimental interest. The theory of homogeneous nucleation was originally developed by Volmer and Weber [35] and Becker and Doring [36]. Subsequent contributions made by others in further development of the theory are well documented by Abraham [37]. For the present purpose, only the basic equations need be given to estimate the supersaturation of a homogeneous solution for the onset of nucleation.

The rate of formation I number of nuclei is an exponential function of the energy of activation ΔF^* for nucleation,

$$I = A_{\circ} \exp(-\Delta F^*/kT) \quad \text{nuclei cm}^{-3}\,\text{sec}^{-1}, \tag{7.55}$$

where $A_{\circ}$ is a constant $\sim 10^{27}$ nuclei cm^{-3} sec^{-1},

$$\Delta F^* = 16\pi\sigma^3/3(\Delta F_{\text{v}})^2 \quad \text{erg nucleus}^{-1}, \tag{7.56}$$

σ is the interfacial energy (using bulk properties) in ergs per square centimeter, the free-energy change

$$\Delta F_{\text{v}} = -(kT/\Omega)\ln(K_{\text{s}}/K_{\circ}) \quad \text{erg cm}^{-3}, \tag{7.57}$$

k is the Boltzmann constant, Ω the molecular volume of the nucleus (using bulk properties of the nucleated phase), and $K_{\text{s}}/K_{\circ}$ the supersaturation ratio relative to $K_{\circ}$ for the bulk phase equilibrium. It should be noted that ΔF^* is not too sensitive to the choice of I, which is usually taken to be somewhere in the range 10–10^3 nuclei cm^{-3} sec^{-1}.

For the condensation of liquid droplets from its vapor phase, the supersaturation ratio $K_{\text{s}}/K_{\circ}$ is the ratio of the vapor pressure of the nucleated liquid over the vapor pressure of the bulk liquid $p_{\text{s}}/p_{\circ}$. The supersaturation ratio $p_{\text{s}}/p_{\circ}$ for condensation of liquid–iron droplets from the vapor phase, calculated from the above equations with the known values of molar volume, surface tension, and the vapor pressure of liquid iron, is shown as an example in Fig. 7.8. The supersaturation ratio decreases with increasing temperature, and becomes unity at the critical temperature. From the empirical relations in Eq. (6.20), the critical temperature of iron is estimated to be 4592 °K which agrees well with the tangential extrapolation of the curve in Fig. 7.8 to $\log(p_{\text{s}}/p_{\circ}) = 0$.

If we are considering, for example, the nucleation of gas bubbles in liquids, the ratio $K_{\text{s}}/K_{\circ}$ is the ratio for the gas solubility in the supersaturated liquid over

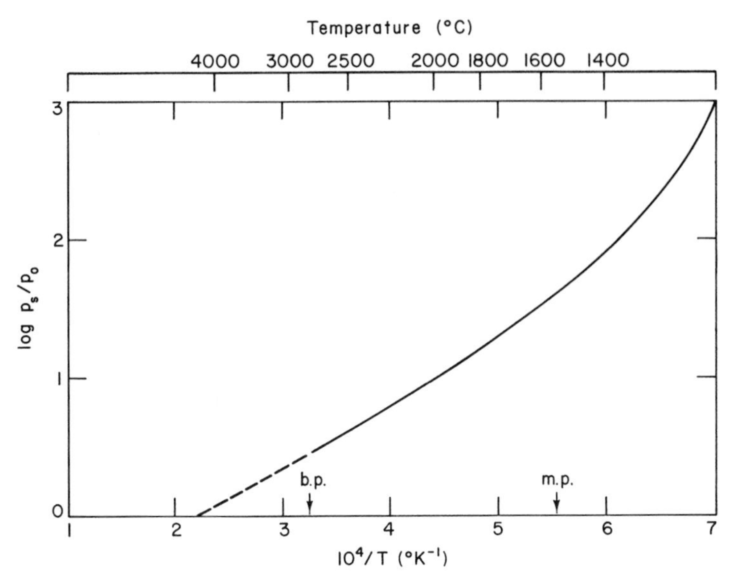

FIG. 7.8 Supersaturation ratio $p_s/p_\circ$ for nucleation of liquid-iron droplets from the vapor phase calculated from Eq. (7.57).

that for the equilibrium at atmospheric pressure. For liquid metals with surface tensions in the range 1000–1600 dyn cm^{-1}, the supersaturation for the nucleation of gas bubbles would be in the range 1×10^4–5×10^4.

As we shall see from the examples given in other chapters, the homogeneous nucleation is not as a rule a rate limiting process in reactions involving the formation of a second phase, because of the presence of impurities (as inclusions) and surfaces, or in the case of solids, strained lattice, dislocations, and grain boundaries.

Next, let us consider the growth of nucleated particles dispersed in a single-phase system. Since the size of nuclei is in the range of a few atom diameters, the growth by collision through Brownian motion is one possible growth mechanism in fluid media. Smoluchowski [38] derived an expression for the decrease in the number of particles resulting from coalescence of colliding particles caused by the Brownian motion; thus, at time t the remaining Z number of particles is

$$Z \simeq 3\eta/8kTt, \tag{7.58}$$

where η is the viscosity of the fluid medium. In liquid metals at elevated temperatures, it would take about 3 hr for Z to be reduced to about 10^7 particles cm^{-3} in which the particle size is within the μm range. This growth mechanism would not apply to most cases of experimental or practical interest.

A realistic growth mechanism is that due to the Ostwald ripening effect; that is, the growth of large particles at the expense of small particles is caused by the difference in the chemical potential of the solute in the small and large particles. Wagner [39] derived the following equation for the time dependence of the mean radius r of an assembly of particles, with an initial mean radius of r_0, in solid or fluid media where the atom movement is due to diffusion only,

$$r^3 - r_0^3 = (8\sigma DcV^2/9RT)t, \tag{7.59}$$

where D is the diffusivity, c the molar concentration of the solute in the medium, and V the molar volume of the particle per mole of the diffusate. For solute diffusivity of about 10^{-5} cm^2 sec^{-1}, the mean particle size would be in the range 1–3 μm after about 15 min.

As another example, let us consider even distribution of nuclei at the onset of a reaction in a homogeneous solution such as precipitation. All the particles are assumed to be of equal size and a state of equilibrium is being maintained between the particle and the solution at the interface. For this limiting case, the growth rate of spherical particles is limited by diffusion of reactant to the surface of the particle. For evenly distributed Z number of particles per unit volume, each particle has its own spherical diffusion zone of radius r_0, represented by the equation

$$r_0 = (3/4\pi Z)^{1/3} = 0.62Z^{-1/3}. \tag{7.60}$$

A simplified form of the rate equation derived by Ham [40] may be used to calculate the rate of diffusion-limited precipitation from a homogeneous solution,

$$\frac{Dt}{r_0^2}\left(\frac{c_0 - c_i}{c_p}\right)^{1/3} = \frac{1}{6}\ln\frac{u^2 + u + 1}{(u-1)^2} - \frac{1}{\sqrt{3}}\tan^{-1}\left(\frac{2u+1}{\sqrt{3}}\right) + \frac{1}{\sqrt{3}}\tan^{-1}\left(\frac{1}{\sqrt{3}}\right), \tag{7.61}$$

where $u = [(c_0 - c_m)/(c_0 - c_i)]^{1/3}$, c_0 is the initial solute concentration in the solution, c_m the average solute concentration at time t, c_i the solute concentration in solution at the interface, and c_p the concentration of diffusate in the particle.

7.4 Heat and Mass Transfer in a Fluid–Film Layer

Rates of interfacial reactions are often influenced, and more often controlled, by heat and/or mass transfer to and from the reaction surface. In this section, we shall point out some basic features of rates of heat and mass transfer in the fluid–film boundary layer next to the surface of a condensed phase.

7.4.1 Heat- and Mass-Transfer Correlations

For the simple case of a fluid flowing over a surface, the mass flux of a solute in dilute solutions through the boundary layer is given by the relation

$$J_M = k_m(c_i - c_0) \quad \text{mole cm}^{-2}\,\text{sec}^{-1}, \tag{7.62}$$

where k_m is the mean mass-transfer coefficient over the surface in centimeters per second, c_i is the molar concentration of the solute at the surface, and c_0 is that in the bulk fluid away from the interface. Similarly, for a steady state heat flux

$$J_H = k_h(T_i - T_0) \quad \text{cal cm}^{-2}\,\text{sec}^{-1}, \tag{7.63}$$

where k_h is the mean heat-transfer coefficient over the surface in cal cm^{-2} sec^{-1} deg^{-1}; subscripts i and zero indicate surface and free stream temperatures.

If the rate is controlled jointly by mass transfer across the boundary layer and a chemical reaction at the surface, the rate constant for the interfacial reaction may be derived from the measured overall rate, provided the mass-transfer coefficient k_m is known from the flow conditions of the experiments. As an example, let us consider a first-order-type reaction for which the rate is represented by

$$\mathscr{R} = -\Phi(c_i - c_e) = k_m(c_i - c_0), \tag{7.64}$$

where c_e is the equilibrium concentration of the reactant. Solving for c_i gives the rate equation in terms of the overall driving force $c_0 - c_e$ for the reaction

$$\mathscr{R} = -k'(c_0 - c_e), \tag{7.65}$$

where

$$k' = \Phi k_m/(\Phi + k_m) \qquad \text{or} \qquad 1/k' = (1/\Phi) + (1/k_m). \tag{7.66}$$

In discussing heat and mass transfer, it is helpful to introduce the concept of the boundary-layer thickness. Fictitious boundary-layer thickness δ is defined such that in terms of mass diffusivity D and thermal conductivity κ, or thermal diffusivity D^T,

$$k_m = D/\delta_m. \qquad k_h = \kappa/\delta_h = D^T c_p \rho/\delta_h. \tag{7.67}$$

In addition, there is the velocity boundary layer which is usually defined in terms of the displacement thickness δ^*

$$\delta^* = 1.73(\nu l/u_0)^{1/2}, \tag{7.68}$$

where ν is the kinematic viscosity of the fluid, l is the distance in the flow direction, and u_0 is the velocity of the free stream. The displacement thickness is that distance by which the external potential field of flow is displaced outwards as a consequence of the decrease in velocity in the boundary layer. The boundary-layer thickness δ over which the potential velocity is attained to within 1% is about three times larger than the displacement thickness δ^*.

The transport coefficients are functions of flow geometry and transport properties of the fluid medium. For details of the boundary-layer theory and its application in the formulation of transport equations, reference may be made to Schlichting [41], Eckert and Drake [42], Levich [43], and to other

TABLE 7.2

Selected Dimensionless Groups for Fluid Dynamics, Heat, and Mass Transfer

Group	Formula	Nomenclature
Fluid Dynamics		
Drag coefficient	$C_d = \frac{g(\rho - \rho_\circ)l}{\rho u^2}$	l = characteristic length; $\rho_\circ$ = density of dispersed phase
Eötvos number	$N_{eo} = \frac{gl^2(\rho - \rho_\circ)}{\sigma} = \frac{N_{we}}{N_{fr}}$	ρ = density of continuous phase; g = gravitational acceleration; u = velocity
Froude number	$N_{fr} = \frac{\rho u^2}{(\rho - \rho_\circ)gl}$	
Morton number	$N_m = \frac{g\eta^4}{\rho\sigma^3}$	η = viscosity
Reynolds number	$N_{re} = \frac{ul}{v}$	v = kinematic viscosity
Weber number	$N_{we} = \frac{u^2 l\rho}{\sigma}$	σ = surface tension
Heat Transfer		
Grashof number	$N_{gr_h} = \frac{gl^3\beta\,\Delta T}{v^2}$	$\beta = \frac{1}{\rho}\left(\frac{\partial\rho}{\partial T}\right)_p$; ΔT = temperature difference
Nusselt number	$N_{nu} = \frac{k_h l}{\kappa}$	k_h = heat-transfer coefficient
Prandtl number	$N_{pr} = \frac{c_p\eta}{\kappa} = \frac{v}{D^T}$	κ = thermal conductivity; c_p = specific heat
Mass Transfer		
Grashof number	$N_{gr_m} = \frac{gl^3\beta'\,\Delta c}{v^2}$	$\beta' = \frac{1}{\rho}\left(\frac{\partial\rho}{\partial c}\right)_T$
Lewis number	$N_{le} = \frac{\rho c_p D}{\kappa} = \frac{D}{D^T}$	D = interdiffusivity; D^T = thermal diffusivity
Schmidt number	$N_{sc} = \frac{\eta}{\rho D} = \frac{v}{D}$	
Sherwood number	$N_{sh} = \frac{k_m l}{D}$	k_m = mass-transfer coefficient

textbooks [1, 2]. A brief statement of some limiting rate equations for heat and mass transfer is adequate for the present purpose.

Selected dimensionless groups for fluid dynamics, and heat and mass transfer are listed in Table 7.2. The characteristic length l is the length of a plate, diameter of a sphere, or a cylinder. Selected correlations for heat and mass transfer for laminar flow are listed in Table 7.3.

TABLE 7.3

Selected Correlations for Heat and Mass Transfer for Laminar Flow

Type of flow	Average Sherwood number $\bar{N}_{sh}$	Average Nusselt number $\bar{N}_{nu}$
Forced convection over a flat plate	$0.664 N_{re}^{1/2} N_{sc}^{1/3}$	$0.664 N_{re}^{1/2} N_{pr}^{1/3}$
Natural convection from a vertical plate	$0.902\left(\frac{N_{gr_m} N_{sc}}{4(0.861 + N_{sc})}\right)^{1/4}$	$0.902\left(\frac{N_{gr_h} N_{pr}^2}{4(0.861 + N_{pr})}\right)^{1/4}$
Forced convection around a sphere	$2 + 0.6 N_{re}^{1/2} N_{sc}^{1/3}$	$2 + 0.6 N_{re}^{1/2} N_{pr}^{1/3}$
Natural convection around a sphere	$2 + 0.6 N_{gr_m}^{1/4} N_{sc}^{1/3}$	$2 + 0.6 N_{gr_h}^{1/4} N_{pr}^{1/3}$
Gas jet impinging on a solid surface[a]	$1.01(d/l)^{1/4} N_{re}^{3/4} N_{sc}^{1/3}$	$1.01(d/l)^{1/4} N_{re}^{3/4} N_{pr}^{1/3}$

[a] The equation derived by Scholtz and Trass [43a] for local mass transfer in a laminar flow is integrated to obtain N_{sh} for a surface of diameter l and a jet nozzle of diameter d. By analogy, a similar correlation is written for N_{nu}.

A few comments should be made on these correlations. As the Reynolds number, or velocity, approaches zero, the average Sherwood number and Nusselt number for spherical shapes approach a constant value of 2 which is the limiting value predicted theoretically for heat and mass transfer in a stagnant medium. The correlation for forced convection derived from the boundary-layer theory is for a well established flow from the leading edge of a flat plate, and does not apply to stagnant fluid media. In experiments with thin strips of finite size there are edge effects, particularly at low flow rates. Experience has shown that [44] the mass transfer measured for short strips at elevated temperatures is about 1.2 times greater than that given by the theoretical correlation for well established flows.

As indicated by the boundary-layer theory, for $N_{pr} \ll 1 \ll N_{sc}$, as in liquid metals, the thermal and diffusion boundary layers are thicker than the velocity boundary layer; therefore, the fluid properties to be used are those for the surface temperature. For $N_{sc} \gg N_{pr} \gg 1$, as in molten glasses and slags, the thermal boundary layer is thinner than the velocity boundary layer; therefore, the viscosity used is that for the temperature of the free stream; however since $\delta_m \ll \delta_h$, the diffusivity to be used is that for the surface temperature. For $N_{pr} \sim N_{sc} \sim 1$, as in gases, the boundary layers essentially coincide; in this case the average film temperature should be used for the transport properties.

7.4.2 Mass Transfer Controlled Rate of Decarburization of Liquid Iron

As an example, let us consider the rate of decarburization of liquid iron in a stream of CO_2 for three different experimental techniques and flow conditions: (i) the metal droplet levitated and heated in an electromagnetic field in a stream of CO_2 [45], (ii) the reacting gas stream impinging on the surface of inductively stirred melt [46], and (iii) the reacting gas flowing parallel to the surface of the melt [44].

$$\underline{C} + CO_2 = 2CO, \tag{7.69}$$

where the underscore indicates carbon dissolved in iron, the rate of decarburization controlled by the countercurrent diffusion of CO_2 and CO is, from Eq. (6.42),

$$J_{CO_2} = -\frac{PD_{12}}{RT}\frac{dx_{CO_2}}{dz} + x_{CO_2}(J_{CO_2} + J_{CO}) \quad \text{mole cm}^{-2}\,\text{sec}^{-1}, \tag{7.70}$$

where D is the CO_2–CO interdiffusivity, P is the total pressure and dx_{CO_2}/dz is the concentration gradient across the boundary layer. Noting that the flux J_{CO} is equal to $-2J_{CO_2}$, integration over the thickness δ of the boundary layer gives

$$J_{CO_2} = -\frac{P}{RT}\frac{D_{12}}{\delta}\ln\frac{1 + x_{CO_2}}{1 + x^*_{CO_2}}, \tag{7.71}$$

where $x^*_{CO_2}$ is that in equilibrium with carbon in the melt. For carbon contents below 0.5% in iron, $x^*_{CO_2}$ is small enough to be neglected; with $P = p_{CO_2} + p_{CO} =$ 1 atm and inserting practical units, the rate of decrease in the carbon content of iron as a function of p_{CO_2} in the gas stream is, from Eq. (7.71),

$$\frac{d[\%C]}{dt} = -\frac{1200}{\rho l}\frac{1}{RT}\frac{D_{12}}{\delta}\ln(1 + p_{CO_2}), \tag{7.72}$$

where ρ is the density and l is one-sixth the diameter of the iron droplet or depth of melt in a crucible.

Baker *et al.* [45] found that for a given temperature and p_{CO_2} in the gas stream, the rate of decarburization of levitated iron droplet ($\sim$0.7 g) is constant as in Eq. (7.72), and the rate constant D/δ obtained is in close agreement with that derived from the correlation in Table 7.3 for forced flow around a sphere.

Swisher and Turkdogan [46] measured the rate of decarburization of iron–carbon melts in CO_2–CO mixtures impinging on the surface of iron–carbon melts with an average velocity of about 200 cm sec^{-1}. As is seen from their

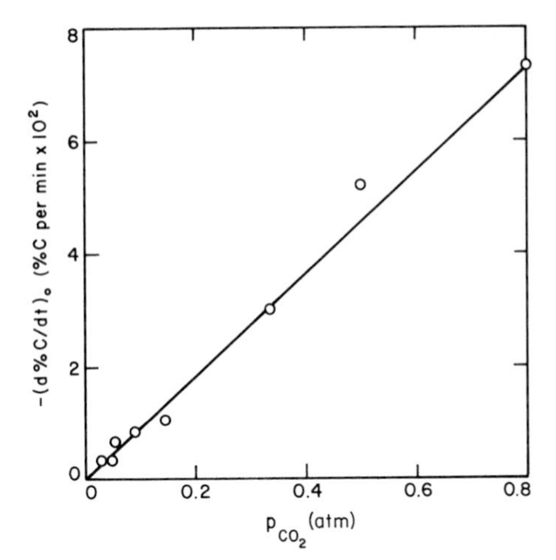

FIG. 7.9 Rate of decarburization of liquid–iron carbon alloys in CO_2–CO mixtures, impinging on the surface of the melt at 200 cm sec^{-1}, as a linear function of $\ln(1 - p_{CO_2})$ in accord with Eq. (7.72) for gas–film mass-transfer control. After Swisher and Turkdogan [46].

results in Fig. 7.9, $-d[\%C]/dt$ is directly proportional to $\ln(1 + p_{CO_2})$ in accord with Eq. (7.72). The slope of the line gives for the mass-transfer coefficient $D/\delta = 2.6$ cm sec^{-1}. For their experimental conditions of $u = 200$ cm sec^{-1} gas velocity, $d = 0.64$ cm jet diameter and $l = 3.18$ cm melt diameter in the crucible, the average mass-transfer coefficient computed from the correlation for N_{sh} in Table 7.3 for laminar flow is 25.8 cm sec^{-1}, which is an order of magnitude greater than the experimental value. This disparity may be attributed to gas turbulence above the melt inside the crucible. In fact, using the correlation for turbulent gas stream impinging on the surface, $\bar{N}_{sh} = dk_m/D = 0.02N_{re}^{0.87}N_{sc}^{0.33}$, gives k_m equal to 3.0 cm sec^{-1}, which agrees well with the experimental value given by the slope of the line in Fig. 7.9. In a subsequent study, Sain and Belton [47] showed that the velocity of gas impingement on the surface has to be greater than 20 m sec^{-1} to overcome the mass-transfer effects in measuring the chemical rate of decomposition of CO_2 on the surface of liquid iron–carbon alloys.

The results obtained by Fruehan and Martonik [44] for the gas stream flowing parallel to the surface of the melt, the rate of decarburization was also shown to be in accord with Eq. (7.72).

Although heat and mass transfer in the gas–film layer are rate-limiting processes, different phenomena occur with vaporization in a reactive gaseous environment or in the presence of a temperature gradient near the vaporizing surface.

7.4.3 Vaporization in Reactive Gases

There are two types of mechanisms of enhanced vaporization in reactive gases: (i) a chemical process involving the formation of volatile species, e.g., oxides, sulfides, halides, carbonyls, and hydroxides, and (ii) a transport process in which the vapor reacts with the gas diffusing toward the vaporizing surface.

The concept of diffusion-limited enhanced vaporization, introduced by Turkdogan *et al.* [48], is illustrated schematically in Fig. 7.10, using as an example the vaporization of iron in a stream of an oxygen–inert gas mixture. In this counterflux–transport process, at some short distance from the surface of the metal, the oxygen and metal vapor react to form a metal oxide mist.

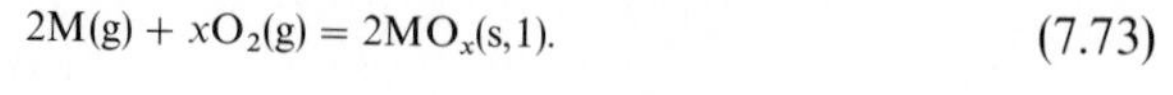

$$2M(g) + xO_2(g) = 2MO_x(s, 1). \tag{7.73}$$

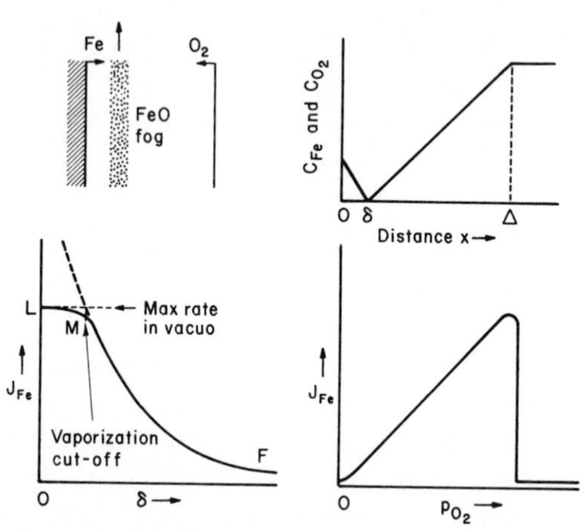

FIG. 7.10 Schematic representation of vaporization of metals (iron as an example) in oxygen-bearing gas mixtures flowing over the surface; counterflux of iron vapor and oxygen resulting in enhanced vaporization of iron. After Turkdogan *et al.* [48].

The formation of an oxide in the gas phase provides a sink for the metal vapor and oxygen resulting in the counterflux of these two gaseous species. Increasing oxygen partial pressure decreases the distance δ through which the metal vapor is diffusing, i.e., the rate of vaporization increases. On further increase in oxygen partial pressure, the flux of oxygen toward the surface of the metal eventually becomes greater than the equivalent counterflux of the metal vapor, resulting in the oxidation of the metal surface and cessation of vaporization. Just before this cutoff, the rate of vaporization will be close to the maximum rate of vaporization as given by Eq. (7.18) for free vaporization in vacuo. This mechanism of enhanced vaporization is the major cause of the formation of metal oxide fumes in smelting and metal refining processes.

As is seen from some examples of the experimental results of Turkdogan *et al.* [48] given in Fig. 7.11, the rates of vaporization of iron and copper increase with increasing partial pressure of oxygen until a maximum rate is obtained, beyond which vaporization almost ceases. The results with various metals summarized in Fig. 7.12 show that the measured maximum rates of vaporization

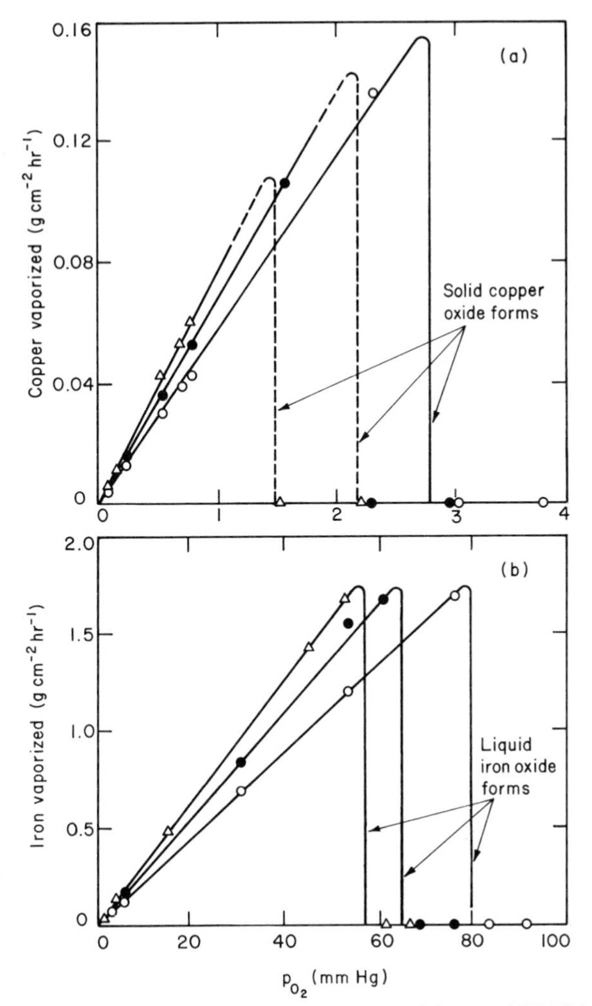

FIG. 7.11 Rates of vaporization of copper at 1200°C and iron at 1600°C in argon–oxygen mixtures at 1 atm pressure as affected by gas velocity parallel to surface and partial pressure of oxygen. Gas velocity: (○) 40, (●) 60, and (△) 80 cm sec^{-1}. After Turkdogan *et al.* [48].

agree well with those calculated for free vaporization. A quantitative analysis of these observations in terms of mass transfer is given in the paper cited [48].

In a subsequent study, Kor and Turkdogan [49] showed that the rate of vaporization of phosphorus from liquid iron can be increased appreciably also by blowing an oxygen–argon mixture on the surface of the melt without oxidizing the surface. In this case, the vaporizing species are Fe, PO, and PO_2, which react with oxygen in the gas phase forming an iron–phosphate mist close to the surface of the melt.

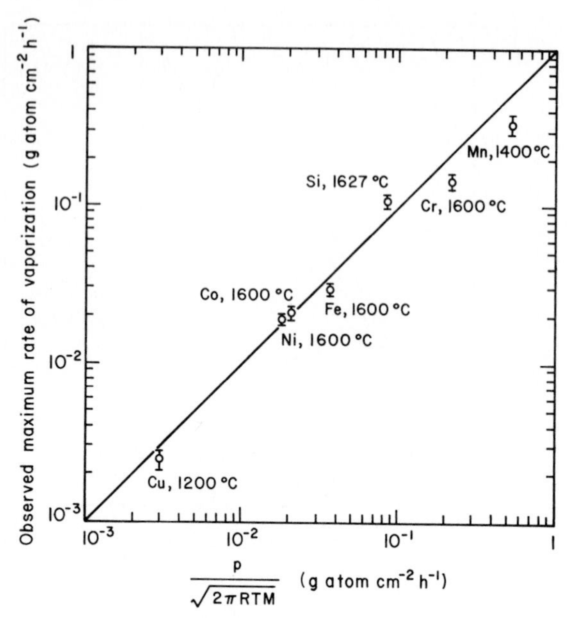

FIG. 7.12 Experimentally observed maximum rate of vaporization of metals in argon–oxygen mixtures compared with rate of free vaporization in vacuo. After Turkdogan *et al.* [48].

7.4.4 Vaporization in a Temperature Gradient

When there is a temperature gradient between a heated object and its immediate fluid surroundings, the vapor condenses at a short distance away from the surface, thus shortening the diffusion distance for the vapor species, and consequently, increasing the rate of vaporization. There are two processes occurring simultaneously: (i) natural convection which brings about the vapor transport and (ii) condensation of the vapor which is a reaction process. As shown by Turkdogan [50], this is a complex phenomenon involving the supersaturation of the vapor for nucleation of metal droplets in a steep temperature gradient, coupled with the effect of the latent heat of condensation on the temperature profile (sometimes referred to as the Lewis effect) and growth of the droplets. High speed motion pictures of a zirconium drop falling in an oxidizing atmosphere and in a temperature gradient [51] illustrate well the condensation occurring in the gas film around the metal drop. A similar phenomenon is observed in the vaporization of Ni_3S_2 droplets in a temperature gradient [52]. Recombination of vaporized sulfur and nickel along with condensation of the nickel vapor within the cooler gas–film layer enhances the rate of vaporization of Ni_3S_2.

Rate equations of various complexity have been derived to formulate this phenomenon in terms of heat and mass transfer and condensation in a supersaturated vapor in the boundary layer [50–54]. Similar phenomena have been proposed for heat and mass transfer over blunt-nosed bodies at hypersonic flight speeds [55, 56].

7.5 Heat Transfer by Radiation

A brief reference should also be made to heat transfer by thermal radiation which is the energy transferred by electromagnetic waves (photons) from the surface of a heated body. The thermal radiation is composed of a continuous spectrum of wavelengths forming an energy distribution as described by *Planck's distribution law*. Integration of Planck's equation gives the total emissive power e of a body, known as the *Stefan–Boltzmann equation*,

$$e = \varepsilon \sigma T^4, \tag{7.74}$$

where σ is the radiation constant, 1.355×10^{-12} cal cm^{-2} sec^{-1} deg^{-4}, and ε the emissivity.

A surface which absorbs all incident radiation and emits radiation of all wavelengths is known as the *blackbody* or *black radiator*. The ratio of the total emissive power of a body to that of the black radiator gives the emissivity. The emissivity varies with temperature, wavelength, direction of the radiation, and surface roughness. The emissivity used in most practical situations is an average emissivity for all wavelengths and all directions. The emissivity increases with increasing surface roughness and often with increasing surface temperature. For example, the emissivity of polished copper is 0.15 at 600 °C and, for oxidized copper, it is about 0.6. Data on emissivity of many materials have been compiled in the references cited [57–59].

The radiant heat-transfer coefficient k_r is defined as

$$k_r = \varepsilon_s \sigma (T_s^4 - T_\circ^4)/(T_s - T_\circ), \tag{7.75}$$

where ε_s and T_s are the emissivity and temperature of the surface and $T_\circ$ is the temperature of the bulk fluid stream. The total heat-transfer coefficient is given by the sum

$$h = k_h + k_r, \tag{7.76}$$

where k_h is the convective heat-transfer coefficient which becomes significant at lower temperatures. Because of the fourth-power temperature dependence, heat transfer by radiation predominates at elevated temperatures.

The polar gases such as CO, CO_2, H_2O, SO_2, and NH_3 interact with thermal radiation. The extent of emission and absorption of radiant energy in a gas layer depends on the emissivity, temperature, pressure, and thickness of the gas layer. This and many other fundamental aspects of thermal radiation cannot be described adequately in this brief mention of heat transfer by radiation; reference should be made to textbooks devoted totally to this subject [57–59].

7.6 Heat and Mass Transfer in the Reaction–Product Layer

Rates of interfacial reactions are seriously impaired by relatively slow diffusion of reactants and slow heat transfer to and from the reaction surface. Because of the theoretical interest, and experimental and practical importance, the Fourier equation for heat conduction and Fick's equation for mass diffusion have been solved for many goemetrical shapes and boundary conditions. Reference may be made, for example, to Crank [60] and to Carslaw and Jaeger [61] for the methods of solution of these equations. In this section we shall discuss only a few special cases of rate control by heat and mass transfer in the reaction–product layer as typical examples.

7.6.1 Special Cases for Spheres, Long Cylinders, and Flat Plates

As an example, let us consider the growth of a reaction product around a spherical particle reacting with a gas phase X_2.

$$\tfrac{1}{2}X_2(g) + \nu A(s) = A_\nu X(s). \tag{7.77}$$

The time and distance dependence of the molar flow F_X of reactant X through the product layer $A_\nu X$ for isothermal conditions is given by

$$(F_X)_{r,t} = -4\pi r^2 D(\partial c/\partial r)_{r,t} \quad \text{mole sec}^{-1}, \tag{7.78}$$

where r is the radial distance from the center of the sphere, D the diffusivity of X, assumed independent of composition, and $\partial c/\partial r$ the concentration gradient of X in the product layer.

As the reaction proceeds, the radial position r_i of the reaction interface between the product layer and the unreacted core moves towards the center of the sphere. When the rate of diffusion is much greater than the rate of movement of the interface, a pseudosteady state $dc/dt \sim 0$ approximation may be made. For the case considered, it is also assumed that the rate of reaction (7.77) is controlled solely by diffusion in the product layer under fixed boundary conditions, $c = c_e$ (equilibrium concentration) at $r = r_i$ and $c = c_\circ$ at $r = r_\circ$. The third assumption is that the radius $r_\circ$ of the sphere (core + shell) remains unchanged. With these boundary conditions, the integration of Eq. (7.78) gives for the molar flow

$$F_X = [4\pi r_0 r_i/(r_0 - r_i)]D(c_0 - c_e). \tag{7.79}$$

For the stoichiometry of reaction (7.77), the following equality of fluxes may be written:

$$\nu F_X/4\pi r_0^2 = -F_A/4\pi r_0^2. \tag{7.80}$$

Noting that the instantaneous rate of reaction is given by

$$F_A = d(\rho \tfrac{4}{3}\pi r_i^3)/dt,$$

where ρ is the molar density of A in the unreacted core, the following is obtained from Eq. (7.80):

$$\nu F_X/4\pi = -\rho r_i^2\, dr_i/dt, \tag{7.81}$$

Combining this with Eq. (7.79) and integrating gives the equation for r_i as a function of reaction time t:

$$2(r_i/r_0)^3 - 3(r_i/r_0)^2 + 1 = (6\nu D/\rho r_0^2)(c_0 - c_e)t. \tag{7.82}$$

The fraction F of the reactant A consumed is given by $r_i/r_0 = (1-F)^{1/3}$. Inserting this is Eq. (7.82) gives for the time dependence of fraction reacted

$$3 - 2F - 3(1-F)^{2/3} = (6\nu D/\rho r_0^2)(c_0 - c_e)t. \tag{7.83}$$

If reaction (7.77) is accompanied by a large enthalpy change and the rate of heat conduction in the product layer is slow, the rate of reaction may be controlled by heat transfer through the product layer, as, for example, in the calcination of limestone. The equation derived for this limiting rate is analogous to Eq. (7.83), thus

$$3 - 2F - 3(1-F)^{2/3} = (6\nu\kappa/\rho\,\Delta H r_0^2)(T_0 - T_i)t, \tag{7.84}$$

where ΔH is the heat of reaction per mole $A_\nu X$, T_0 the surface temperature at $r = r_0$, T_i the interface and core temperature at $0 < r \leqslant r_i$, and κ the thermal conductivity of the product layer.

For flat plates and long cylinders, the following equations are derived for the rate controlled by diffusion in the product layer.

For long cylinders of radius r_0,

$$F + (1-F)\ln(1-F) = (4\nu D/\rho r_0^2)(c_0 - c_e)t. \tag{7.85}$$

For flat plates of thickness l_0, reacting on one surface,

$$F^2 = (2\nu D/\rho l_0^2)(c_0 - c_e)t. \tag{7.86}$$

Often, the overall rate is controlled by both diffusion in the product layer and reaction at the interface of the unreacted core with the product layer. In their book on gas–solid reactions, Szekely *et al.* [62] give general rate equations for overall rates controlled jointly by heat transfer, mass transfer, and chemical reactions involving solid and porous materials. For the case of an isothermal reaction, Eq. (7.77), between a gas and a solid spherical particle with a growing product layer around the unreacted core, the dual rate control is expressed by a general equation

$$t^* = q(F) + \sigma^2 g(F), \tag{7.87}$$

where

$$t^* = (\nu\Phi/\rho r_0)(c_0 - c_e)t \qquad \text{for} \quad 1 - \theta \to 1, \tag{7.88}$$

$$q(F) = 1 - (1-F)^{1/3}, \tag{7.89}$$

$$\sigma^2 = \Phi r_0/6D, \tag{7.90}$$

$$g(F) = 3 - 2F - 3(1-F)^{2/3}. \tag{7.91}$$

For large values of Φ/D, i.e., for fast chemical reaction, Eq. (7.87) reduces to (7.82). For small values of Φ/D, as for slow chemical reaction and fast diffusion in the porous product layer, or in gasification of solids, Eq. (7.87) reduces to

$$1 - (1 - F)^{1/3} = (v\Phi/\rho r_0)(c_0 - c_e)t \tag{7.92}$$

for chemical reaction control at the receding interface.

Aside from solving diffusion equations for various geometrical shapes and boundary conditions, due consideration should also be given to the physico-chemical aspects of mass transfer in the product layer.

7.6.2 Scale Formation—Wagner Equation

The parabolic rate law, which was originally derived by Tammann [63], came about from studies of rates of scale formation on metals. Subsequently, Wagner [64] derived a general form of the parabolic rate law in terms of the self-diffusivity and the thermodynamic activity of the diffusing species in the solid scale.

For a brief account of the basic concept, let us consider a metal A reacting with a gas X_2 to form a scale of nominal composition AX. For simplicity, let us assume that the flux J_A of component A in the scale is much greater than the flux J_X of component X. For this limiting case, Eq. (3.145) for binary diffusivity is reduced to

$$D = -c_X \bar{V}_X J_A/(\partial c_A/\partial z). \tag{7.93}$$

The total quantity Q_A of component A per unit area in the scale of thickness Z at time t is

$$Q_A = \bar{c}_A Z, \tag{7.94}$$

where $\bar{c}_A$ is the average concentration of A. For small compositional changes across the scale, the flux J_A at time t is essentially independent of position along the diffusion path. For this pseudosteady state, the flux is given by

$$J_A = dQ_A/dt = \bar{c}_A\, dZ/dt. \tag{7.95}$$

Introducing a parabolic rate constant $\lambda = d(z^2)/dt$, and inserting $dZ/dt = \lambda/2Z$ in Eq. (7.95) and combining with Eq. (7.93) gives

$$\lambda d(z/Z) = -(2D/c_X \bar{V}_X)\, dc_A/\bar{c}_A. \tag{7.96}$$

Upon integration across the scale layer from $z/Z = 0$ (scale–gas interface) with $c_A = c_A''$ to $z/Z = 1$ (scale–metal interface) with $c_A = c_A'$, and inserting the approximation $\bar{c}_A \simeq c_A$, the expression obtained for the parabolic rate constant is

$$\lambda = 2 \int_{c_A''}^{c_A'} \frac{D}{c_X \bar{V}_X}\, d \ln c_A. \tag{7.97}$$

For the simplified case of $D_A^* \gg D_X^*$, Darken's equation (3.148) reduces to

$$D = c_X \bar{V}_X D_A^* \, [d(\ln a_A)/d(\ln c_A)]. \tag{7.98}$$

Substituting this in Eq. (7.97) gives for the parabolic rate constant

$$\lambda = 2 \int_{a''_A}^{a'_A} D_A^* \, d(\ln a_A). \tag{7.99}$$

By invoking the Gibbs–Duhem relation, relation (17) in Table 1.1, λ may be given in terms of the activity of component X,

$$\lambda = 2 \int_{a_X}^{a''_X} \frac{x_X}{x_A} D_A^* \, d(\ln a_X), \tag{7.100}$$

where a's are the activities and x's are the atom fractions in the scale. If both diffusivities D_A^* and D_X^* are of the same order of magnitude, the rate equation in a complete form is given by the equation

$$\lambda = 2 \int_{a'_X}^{a''_X} \left(\frac{x_X}{x_A} D_A^* + D_X^* \right) d(\ln a_X). \tag{7.101}$$

A different approach was made by Wagner [64] in the original derivation of this equation; in addition, the parabolic rate constant used by Wagner, *rational rate constant* k_r, is in units of gram-equivalents per centimeter per second. The conversion from λ [cm^2 (scale) sec^{-1}] to k_r is given by

$$k_r = (\rho/M)\lambda, \tag{7.102}$$

where ρ is the density of the scale and M the average molecular weight of the scale per gram-atom of component X.

7.6.3 Subscale Formation

Many have contributed to the study of subscale formation (oxides, sulfides, nitrides, carbides, etc.) in metal alloys, the contributions made by Wagner being the most outstanding. When an alloy is heated, for example, in an oxidizing gas, the oxygen diffuses into the metal, reacts with the less-noble alloying element, and forms a subscale consisting of minute oxide particles embedded in the metal matrix. As discussed in the papers cited, for example, Rapp [65] and Hepworth *et al.* [66], many measurements have been made of the permeability and diffusivity of interstitial solutes in metals by determining the rate of growth of the subscale.

The diffusivities of interstitial solutes, e.g., O, N, and C, are much higher than those of the substitutional alloying elements. Therefore, for the case of almost complete reaction of the solute in the subscale, the isothermal rate equation derived for parabolic growth of the subscale may be given in a simplified form:

$$\xi^2 = [2M_A/rM_X(\%\,A)][(\%\,X)'' - (\%\,X)']D_X t, \tag{7.103}$$

where ξ is the thickness of the subscale at time t, M_A the atomic weight of the less-noble alloying element, M_X the atomic weight of the diffusing reactant, %X the concentration of the diffusate (O, N, C, etc.) in the metal, double prime at the surface and prime at the subscale–metal interface, which assumed to remain constant, D_X the diffusivity of the reactant, and r the X/A atom ratio in the reaction product in the subscale

The amount of the diffusate X transported across the unit area of the metal is, from mass balance,

$$n = (\rho/M_A)r(\%\,A/100)\xi \quad \text{mole cm}^{-2},$$

where ρ is the density of the parent metal. Substituting this in Eq. (7.103) gives

$$n^2 = 2 \times 10^{-4}(\rho^2 r/M_A M_X)(\%A)[(\%\,X)'' - (\%\,X)']D_X t. \tag{7.104}$$

We see from Eqs. (7.103) and (7.104) that as the concentration of the less-noble alloying element A increases, the thickness ξ of the subscale decreases but the amount of X transported increases. Therefore, when the concentration of A is sufficiently high, there will be a transition from the subscale to the surface scale formation. Wagner's theoretical analysis [67] of scale–subscale transition, based on the concept of interface instability, leads to the following criteria for this transition:

scale formation when

$$\frac{D_A(\%\,A)}{D_X(\%\,X)''} > \frac{\pi}{2r}\frac{M_A}{M_X}, \tag{7.105a}$$

subscale formation when

$$\frac{D_A(\%\,A)}{D_X(\%\,X)''} < \frac{\pi}{2r}\frac{M_A}{M_X}. \tag{7.105b}$$

If the alloying element is more noble than the base metal, the oxidation of the latter results in the enrichment and ultimate precipitation of the alloying element at the scale–metal interface. This type of subscale formation has been observed in the oxidation of iron containing copper, tin [68], or sulfur [69]. However, if the solute diffusivity in the oxide is relatively high and the rate of scale growth is slow, there will be sufficient time for the solute to diffuse through the thin scale into the gas phase, and subscale will not form. Such a situation may occur in the oxidation of iron containing sulfur [70].

Because of the limited coverage that could be given here to the discussion of heat and mass transfer in the reaction–product layer, the examples chosen in this section had to be confined to simple systems. The solution of heat and mass-transfer problems for systems of practical interest is much more complex, as, for example, in the solidification of alloys, attack of molten slags on the refractory lining of smelting- and metal-refining furnaces, high temperature corrosion of jet turbine blades, heat and mass transfer in fuel elements of nuclear reactors, and so on.

7.7 Reaction of Gases with Porous Media

The rate of a heterogeneous reaction between a fluid and a porous medium is much affected by the counterdiffusive flow of fluid reactants and products through the pores of the medium. The mathematical analysis of Thiele [71] for mixed rate control. involving reaction on the surface of the pore walls and gas diffusion in the pores of the medium, formed the basis for the subsequent earlier studies of reaction of gases with porous materials. References to early work are given in the publications of, for example, Wheeler [72, 73] and Petersen [74, 75].

Within the context of Chapter 7, we shall consider three types of reaction systems that are frequently encountered in experiments and in industrial processes: porous catalysts, gasification of porous materials, and the formation of a porous reaction product layer.

7.7.1 Irreversible Reactions in Porous Catalysts

In order to highlight the kinetic parameters and properties of the system that have the greatest effect on the overall rates of reactions in porous catalysts, it is expedient to make certain simplifying assumptions. It is assumed that (i) there is no pressure buildup in the porous pellet, (ii) the reaction is irreversible, (iii) the mass transfer is by diffusion only, no forced flow, (iv) the pore structure is uniform and remains unchanged, and (v) the temperature of the system is uniform and constant. For the case considered, the mass balance within the pellet is represented by the equation

$$D_e\, d^2c/dz^2 - \Phi'\rho S c^n = 0, \tag{7.106}$$

where D_e is the effective gas diffusivity, c the molar concentration of the reacting gas, z the diffusion distance normal to the surface of porous pellet, ρ the density of porous pellet, S the pore surface area per unit mass of pellet, and Φ' the apparent rate constant, mole per unit area of the pore wall, for $1 - \theta \to 1$.

For a spherical pellet of radius r, and the boundary conditions $c = c_0$ at $z = r$ (the surface of the pellet) and $c = dc/dz = 0$ at $z = 0$ (the center of the pellet), the solution of Eq. (7.106) gives for the rate of reaction per pellet

$$\mathscr{R} = 4\pi r^2 \left(\frac{2}{n+1} \Phi' \rho S D_e \right)^{1/2} c_0^{(n+1)/2} \left[\frac{1}{\tanh 3\phi} - \frac{1}{3\phi} \right], \tag{7.107}$$

where ϕ is the Thiele's dimensionless parameter defined by

$$\phi = \tfrac{1}{3} r \left(\tfrac{1}{2}(n+1)(\Phi'\rho S/D_e) c_0^{n-1}\right)^{1/2}. \tag{7.108}$$

When ϕ is small, e.g., $\phi < 0.2$, as with (i) decreasing particle size, (ii) decreasing temperature (i.e., decreasing Φ'/D_e), and (iii) decreasing pressure (i.e., increasing D_e), the dimensionless parameter in Eq. (7.107) approaches the value of $\phi c_0^{(n-1)/2}$; for this limiting case the rate equation (7.107) simplifies to

$$\mathscr{R} = \tfrac{4}{3}\pi r^3 \Phi' \rho S c_0{}^n \quad \text{mole sec}^{-1}. \tag{7.109}$$

That is, the rate of reaction for given gas composition and temperature is proportional to the mass of the porous pellet. This is the limiting case for almost complete pore diffusion in pellets of any geometry.

When ϕ is large, e.g., $\phi > 2$, as with (i) increasing particle size, (ii) increasing temperature (i.e., increasing Φ'/D_e) and (iii) increasing pressure (i.e., decreasing D_e), the dimensionless parameter in Eq. (7.107) approaches unity, and the rate equation reduces to

$$\mathcal{R} = 4\pi r^2([2/(n+1)]\Phi'\rho S D_e)^{1/2} c_0^{(n+1)/2} \tag{7.110}$$

For this limiting mixed control, the reaction is confined to the pore mouths at the outer surface of the pellet; hence, the rate per unit area, $\mathcal{R}/4\pi r^2$, given by Eq. (7.110) applies to pellets of any geometrical shape. Although the rate is the nth-type order for a plane surface of a solid particle or for the surface of the pore walls, for the limiting mixed control, the apparent order of the reaction is $\frac{1}{2}(n+1)$. Another important feature is that, since D_e does not change much with temperature, the apparent heat of activation for the reaction in the region of limiting mixed control is about one-half of that for the case of almost complete pore diffusion when the rate is controlled by reaction on the pore walls.

7.7.2 Gasification of Porous Solids

There are many reactions of practical importance classified as gasification reactions, such as combustion of solid fuels, methane generation by C–H_2 reaction, and the formation of volatile suboxides, sulfides, and halides. Let us consider the gasification of a spherical particle of porous solid A with a gaseous reactant X_2 to form a gaseous product A_vX.

$$\tfrac{1}{2}X_2(g) + vA(s) = A_vX(g). \tag{7.111}$$

In order to highlight the salient features of the rate phenomena in the gasification of porous solids, let us neglect the reverse reaction and also assume that the flow conditions are such that the transport of gaseous species to and from the surface of the pellet is not rate determining. It is also assumed that the temperature and pressure in the system is constant and the pore structure remains unchanged during gasification. For the case considered, Eq. (7.107) gives the instantaneous rate of gasification of a porous solid of radius r at time t.

For the limiting case of complete gas diffusion in the porous pellet, the mass fraction F of pellet gasified at time t is readily derived from Eq. (7.109), giving

$$\ln(1-F) = -v\Phi' S c_0^n t. \tag{7.112}$$

That is, for given gas composition and temperature, $\ln(1-F)$ is a linear function of time and the rate $\partial[\ln(1-F)]/\partial t$ is independent of the particle size. This is the idealized limiting case for almost uniform internal "burning" without change in the outer dimensions of the pellet.

When the rate of gas diffusion in the pellet is slow, the gasification is confined to the outer surface of the pellet as described by Eq. (7.110). Taking into account the shrinking size of the spherical pellet during gasification, integration of Eq. (7.110) gives

$$1 - (1 - F)^{1/3} = (v/\rho r_0)([2/(n+1)]\Phi'\rho S D_e)^{1/2} c_0^{(n+1)/2} t, \qquad (7.113)$$

where r_0 is the initial radius of the porous pellet.

It follows from Eqs. (7.112) and (7.113) that, in terms of mass fraction gasified w/w_0, the initial rate of gasification $[d(w/w_0)/dt]_0$ is independent of the particle size for uniform internal gasification; for external gasification $[d(w/w_0)/dt]_0$ is inversely proportional to the particle diameter. With large particles and low velocity gas flows, the rate is ultimately controlled by mass transfer in the gas–film boundary layer, for which the correlation in Table 7.3 for Sherwood number gives $[d(w/w_0)/dt]_0$ inversely proportional to the square of the particle diameter. In the light of these three limiting-rate laws, the initial rate of isothermal gasification will change with particle size as depicted schematically in Fig. 7.13a.

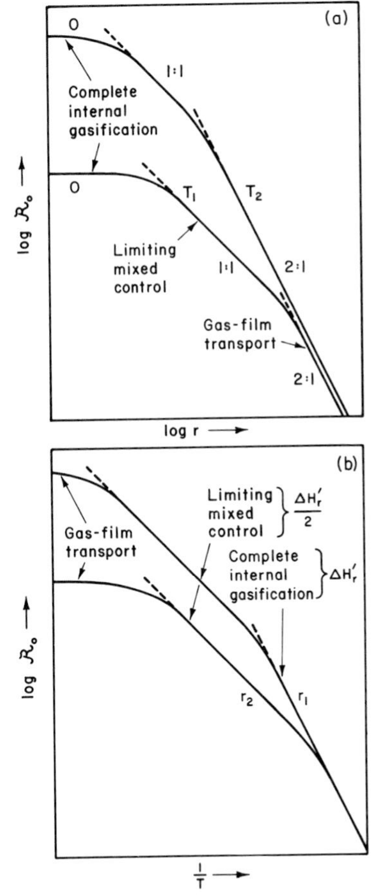

FIG. 7.13 Schematic representation of the effect of (a) pellet size and (b) temperature on the initial rate of gasification of porous materials of spherical shape. For (a), $T_1 < T_2$, and for (b), $r_1 < r_2$.

In addition, in accord with deductions made from Thiele's dimensionless parameter ϕ, the general form of the temperature dependence of the initial rate for a given particle size will be as shown in Fig. 7.13b; $\Delta H_r'$ is the apparent heat of activation for the chemical reaction on the pore walls. The higher the temperature, the smaller the particle size for the onset of complete internal gasification.

The effects of temperature and particle size on the initial rate of gasification of electrode graphite spheroids in a stream of carbon dioxide, measured by Turkdogan *et al.* [76] are shown in Fig. 7.14. In this log $\mathscr{R}_{\circ}$ versus log diameter plot, the curves approach the limiting slope 3:1 for small particles, indicating

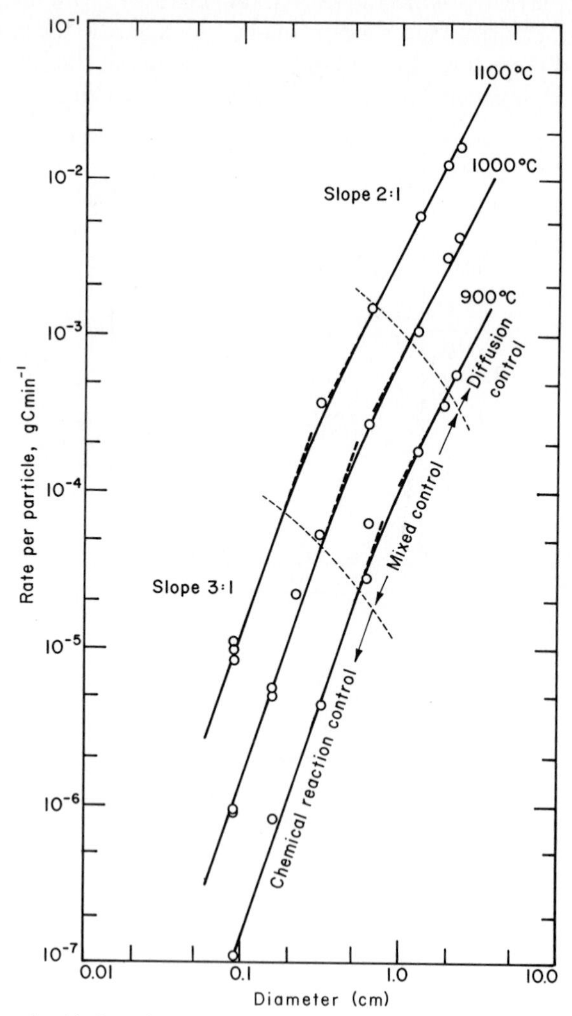

FIG. 7.14 Rate of oxidation of electrode graphite as a function of particle diameter for 1 atm CO_2. After Turkdogan *et al.* [76].

almost complete internal burning as in Eq. (7.109); and the other limiting slope 2:1 for large spheres, indicating external burning as in Eq. (7.110) for the limiting mixed control. The changeover from the external to internal burning, with an intermediate region for mixed control, occurs at particle sizes decreasing with increasing temperature. This is as would be predicted from Thiele's parameter ϕ, Eq. (7.108).

Within the region of molecular diffusion in the pores, the effective diffusivity decreases with increasing pressure. For example, as seen from the data of Turkdogan and Vinters [77] in Fig. 7.15 for the oxidation of electrode graphite at 1000°C, the rate per unit mass is independent of the particle size over the range 0.9–3 mm diameter at 1 atm CO_2, while at 5 and 30 atm CO_2, the rate decreases with increasing particle size because of lack of pore diffusion at higher pressures.

The pore surface area is another important property of porous materials; in general, the higher the pore surface area, the higher would be the reactivity of the material. However, a study of pore characteristics of various types of carbon [78] has indicated that only a certain fraction of the total pore surface area can participate effectively in the internal oxidation of carbon. For an electrode graphite with S equal to 1 $m^2 g^{-1}$, the rate of internal oxidation of granules in 1 atm CO_2 at 1100 °C is about 0.01 fractional mass loss per minute. For coconut charcoal with S equal to 1000 $m^2 g^{-1}$, the rate is 0.4 min^{-1}, instead of the 10 min^{-1} that would be expected from a 1000-fold increase in the pore surface area relative to graphite.

Because of variations in pore size and faster gas diffusion in larger pores, most of the reaction will occur on the walls of larger pores. Consequently, only a fraction of the total pore surface area is expected to be available for the reaction; also, the larger the pore surface area, the smaller the fraction of the total pore wall utilized in the reaction. Because of these complexities and uncertainties inherent in porous materials, the apparent rate constant Φ' derived from

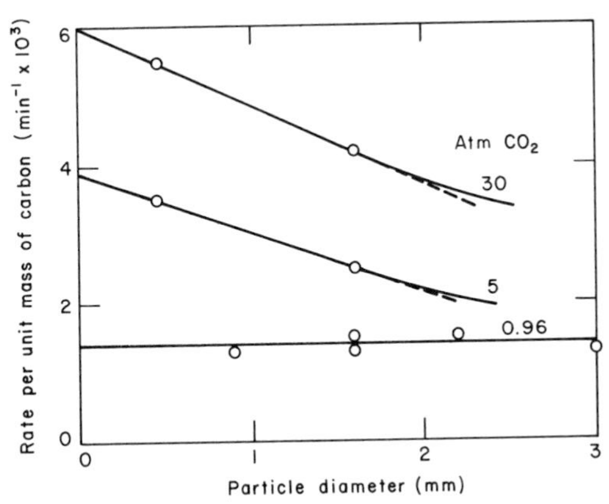

FIG. 7.15 Effect of particle size on the rate of oxidation of electrode graphite in CO_2 at 1000 °C and indicated pressures. After Turkdogan and Vinters [77].

the measurements, within the regime of almost complete internal gasification, may not be amenable to interpretation in terms of reaction kinetics. Nevertheless, an equation such as (7.107) does provide general guidance in characterizing the reactivity of porous materials in terms of the specific properties of the system: temperature, pressure, and gas composition.

In the gasification of carbon in CO_2–CO mixtures, there is the added complication arising from strong chemisorption of CO on the pore walls. The rate equations have been derived for mixed control involving pore diffusion, forward and reverse reactions, chemisorption of CO [79], and nonisothermal gasification [80] in CO_2–CO mixtures.

7.7.3 Porous Product Layer

In the dehydration, calcination, and reduction, a porous reaction product is formed. If the initial porosity of the material is negligibly small, as in calcite or fused oxides, a layer of porous reaction product will form around the unreacted core, for which the rate equations are similar to those given in Section 7.6.1. For example, for the reduction of a dense spherical oxide particle, Eq. (7.87) will apply for the dual rate control. In the early stages of the reaction, when the porous product layer is thin, the rate will be controlled essentially by the interfacial chemical reaction, for which Eq. (7.92) applies. Ultimately, when the product layer is sufficiently thick, the rate will be controlled by gas diffusion in the pores of the product layer. For equimolar countercurrent diffusion at constant pressure and temperature, Eq. (7.83) gives the limiting rate of reaction for a spherical particle. The diffusivity D in this case is the effective gas diffusivity D_e in the porous layer. If the rate is controlled by heat transfer through the porous layer, the rate for a spherical particle is that given by Eq. (7.84) in which κ is replaced by the effective thermal conductivity κ_e of the porous layer.

If the material to be calcined or reduced is porous, the reaction will occur within the pores, with the deposition of the reaction product on the pore walls as in the reduction of porous oxides and sulfides. The characteristic features of reduction of porous metal oxides are discussed in Chapter 8; for the present we need to consider only the general form of the rate equations for simple cases.

If the effective gas diffusivity and the pore surface area are not affected much with the deposition of the reaction product, which is also porous, the isothermal rate of reaction per pellet is that given by Eq. (7.107). With small particles of porous oxides or sulfides at low temperatures, the internal reduction will predominate; for this limiting case, Eq. (7.109) will apply. With large particles of low porosity and at high temperatures, the initial rate will be that given by Eq. (7.110) for the limiting mixed control. This situation leads to the formation of a porous product layer and the rate is ultimately controlled by gas diffusion in the porous product layer, as given by Eq. (7.83) for a spherical particle and equimolar countercurrent diffusion.

These general considerations may be summarized by stating that there are three limiting rate processes pertinent to the reaction of gases with porous materials forming a porous reaction product: uniform internal reaction, limiting mixed control, and gas diffusion in the porous product layer. When the reaction is controlled solely by any one of these reaction steps, then the time t of reaction would be related to the particle (spheroidal) diameter d in one of the following ways:

Uniform internal reaction. t is independent of d.
Limiting mixed control. t is proportional to d.
Diffusion in porous layer. t is proportional to d^2.

The t–d relation for these individual limiting rate processes are shown schematically by dotted curves in Fig. 7.16. The curve drawn represents the experimental observation. Depending on temperature, gas composition, particle size, and porosity of the material, there will be transition from one limiting rate controlling process to another as the reaction progresses. The interpretation of the rate data within the region of mixed control is subject to the choice of the mathematical model for the overall reaction.

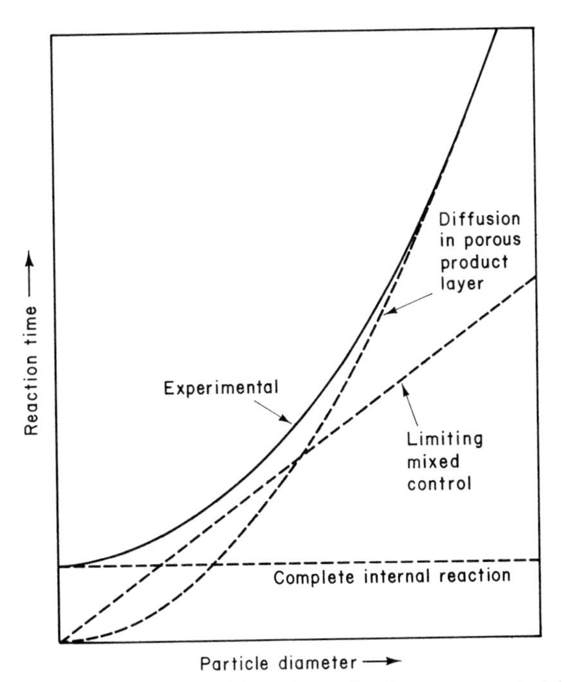

FIG. 7.16 Schematic representation of time of reaction for porous materials as a function of particle size for three limiting rate controlling processes.

Numerous rate equations have been derived for various models of reaction of gases with porous materials. A generalized model of a special type is the so-called *grain model* proposed by Sohn and Szekely [81]. The porous pellet is considered as an ensemble of solid grains of equal size; the solid grains may be spherical, cylindrical or flat platelike. In the schematic representation of this

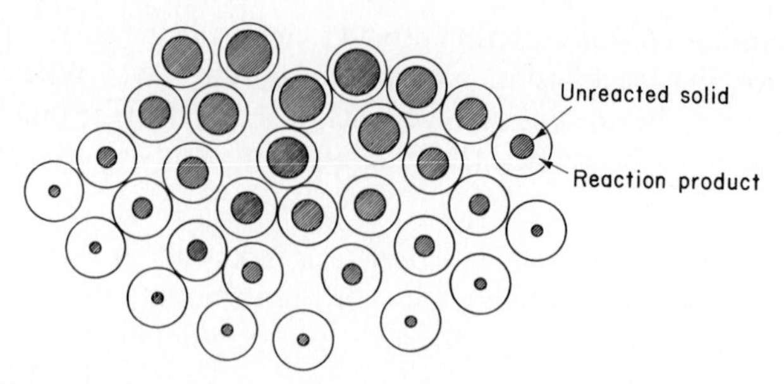

FIG. 7.17 Schematic representation of grain model for reaction of gases with porous material consisting of solid spheroidal grains of uniform radius r_g.

model in Fig. 7.17, the spherical pellet of radius r_0 is considered to be made up of spherical solid grains of radius r_g. Because of the gas concentration gradient in the pellet, the extent of reaction in the grains decreases with the radial distance from outside to the center of the pellet. On the assumption that within each grain the reaction front retains its original geometrical shape, the mass balance within the pellet is represented by the equation

$$D_e\, d^2(c_X - c_Y/K)/dz^2 - \mathscr{R}_l(1 + 1/K) = 0, \tag{7.114}$$

where D_e is the effective gas diffusivity in the pellet, K is the equilibrium constant for the gas–solid reaction, c is the local concentration of gas (X reactant, Y product) at the radial distance z, and $\mathscr{R}_l$ is the local rate of reaction.

On the assumption that diffusion in the product layer around the grain is not rate limiting, for a first-order-type reaction, the local rate on the surface of a spherical grain is represented by

$$\mathscr{R}_l = [3(1 - \varepsilon)\Phi/r_g{}^3]r_z{}^2(c_X - c_Y/K), \tag{7.115}$$

where ε is the initial porosity of the pellet, Φ the specific rate constant, r_g the radius of the grain, and r_z the radial distance of the reaction front in the grain at the radial distance z in the pellet.

Using an expression similar to Thiele's modulus ϕ, Eq. (7.108), Sohn and Szekely [81] showed that for $\phi < 0.2$ there is essentially uniform reaction of grains across the pellet for which the rate equation simplifies to that for gasification of spheroidal solids, which is similar to Eq. (7.92) and differs from Eq. (7.112) for uniform internal reaction throughout the porous pellet and porous grains. The fraction reacted F as a function of time is different for these two models of "uniform" reaction in porous pellets, particularly at longer reaction times.

For $\phi > 2$, integration of Eq. (7.114) gives an expression identical in form to Eq. (7.113) for the limiting mixed control that is appropriate for the initial stages of the reaction which is confined to the pore mouths at the surface of the pellet.

When the grains on the periphery of the pellet have completely reacted, we have to consider two zones in the pellet: (i) gas diffusion through the reacted outer layer and (ii) partial diffusion and reaction in the core of the pellet, for which the appropriate rate equation has been derived by Szekely *et al.* [62, 81]. Further discussion of the kinetics of reduction of metal oxides is deferred to Chapter 8.

7.8 Mass Transfer across Stirred Liquid Interfaces

Gas evolution in smelting and metal refining processes brings about effective mixing in the metal and slag phases, and also enhances mass transfer across the slag–metal interface. The discussion here is confined to the rate of mass transfer across liquid interfaces stirred with gas bubbles.

The surface renewal or penetration theory of Higbie [82] has been used extensively in the derivation of rate equations for mass transfer in stirred liquids. In terms of the mass transfer coefficient k_m, Higbie's penetration theory is reduced to the following form for low concentrations of the diffusate:

$$k_m = 2(D/\pi t_c)^{1/2}, \tag{7.116}$$

where D is the diffusivity and t_c is the time of contact of the liquid element with the surface layer. For a bubble of diameter d approaching the surface with velocity u_b, the contact time would be

$$t_c = d/u_b. \tag{7.117}$$

Richardson and co-workers [83–86] studied the effect of gas bubbling on the mass transfer in several two-liquid systems: aqueous solutions–molten metals and molten metals–molten salts. For the case of no dispersion of the two liquids, no solute-induced interfacial turbulence, and interfacial reaction equilibrium, the mass transfer of a solute from one phase to another is found to obey approximately the relation

$$k_m^2 = bD\dot{Q}, \tag{7.118}$$

where b is the proportionality factor and $\dot{Q}$ is the volume flux of gas bubbles across the interface. This relation is as would be expected from Higbie's surface renewal model. From experimental work with various liquids, it was found [83–86] that for liquid depths more than about 5 cm, the proportionality factor b is about 120 cm^{-1} for liquid metals and about 50 cm^{-1} for slags.

As an example, let us consider the decarburization of liquid iron with iron oxide in the slag. In this slag–metal system, oxygen is transferred from slag to metal and reacts with carbon in the metal to form CO bubbles. For the overall reaction

$$\underline{C} + \underline{O} = CO \tag{7.119}$$

the isothermal mass-transfer controlled rate is given by the equation

$$\dot{n}/A = k_m'(c_i' - c') = k_m''(c'' - c_i''), \tag{7.120}$$

where $\dot{n}$ is the rate of oxygen transfer to metal $\equiv$ −rate of decarburization, A is the static slag–metal interfacial area, k_m is the mass transfer coefficient, c_i is the iron oxide content at the interface, and c is the iron oxide content of bulk substance, and the prime and double prime refer to metal and slag, respectively.

Assuming that there is equilibrium distribution of iron oxide between the slag and metal at the interface, i.e., $c_i'/c_i'' = k_0$ is constant for a given temperature, Eq. (7.120) is transformed to

$$\dot{n}/A = k_m{}^*(k_0 c'' - c'), \tag{7.121}$$

where $k_m{}^*$ is the overall mass-transfer coefficient given by

$$1/k_m{}^* = (1/k_m') + (k_0/k_m''). \tag{7.122}$$

Noting that $\dot{n}/A \propto \dot{Q}$, Eqs. (7.118) and (7.121) give

$$\dot{n}/A \propto D_O(k_0 c'' - c')^2, \tag{7.123}$$

where D_O is the diffusivity of oxygen in liquid iron. This relation indicates the autocatalytic nature of decarburization of the metal by the slag. That is, the mass-transfer coefficient for oxygen is a function of the rate of CO evolution such that doubling the oxygen concentration driving force increases the rate of decarburization by fourfold.

The autocatalytic decarburization of liquid iron with iron oxide in the slag is substantiated by the experimental work of Philbrook and Kirkbride [87], who found that the rate of decarburization is proportional to the square of the iron oxide content of the slag. Also, as shown by Subramanian and Richardson [84], the rate of decarburization derived from plant data [88] on open-hearth steelmaking agrees well with that calculated from Eq. (7.123).

As discussed in Sections 4.2 and 4.3, the interface turbulence and emulsification of liquids are much enhanced when the surface-active elements participate in interfacial reactions. Reference may be made to a textbook by Davies [89] for numerous other examples of mass transfer as affected by interfacial turbulence and rate phenomena in stirred liquids.

7.9 Gas Bubbles in Liquids

Much use is made of gas–liquid interaction in the processing of chemicals and refining of liquid metals. Consequently, a great deal of experimental and theoretical work has been done on the rate phenomena involving gas bubbles in liquids such as the dynamics of the formation of bubbles at submerged orifices and gas jet streams, nucleation and growth of bubbles, and mass transfer in the swarm of bubbles.

7.9.1 Formation of Bubbles at Submerged Orifices

Much work has been done over the years on the dynamics of the formation of gas bubbles at submerged orifices. Despite all these efforts, however, there are disparities in the results of experimental and theoretical work of different investigators. Some of these disparities and the reasons therefor are pointed out in the following brief outline of the salient features of the dynamics of bubble formation at submerged nozzle orifices.

At low gas flow rates, the inertial and viscous forces have negligible effects on the formation of single bubbles at submerged orifices. The bubble is released when the upward buoyancy force exceeds the downward force due to the surface tension of the liquid at the periphery of the orifice (inner circumference of the nozzle),

$$V_d = 2\pi a_\circ \sigma/\rho g, \tag{7.124}$$

where V_d is the bubble volume at detachment, $a_\circ$ the orifice radius, σ the surface tension, ρ the liquid density, and g the gravitational acceleration.

This limiting bubble volume, independent of gas flow rate, has been substantiated experimentally for aqueous solutions by van Krevelen and Hoftijzer [90] and by many others. If the liquid does not wet the nozzle material, as with liquid metals, the bubble spreads across the nozzle tip and Eq. (7.124) does not hold. For example, Sano and Mori [91] found that the bubble volumes at low gas flows in liquid mercury and silver approximate Eq. (7.124) when the outside nozzle diameter is used in the equation, instead of the orifice diameter. Such side effects may be circumvented by employing knife-edge vertical orifices away from the container walls.

Beyond a certain flow rate, the bubble volume at detachment increases with increasing gas flow rate. In the simple dynamic model considered by Davidson and Schuler [92] for a point source orifice, the buoyancy force of the growing bubble is assumed to be balanced by the inertial force to accelerate the liquid away from the expanding gas–liquid interface. The bubble detachment occurs when the upward velocity of the bubble center ds/dt exceeds the velocity of the bubble growth $da/dt = G/4\pi a^2$, where G is the constant gas flow rate and a is the bubble radius. For the limiting case of point source orifice in inviscid liquids, Davidson and Schuler derived the following theoretical equation for bubble volume at detachment for isothermal and constant flow conditions:

$$V_d = 1.378 G^{1.2} g^{-0.6} = 0.022 G^{1.2} \quad \text{cm}^3, \tag{7.125}$$

where the flow rate G is in cubic centimeters per second. The ratio G/V_d is the bubble frequency f,

$$f = 45.45 G^{-0.2}, \tag{7.126}$$

which is the parameter measured experimentally.

This equation has been tested time and again; some fit Eq. (7.126) well and some differ vastly. In particular, we may refer to the empirical correlations obtained by Leibson *et al.* [93],

$$V_d = 0.148 a_\circ^{0.5} G \quad \text{cm}^3, \tag{7.127}$$

and by Davidson and Amick [94],

$$V_d = 0.110(a_o^{0.5}G)^{0.87} \quad \text{cm}^3. \tag{7.128}$$

In an attempt to resolve, at least approximately, the apparent disparity between the theoretical equation (7.125) and the empirical relations (7.127) and (7.128), let us qualify the range of applicability of the theoretical model of Davidson and Schuler. Experimental data of different origin are shown in Fig. 7.18 for bubble frequency as a function of gas flow rate for orifice radii of 0.033–0.64 cm. For a given orifice size, the bubble frequency increases with increasing gas flow rate, ultimately reaching the dotted curve calculated from Eq. (7.126).

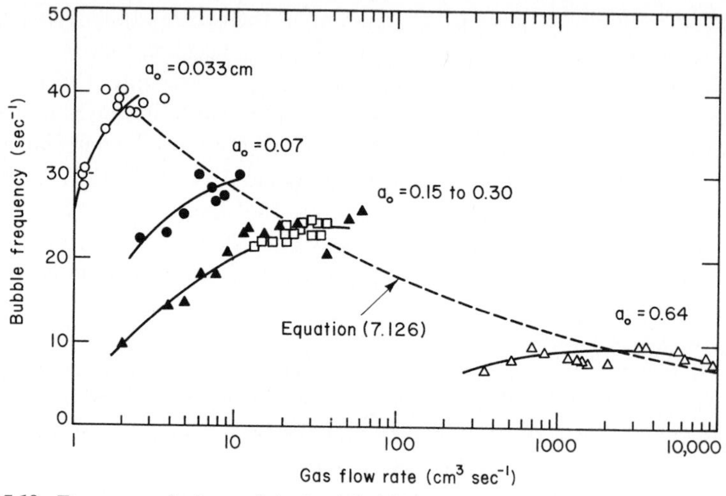

FIG. 7.18 Frequency of release of single air bubbles in water as a function of gas flow rate for indicated orifice radii. (○), (□) Davidson and Schuler [92]; (●), (▲) Ramakrishnan *et al.* [95]; (△) Walters and Davidson [98].

One criterion of the Davidson–Schuler model is that the rate of gas flow through the orifice must not be affected by the detachment of the bubble. That is, at the instant of detachment, there must remain a residual bubble at the orifice so that the subsequent bubble may grow without delay. At low gas flow rates the pressure in the liquid below the detached bubble is greater than the pressure in the orifice, resulting in liquid drainage, hence delaying the initiation of the bubble at the orifice. As shown by McCann and Prince [96], the smaller the orifice diameter the lower the critical gas flow rate below which weeping occurs. For this reason, therefore, the smaller the orifice diameter, the lower the critical gas flow rate below which the bubble frequency would decrease with decreasing flow rate; this may account for the experimental observations in Fig. 7.18.

The critical flow rate above which Eq. (7.126) is applicable for a given orifice size may be evaluated by considering the theoretical and experimental findings of Wraith and Kakutani [97]. They derived an equation of motion for the bubble in inviscid liquids by considering the velocity potential and pressure field around the growing bubble in the liquid. From theoretical calculations, it was found that the relative pressure at the orifice reaches a maximum when the ratio of the distance of the bubble center to the bubble radius s/a has risen to about 1.5. This is taken to be the criterion for the cutoff of the gas supply to the bubble by closing off the neck between the orifice and the bubble.

Let us now assume that for a given orifice radius $a_\circ$, the critical flow rate, below which the weeping occurs, is that for s equal to $a + a_\circ$. With s/a equal to 1.5 for bubble detachment, the bubble radius for this limiting case would be twice the orifice radius. Now, substituting $V_d = \frac{4}{3}\pi(2a_\circ)^3$ in Eq. (7.125) gives the critical flow rate G_c and the critical (maximum) bubble frequency f_c

$$G_c = 449a_\circ^{2.5}, \qquad f_c = 13.40a_\circ^{-0.5}. \tag{7.129}$$

With this definition of maximum bubble frequency, the bubble volume for flow rates above G_c is, from the above equation,

$$V_d = 0.075a_\circ^{1/2}G \quad \text{cm}^3. \tag{7.130}$$

This semiempirical modification of the Davidson–Schuler model for constant flow rates gives an equation that is similar to the empirical relations (7.127) and (7.128).

To test Eq. (7.130), typical examples of experimental data are given in Fig. 7.19 from the work of Davidson and Schuler [92], Walters and Davidson [98], and Wraith [99] for the formation of single bubbles at constant air flow rates in water over wide ranges of orifice sizes and flow rates. The modified Davidson–Schuler equation is well substantiated with the experimental data, which may be represented by the equation

$$V_d = 0.11a_\circ^{1/2}G \quad \text{cm}^3. \tag{7.131}$$

For orifice radius 0.033 cm, the limiting bubble volume at low gas flow rates is 0.015 cm^3, as given by Eq. (7.124). Departure of these data points in Fig. 7.19 from the linear relation is evidently due to the surface-tension effect on the bubble volume at detachment.

The bubble volumes measured, under essentially constant flow conditions by Sano and Mori [91] with liquid mercury and silver and by Irons and Guthrie [100] with graphite-saturated liquid iron are in substantial agreement with Eq. (7.131) at high gas flow rates, e.g., 100–1000 cm^3 sec^{-1} at liquid metal temperatures, provided $a_\circ$ is substituted by the nozzle radius, because of nonwetting of the orifice by liquid metals.

It should be noted that the antechamber volume below the orifice should be small to maintain constant gas flow through the orifice. With large chamber volumes, the bubble size at detachment is larger than that predicted by the Davidson–Schuler equation or the modification thereof. Also, measurements made at elevated temperatures with liquids in small containers are subject to errors because of the nonisothermal conditions at the orifice, wall effect with

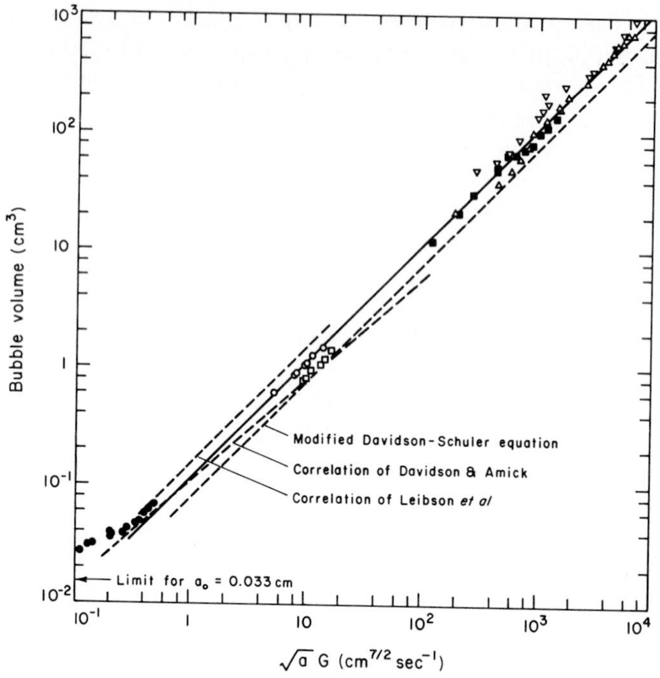

FIG. 7.19 Volume of bubble at detachment as a function of the product of the gas flow rate and the square root of orifice radius for air–water systems at room temperature and for conditions of constant flow rate through the orifice; from the data of Davidson and Schuler (for orifice radius of (●) 0.033, (○) 0.15, and (□) 0.25 cm) [92], Walters and Davidson ((▽) orifice radius of 0.64 cm) [98], and Wraith (orifice radius of (■) 0.337 and (△) 0.635 cm) [99].

small containers, and experimental difficulties in measuring the bubble frequency, particularly in liquid metals.

We should mention in passing that Davidson and Schuler extended their theoretical formulation of the dynamics of bubble formation to viscous liquids and showed that the bubble volume increases with increasing viscosity. For example, an increase in kinematic viscosity from 0.01 $cm^2\,sec^{-1}$ for water up to 5 $cm^2\,sec^{-1}$ for a viscous liquid doubles the bubble volume.

With further increase in gas flow rate, doublets and triplets are formed at the orifice, and the bubble volume at detachment is no longer described by Eq. (7.131), or by any empirical relation. At sufficiently high gas flow rates the swarm of bubbles are formed above the orifice plate. Within this flow regime the bubbles are essentially of uniform size, the average bubble radius in water being 0.45 cm. This finding of Leibson *et al.* [93] has been substantiated by many subsequent studies.

7.9.2 Mass Transfer to and from Gas Bubbles

The motion of gas bubbles in liquids is characterized by the dimensionless parameters, Reynolds number N_{re}, Eötvos number N_{eo}, and Morton number

N_m defined in Table 7.2. Depending on the bubble size and the properties of the liquid (density, viscosity, and surface tension), the bubbles in motion in liquids acquire spherical, ellipsoidal, or spherical-cap shapes. For all shapes, the bubble size is defined in terms of the bubble diameter of an equivalent sphere, that is, the diameter of a sphere with the same volume as the nonspherical bubble. From a study of the available data on the velocity of rise of bubbles and their shapes, a correlation shown in Fig. 7.20 was obtained by Grace [101] for Newtonian liquids with Morton numbers of 10^{-14}–10^8.

For N_{re} less than about 2 and for all values of N_{eo}, bubbles are spherical and their terminal velocities are within the range limited by Stokes' law and Hadamard–Rybczynski equation. The Stokes' law applies to small bubbles that behave like rigid spheres for which the terminal velocity u_t is given by the

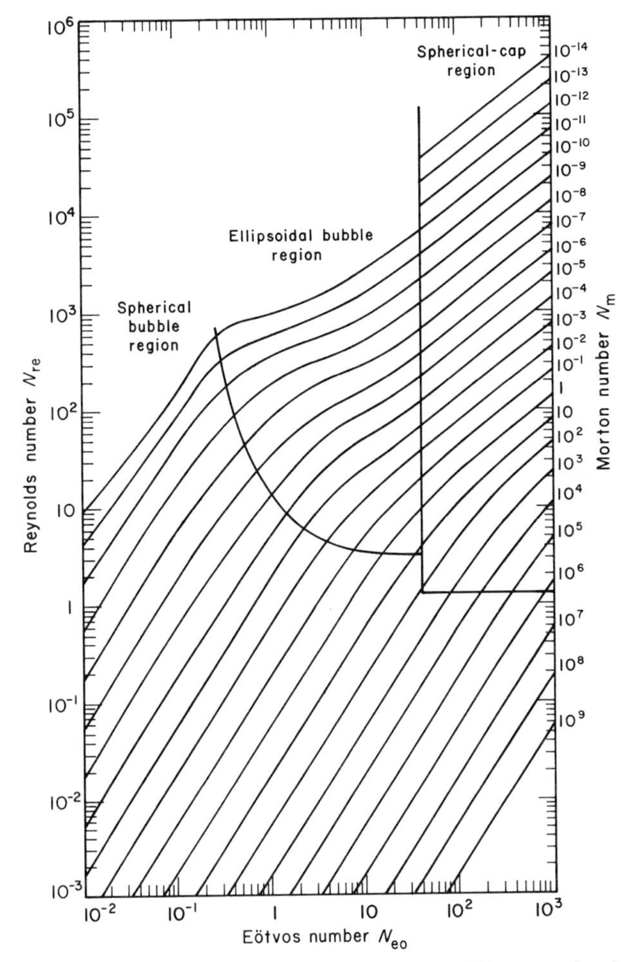

FIG. 7.20 Generalized correlation of Reynolds number versus Eötvos number for single bubbles rising in infinite Newtonian liquids. After Grace [101].

equation

$$u_t = gd^2 \, \Delta\rho/18\eta, \tag{7.132}$$

where d is the bubble diameter, $\Delta\rho$ is the density difference between liquid and gas, and η is the viscosity of the liquid. For larger bubbles with a mobile surface, there is momentum transfer across the interface and gas circulation within the bubble; for such bubbles the Hadamard–Rybczynski equation gives the terminal velocity that is 50% greater than the limiting Stokes' law.

For N_{re} more than about 100 and N_{eo} more than about 40, the bubbles acquire spherical-cap shape. The straight lines drawn in Fig. 7.20 for this region are in accord with the terminal velocity derived by Davies and Taylor [102]. They found that the equation for potential flow, around a sphere in the region close to the forward stagnation point, could be combined with *Bernouilli's equation* to give the following expression for the velocity of rise of spherical-cap-shape bubbles:

$$u_t = 0.72(gd)^{1/2}. \tag{7.133}$$

Experimentally measured velocities of rise of spherical-cap-shape air bubbles in water and mercury [103] agree well with Eq. (7.133) for equivalent bubble diameters of 1–1.5 cm. For larger bubbles, the measured velocities are lower than predicted from theory; this discrepancy is attributed to the effect of the container wall.

Combining Eq. (7.116) with (7.117), substituting u_t for u_b, and combining with Eq. (7.133), gives the mass-transfer coefficient controlled by surface renewal on spherical-cap-shape bubbles,

$$k_m = 0.975 D^{1/2}(g/d)^{1/4}. \tag{7.134}$$

In terms of the average Sherwood number, Eq. (7.116) and (7.117) give for mass transfer on a mobile bubble surface

$$\bar{N}_{sh_m} = 1.13 N_{re}^{1/2} N_{sc}^{1/2}. \tag{7.135}$$

Adsorption of surface-active species on the bubble surface retards surface renewal and circulation within the bubble or liquid drop. The bubble surface becomes immobile when there is strong adsorption which affects the velocity of the moving bubble or liquid drop and its mass-transfer characteristics [104]. The bubble with an immobile surface behaves like a rigid sphere for which the Sherwood number is that given in Table 7.3:

$$\bar{N}_{sh_i} = 2 + 0.6 N_{re}^{1/2} N_{sc}^{1/3}. \tag{7.136}$$

Noting that N_{sc} is about 10^3 for inviscid liquids, we see from the above equations that for bubbles with $d > 1$ mm, the mass-transfer coefficient for a mobile surface is about six times that for the immobile surface.

In a study of absorption of CO_2 from large bubbles (8–42 mm) in water, Baird and Davidson [105] found the rate of absorption to be about 50% greater than expected from the penetration theory, Eq. (7.134). This was attributed to the additional absorption of CO_2 from the turbulent, rear surface of the bubble. With an addition of 0.1% hexanol, the turbulence in the bubble wake was suppressed and the measured mass transfer was found to be in close agreement with Eq. (7.134). Guthrie and Bradshaw [106] measured the rate of solution of oxygen in liquid silver by a single-bubble technique. The mass-transfer coefficient k_m is about 20% lower than that given by Eq. (7.134), possibly because of the partial immobility of the bubble surface by the adsorption of oxygen on the bubble surface.

There are many applications of gas purging in the treatment of liquids at elevated temperatures, as, for example, in the deoxidation of liquid copper with carbon monoxide, removal of hydrogen from liquid metals by purging with argon, steelmaking with oxygen bottom blowing, and air bottom blowing in copper converters. For a mass-transfer controlled rate, involving solute X, we may write for the rate of change of solute concentration in a cylindrical column of liquid

$$A\,d(\%\mathrm{X})/dt = -(S_b k_m/H)[\%\mathrm{X} - (\%\mathrm{X})_e] \quad \mathrm{cm^2\,sec^{-1}}, \tag{7.137}$$

where A and H are the cross-sectional area and height from the orifice tip, respectively, of cylindrical liquid column, S_b is the total bubble surface area in the liquid column, and $\% X_e$ the solute concentration in equilibrium with the gas at the bubble surface.

For the limiting case of single-bubble formation at a submerged orifice, and with the assumption of negligible change in the bubble volume during ascent, the isothermal mass-transfer parameter $S_b k_m/H$ may be calculated from the equations already given. The total bubble surface area in the liquid column is

$$S_b = \pi d^2 f t_r, \tag{7.138}$$

where $f = G/V$ is the bubble frequency and $t_r = H/u_t$ is the time of residence of bubbles in the liquid column. Combining Eqs. (7.133), (7.138), and substituting Eq. (7.131) for the bubble volume gives for the condition of isothermal, constant rate of gas flow

$$S_b = (6H/0.72)(\pi G/0.66g\sqrt{a_\circ})^{1/2} = 0.58H(G/\sqrt{a_\circ})^{1/2} \quad \mathrm{cm^2} \tag{7.139}$$

for H and $a_\circ$ in centimeters and G in cubic centimeters per second. Combining Eqs. (7.131) and (7.134) gives for the mass-transfer coefficient for bubbles formed at a submerged orifice

$$\begin{aligned} k_m &= 0.975D^{1/2}g^{1/4}(0.66/\pi)\sqrt{a_\circ G}]^{-1/12} \\ &= 6.214D^{1/2}(\sqrt{a_\circ}G)^{-1/12} \quad \mathrm{cm\,sec^{-1}}. \end{aligned} \tag{7.140}$$

These equations give for the surface mass-transfer coefficient for single gas bubbles in inviscid liquids

$$S_b k_m/H = 3.60 D^{1/2} a_o^{-7/12} G^{5/12} \quad \text{cm}^2\ \text{sec}^{-1}. \tag{7.141}$$

As noted earlier, ranges of flow rates for the single-bubble regime depend on the orifice size; the smaller the orifice size, the lower is the range of the flow rate. Based on the studies of single-bubble formation, of which data in Fig. 7.19 are typical examples, Eq. (7.141) may be applicable for the following estimated ranges of gas flow rates in aqueous solutions.

a_0 (cm)	0.05	0.1	0.5	1.0
G ($\text{cm}^3\ \text{sec}^{-1}$)	0.5–1	10–30	100–1000	1000–10,000

It should also be noted that Eq. (7.141) is for bubbles with a mobile surface, i.e., in the absence of surface-active species. However, with the information already given, $S_b k_m/H$ is readily calculated for bubbles with an immobile surface. If there is strong chemisorption, then the rate may be controlled by a surface reaction, as, for example, in the denitrogenation of liquid iron at high oxygen levels with an argon purge.

In most applications of gas injection into liquids, we are concerned with rates of reactions in the swarm of bubbles. Mass-transfer phenomena in the swarm of bubbles have been studied extensively with aqueous solutions; accumulated knowledge is well documented in a review paper by Calderbank [107]. For an average bubble velocity u_b, the liquid-phase mass-transfer coefficient for the the surface-renewal regime is given by

$$k_m = 1.28(D u_b/d)^{1/2}. \tag{7.142}$$

If we take u_b equal to the terminal velocity u_t and combine Eqs. (7.133) and (7.142), the expression for k_m is the same as (7.134) except for the proportionality factor, which now becomes 1.086. However, in the swarm of bubbles, u_b is greater than u_t; the bubble velosity is a function of the superficial gas velocity and the fractional holdup of gas:

$$u_b = u_s/\varepsilon, \tag{7.143}$$

where ε is the ratio of the volume of gas bubbles to the volume of gas–liquid emulsion and u_s the superficial gas velocity, equal to G/A.

Yoshida and Akita [108] have measured the increase in the fractional holdup of gas in nonfoaming, inviscid aqueous solutions with increasing superficial gas velocity up to about 60 cm sec^{-1}. From their experimental results, the following relation is obtained for $u_s > 3$ cm sec^{-1}:

$$\log(1 - \varepsilon)^{-1} = 0.146 \log(1 + u_s) - 0.06. \tag{7.144}$$

At high superficial gas velocities in nonfoaming liquids, the limiting value of ε in most practical situations is about 0.5. Yoshida and Akita also measured the volumetric liquid-phase mass-transfer coefficient in the swarm of bubbles. Since the bubble interfacial area $S_\circ$ per unit volume of the aerated liquid is

$$S_\circ = 6\varepsilon/d \text{ cm}^2/\text{cm}^3 \text{ of aerated liquid} \qquad (7.145)^*$$

the volumetric mass-transfer coefficient from Eqs. (7.142) and (7.143) is given by

$$S_\circ k_m = 7.68(\varepsilon D u_s)^{1/2} d^{-3/2}. \qquad (7.146)$$

As mentioned earlier, Leibson *et al.* [93] found that in the swarm of bubbles formed in water with air injection at high velocities, the average bubble diameter is about 0.45 cm. The values of $S_\circ k_m$ as a function of u_s, calculated for $D = 2 \times 10^{-5}$ cm^2 sec^{-1} and $d = 0.45$ cm, agree well with the mass-transfer measurements in water [108].

Because of the experimental difficulties, no reliable data are available for the average bubble size and the mass-transfer coefficient for the swarm of bubbles in liquid metals. To evaulate $S_b k_m/H$ for the swarm of bubbles in liquid metals, we must first estimate the average bubble size.

Relatively large bubbles in motion are subject to deformation and ultimately to fragmentation into smaller bubbles. The drag force exerted by the liquid on a moving bubble induces rotational and probably turbulent motion of the gas within the bubble. This motion creates a dynamic pressure on the bubble surface, and when this force exceeds the surface tension bubble breakup occurs. Because of the large difference between the densities of the gas and the liquid, the energy associated with the drag force is much greater than the kinetic energy of the gas bubble. Therefore, the gas velocity in the bubble will be similar to the bubble velocity. On the basis of this reasoning, Levich [43] derived the following equation for the critical bubble size as a function of bubble velocity:

$$d_c = (3/C_d \rho_g \rho_1{}^2)^{1/3}(2\sigma/u_b{}^2), \qquad (7.147)$$

where C_d is the drag coefficient, which may be set equal to unity. Although this equation has not been tested experimentally, it is anticipated that Eq. (7.147) will not hold at high bubble velocities and for the corresponding small bubbles, because the gas circulation within the bubble diminishes below a certain bubble size.

Because of the relation in Eq. (7.133), the bubble velocity u_b may be assumed to be proportional to the square root of d_e of the equivalent bubble. With this in mind, we may now assume a *similarity relation* for liquids 1 and 2, based on Eq. (7.147),

$$d_1{}^2 \rho_1^{2/3}/\sigma_1 \simeq d_2{}^2 \rho_2^{2/3}/\sigma_2, \qquad d_1/d_2 \simeq (\rho_2/\rho_1)^{1/3}(\sigma_1/\sigma_2)^{1/2}. \qquad (7.148)$$

* It is sometimes convenient to use bubble interfacial area S relative to the unit mass of the quiescent liquid, i.e.,

$$S = \frac{S_0}{(1-\varepsilon)\rho} = \frac{6\varepsilon}{1-\varepsilon}\frac{1}{\rho d} \text{ cm}^2/\text{g liquid}, \qquad (7.145\text{a})$$

where ρ is the liquid density.

For the swarm of air bubbles in water, d_2 is about 0.45 cm, and inserting the appropriate values of ρ_2 and σ_2 for water, we now obtain an estimate of the average bubble size for the swarm of bubbles in any inviscid liquid, free of surface-active solutes,

$$d \simeq 0.053\rho^{-1/3}\sigma^{1/2}. \tag{7.149}$$

Because the immobile surface retards the transfer of momentum across the bubble surface and hence impedes the fragmentation of the bubbles, larger bubble size is expected for the swarm of bubbles when the liquid contains surface-active solutes.

For liquid steel with $\sigma = 1500$ dyn cm^{-1} and $\rho = 7$ g cm^{-3}, the average bubble diameter is estimated to be about 1.1 cm in the swarm of bubbles. For the reason stated above, Eqs. (7.147) and (7.149) will not hold for liquids containing surface-active solutes. Hence, for the swarm of bubbles in liquid iron with high concentrations of oxygen or sulfur, the average bubble size will be larger than that estimated from Eq. (7.149).

The superficial gas velocity in metal-refining processes is in the range 100–400 cm sec^{-1}, and the melt swells to about one-third the bath depth, for which ε is about 0.25. For an average solute diffusivity of 5×10^{-5} cm sec^{-1} and an estimated bubble diameter of 1.1 cm, and also noting that $S_b k_m/AH$ is equal to $S_o k_m$, the following approximate rate equation is obtained from Eqs. (7.137) and (7.146) for the rate of reaction of a gas stream with liquid steel, controlled by liquid-phase mass transfer in the dispersed gas jet stream:

$$d(\%X)/dt \simeq -0.023\sqrt{u_s}[\%X - (\%X_e)]. \tag{7.150}$$

Integration gives

$$\log \frac{\%X - (\%X)_e}{(\%X)_0 - (\%X)_e} \simeq -0.01\sqrt{u_s}t, \tag{7.151}$$

where $(\%X)_0$ is the initial concentration of the reactant in the metal; u_s is in centimeters per second and the reaction time is in seconds. The above equation is for a mobile bubble surface; if there is strong adsorption, the estimated rate could be about an order of magnitude lower than given by Eq. (7.151).

It can be shown from the material balance and Eq. (7.151) that in pneumatic metal-refining processes, the rates of gas–metal reactions are controlled by the liquid-phase mass transfer, only when the concentration difference $\%X - (\%X)_e$ is less than about 0.1%. At higher concentrations of the reactant, the gas blown into the reactor is consumed almost completely to satisfy the bulk gas–metal equilibrium. For this limiting case of gas starvation, the rate of reaction is proportional to the rate of gas flow, or superficial gas velocity u_s; while for the mass-transfer control, the rate is proportional to the square root of u_s.

In laboratory experiments with liquid metals in small containers, the characteristics of the gas stream in the melt is ill defined. Therefore, experimentally measured overall rate parameter $S_b k_m/H$ cannot be extrapolated with confidence to high gas flow rates of practical interest.

7.9.3 Nucleation and Growth of Bubbles

As pointed out in Section 7.3, the theory of homogeneous nucleation predicts a supersaturation of 1×10^4–5×10^4 atm for nucleation of gas bubbles in pure liquid metals with surface tension in the range 1000–1600 dyn cm^{-1}. Yet, in experiments done with levitated liquid drops of iron–carbon alloys not in contact with refractory surfaces and in an oxidizing atmosphere, the CO gas bubbles could be initiated with supersaturations in the range 20–100 atm [109, 110]. Based on these observations, it has been suggested [109] that the theory of homogeneous nucleation in its present form may not be applicable to liquids of high surface tension because of possible clustering of solute atoms in the proximity of holes in the liquid structure. An alternative explanation [110] is that the turbulent eddies in the levitated melt, induced by the applied electromagnetic field for levitation and melting, may develop vortices with negative pressures, hence lowering the supersaturation needed for the nucleation of bubbles. However, it is well to bear in mind that the iron–carbon alloys used in these experiments will contain trace amounts of impurities such as alumina and zirconia which may stimulate premature nucleation of CO bubbles in levitated melts. We cannot therefore conclude unequivocally that the theory of homogeneous nucleation in its present form would not apply to liquid metallic solutions.

Bubble nucleation at gas-filled crevices on the container wall is of more practical importance. It has long been recognized that bubble nucleation is much easier in melts held in refractory crucibles with a rough surface than in crucibles with a glazed surface [111]. For a bubble to grow and detach from a gas-filled crevice, the pressure P_b in the bubble must be less than that in equilibrium with the concentration of dissolved gas in the metal. As an example, let us consider the growth of CO bubbles in liquid Fe–C–O alloys for which the criterion for bubble growth is

$$p_{CO} > P_b = P_0 + (2\sigma/a), \tag{7.152}$$

where p_{CO} is the pressure of CO in equilibrium with the concentrations of carbon and oxygen in solution in iron, P_0 is the static pressure over the bubble, and a is the bubble radius. It is implicit from Eq. (7.152) that for the bubble to grow at the mouth of a cylindrical crevice, the radius of the crevice must be greater than the critical minimum a_{min}.

$$a_{min} > 2\sigma/(p_{CO} - P_0). \tag{7.153}$$

By the same token; there is a maximum crevice radius above which the crevice cannot sustain the residual gas pocket

$$a_{max} < -(2\sigma/P_0)\cos\theta, \tag{7.154}$$

where θ is the contact angle, usually $>90°$, between the melt and the refractory surface. The validity of these equations has been tested experimentally by Alexander *et al.* [112] for CO evolution from liquid steel held in refractory crucibles of varying roughness with pore sizes in the range 0.01–0.1 mm.

The bubble is released when the volume has grown to the critical size V_d of Eq. (7.124). Subsequent growth of the bubble as it ascends in the melt may be calculated from the equation derived by Bradshaw and Richardson [113] for liquid-phase diffusion control,

$$P\,dV/dP + V = (S_a k_m/\rho g u_t)[c_X - (c_X)_e]RT, \tag{7.155}$$

where S_a is the surface area of the bubble, c_X the molar concentration of the diffusate, and $(c_X)_e$ that for the gas–metal equilibrium; other symbols have already been defined. By substituting the appropriate equations for S_a, k_m, and u_t in terms of the bubble volume V, Davenport [114] solved Eq. (7.155) for the growth of CO bubbles during ascent in liquid steel saturated with CO. He found that the volume of the bubble at a given distance of rise is essentially independent of the initial volume at the nucleation site. The calculated CO volumes of 5 and 13 cm^3 at 36 and 72 cm rise, respectively, agree well with bubble sizes of 5–10 cm^3 observed [115] in open-hearth steelmaking with a bath depth of about 40 cm. This mechanism of growth of single bubbles during gas evolution does not apply to the swarm of bubbles formed at high rates of gas evolution, as in pneumatic metal-refining processes.

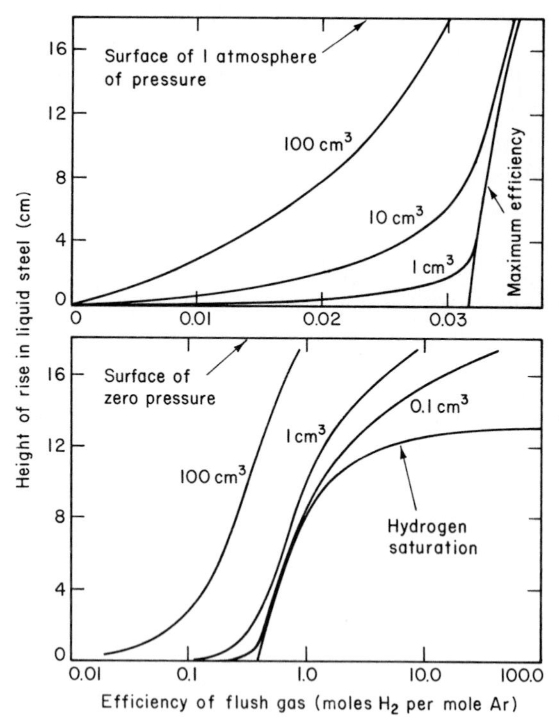

FIG. 7.21 Calculated efficiency of usage of flush gas in removal of hydrogen from steel (initially containing 5 ppm H) as a function of depth of steel and initial volume of injected gas bubbles. After Davenport [114].

7.9.4 Gas Purging

Equations (7.153) and (7.154) indicate that the bubbles will not grow at crevices unless the concentration of the dissolved gas in the liquid is sufficiently high to sustain the required excess pressure in the bubble under a given static pressure. Therefore, gas purging is necessary for desorption of trace amounts of dissolved gases.

Based on Eq. (7.155), Davenport [114] calculated the effect of pressure over the liquid column on the efficiency of desorption by gas purging, using as an example, hydrogen removal from liquid steel by purging with argon. As is seen from the results of his calculations in Fig. 7.21, the efficiency of usage of flush gas increases considerably when there is vacuum over the melt. There is also a marked increase in the efficiency of degassing with decreasing size of the injected inert-gas bubbles. Similar calculations have been made by Papamantellos *et al.* [116] by modifying the rate equation (7.155) with the inclusion of slow chemical reaction at the bubble surface.

7.9.5 Stream Degassing

In some applications the liquid is degassed by breakup of the liquid stream in a "vacuum" chamber at pressures less than 1 mm Hg. The efficiency of degassing in such a process depends on the extent of stream breakup.

Olsson and Turkdogan [117] showed that effective stream breakup may be achieved, even at moderate chamber pressures, by injecting a small quantity of an inert gas into the liquid stream just before passage through a nozzle to a chamber at a lower pressure. In such an application of stream breakup to liquid steel, the metal droplets are coalesced on a conical-shape spray collector which directs the degassed liquid steel into a casting mold in the low pressure chamber. The effect of gas injection on breakup of the stream of liquid steel is shown in Fig. 7.22. With argon injection at a rate of 250 $cm^3 \, sec^{-1}$ (STP) into a metal stream flowing at 400 $g \, sec^{-1}$, the average particle diameter of the metal droplets was about 100 μm.

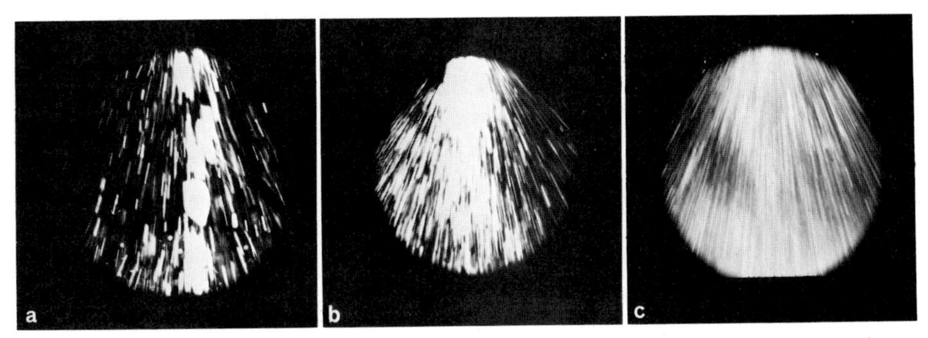

FIG. 7.22 High-speed photographs of liquid steel sprays with stream breakup by gas injection, chamber pressure from 5 to 12 mm Hg; (a) no argon injection, (b) 2 $l \, min^{-1}$, (c) 15 $l \, min^{-1}$. After Olsson and Turkdogan [117].

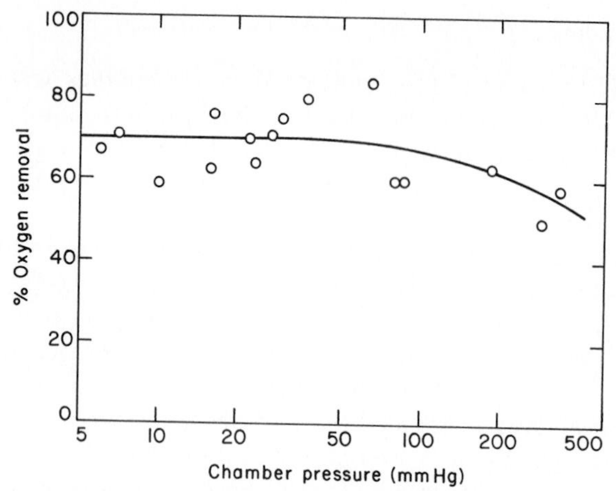

FIG. 7.23 Carbon deoxidation accompanying stream breakup at indicated chamber pressures. After Olsson and Turkdogan [117].

In laboratory experiments with 15 kg melts, the carbon deoxidation achieved by stream breakup of 0.1% C steel is shown in Fig. 7.23. The flight time of metal droplets in the lower pressure chamber was about 0.08 sec. Within this time of reaction about 70% of the oxygen was removed from the steel as CO at chamber pressures up to 200 mm Hg. If the rate of reaction is controlled by diffusion of oxygen to the surface of the metal droplets, about 80% oxygen removal would be expected (for oxygen diffusivity of 5×10^{-5} $cm^2 sec^{-1}$, from 100-μm-diameter steel droplets in a flight time of 0.08 sec; this is close to the value obtained experimentally.

References

1. R. B. Bird, W. E. Stewart, and E. N. Lightfoot, "Transport Phenomena." Wiley, New York, 1960.
2. J. Szekely and N. J. Themelis, "Rate Phenomena in Process Metallurgy." Wiley, New York, 1971.
3. H. Eyring, *J. Chem. Phys.* **3**, 107 (1935).
4. S. Glasstone, K. J. Laidler, and H. Eyring, "The Theory of Rate Processes." McGraw-Hill, New York, 1941.
5. H. Eyring, D. Henderson, and W. Jost (ed.), "Physical Chemistry, an Advanced Treatise," Vols. VIA, VIB, VII. Academic Press, New York, 1975.
6. L. S. Darken and E. T. Turkdogan, *in* "Heterogeneous Kinetics at Elevated Temperatures" (G. R. Belton and W. L. Worrell, eds.), p. 25. Plenum, New York, 1970.
7. C. Wagner, *Adv. Catal.* **21**, 323 (1970).
8. G. R. Belton, *in* "Physical Chemistry in Metallurgy—Darken Conference" (R. M. Fisher, R. A. Oriani, and E. T. Turkdogan, eds.), p. 93. U.S. Steel, Monroeville, Pennsylvania, 1976.
9. H. J. Grabke and G. Horz, *Annu. Rev. Mater. Sci.* **7**, 155 (1977).
10. R. D. Pehlke and J. F. Elliott, *Trans. Metall. Soc. AIME* **227**, 844 (1963).
11. R. G. Mowers and R. D. Pehlke, *Metall. Trans.* **1**, 51 (1970).
12. H. Schenck, M. G. Frohberg, and H. Heinemann, *Arch. Eisenhuettenwes.* **33**, 593 (1962).

13. T. Fuwa, S. Ban-ya, and T. Shinohara, *Tetsu To Hagane* **54**, S436 (1968).
14. K. Mori and K. Suzuki, *Trans. Iron Steel Inst. Jpn.* **10**, 232 (1970).
15. K. Narita, S. Koyama, T. Makino, and M. Okamura, *Trans. Iron Steel Inst. Jpn.* **12**, 444 (1972).
16. M. Inouye and T. Choh, *Trans. Iron Steel Inst. Jpn.* **8**, 134 (1968); **12**, 189 (1972).
17. E. T. Turkdogan and P. Grieveson, *J. Electrochem. Soc.* **114**, 59 (1967).
18. H. J. Grabke, *Ber. Bunsenges. Phys. Chem.* **72**, 541 (1968).
19. E. D. Hondros, *Met. Sci. J.* **1**, 36 (1967).
20. H. J. Grabke, *Arch. Eisenhuettenwes.* **44**, 603 (1973).
21. F. Portele, *Z. Naturforsch., Teil A* **24**, 1268 (1969).
22. P. Grieveson and E. T. Turkdogan, *Trans. Metall. Soc. AIME* **230**, 407, 1604 (1964).
23. H. J. Grabke and G. Tauber, *Arch. Eisenhuettenwes.* **46**, 215 (1975).
24. D. E. Rosner, H. M. Chung, and H. H. Feng, *J. Chem. Soc., Faraday Trans. I* **72**, 842, 858 (1976); *Metall. Trans.* **5**, 2305 (1974).
25. T. P. Floridis, *Trans. Metall. Soc. AIME* **215**, 870 (1959).
26. V. D. Sehgal and A. Mitchell, *J. Iron Steel Inst., London* **202**, 216 (1964).
27. R. Ohno and T. Ishida, *J. Iron Steel Inst., London* **206**, 904 (1968).
28. V. D. Sehgal, *J. Iron Steel Inst., London* **207**, 95 (1969).
29. R. J. Fruehan and E. T. Turkdogan, *Metall. Trans.* **2**, 895 (1971).
29a. R. J. Fruehan, *Metall. Trans. B* **9**, 287 (1978).
30. G. R. Belton, R. J. Fruehan, and E. T. Turkdogan, *Metall. Trans.* **3**, 596 (1972).
31. T. B. King and S. Ramachandran, *in* "Physical Chemistry of Steelmaking" (J. F. Elliott, ed.), p. 125. MIT Press, Cambridge, Massachusetts, 1958.
32. A. Nilas and M. G. Frohberg, *Arch. Eisenhuettenwes.* **41**, 951 (1970).
33. K. M. Goldman, G. Derge, and W. O. Philbrook, *Trans. Metall. Soc. AIME* **200**, 534 (1954).
34. E. T. Turkdogan and P. M. Bills, cited in E. T. Turkdogan and M. L. Pearce, *Trans. Metall. Soc. AIME* **227**, 940 (1963).
35. M. Volmer and A. Weber, *Z. Phys. Chem., Stoechiom. Verwandschaftsl.* **119**, 227 (1926).
36. R. Becker and W. Doring, *Ann. Phys.* (*Leipzig*) **24**, 719 (1935).
37. F. F. Abraham, "Homogeneous Nucleation Theory." Academic Press, New York, 1974.
38. M. V. Smoluchowski, *Phys. Z.* **17**, 585 (1916).
39. C. Wagner, *Z. Elektrochem.* **65**, 581 (1961).
40. F. S. Ham, *J. Phys. Chem. Solids* **6**, 335 (1958).
41. H. Schlichting, "Boundary Layer Theory." McGraw-Hill, New York, 1960.
42. E. R. G. Eckert and R. M. Drake, Jr., "Heat and Mass Transfer." McGraw-Hill, New York, 1959.
43. V. G. Levich, "Physicochemical Hydrodynamics." Prentice-Hall, Englewood Cliffs, New Jersey, 1962.
43a. M. T. Scholtz and O. Trass, *AIChE J.* **9**, 548 (1963).
44. R. J. Fruehan and L. J. Martonik, *Metall. Trans.* **5**, 1027 (1974).
45. L. A. Baker, N. A. Warner, and A. E. Jenkins, *Trans. Metall. Soc. AIME* **230**, 1228 (1964).
46. J. H. Swisher and E. T. Turkdogan, *Trans. Metall. Soc. AIME* **239**, 602 (1967).
47. D. R. Sain and G. R. Belton, *Metall. Trans. B* **7**, 235 (1976).
48. E. T. Turkdogan, P. Grieveson, and L. S. Darken, *J. Phys. Chem.* **67**, 1647 (1963).
49. G. J. W. Kor and E. T. Turkdogan, *Metall. Trans. B* **6**, 411 (1975).
50. E. T. Turkdogan, *Trans. Metall. Soc. AIME* **230**, 740 (1964); E. T. Turkdogan and K. C. Mills, *Trans. Metall. Soc. AIME* **230**, 750 (1964).
51. L. S. Nelson and H. W. Levine, *in* "Heterogeneous Kinetics at Elevated Temperatures" (G. R. Belton and W. L. Worrell, eds.), p. 503. Plenum, New York, 1970.
52. Y. Fukunaka and J. M. Toguri, *Metall. Trans. B* **9**, 33 (1978).
53. A. W. D. Hills and J. Szekely, *J. Chem. Eng. Sci.* **19**, 79 (1964).
54. D. E. Rosner, *Int. J. Heat Mass Transfer* **10**, 1267 (1967); D. E. Rosner and M. Epstein, *Int. J. Heat Mass Transfer* **13**, 1393 (1970).

55. L. Lees, *Jet Propul.* **26**, 259 (1956).
56. J. A. Fay and F. R. Riddell, *J. Aeronaut. Sci.* **25**, 73 (1958).
57. E. M. Sparrow and R. C. Cess, "Radiation Heat Transfer." Brooks/Cole, Belmont, California, 1966.
58. H. C. Hottel and A. F. Sarofim, "Radiative Transfer." McGraw-Hill, New York, 1967.
59. T. J. Love, "Radiative Heat Transfer." Merrill, Columbus, Ohio, 1968.
60. J. Crank, "Mathematics of Diffusion." Oxford Univ. Press, London and New York, 1956.
61. H. S. Carslaw and J. C. Jaeger, "Conduction of Heat in Solids." Oxford Univ. Press (Clarendon), London and New York, 1959.
62. J. Szekely, J. W. Evans, and H. Y. Sohn, "Gas–Solid Reactions." Academic Press, New York, 1976.
63. G. Tammann, *Z. Anorg. Allg. Chem.* **111**, 78 (1920).
64. C. Wagner, *Z. Phys. Chem., Abt. B* **21**, 25 (1933); *in* "Atom Movements," p. 153. Am. Soc. Met., Cleveland, Ohio, 1950.
65. R. A. Rapp, *Corrosion* (*Houston*) **21**, 382 (1965).
66. M. T. Hepworth, R. P. Smith, and E. T. Turkdogan, *Trans. Metall. Soc. AIME* **236**, 1278 (1966).
67. C. Wagner, *J. Electrochem. Soc.* **103**, 571 (1956).
68. R. G. Olsson, B. B. Rice, and E. T. Turkdogan, *J. Iron Steel Inst., London* **207**, 1607 (1969).
69. E. T. Turkdogan and G. J. W. Kor, *Metall. Trans.* **2**, 1561 (1971).
70. R. H. Tien and E. T. Turkdogan, *Met. Sci.* **9**, 240 (1975).
71. E. W. Thiele, *Ind. Eng. Chem.* **31**, 916 (1939).
72. A. Wheeler, *Adv. Catal.* **3**, 249 (1951).
73. A. Wheeler, *Catalysis* **2**, 105 (1955).
74. E. E. Petersen, *AIChE J.* **3**, 443 (1957).
75. E. E. Petersen, "Chemical Reaction Analysis." Prentice-Hall, Englewood Cliffs, New Jersey, 1965.
76. E. T. Turkdogan, V. Koump, J. V. Vinters, and T. F. Perzak, *Carbon* **6**, 467 (1968).
77. E. T. Turkdogan and J. V. Vinters, *Carbon* **7**, 101 (1969).
78. E. T. Turkdogan, R. G. Olsson, and J. V. Vinters, *Carbon* **8**, 545 (1970).
79. R. H. Tien and E. T. Turkdogan, *Carbon* **8**, 607 (1970).
80. R. H. Tien and E. T. Turkdogan, *Carbon* **10**, 35 (1972).
81. H. Y. Sohn and J. Szekely, *Chem. Eng. Sci.* **27**, 763 (1972).
82. R. Higbie, *Trans. Inst. Chem. Eng.* **31**, 365 (1935).
83. W. F. Porter, F. D. Richardson, and K. N. Subramanian, *in* "Heat and Mass Transfer in Process Metallurgy" (A. W. D. Hills, ed.), p. 79. Inst. Min. Metall., London, 1967.
84. K. N. Subramanian and F. D. Richardson, *J. Iron Steel Inst., London* **206**, 576 (1968).
85. J. K. Brimacombe and F. D. Richardson, *Inst. Min Metall., Trans., Sect. C* **80**, 140 (1971); **82**, 63 (1973).
86. F. D. Richardson, D. G. C. Robertson, and B. B. Staples, *in* "Physical Chemistry in Metallurgy—Darken Conference" (R. M. Fisher, R. A. Oriani, and E. T. Turkdogan, eds.), p. 25. U.S. Steel, Monroeville, Pennsylvania, 1976.
87. W. O. Philbrook and L. D. Kirkbride, *Trans. AIME* **206**, 351 (1956).
88. T. E. Brower and B. M. Larsen, *Trans. AIME* **172**, 137, 164 (1947).
89. J. T. Davies, "Turbulence Phenomena." Academic Press, New York, 1972.
90. D. W. van Krevelen and P. J. Hoftijzer, *Chem. Eng. Prog.* **46**, 29 (1950).
91. M. Sano and K. Mori, *Trans. Jpn. Inst. Met.* **17**, 344 (1976).
92. J. F. Davidson and B. O. G. Schuler, *Trans. Inst. Chem. Eng.* **38**, 144, 335 (1960).
93. I. Leibson, E. G. Holcomb, A. G. Cacuso, and J. J. Jacmic, *AIChE J.* **2**, 296 (1956).
94. L. Davidson and E. H. Amick, *AIChE J.* **2**, 337 (1956).
95. S. Ramakrishnan, R. Kumar, and N. R. Kuloor, *Chem. Eng. Sci.* **24**, 731 (1969).
96. D. J. McCann and R. G. H. Prince, *Chem. Eng. Sci.* **24**, 801 (1969); **26**, 1505 (1971).
97. A. E. Wraith and T. Kakutani, *Chem. Eng. Sci.* **29**, 1 (1974).

98. J. K. Walters and J. F. Davidson, *J. Fluid Mech.* **17**, 321 (1963).
99. A. E. Wraith, *Chem. Eng. Sci.* **26**, 1659 (1971).
100. G. A. Irons and R. I. L. Guthrie, *Metall. Trans. B* **9**, 101 (1978).
101. J. R. Grace, *Trans. Inst. Chem. Eng.* **51**, 116 (1973).
102. R. M. Davies and G. I. Taylor, *Proc. R. Soc. London, Ser. A* **200**, 375 (1950).
103. W. G. Davenport, F. D. Richardson, and A. V. Bradshaw, *J. Iron Steel Inst., London* **205**, 1034 (1967).
104. W. S. Huang and R. C. Kintner, *AIChE J.* **15**, 735 (1969).
105. M. H. I. Baird and J. F. Davidson, *Chem. Eng. Sci.* **17**, 87 (1962).
106. R. I. L. Guthrie and A. V. Bradshaw, *Trans. Metall. Soc. AIME* **245**, 2285 (1969).
107. P. H. Calderbank, *in* "Mixing" (V. W. Uhl and J. B. Gray, eds.), Vol. 2, p. 2. Academic Press, New York, 1967; *Chem. Eng.* (*London*) No. 212, CE209 (1967).
108. F. Yoshida and K. Akita, *AIChE J.* **11**, 9 (1965).
109. P. A. Distin, G. D. Hallett, and F. D. Richardson, *J. Iron Steel Inst., London* **206**, 821 (1968).
110. D. G. C. Robertson and A. E. Jenkins, *in* "Heterogeneous Kinetics at Elevated Temperatures" (G. R. Belton and W. L. Worrell, eds.), p. 393. Plenum, New York, 1970.
111. F. Korber and W. Oelsen, *Mitt. Kaiser-Wilhelm-Inst. Eisenforsch. Duesseldorf* **17**, 39 (1935).
112. J. Alexander, G. S. F. Hazeldean, and M. W. Davies, "Chemical Metallurgy of Iron and Steel," p. 107. Iron Steel Inst., London, 1973.
113. A. V. Bradshaw and F. D. Richardson, *in* "Vacuum Degassing of Steel," Spec. Rep. No. 92, p. 24. Iron Steel Inst., London, 1965.
114. W. G. Davenport, *Can. Metall. Q.* **7**, 127 (1967).
115. F. D. Richardson, *Iron Coal Trades Rev.* **183**, 1105 (1961).
116. D. Papamantellos, K. W. Lange, K. Okohira, and H. Schenck, *Metall. Trans.* **2**, 3135 (1971).
117. R. G. Olsson and E. T. Turkdogan, *J. Iron Steel Inst., London* **211**, 1 (1973).

CHAPTER 8

CALCINATION, REDUCTION, ROASTING, AND SINTERING

8.1 Introduction

Raw materials, recycled process materials, and waste by-products are treated by calcination, reduction, or roasting in the preparation of materials for a wide variety of manufacturing processes in the chemical and metallurgical industries. A feature common to calcination, reduction, and roasting is that they are gas–solid reactions often involving porous reaction products. The calcination is thermal decomposition of substances such as hydroxides, carbonates, nitrates, and sulfates to metal oxides; there is also the reduction calcination as in the thermal decomposition of phosphates, sulfates, and carbonates by carbon. The physical chemistry of reduction discussed here concerns the gaseous reduction of iron oxides. The oxidation of sulfide–ore concentrates to metal oxides or sulfates constitutes the subject of roasting. In all these processes, the rate of the reaction is much affected by the structure of the porous reaction product; sintering that occurs as the reaction progresses may have a decisive influence on the pore structure.

The physical chemistry of reactions pertaining to these processes is presented with reference to particular systems and reactions of major industrial importance. The plant engineering and the operational details of these processes are not included in this discussion of the subject.

8.2 Calcination

8.2.1 Thermal Decomposition of Limestone

Of the many calcination processes, more in-depth study has been made of the calcination of calcium carbonate because of the extensive usage of burnt lime for a variety of purposes in the diversified areas of farming, building, chemical, and metallurgical industries.

There are three types of calcination processes: (1) calcination of large pieces of hard limestone in a gas-fired shaft furnace; (2) gas, oil, or coal fired rotary kilns for the calcination of mixed size limestone that is prone to crumble; and (3) calcination of granulated limestone or powdered calcium carbonate by-product in a fluidized bed.

The calcination is simply a thermal decomposition of calcium carbonate

$$CaCO_3 = CaO + CO_2 \tag{8.1}$$

for which the temperature dependence of the equilibrium CO_2 pressure is represented by the equation [1. 2]

$$\log p_{CO_2} \quad (\text{atm}) = -(8427/T) + 7.169. \tag{8.2}$$

This equilibrium relation gives for the enthalpy of dissociation $\Delta H = 38.55$ kcal mole^{-1} $CaCO_3$.

The first experimental clue to the rate controlling step in the calcination of limestone may be traced back to the work of Furnas [3] in 1931. He found that the temperature at the center of a block of limestone remains unchanged during calcination, until the calcination front reaches the center of the block. Subsequently, Azbe [4, 5] made the first substantial contribution to the study of calcination. Although Azbe did not study the fundamental aspects of the rate phenomena, he did demonstrate that the time of calcination obeys the parabolic relation and the rate is controlled primarily by heat transfer through the calcined layer. However, many subsequent investigators [6–9] have claimed that in the calcination of various types of calcium carbonate of numerous shapes the thickness of the calcined layer increases linearly with the reaction time, and suggested that the rate of decomposition is controlled by a chemical reaction at the calcination front. Most of these conflicting views on the rate controlling mechanisms of calcination have been resolved in later studies by Philbrook and Nateson [10], Hills [11], and Turkdogan *et al.* [12].

The variation of the temperature profile during calcination in air of a 10-cm-diameter cylinder of limestone, measured by Turkdogan *et al.* [12] is shown in Fig. 8.1; the furnace temperature was rising from 1050 to 1200 °C. For a period of calcination from 100 to 300 min, the temperature across the unburnt core is essentially uniform, within a range of about 10 °C. The small increase in the core temperature of a few degrees over a period of hours may be attributed partly to the rise of the furnace temperature. On the 250-min isochronal, when the unburnt core has shrunk to less than 3 cm diameter and the furnace temperature is nearly 1200 °C, the core temperature is only about 887 °C.

The variation of surface and center temperatures during calcination is shown in Fig. 8.2 for a sphere of limestone, 8.3 cm diameter. In the latter half of the calcination, the center temperature T_i remains unchanged at 897 °C and the surface temperature T_0 rises only a few degrees; the temperature difference $\Delta T = T_0 - T_i$ is essentially constant at about 425 °C. At the end of calcination, the center temperature rose rapidly, approaching the surface temperature of 1330 °C.

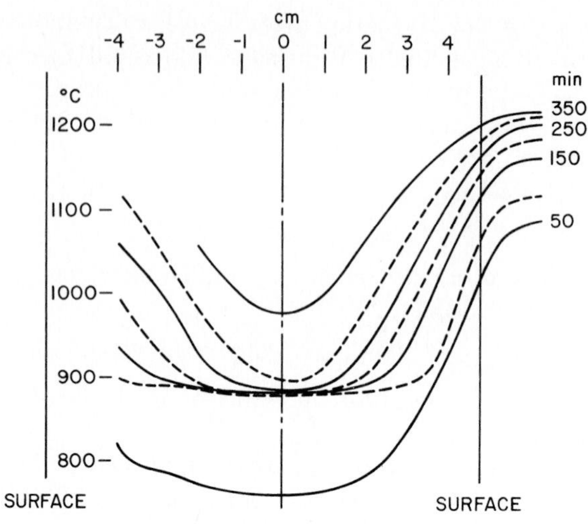

FIG. 8.1 Isochronal temperature distributions in limestone calcining in air as furnace temperature increases from 1050 to 1200 °C. After Turkdogan *et al.* [12].

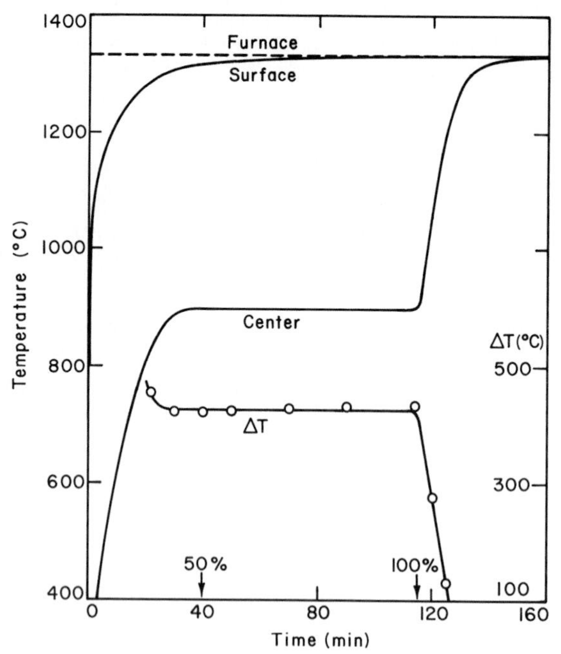

FIG. 8.2 Variation of surface and center temperatures during calcination of 8.3-cm-diameter limestone sphere in argon at 1 atm and furnace temperature of 1332 °C. After Turkdogan *et al.*[12].

Assuming that the interface temperature at the receding calcination front is essentially the same as the center temperature for the period of the calcination when ΔT remains almost constant, and the initial radius r_0 of the spherical particle remains unchanged, Eq. (7.84) gives the fraction F of the carbonate decomposed as a function of time for the rate controlled by heat transfer through the porous lime shell,

$$Y = 3 - 2F - 3(1 - F)^{2/3} = (6\kappa_e/\rho\,\Delta H r_0^2)(T_0 - T_i)t + C, \tag{8.3}$$

where κ_e is the effective thermal conductivity of the porous lime layer, ΔH the enthalpy of decomposition, ρ the bulk density of the limestone in mole per cubic centimeter and C is a constant (a negative number) that takes account of all early time departures from the assumed boundary conditions from which Eq. (8.3) is derived. The value of C depends on particle size and temperature. As is seen from the example in Fig. 8.3 for a 14-cm-diameter limestone sphere calcined in air at a furnace temperature of 1135 °C, the parameter $Y = 3 - 2F - 3(1 - F)^{2/3}$ is a linear function of the time after heat-up, corresponding to the period before about 50% calcination. From Eq. (8.3) the rate constant is represented by

$$(dY/dt)r_0^2 = (6\kappa_e/\rho\,\Delta H)\,\Delta T. \tag{8.4}$$

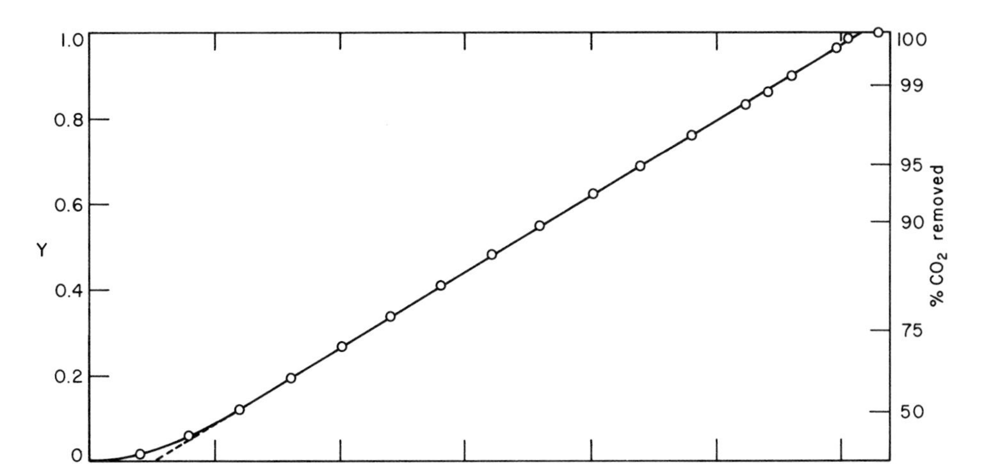

FIG. 8.3 Calcination of 14-cm-diameter (3.58 kg) limestone spheroid in air at 1135 °C; $Y = 3 - 2F - 3(1 - F)^{2/3}$, where F is fraction of CO_2 removed. After Turkdogan *et al.* [12].

The applicability of this limiting rate law to calcination controlled by heat transfer in the product layer is demonstrated in Fig. 8.4 by the direct proportionality of $(dY/dt)r_0^2$ to ΔT. For known values of $\rho = 2.47\ \text{g cm}^{-3}$ and $\Delta H = 38.55$ kcal/mole $CaCO_3$, the slope of the line gives for the effective thermal conductivity $\kappa_e = 0.00126\ \text{cal cm}^{-1}\ \text{sec}^{-1}\ \text{deg}^{-1}$.

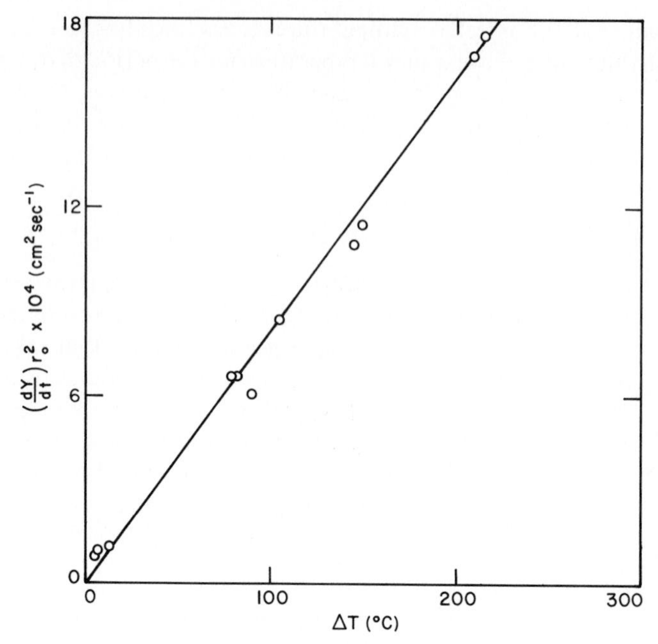

FIG. 8.4 Variation of the rate of calcination $(dY/dt)r_0^2$ with the temperature difference across the calcined layer. After Turkdogan *et al.* [12].

Woodside and Messmer [13] proposed a geometric mean equation to describe the effect of porosity on the effective thermal conductivity at temperatures where the radiation effects are negligible as in small pores at temperatures below 1800 °C,

$$\kappa_e = (\kappa_g)^{\varepsilon}(\kappa_s)^{1-\varepsilon}, \tag{8.5}$$

where κ_g is the thermal conductivity of the gas in the pores, κ_s the thermal conductivity of the solid skeleton of the porous material, and ε the volume fraction of pores. From known values of $\kappa_g = 1 \times 10^{-4}$ cal cm^{-1} sec^{-1} deg^{-1} for CO_2 and $\kappa_s = 0.03$ cal cm^{-1} sec^{-1} deg^{-1} for dense calcium oxide [14] at about 1000 °C, the effective thermal conductivity of burnt lime with ε equal to 0.59 is estimated to be 0.0011 cal cm^{-1} sec^{-1} deg^{-1}, which agrees well with that derived from the rate data in Fig. 8.4.

From Eq. (6.42) for molecular pore diffusion at essentially constant pressure with $J_1 = J_{CO_2}$ and $J_2 = 0$, and Eq. (7.83) for the shrinking core model, the rate equation derived for diffusion of CO_2 in the calcined layer is

$$Y = 3 - 2F - 3(1-F)^{2/3} = \frac{6D_eP}{\rho r_0^2 RT} \ln \frac{P - p_{CO_2}^0}{P - p_{CO_2}^i} + C, \tag{8.6}$$

where D_e is the effective diffusivity of CO_2-inert furnace gas mixtures, and the superscripts 0 and i indicate, the partial pressure of CO_2 at the surface of the pellet and at the calcination front, respectively; P is the total pressure.

Combining Eqs. (8.3) and (8.6) gives

$$\ln \frac{P - p^0_{CO_2}}{P - p^i_{CO_2}} = \frac{RT}{P\,\Delta H} \frac{\kappa_e}{D_e} (T_0 - T_i). \tag{8.7}$$

For known values of κ_e, D_e, P, and $p^0_{CO_2}$, the partial pressure of CO_2 and the corresponding equilibrium temperature at the calcination front can be calculated from Eqs. (8.2) and (8.7) for any given surface temperature. For the simple calcination model considered, with the assumption of equilibrium at the calcination front, the values of $p^i_{CO_2}$ and T_i so obtained are independent of particle size and time of calcination, and the steady state ΔT is a simple function of the surface temperature T_0. The value of the steady state ΔT thus calculated for the experiment cited in Fig. 8.2 is within a few degrees of that measured.

It should be noted that constant pressure assumed in deriving Eq. (8.7) is a slight approximation. As discussed in Section 6.3.1, mass flow in porous media induces a balancing pressure gradient in the flow direction. The steady state excess internal pressure measured [12] during the calcination of limestone increases from about 0.008 atm at the calcination temperature of 800 °C to about 0.08 atm at 1200 °C.

The effects of time and temperature on the pore surface area of burnt lime are shown in Fig. 8.5; the initial pore surface area of Michigan limestone used in these experiments was 0.07 $m^2\,g^{-1}$. The soft burnt lime (low calcination temperature) being more reactive than the hard burnt lime (high calcination temperature) is self-evident from the time and temperature effect on the pore surface area. The variation of the pore volume of burnt lime with the time and

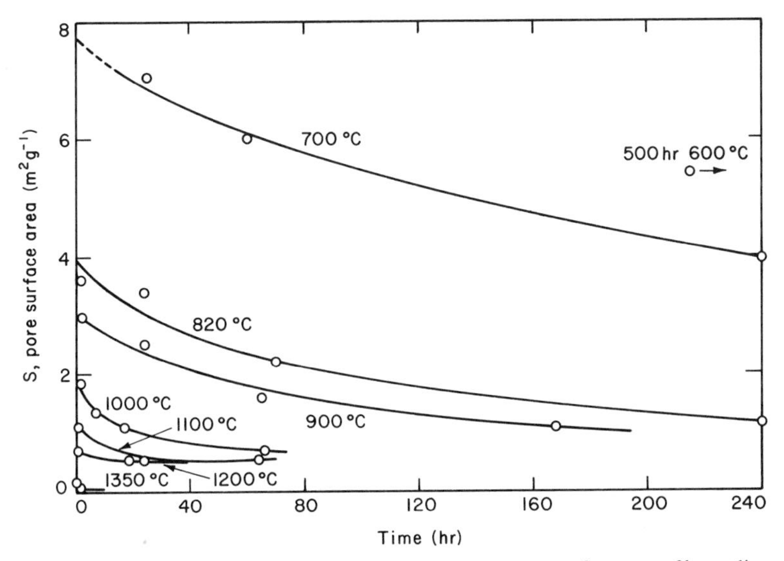

FIG. 8.5 Effect of heat-treatment time and temperature on pore surface area of burnt lime. After Turkdogan *et al.* [12].

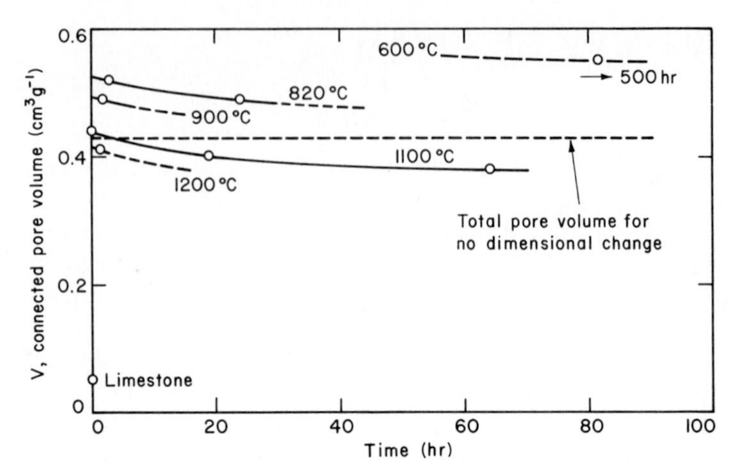

FIG. 8.6 Effect of heat-treatment time and temperature on connected pore volume of burnt lime. After Turkdogan *et al.* [12].

temperature of heat treatment subsequent to calcination is shown in Fig. 8.6. The dotted line is the calculated total pore volume of 0.43 $cm^3\,g^{-1}$ when there is no shrinkage or swelling upon calcination of limestone having a bulk density of 2.47 $g\,cm^{-3}$, corresponding to the initial pore volume of 0.04 $cm^3\,g^{-1}$. The data in Fig. 8.6 show marked volume expansion at lower calcination temperatures. At and above 1100 °C, there is shrinkage which increases with time of heating of the burnt lime. The extent of shrinkage upon calcination depends on the type of carbonate dissociated. For example [12], the calcium oxide produced by calcination of Iceland spar and heat treated at 1100 °C for 20 hr has a pore volume of 0.21 $cm^3\,g^{-1}$, corresponding to 27% volume shrinkage. Similar observations were made by Fischer [15].

For some applications, burnt lime low in sulfur is required, e.g., 0.03% S (maximum) in burnt lime used in steelmaking. Limestone invariably contains some sulfur, primarily in the form of iron pyrite; depending on the geographical location, the sulfur content varies over a wide range from 0.02% to 0.3%. To produce burnt lime low in sulfur, particular attention must be paid to the gas composition in the calcination furnace atmosphere.

During heating of limestone to the calcination temperature, the pyrite dissociates to pyrrhotite and sulfur vapor which may react with $CaCO_3$ to form CaS, SO_2, and CO. At the calcination front, i.e., $CaCO_3$–CaO interface, S_2 may react with CO_2 to form SO_2 and CO. As SO_2 diffuses through the pores of the lime shell together with CO_2, some $CaSO_4$ may also form. The pyrrhotite formed during heating may react with $CaCO_3$ to form CaS in the core of the limestone. At the calcination front, sulfur may be removed by the reaction

$$CaS + 3CO_2 = CaO + SO_2 + 3CO \tag{8.8}$$

and some of the sulfur may be converted to $CaSO_4$,

$$CaS + 4CO_2 = CaSO_4 + 4CO. \tag{8.9}$$

Towards the end of calcination, almost all the residual sulfur is presumably in the form of $CaSO_4$, for which the following equilibrium may be written

$$CaSO_4 + CO = CaO + SO_2 + CO_2. \tag{8.10}$$

The reactions (8.8) and (8.10) are the key reactions for desulfurization during and subsequent to calcination of limestone. The temperature dependence of the equilibrium constants of these reactions may be represented by the following equations [16]:

for reaction (8.8),

$$\log p_{SO_2}(p_{CO}/p_{CO_2})^3 = -(20{,}000/T) + 9.270; \tag{8.11}$$

for reaction (8.10),

$$\log p_{SO_2}(p_{CO_2}/p_{CO}) = -(9617/T) + 8.021, \tag{8.12}$$

where the gas partial pressures are in atmospheres.

It follows from the sulfide–sulfate equilibrium diagram in Fig. 8.7, drawn for a total pressure of 1 atm, that burnt lime low in residual sulfur may be obtained by appropriate adjustment of the oxygen potential of the furnace atmosphere during calcination. The invariant for the coexisting three condensed phases CaO, CaS, and $CaSO_4$, shown by the dotted curve, gives the optimum oxygen activity (CO_2/CO ratio) for desulfurization of burnt lime.

The effect of gas composition on the rate of desulfurization of limestone spheroids, 3 cm in diameter, during and after calcination is shown in Fig. 8.8

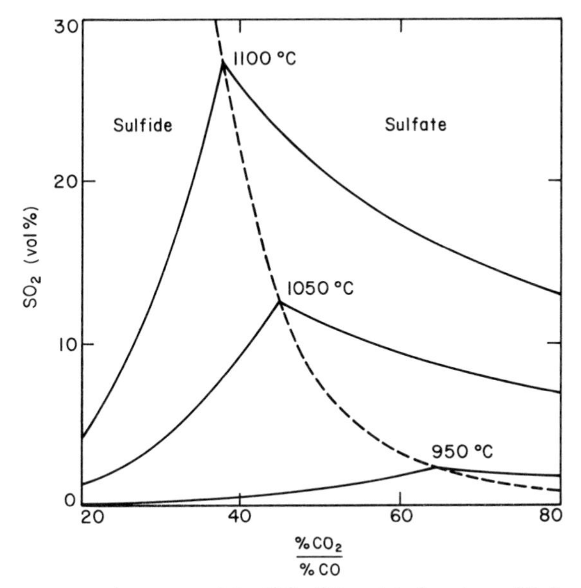

FIG. 8.7 SO_2 content and oxygen activity (CO_2/CO ratio) of gas in equilibrium with CaO–CaS and CaO–$CaSO_4$ at indicated temperatures and total pressure of 1 atm.

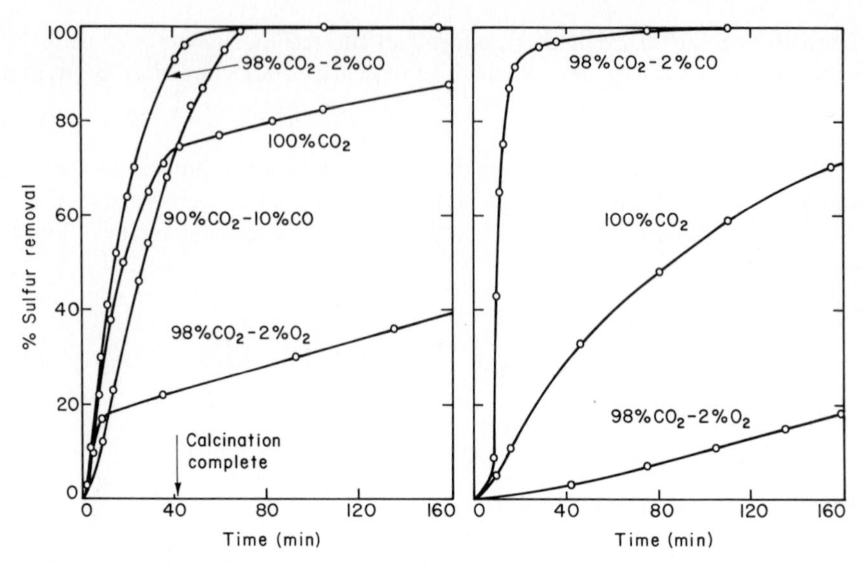

FIG. 8.8 Desulfurization of burnt lime at 1100 °C in indicated gas compositions (a) during and subsequent to calcination and (b) after calcination of limestone in oxygen at 1 atm. After Turkdogan and Rice [17].

[17]. During calcination (Fig. 8.8a), enhanced desulfurization is obtained with CO_2–CO mixtures containing 2% < CO < 10%. This is in general accord with that anticipated from the equilibrium data in Fig. 8.7. The gas composition has a more marked effect on the rate of desulfurization after calcination (Fig. 8.8b). When limestone is calcined in air essentially all the sulfur is converted to $CaSO_4$, and hence subsequent desulfurization is much faster in a 2% CO–CO_2 mixture. We see from these examples that low residual levels of sulfur in burnt lime can be achieved by controlling the SO_2 content and the oxygen potential of the furnace atmosphere, particularly in the exit part of the calcination furnace or kiln where the burnt lime is cooled.

The properties of lime, in particular its reactivity, are affected by calcination conditions. The classification of burnt lime for its reactivity depends on the nature of the reaction involved. As an example, we may consider the injection of powdered limestone into hot flue gases for the removal of sulfur oxides. With this application in mind, Mullins and Hatfield [18] developed a method of characterizing the effects of calcination conditions on the properties of lime such as the extent of shock calcination, density, pore volume, crystallite size, and SO_2-absorbing capacity as functions of the independent variables: time and temperature of calcination and particle size. By using a quadratic formalism involving these three independent variables, the measured properties were extrapolated to short reaction times pertinent to the injection process. Their results reproduced in Fig. 8.9 show systematic variations in properties with time and temperature of calcination and particle size.

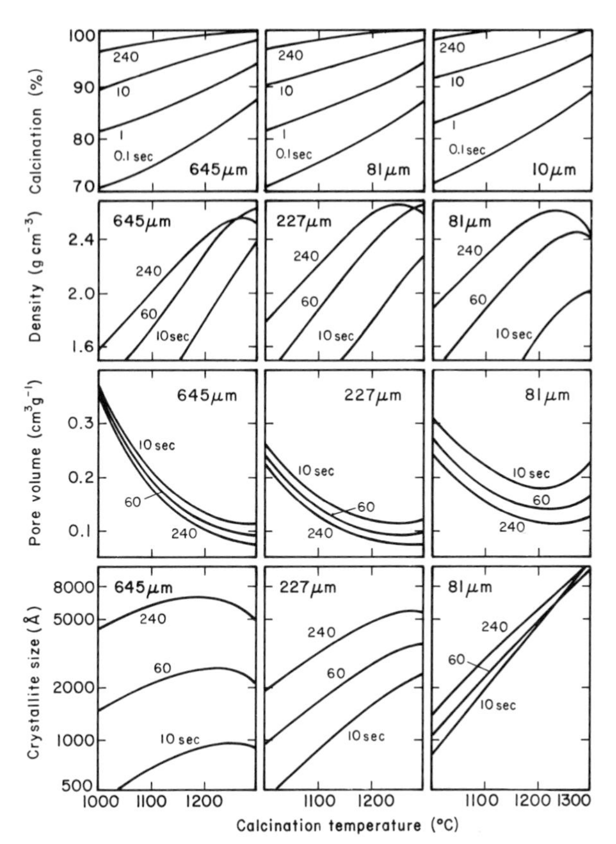

FIG. 8.9 Effect of initial particle size, time, and temperature on the extent of calcination and properties of calcines. After Mullins and Hatfield [18]. Reprinted by permission of the American Society for Testing and Materials, copyright 1970.

Response to SO_2-absorbing capacity of the calcines shown in Fig. 8.10 for sulfating tests at 920–940 °C for 30 min in a gas stream containing 4% SO_2, 4% O_2, and 92% N_2 is consistent with the variations of density, pore volume, and crystallite size of the calcines with the conditions of shock calcination. The effect of calcination time on SO_2 absorption is greater at lower than at higher calcination temperatures. The lower SO_2-absorbing capacity at higher calcination temperatures is consistent with higher density and lower pore volume of the calcine.

For many applications, the reactivity of burnt lime is evaluated by measuring either the rise of temperature during slaking in water or the rate of dissolution in a hydrochloric acid solution. There are, however, conflicting views on the validity of such tests as a measure of reactivity of burnt lime used in pneumatic steelmaking processes [19], because of recrystallization of lime upon reburning at steelmaking temperatures and also the variation of the rate of dissolution of lime with the composition of the slag.

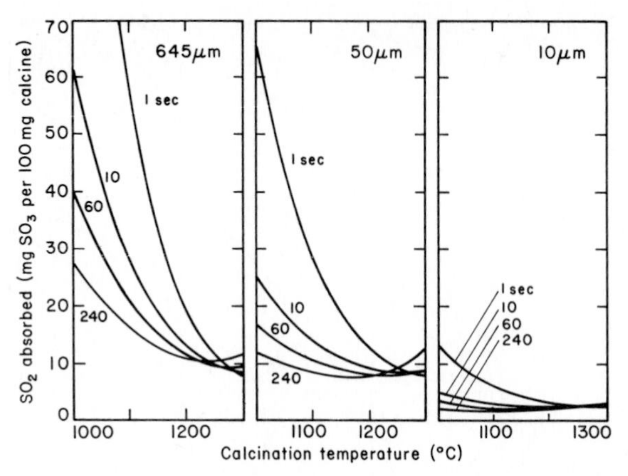

FIG. 8.10 Effect of initial particle size, time, and temperature of calcination on absorption of sulfur dioxide by calcines sulfated 30 min at 920–940 °C. After Mullins and Hatfield [18]. Reprinted by permission of the American Society for Testing and Materials, copyright 1970.

8.2.2 Reduction Calcination

There are numerous industrial applications of reduction calcination as, for example, in the production of phosphorus by fusion of a mixture of calcium phosphate, silica, and carbon; the production of sodium by fusion of sodium carbonate with carbon; and the production of sulfur or sulfuric acid from gypsum by calcination with carbon. Reduction calcination of gypsum or anhydrite is particularly interesting because of the occurrence of a series of coupled reactions during calcination.

Depending on the relative amount of carbon used, the calcium sulfate may be reduced either to SO_2 and CaO, to CaS, or to a mixture thereof.

$$CaSO_4 + \tfrac{1}{2}C = CaO + SO_2 + \tfrac{1}{2}CO_2, \tag{8.13}$$

$$CaSO_4 + 2C = CaS + 2CO_2. \tag{8.14}$$

Turkdogan and Vinters [20] studied the reduction of calcium sulfate with various types of carbon, such as electrode graphite, coconut charcoal, coal char, and coal. As is seen from an example of their results in Fig. 8.11 for temperatures of 900–1100 °C, the relative amounts of SO_2 and CaS formed depend on the amount of carbon (coal in this case) reacted with the calcium sulfate. It should be noted that the preferential conversion to SO_2 or CaS occurs at carbon additions corresponding closely to the respective stoichiometric amounts for reactions (8.13) and (8.14); in calculating these critical amounts, due allowance is made for the volatile and ash contents of the coal.

Since the solid–solid reactions are inherently slow, reactions (8.13) and (8.14) must occur via the intermediate gaseous reaction products CO, CO_2, and SO_3. In the reduction of calcium sulfate with a small amount of carbon, corresponding to the stoichiometry of reaction (8.13), almost all the carbon is used up to form CaS, which subsequently reacts with the residual $CaSO_4$ to produce CaO and

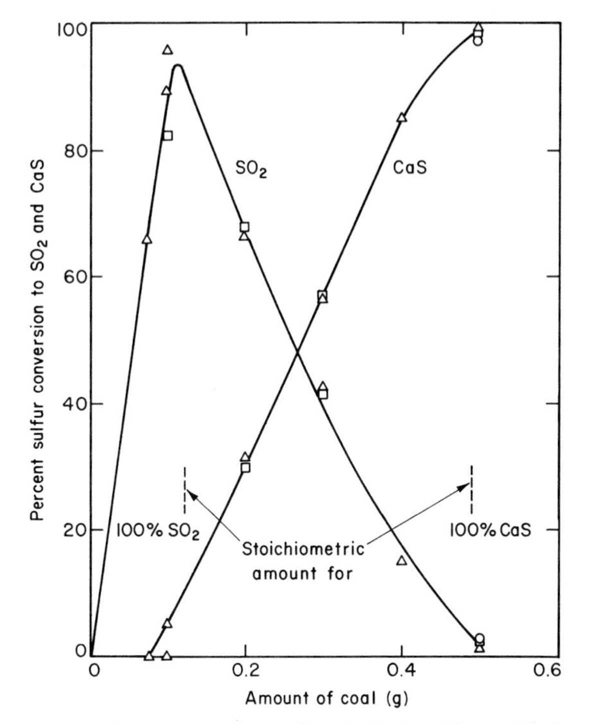

FIG. 8.11 Conversion of sulfur in calcium sulfate (1.65 g) to SO_2 and CaS at temperatures of (○) 900, (△) 1000, and (□) 1100 °C as a function of amount of coal in the mixture. After Turkdogan and Vinters [20].

SO_2. Since most of the SO_2 evolution occurs in the absence of carbon via the reaction

$$3CaSO_4 + CaS \rightarrow 4CaO + 4SO_2, \tag{8.15}$$

the rate of reduction should be independent of the type of carbon used. This is substantiated by the experimental results obtained (Fig. 8.12) with different types of carbon using mixtures with molar ratio $CaSO_4/C$ equal to about 2/1.

This solid–solid reaction presumably occurs via the gaseous species SO_3 thus

$$3CaSO_4 \rightarrow 3CaO + 3SO_3, \qquad CaS + 3SO_3 \rightarrow CaO + 4SO_2. \tag{8.16}$$

The rates of these reactions are apparently fast, and it is believed that under the experimental conditions employed, the rate of the overall reaction (8.15) is controlled by gas diffusion through the interparticle pores within the powder mixture, where the gas is essentially saturated with SO_2 in accord with reaction (8.15). This argument is substantiated by the finding that the temperature dependence of the rate derived from the data in Fig. 8.12 is almost identical to the enthalpy change for reaction (8.15), $\Delta H^\circ = 55.9$ kcal mole^{-1} SO_2.

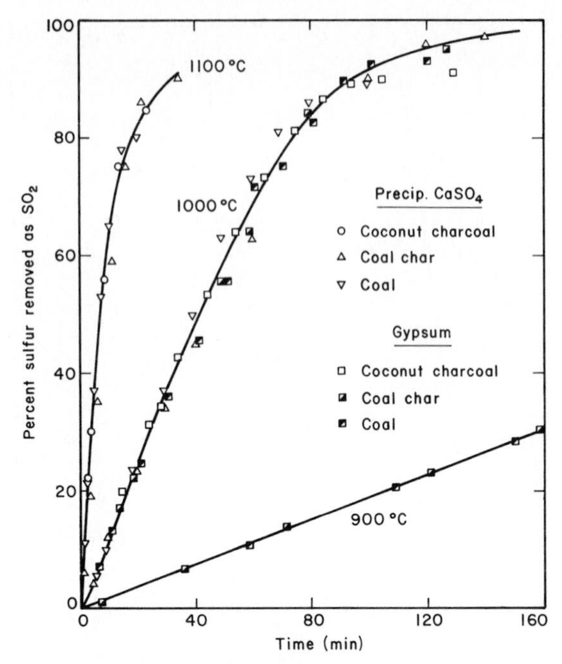

FIG. 8.12 Decomposition of $CaSO_4$ by carbonaceous materials using mixtures with a molar ratio $CaSO_4/C$ of about 2/1. After Turkdogan and Vinters [20].

In the presence of excess carbon, the following sequence of reactions may be considered:

$$CaSO_4 + 4CO \rightarrow CaS + 4CO_2, \tag{8.17}$$

$$CO_2 + C \rightarrow 2CO. \tag{8.18}$$

The clue to the rate controlling reaction is provided by the experimental data in Fig. 8.13 for various types of carbon using mixtures with the molar ratio $CaSO_4/C$ of about 1/2. As the reactivity of carbon increases from a low reactivity for graphite to a high reactivity for coconut charcoal, there is less SO_2 evolution and more CaS formation. That is, the overall rate of conversion of $CaSO_4$ to CaS is controlled by the reaction of CO_2 with carbon to produce CO which is rapidly consumed in reaction (8.17). Under these conditions, the ratio p_{CO_2}/p_{CO} in the exit gas would be high, and close to the equilibrium value for reaction (8.17) as verified experimentally (~ 54 at 1000 °C).

These observations suggest that a fluidized-bed-type reactor would be more efficient in producing SO_2 by the reduction calcination of gypsum. On the other hand, if calcium sulfide is the desired end product, a coal-fired rotary kiln, with coal as a reductant, would be more suitable.

In relation to the reduction calcination of gypsum, it is appropriate to consider the chemistry of reactions in the production of matte in a low-shaft

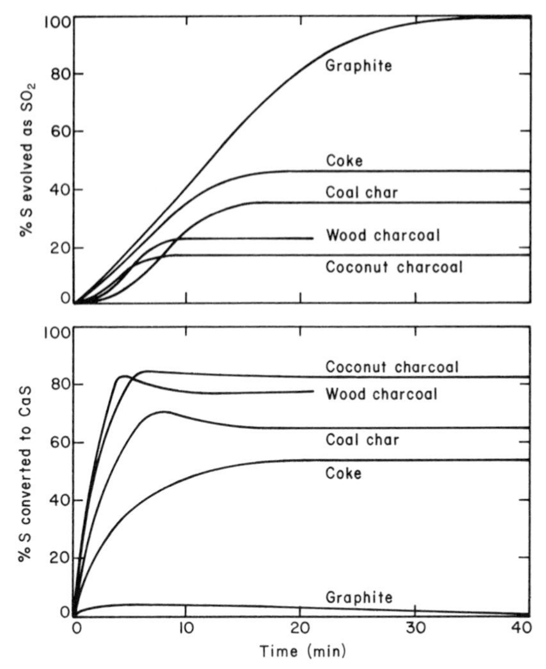

FIG. 8.13 Relative amounts of sulfur in $CaSO_4$ converted to SO_2 and CaS when a mixture with a molar ratio $CaSO_4/C$ of about 1/2 is heated in helium at 1100 °C. After Turkdogan and Vinters [20].

blast furnace with nickel-bearing garnierite ores. In this process the charge consists of ore, coke, gypsum, limestone, and recycled converter slag. The SO_2 produced by the reaction of gypsum with carbon, or carbon monoxide, forms sulfides from the oxides of iron, nickel, cobalt, etc. The overall reaction is represented by

$$CaSO_4 + MO + 2C \rightarrow MS + CaO + 2CO_2, \tag{8.19}$$

where MO is the metal oxide and MS is the matte. It follows from the foregoing considerations of the $CaSO_4$–C reactions that the amount and the reactivity of the carbon used in the production of matte by this process should be adjusted such that little or no calcium sulfide is formed; consequently, almost all the SO_2 evolved is consumed in the conversion of mineral oxides to their respective sulfides. That is, best results are expected in this smelting process with carbon of low reactivity and with an amount sufficient to satisfy the minimum stoichiometric requirement for reaction (8.13).

8.3 Gaseous Reduction of Iron Oxides

Reduction of metal oxides has been studied for many decades, greater emphasis being on the reduction of iron oxides because of its major industrial importance. The reduction of iron ores is well documented in a textbook by von Bogdandy and Engell [21], covering both the physicochemical principles of the iron ore reduction and the process technology of ironmaking. These authors also give a comprehensive, critical review of all the work done on reduction of iron ores prior to the mid-1960s. The present discussion is confined to recent developments in the understanding of the mechanisms of gaseous reduction of iron oxides with particular emphasis on the effect of pore structure on the kinetics of reduction. The knowledge acquired on this subject is considered applicable also to many other gas–solid reactions involving porous materials.

8.3.1 Volume Change during Reduction

At room temperature the specific volumes of hematite, magnetite, wustite, and iron are, respectively, 0.146, 0.193, 0.175, and 0.127 $cm^3 g^{-1}$. Therefore, volume expansion in the early stages of reduction followed by some shrinkage in the final stages of reduction is anticipated. However, because of limited sintering of the reduced oxide and the metal, the extent of swelling or shrinkage varies from one type of iron oxide pellet to another. This is manifested by variations in the experimental observations of different investigators [22–25].

One cause of pellet swelling is attributed by Wenzel *et al.* [26] and Granse [27] to the directional crystallization of metallic iron, the so-called *iron whiskers*; similar observations were made by Moon and Walker [28]. However, such whiskers are not always observed even in partially reduced iron oxides. In general, the swelling during reduction in CO-rich gas is much greater than in H_2-rich gas. It is an established fact that metal dusting occurs during carbon deposition in CO-containing gas mixtures; reduction of iron oxides under these conditions will certainly cause swelling and disintegration of the reduced pellets. However, it is difficult to explain swelling that may occur during reduction in $CO–CO_2$ mixtures even when there is no carbon deposition.

The cause and effect of swelling or shrinkage accompanying reduction have not, as yet, been resolved. Observations made by Turkdogan and Vinters [29] reveal some unpredictable behavior, as demonstrated by the examples in Fig. 8.14. The pellets made from reagent-grade Fe_2O_3, shown in Fig. 8.14a, undergo dramatic dimensional changes upon sintering and reduction. An air-dried green pellet, 14 mm in diameter, shrinks ($\sim 63\%$) to 10 mm diameter upon sintering in air for 40 min at 1300 °C. When the pellet is reduced in hydrogen at 900 °C, there is further shrinkage (~ 38 vol %). However, when the pellet is reduced in $CO/CO_2 = 9/1$, there is extensive swelling and distortion such that the size of the reduced pellet exceeds the initial size of the pellet in the green state.

The shrinkage resulting from crystal growth during sintering and H_2 reduction is not surprising. However, swelling accompanying reduction in $CO–CO_2$ mixtures cannot readily be explained. With the gas composition employed, there was no carbon deposition; the iron formed upon reduction contained

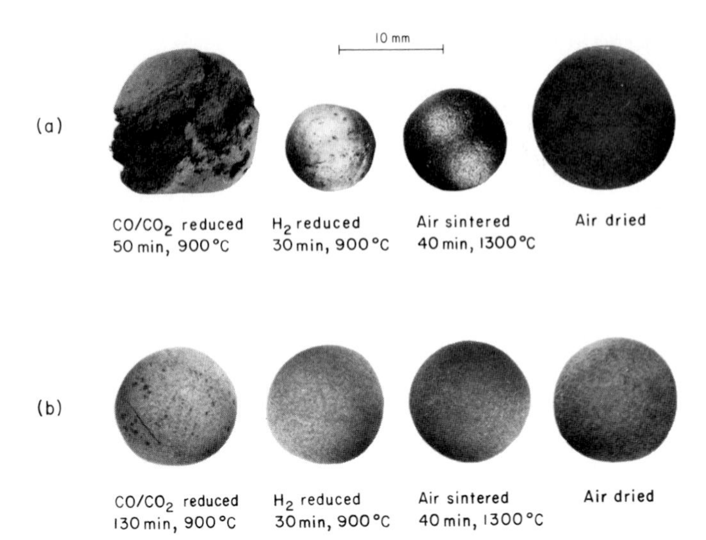

FIG. 8.14 Volume change during sintering and reduction of iron oxides in hydrogen and carbon monoxide (+CO_2 to suppress carbon deposition) for (a) reagent-grade Fe_2O_3 pellet and (b) Venezuelan ore pellet. After Turkdogan and Vinters [29].

0.2% carbon (equilibrium value at 900 °C is 0.3%). In contrast to this behavior, there were only small dimensional changes (<15 vol%) when pellets made from Venezuelan ore were sintered and reduced in H_2 or in $CO/CO_2 = 9/1$ (Fig. 8.14b). Similar behavior was observed with a number of commercial pellets and with pellets made from a high grade taconite (Fe_3O_4) ore concentrate. However, at reduction temperatures above 900 °C, there was some swelling (~25 vol%) when these pellets were reduced in CO–CO_2 mixtures (without carbon deposition).

In the presence of silica (up to 5%), reagent-grade Fe_2O_3 pellets do not swell when reduced in CO (with CO_2 added) [29]. When lime was added, up to 4%, the swelling upon reduction in CO was less than that observed in the absence of lime. The addition of lime and/or silica to pellets made from iron oxide ores had no perceptible effect on the swelling of the pellets. These observations [29], however, differ from those of Burghardt *et al.* [25] and of Moon and Walker [28], who found that a small amount of lime addition to hematite ore pellets (<0.1% CaO) brought about considerable swelling during reduction. On the other hand, Granse [27] and Hoffman *et al.* [30] found that about 1% CaO addition to hematite pellets suppressed swelling during reduction. These variations in the observed effect of lime on swelling may be due, in part, to the presence or absence of other impurities in the ore, such as alkalies.

vom Ende and co-workers [31] found that in the presence of alkalies (0.1–1% Na and K added as bicarbonates) there was extensive swelling during reduction in H_2 or CO. The effect of alkalies becomes more severe with increasing CaO/SiO_2 ratio in the pellet. Conversely, adequate silica additions were found to counteract the disintegration of the pellets.

8.3.2 Pore Structure of Reduced Iron

The pore structure of reduced iron is much affected by the nature of the oxide and reduction conditions; also the reducibility is much affected by the pore structure of the oxide and of the reaction product.

In the very early stages of reduction of dense wustite at a slow rate, a dense layer of iron is formed, provided the iron layer is less than about 5 μm. As demonstrated by Kohl and Engell [32], the rate of reduction under these conditions is controlled by diffusion of oxygen through the iron layer. Because of insufficient plastic flow of the material and the development of gas pressure at the iron–wustite interface, pores begin to form when the iron layer is more than a few microns thick.

The effect of reduction temperature on the pore structure of reduced iron as viewed by scanning electron microscopy is shown in Fig. 8.15 for Venezuelan ore reduced in hydrogen at temperatures of 600–1200 °C [33]. The higher the reduction temperature, the coarser the pore structure; the same trend is observed when the oxide is reduced in CO. However, the pore structure obtained for any given temperature in CO reduction is always coarser than that obtained in H_2 reduction. This is consistent with the general tendency to swelling when the ore is reduced in CO. In all cases of reduction of metal oxides and thermal decomposition of carbonates, sulfates, nitrates, etc., the reaction product is porous, and the pore structure becomes coarser with increasing reaction temperature.

The top left micrograph in Fig. 8.15 shows the fracture surface of lump Venezuelan ore which consists of agglomerates of spheroidal ore grains of

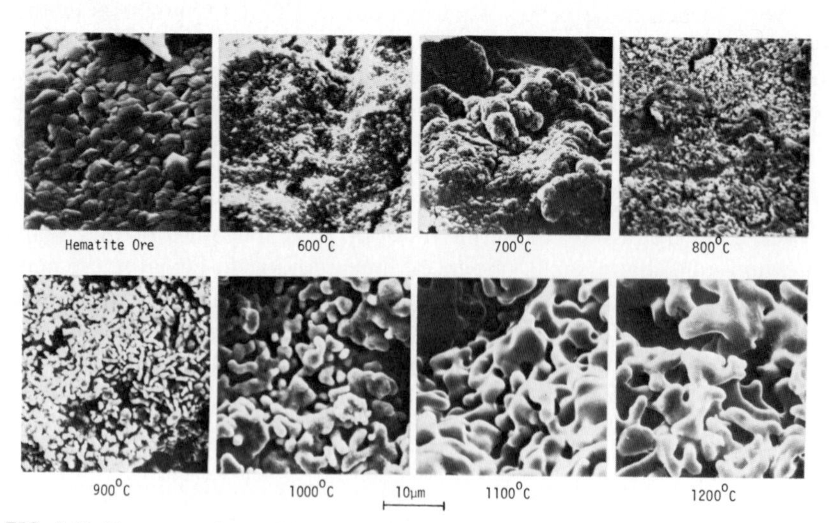

FIG. 8.15 Fracture surfaces of lump Venezuelan ore and porous iron formed by reduction in hydrogen at indicated temperatures as viewed in the scanning electron microscope. After Turkdogan and Vinters [33].

2–5 μm diameter. Although each grain appears to be relatively dense, fine pores are present within the grains as indicated by the relatively large pore surface area, 14 $m^2 g^{-1}$, of the ore. This is much greater than the calculated intergranular surface area of 0.25–0.62 $m^2 g^{-1}$ for 5–2-μm-diameter granules with a bulk density of 4.86 $g cm^{-3}$.*

In addition to microscopic examination, the characteristics of porous media should be qualified by measurements of pore volume, pore surface area, and effective gas diffusivity. In the reduction experiments made with hard lump ore (Venezuelan crustal ore), containing 29% oxygen primarily as Fe_2O_3, the average bulk density of the ore was 3.6 $g cm^{-3}$. The connected pore volume in the ore was about 0.1 $cm^3 g^{-1}$, and upon complete reduction, the average connected pore volume of iron increased to about 0.26 $cm^3 g^{-1}$, which is essentially the same as the total pore volume when there is little or no swelling or shrinkage.

The effect of reduction temperature and gas composition on the pore surface area of the reduced hematite ore is shown in Fig. 8.16 [34]. The pore surface area was measured with the *BET* method from N_2 adsorption at $-195\,°C$. Curve b is for iron and wustite formed by H_2–H_2O reduction of hematite ore B, and curve c is for CO–CO_2 reduction. When hematite is reduced in CO–CO_2 mixtures, the pore surface area is about two-thirds of that for reduction in H_2–H_2O mixtures. This is consistent with the coarser pore structure of CO-reduced iron observed under the microscope. The higher the reduction temperature, the lower the pore surface area as would be anticipated from the micrographs in Fig. 8.15. It should be noted from Fig. 8.16 that when hematite is reduced to wustite, the pore surface area obtained is essentially the same as for iron, indicating that the pore structure of iron is determined by the pore structure of wustite, which is a transient reaction product in the reduction of hematite to iron.

The pore surface area of the iron formed by gaseous reduction depends on the initial pore surface area of the iron oxide. This is demonstrated in Fig. 8.17; for a given type of iron oxide, such as natural hematite ore or sintered hematite pellets, the pore surface area of iron formed upon reduction decreases with decreasing initial pore surface area of the oxide. However, there is considerable overlap. For example, a sintered oxide pellet with $S = 0.1\ m^2 g^{-1}$ and a hematite ore with $S = 10\ m^2 g^{-1}$, when reduced, both give porous iron with $S \sim 1\ m^2 g^{-1}$. That is, the pore surface area of the reduction or calcination product may be larger or smaller than the pore surface area of the unreacted material.

In Section 6.3.3b, we showed that gas diffusion in reduced porous iron may be described in terms of an ideal pore structure such that both molecular and Knudsen diffusion processes occur mainly via the macropores that constitute

* For fcc packing, the fractional volume of the intergranular space is $\varepsilon = 0.26$. Therefore, for a bulk density of $\rho'' = 3.6\ g cm^{-3}$, the bulk density of the granules would be $\rho' = \rho''/(1 - \varepsilon) = 4.86$ g cm^{-3}. This gives for the intergranular surface area $S = 6/\rho' d$, where d is the diameter of the spherical granules.

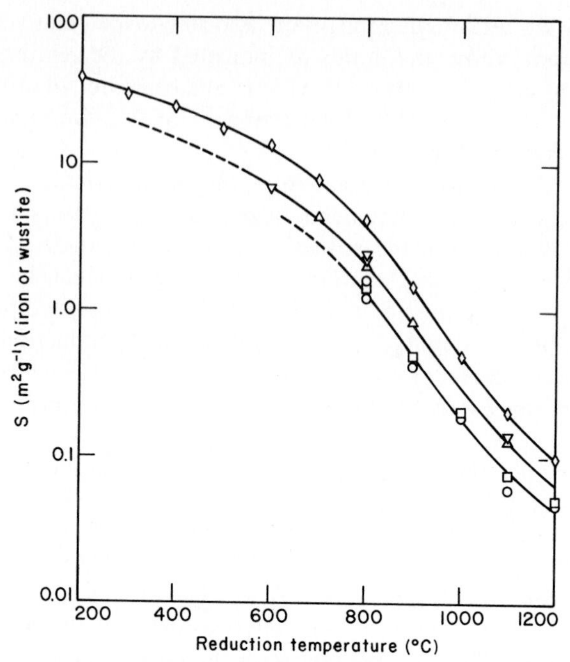

FIG. 8.16 Internal pore surface area of iron and wustite formed by reduction of Venezuelan ores A and B as a function of reduction temperature for (a) ore A reduced to iron (◇) in H_2, (b) ore B reduced to iron (▽) in H_2 and to wustite (△) in an H_2–H_2O mixture, and (c) ore B reduced to iron (□) in a CO–CO_2 mixture and to wustite (○) in a CO–CO_2 mixture. After Turkdogan and Vinters [34].

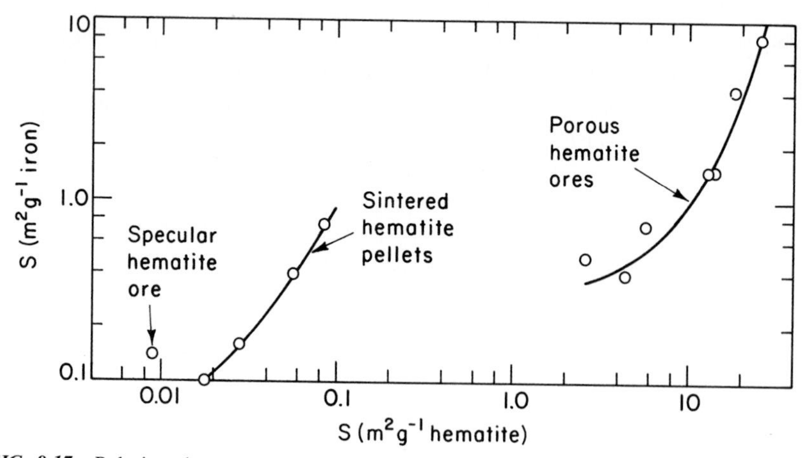

FIG. 8.17 Relations between the internal pore surface area of H_2-reduced iron at 800 °C and the pore surface area of the corresponding hematite ores or sintered pellets. After Turkdogan and Vinters [34].

most of the pore volume. The variation of the gas diffusivity ratio D_e/D_{12} in porous iron with the reduction temperature is shown in Fig. 8.18 [33, 35]; the values derived from the rate of reduction of the oxide are discussed later. As the reduction temperature decreases, the pore structure becomes finer, and Knudsen diffusion predominates, hence the effective diffusivity decreases, as given by Eq. (6.46). At higher reduction temperatures resulting in coarser pore structure, the molecular diffusion predominates, and the diffusivity ratio approaches the theoretical value of $D_e/D_{12} = \varepsilon/2$ for molecular diffusion in an ideal pore structure.

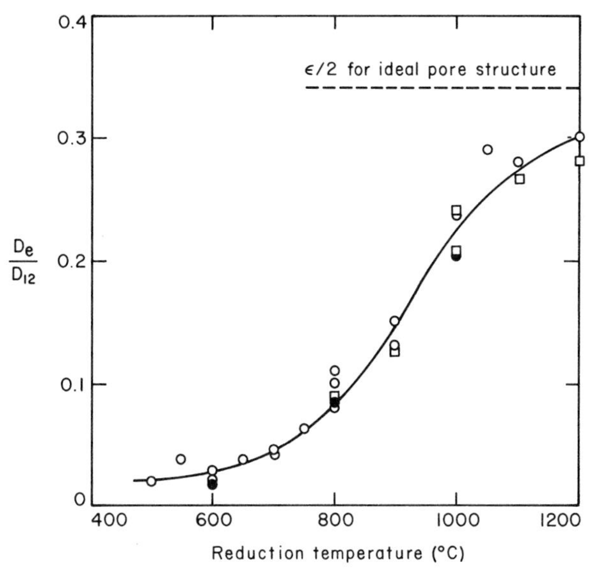

FIG. 8.18 Diffusivity ratio D_e/D_{12} for porous iron as a function of temperature of reduction in hydrogen: (●) direct measurements, (□) calculated from pore structure, (○) derived from reduction data. After Turkdogan *et al.* [33, 35].

8.3.3 Modes of Reduction

The formation of product layers during the reduction of dense sintered hematite pellets or natural iron ore particles is a well-known phenomenon that has been observed by many investigators. The polished section of a partially reduced hematite pellet shown in Fig. 8.19a is a typical example of the layer formation. At higher magnifications the interfaces are seen to be diffuse, particularly near the iron–wustite interface, indicating that some internal reduction occurs ahead of the advancing interface. With pellets or lump ores of high porosity, permitting easy gas diffusion in the pores, there is rapid reduction of hematite to wustite at the internal portion of the particle without the formation of distinct product layers, as in Fig. 8.19b.

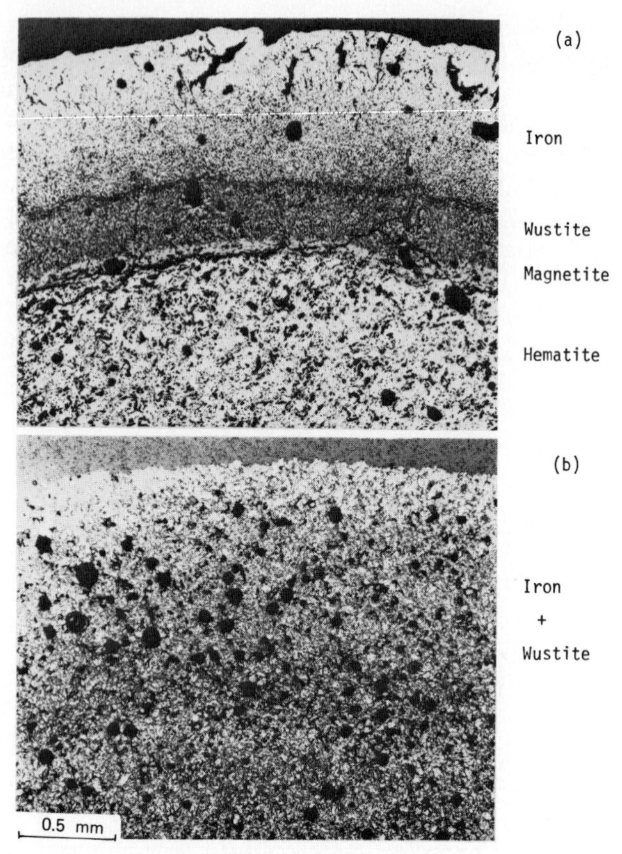

FIG. 8.19 Polished sections of sintered hematite pellets reduced 30% in hydrogen at 900 °C for (a) dense-Fe_2O_3 pellets showing product layers and (b) porous ore pellets showing no product layers.

The greater the driving force for reduction and the faster the rate of the chemical reaction, the more pronounced is the formation of the product layers. For example, as demonstrated in Fig. 8.20 by the photomicrographs of polished sections of sintered hematite pellets reduced 70% at 900 °C, the product layers are sharp when reduced in 100% H_2 (left picture), but diffuse when reduced in a 90% CO + 10% CO_2 mixture (right picture). The rate of chemical reaction in this CO–CO_2 mixture is about one-thirtieth the rate in 100% H_2, hence there is more time for the gas to diffuse in the pores, and consequently, there is more internal reduction resulting in diffuse product layers.

There are also variations in the mode of internal reduction, depending on the type of iron oxide and reduction temperature. Two typical examples are shown in Fig. 8.21 for samples reduced about 70% in a 90% CO + 10% CO_2 mixture. In both cases there is essentially uniform internal reduction, at least on macro scale. The one on the left is for a porous sintered hematite pellet

FIG. 8.20 Sintered hematite pellets (10 mm diam) reduced 70% at 900 °C; distinct layer formation when reduced in hydrogen (left); and extensive internal reduction, hence diffuse layers, when reduced in a 90% CO + 10% CO_2 mixture (right).

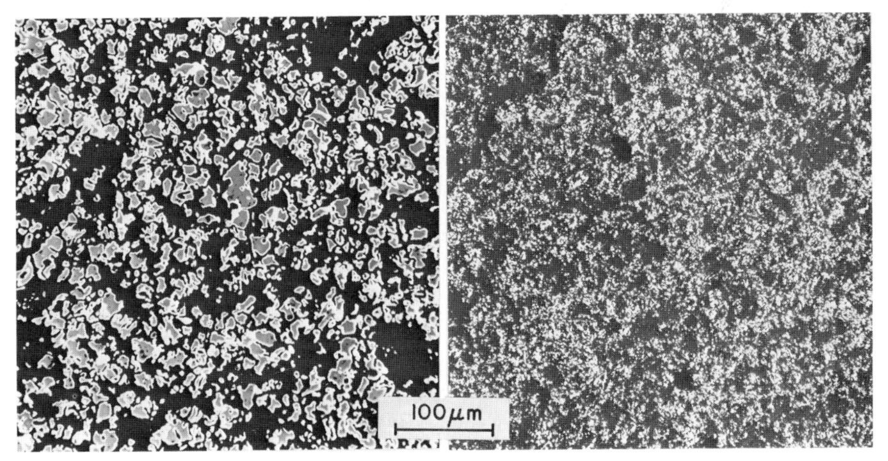

FIG. 8.21 Polished sections of 70% reduced (in CO–CO_2) porous hematite ore pellets showing variations in the mode of internal reduction: (left) wustite grains coated with a layer of iron; (right) random distribution of iron flakes within the wustite matrix.

(~10 mm diam) reduced at 1000 °C; the structure resembles the grain model for uniform internal reduction (Fig. 7.17). The one on the right in Fig. 8.21 is for a sample of porous hematite ore (~4 mm diam) showing random distribution of flakes of iron within the wustite matrix; this does not resemble the grain model.

8.3.4 Rate of Reduction

The kinetics of the formation and dissociation of H_2O and CO_2 on surfaces of dense iron and wustite have been measured by numerous techniques. It is generally agreed that this is a first-order-type reaction relative to the partial pressure of the reacting gas

$$FeO + H_2 \rightarrow Fe + H_2O, \tag{8.20}$$

$$FeO + CO \rightarrow Fe + CO_2. \tag{8.21}$$

From a critical evaluation of the available experimental data [34], the following equations are derived for the temperature dependence of the rate of formation of H_2O from H_2 and of CO_2 from CO on the surface of iron and wustite, at oxygen activities corresponding to the iron–wustite equilibrium:

for H_2 on iron surface

$$\log \Phi_{H_2} = -(4538/T) - 1.020; \tag{8.22a}$$

for H_2 on wustite surface

$$\log \Phi_{H_2} = -(4824/T) - 2.077; \tag{8.22b}$$

for CO on iron surface

$$\log \Phi_{CO} = -(8774/T) + 1.273; \tag{8.23a}$$

for CO on wustite surface

$$\log \Phi_{CO} = -(8774/T) + 0.291. \tag{8.23b}$$

In these equations the rate constants are in units of mole $cm^{-2}\,sec^{-1}\,atm^{-1}$. For both gases, the rate of reaction on the surface of iron is about an order of magnitude greater than on the surface of wustite.

Turkdogan and Vinters [33, 34] made an extensive study of the rate of reduction of Venezuelan ore, over wide ranges of particle size and temperature, in H_2, a CO–CO_2 mixture with $CO/CO_2 = 9$, and mixtures thereof. Because of the self-consistency of the rate data, the following discussion of the rate phenomena is based primarily on their observations. After dehydration at 600 °C, the average composition of the hematite ore B used was $97 \pm 1\%$ Fe_2O_3, 1%–2% SiO_2, and <1% (CaO + MgO + MnO). The ore had an average pore surface area of $14\,m^2\,g^{-1}$ and about 30% porosity. The rate measurements were made with sized particles over the range 0.4–4 mm and spheroidal single particles in the range 7–15 mm diam. Sufficiently high gas flows were used to avoid the impedence of the rate of reduction by the gas–film mass transfer of the reactants to and from the particles. However, as pointed out later, the gas flows may not have been fast enough to overcome convective mass-transfer effect in the initial stages of the reduction of large oxide particles.

Typical examples of the rate of reduction of granules in H_2 and CO are shown in Fig. 8.22, where F is the fraction of oxygen removed. For reduction in H_2 at temperatures above 900 and below 500 °C, $\log(1 - F)$ is a linear function of time up to about 96%–98% oxygen removal. At intermediate temperatures, the plots are linear up to about 80% oxygen removal, beyond which reduction is sluggish; this anomalous behavior is discussed later. For large

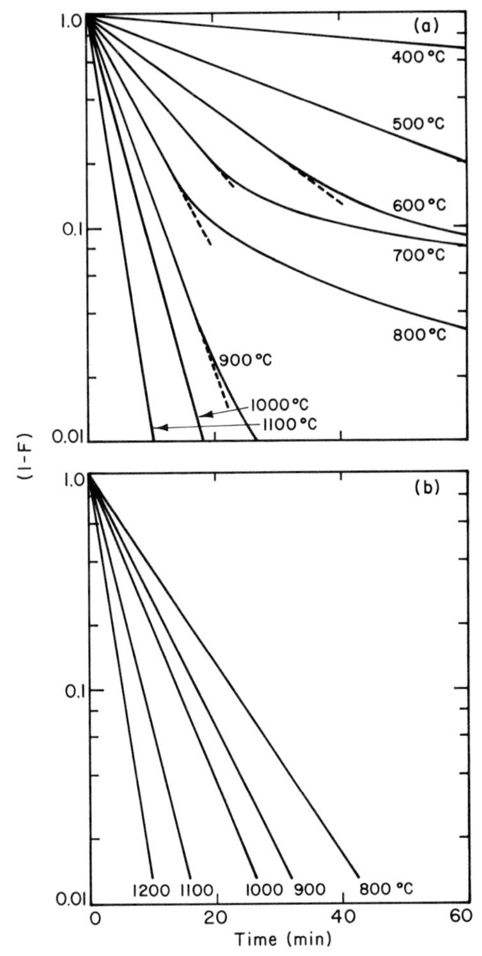

FIG. 8.22 Rate data for reduction of Venezuelan ore granules (a) in hydrogen (800 μm diam) and (b) in a 90% CO + 10% CO_2 mixture (400 μm diam) at indicated temperatures. After Turkdogan and Vinters [34].

particles $\log(1 - F)$ versus t plots are curved, concave downward, because of the diffusional effects discussed later.

For the limiting case of almost complete gas diffusion in the pores of the ore granules, there is essentially uniform internal reduction. Conversion of hematite to wustite and nucleation of iron occurs rapidly in the early stages of the reaction. Therefore, the rate controlling reaction to be considered is the reduction of wustite to iron. The rate is presumably controlled by reaction of H_2, or CO, with oxygen on the surface of a thin layer of iron on the pore walls of wustite, followed by rapid oxygen transfer from wustite to iron at the interface on the pore walls and oxygen diffusion through a few atoms thick iron layer. There are, of course, other plausible atomistic models to be postulated. Whatever the atomistic reaction model, using an equation similar to (7.112), the overall rate of internal reduction may be represented by the equation

$$\ln(1 - F) = - S'\Phi_i[p_i - (p_i)_e]t, \tag{8.24}$$

where Φ_i is the rate constant, p_i the partial pressure of H_2 or CO, and $(p_i)_e$ is the corresponding equilibrium value for coexistence with iron and wustite; S' is the effective pore surface area per unit mass of wustite. Because of incomplete gas diffusion in micropores, the effective area of the pore surface where the reaction is occurring could be less than the total pore surface area.

The rate of reduction, $d\ln(1 - F)/dt$, for a given gas composition and temperature is shown in Fig. 8.23 as a function of particle size. The extrapolation to zero particle size gives the truer estimate of the rate of internal reduction. Because of variations in the pore size and faster gas diffusion in larger pores,

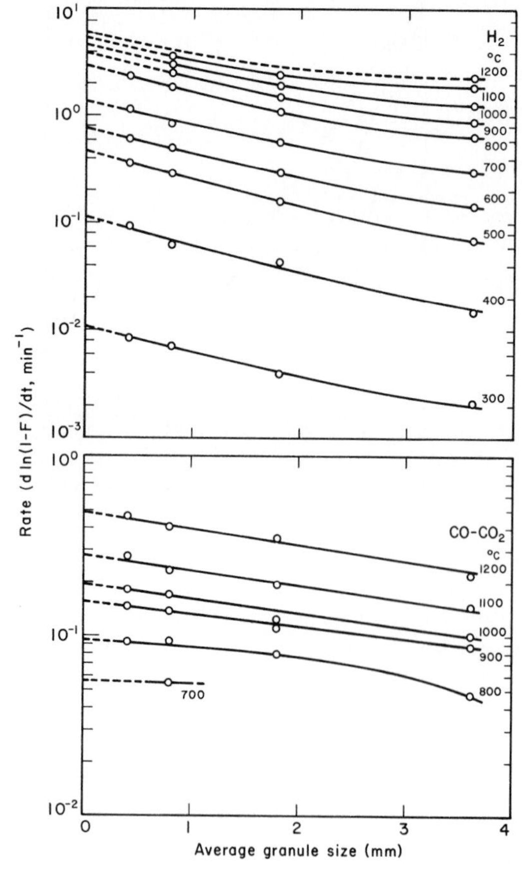

FIG. 8.23 Effect of granule size on the rate of internal reduction of Venezuelan ore B in hydrogen and a 90% CO + 10% CO_2 mixture. After Turkdogan and Vinters [34].

most of the reaction will occur on the walls of larger pores. The fraction of the total pore surface area Θ_S utilized in internal reduction may be estimated from the rate of internal reduction at zero particle size, $[d\ln(1-F)/dt]_0$, and the values of the total pore surface area S, from Fig. 8.16, and the rate constant Φ_i, from Eq. (8.22) or (8.23):

$$\Theta_S = [d\ln(1-F)/dt]_0/68.9S\Phi_i\Delta p_i, \tag{8.25}$$

where 68.9 is the molecular weight of wustite per gram-atom of oxygen.

If Φ_i is that for the iron surface from the data presented, Θ_S is estimated to be about 0.01–0.02. If the rate controlling reaction is occurring on the wustite surface of the pore walls (the slower rate), then Θ_S would be about 0.1–0.2. It is more reasonable to suppose that greater percentages of the pore walls are participating in the reaction, which is perhaps controlled jointly by oxygen diffusion through the thin layer of iron on the pore walls of wustite and the chemical reaction on the surface of the iron in the porous medium consisting of iron and wustite.

It is all too clear from the foregoing discussion that the rate of gas–solid reactions involving porous media cannot be interpreted rigorously in terms of a particular reaction or pore–structure model. Moreover, the rate constant measured for the so-called uniform internal reduction of a particular porous metal oxide may not give a correct estimate of the rate for an oxide of different pore surface area. This is demonstrated by the data in Fig. 8.24, which shows a nonlinear increase in the rate of internal reduction of various types of hematite granules with the pore surface area of iron, or wustite formed upon reduction. This nonlinear relation is similar to the rate of internal burning of carbon as a function of pore surface area, discussed in Section 7.7.2.

There is also anomalous sluggish reduction of iron oxide granules within the temperature range 500–800 °C, as indicated by the data in Fig. 8.22. A number of investigators have observed an unusual temperature effect on the rate of

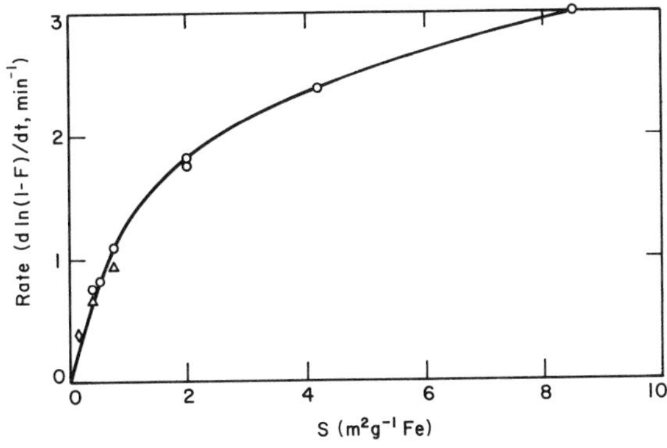

FIG. 8.24 Rate of reduction at 800 °C and 1 atm hydrogen of different types of hematite ore granules (800 μm) as a function of pore surface area of iron (or wustite) formed. After Turkdogan and Vinters [34].

reduction of iron oxide granules. For example, with mill scale (Fe_3O_4) granules in size ranges of ~40 and 110 μm, Smith [36] observed maximum and minimum rates at low and high temperatures as shown in Fig. 8.25. The temperature for maximum rate decreases from 700 to 650 °C with the progress of reduction. Edstrom [22] also found that the rate of reduction of natural hematite and magnetite (4 mm diam) in hydrogen increases with increasing temperature, reaching a maximum at about 550 °C, then decreases, reaching a minimum at about 650 °C, and subsequently increases again with increasing temperature. Similar observations were made by others [37, 38]. So far, no satisfactory explanation has been given of this unusual but well proven reduction behavior of iron oxide granules.

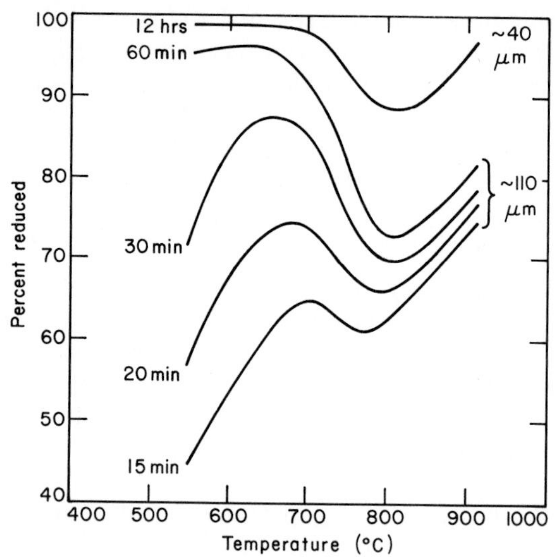

FIG. 8.25 Anomalous temperature effect on the rate of reduction of mill scale granules in hydrogen. After Smith [36].

In the reduction of dense lump ores and sintered ore pellets, concentric layers of porous magnetite, wustite, and iron are formed around the hematite core of the particle as in Fig. 8.19. When the thickness of the iron layer is 1–1.5 mm, the subsequent rate of reduction is controlled primarily by gas diffusion through the porous iron layer. For this limiting case, the isothermal rate for a spherical particle is given by Eq. (7.83), provided there is little or no volume change during reduction,

$$3 - 2F - 3(1 - F)^{2/3} = (6D_e/\rho r_0^2)\{[p_i - (p_i)_e]/RT\}t + C, \qquad (8.26)$$

where r_0 is the particle radius, D_e effective gas diffusivity, and C a constant (a negative number) that takes account of all early time departures from the assumed boundary conditions. The value of C depends on particle size, gas composition, and temperature. The higher the temperature, the larger the particle size, and the greater the driving force for the reaction, the closer is the value of C to zero.

The rate data plotted in accord with Eq. (8.26) usually give elongated S-shaped curves as in Fig. 8.26 for 15-mm-diameter spheroids of Venezuelan ore reduced in H_2 at 1 atm [33]. From about 50% to 95% or 99% reduction, data are well represented by straight lines. The effective gas diffusivity D_e for H_2–H_2O derived from the slopes of these lines shown in Fig. 8.18 as the ratio D_e/D_{12} are seen to be consistent with those measured directly or estimated from the pore structure. The plot of $\ln D_e$ versus $1/T$ is essentially a straight line, the slope of which gives about 15 kcal for the apparent heat of activation for reduction in H_2. This apparent temperature effect on the effective gas diffusivity is due primarily to the coarsening of the pore structure of reduced iron with increasing temperature of reduction.

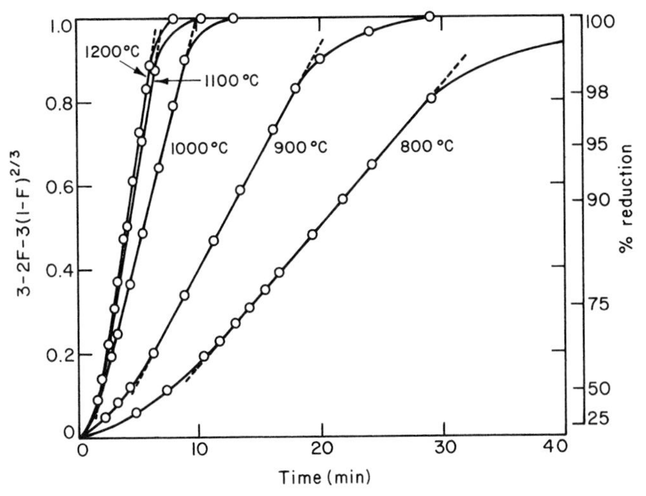

FIG. 8.26 Diffusion plot of reduction data for 15-mm-diameter spheroidal hematite ore and 1 atm hydrogen. After Turkdogan and Vinters [33].

For reduction in H_2–CO–CO_2 mixtures, with $CO/CO_2 = 9$ to suppress carbon deposition, the effect of gas composition on the time for the reaction to achieve 50%, 75%, 90%, and 95% reduction at 900 °C is shown in Fig. 8.27 for hematite and magnetite ore pellets sintered in air at 1300 °C. The rate beyond about 50% reduction could be interpreted [29] in terms of the countercurrent diffusion of H_2 + CO and $H_2O + CO_2$ through the porous iron layer, giving D_e that is a linear function of the average concentrations of $H_2 + H_2O$ and $CO + CO_2$ in the porous iron layer.

MeKewan [39, 40] measured the rate of reduction of dense pure Fe_2O_3 sintered spheroidal pellets ($\rho = 5\ \mathrm{g\,cm^{-3}}$) in hydrogen at pressures of 0.25–40 atm. Although he interpreted his data in terms of rate controlling chemical

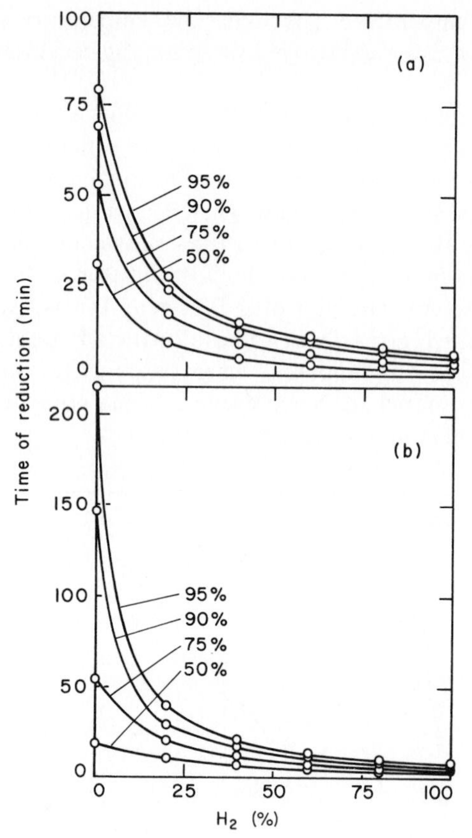

FIG. 8.27 Effect of gas composition (H_2–CO_2/CO = 9 mixtures at 1 atm) on time of reduction at 900 °C of (a) hematite ore and (b) magnetite ore pellets (10–12 mm diam) sintered in air at 1300 °C. After Turkdogan and Vinters [29].

reaction at the receding iron–wustite interface, the replot of his data in Fig. 8.28 in accord with Eq. (8.26) indicates that beyond about 50% reduction, the rate may be controlled by countercurrent diffusion of H_2 and H_2O in the porous iron layer. The effective gas diffusivities calculated from the slopes of the lines for the range 50%–95% reduction shown in Fig. 8.29 as a function of total pressure are seen to agree well with those derived from direct measurements of D_e [35].

The interpretation of the initial rate of reduction of large hematite particles (5–15-mm diam) is more complex. If we assume that for the experimental conditions employed the convective mass transfer through the gas–film boundary layer was not rate limiting, then the initial rate of reduction would be controlled jointly by partial gas diffusion and the chemical reaction in the pore mouths near the surface of the particle. From Eq. (7.110) for this limiting mixed

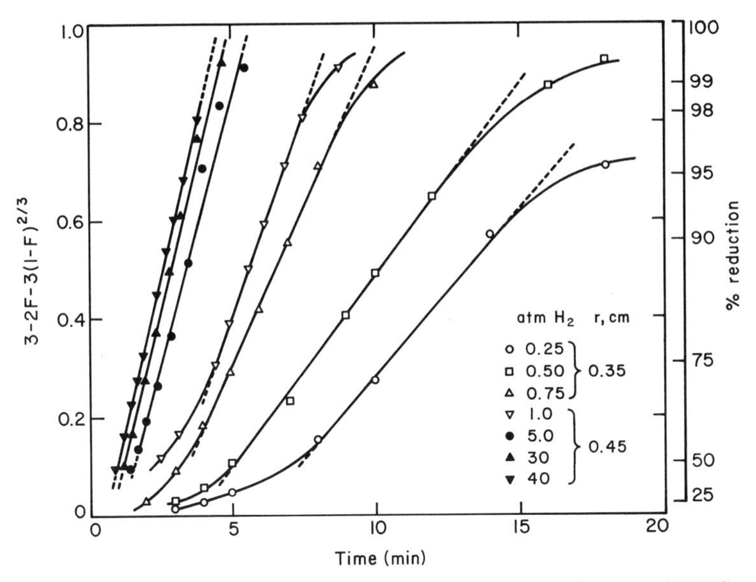

FIG. 8.28 Diffusion plot for reduction of dense Fe_2O_3 pellets in hydrogen at 1000 °C and indicated pressures, using data of McKewan [39].

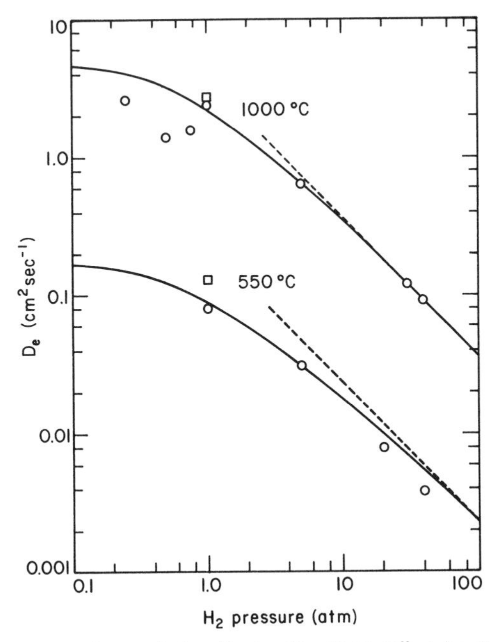

FIG. 8.29 Pressure dependence of the effective H_2–H_2O diffusivity D_c for porous iron (○) derived from rate of reduction in hydrogen (Fig. 8.28) compared with (□) directly measured values of D_c at 550 and 1000 °C.

control, the initial rate of reduction may be represented by the equation

$$\left(\frac{dF}{dt}\right)_\circ = \frac{3}{r_0}\left(\frac{S'\Phi' D_e}{\rho}\right)^{1/2}\left(1+\frac{1}{K}\right)^{1/2}\frac{p^\circ_{H_2} - P/(1+K)}{RT}, \tag{8.27}$$

where Φ' is the specific rate constant, in centimeters per second, for the chemical reaction, i.e., $\Phi' = RT\Phi_{H_2}$ of Eq. (8.24), and K is the equilibrium constant $(p_{H_2O}/p_{H_2})_e$ for coexisting iron and wustite. The apparent rate constants $(S'\Phi' D_e)^{1/2}$ derived by Turkdogan and Vinters [33] from the initial rates should be compared with those calculated from $S'\Phi'$ for internal reduction extrapolated to zero particle size (Fig. 8.23) and the values of D_e in Fig. 8.18. The experimental values of $(S'\Phi' D_e)_e^{1/2}$ are one-half to one-third the calculated values of $(S'\Phi' D_e)_c^{1/2}$. Moreover, the apparent heat of activation derived from the temperature dependence of $(S'\Phi' D_e)_e^{1/2}$ is about 12 kcal, while that derived for $(S'\Phi' D_e)_c^{1/2}$ is about 18 kcal.

In a subsequent interpretation of the initial rate of reduction, Hills [41] showed that the convective mass transfer through the gas–film boundary layer does, in fact, impede the initial rate of reduction. It is probably for this reason that the oversimplified interpretation of the initial rate in terms of the so-called limiting mixed control is two to three times lower than that calculated from independently measured values of $S'\Phi'$ and D_e.

Generalized forms of rate equations of varying degrees of complexity have been derived by many for gaseous reduction of iron oxide particles through mathematical modeling, of which [42–44] are typical examples. Of necessity, many simplifying assumptions are made in mathematical modeling of rates of reactions involving chemical reaction and mass transfer. Noting from the earlier discussion that even for the simpler limiting cases the interpretation of the rate data is not rigorous, we should not overestimate the usefulness and reliability of mathematical formulations of the rates of reactions of gases with porous materials.

8.3.5 Reduction of Metal Oxides with Carbon

In compacted powder mixtures of metal oxides and carbon, the consecutive reduction and oxidation reactions occur via the intermediate reaction products CO and CO_2; thus

$$MO + CO = M + CO_2, \tag{8.28}$$

$$CO_2 + C = 2CO. \tag{8.29}$$

The excess gas generated diffuses out of the system through the interparticle pores where the CO_2/CO ratio can vary over a wide range, depending on the type of metal oxide and carbon in the mixture. For example, in the reduction of higher oxides of iron [45] and manganese [46] and lead oxide [47] by carbon, the reaction product is essentially pure CO_2. For the wustite–carbon mixtures [45, 48] the CO_2/CO ratio in the gas evolved is initially close to the value for the iron–wustite equilibrium, but decreases with the progress of reduction. However, indications are that in all the systems studied the overall rate of reduction of the oxide is controlled primarily by the rate of oxidation of carbon in the CO_2–CO mixture that prevails within the interparticle pores.

In the presence of certain metals and metal oxides, the oxidation of certain types of carbon in CO_2 is catalyzed; this catalytic effect is catastrophic in the reduction of, for example, oxides of lead, zinc, copper, and nickel with powdered electrode graphite. Yet, in experiments with iron oxide–carbon mixtures [45], there is no evidence of catlytic oxidation when coconut charcoal is used; with graphite, there is some catalytic effect. Because of this unpredictable catalytic reaction in the mixture, equations derived through mathematical modeling [49, 50] to describe the overall rate of the reaction are of limited value, and may be applicable only to systems where reactions are not catalyzed.

If the furnace atmosphere is a neutral gas, the rate of reduction of metal oxides by carbon decreases rapidly with the progress of the reaction, because of the dilution of CO_2 in the interparticle pores of the bed by back diffusion of the furnace atmosphere [50]. With a CO–CO_2 mixture in the furnace atmosphere, extensive reduction of the oxide by carbon is possible even when the ratio CO_2/CO in the furnace atmosphere is oxidizing to iron.

8.3.6 Iron Ore Reduction in Practice

In discussing the physical chemistry of reduction of iron oxides, a brief reference should be made to current industrial processes of ore reduction. The blast furnace is, of course, the primary process for the reduction and smelting of iron ore to produce carbon-saturated *hot metal* which is subsequently refined to the desired composition in steelmaking furnaces. There are numerous other processes, known as *direct-reduction processes*, where iron ore granules, pellets, or lump ore are reduced without melting to produce *sponge iron* as a feed material primarily for the electric-furnace steelmaking.

Basically there are four types of direct-reduction processes: (1) fluidized bed, (2) retort furnace, (3) shaft furnace, and (4) rotary kiln. In processes (1)–(3), reformed natural gas is the reductant; in the rotary kiln process the ore is reduced by coal. For technological details on these processes, reference should be made to textbooks on ironmaking such as, that by von Bogdandy and Engell [21].

In the fluidized bed practice, the reformed natural gas (70%–75% CO, 20%–23% H_2, remainder CH_4, CO_2, and H_2O) is preheated to about 825–875 °C and the reduction bed temperature is maintained at temperatures of 700–775 °C. Although iron ore granules reduce sluggishly (Figs. 8.22 and 8.25) within this temperature range, the lower and upper limits of the reduction temperature are dictated for other reasons. At temperatures below 700 °C, there is excessive carbon deposition from CO–H_2 mixtures; in fact, under normal operating conditions the sponge iron contains 1–1.5% C. At temperatures above 800 °C, there is defluidization of the bed due to sticking and sintering of iron particles.

In the shaft furnace practice, iron ore pellets and lump ore are reduced at about 900–950 °C in a countercurrent flow of reformed natural gas, preheated to about 950–1000 °C. Aside from high reducibility, the pellets should have high resistance to breakage and sticking.

With reference to the blast furnace, we should consider also the high temperature reducibility of the ore or pellet. In the blast furnace, most of the reduction is done in the stack at temperatures of 700–1100 °C; the residual iron oxide, 5%–10% wustite, is reduced in the bosh of the blast furnace at temperatures of 1200–1400 °C. Experience has shown that the composition of the ore or the pellet, in terms of CaO/SiO_2 ratio, has little effect on the reducibility at temperatures below 1100 °C. This is no longer the case at higher temperatures, as demonstrated by the example in Fig. 8.30 [51], which shows the dramatic effect of lime addition on increasing the high temperature reducibility of the acidic iron ore pellet (~5% SiO_2). Addition of magnesia or dolomite to the pellet has a similar beneficial effect, but to a lesser extent.

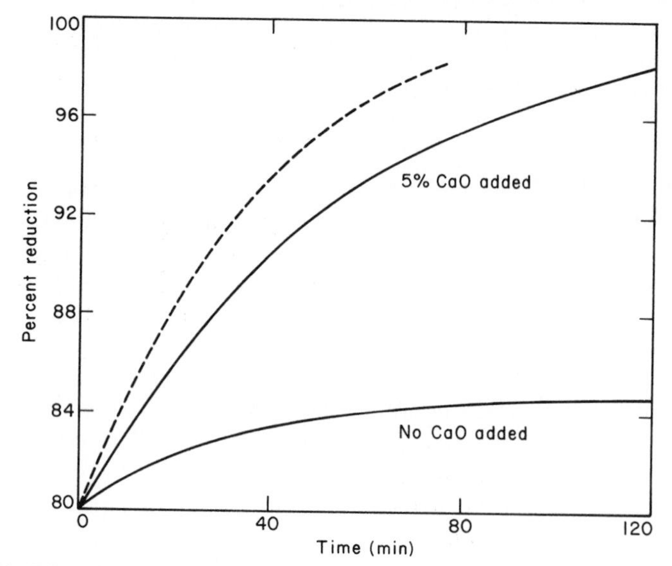

FIG. 8.30 Effect of lime addition on reducibility of sintered hematite ore pellets at 1300 °C——— subsequent to 80% reduction at 1000 °C (– – –) in a 90% CO + 10% CO_2 mixture at 1 atm. After Turkdogan [51].

When the temperature of a partially reduced acidic pellet is increased to 1150 °C or higher, liquid iron silicate forms in the pore structure as in Fig. 8.31a; wustite dendrites are crystallized from the liquid phase during cooling of the sample. Because of lack of gas diffusion into the pellet impregnated with a liquid silicate, the reduction of the residual iron oxide becomes sluggish. Furthermore, the presence of a liquid phase in the pores facilitates grain growth of the reduced iron, which forms a dense layer around the pellet, so that further reduction is

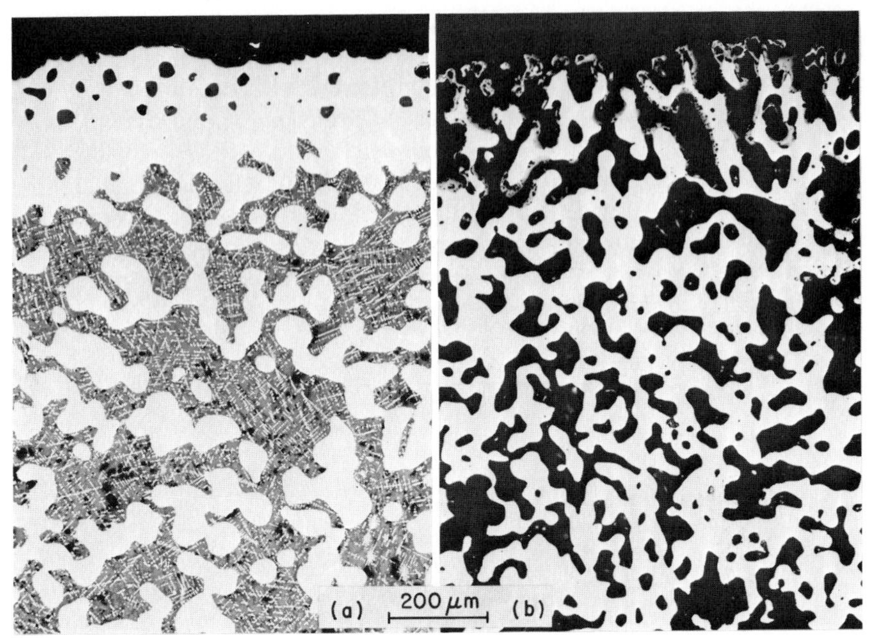

FIG. 8.31 Photomicrographs of polsihed sections of pellets after reduction at 1300 °C in 90% CO–10% CO_2: (a) liquid phase within pore structure of acidic pellet; (b) open pore structure, no liquid phase present, in basic pellet with $CaO/SiO_2 \sim 1.0$. After Turkdogan [51].

not possible until the temperature is high enough to melt the iron shell. This situation leads to early slag formation in the upper part of the bosh with adverse effects on the performance of the blast furnace.

In basic pellets with a CaO/SiO_2 ratio of about 1, there is no liquid phase and the pores remain open as in Fig. 8.31b. Apparently, during sintering of the basic pellet a calcium silicate of high melting point is formed, and so at the final reduction temperature of 1300°C no liquid phase can form; hence the residual iron oxide is rapidly reduced.

It follows from the foregoing considerations that in evaluating the reducibility characteristics of iron ore pellets or sinters, the reduction tests should be made both at low and high temperatures. There are, of course, many other characteristics of the burden that the blast-furnace operator is keenly aware of, such as swelling, strength, breakage, gas permeability, softening temperature, and so on. It should be recognized that simple reduction experiments with single pellets give useful but only limited information on the reducibility pertinent to the conditions in the blast furnace.

8.3.7 Reoxidation of Reduced Iron

The sponge iron produced by direct reduction in a fluidized bed or a shaft furnace is prone to reoxidation in air even when discharged after cooling in a protective atmosphere. The pyrophoric nature of reduced iron is due to its large pore surface area; the lower the reduction temperature, and hence the higher the pore surface area, the greater the extent of reoxidation.

Reoxidation of reduced iron in air is accompanied by a rapid rise in temperature. This is shown in Fig. 8.32 for a completely reduced iron ore pellet (~12 mm diam) at 700 °C, and reoxidized in air at indicated furnace temperatures [52]. The lower the furnace temperature, the greater the extent of reoxidation and greater the accompanying temperature rise that occurs in the 2–3-min reaction time. Even when the sponge iron is hot pressed into briquettes and discharged after cooling, they are susceptible to reoxidation.

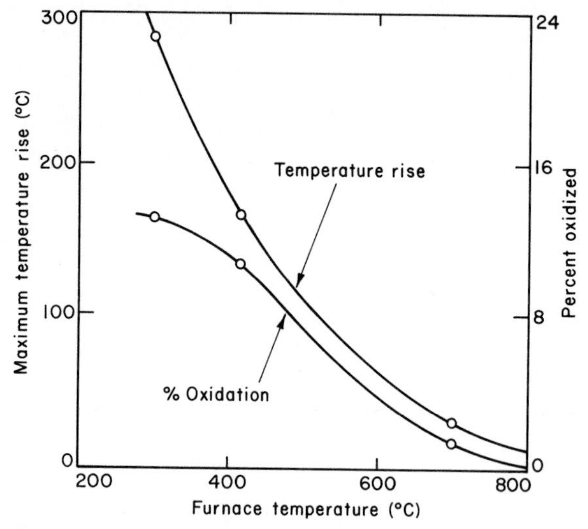

FIG. 8.32 Maximum temperature rise and corresponding percentage of oxidation in air as a function of furnace temperature for pellets previously reduced at 700 °C. After Turkdogan and Vinters [52].

This problem has been resolved in practice by cooling the sponge iron in a slightly oxidizing atmosphere to passivate the product against reoxidation in air. The stability of passivated reduced iron depends on the temperature of (1) reduction, (2) passivation, and (3) reoxidation. The stability of passivated reduced iron is usually evaluated by measuring the extent of reoxidation in 60 min in air at 230 °C. The results of stability tests in Fig. 8.33 [52] are for pellets reduced completely at 700 and 800 °C and passivated by 2% preoxidation in a 1% air–nitrogen mixture at indicated temperatures. The pellet reduced at 800 °C is passivated to reoxidation more effectively than that reduced at 700 °C. Also, the reduced pellet is more resistant to reoxidation when passivated at 200°C.

Since water also reacts with reduced iron, precautions are taken in practice to protect even the passivated reduced iron from the rain. As an added pre-

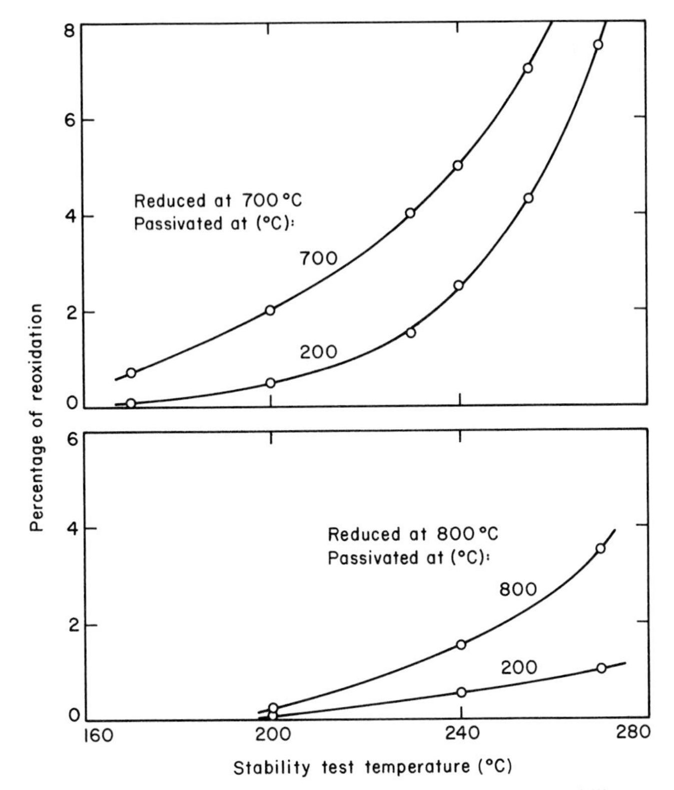

FIG. 8.33 Extent of reoxidation in 60 min in air as a function of the stability test temperature for pellets reduced at 700 and 800 °C and passivated by preoxidation (~2%) in 1% air–nitrogen at indicated temperatures. After Turkdogan and Vinters [52].

caution, the product is not stockpiled on large heaps to avoid excessive temperature rise in the stockpile because of the possibility of some reoxidation.

8.4 Roasting

To facilitate the extraction of metals from their sulfide ores, the sulfide–ore concentrates are roasted by air oxidation to convert the sulfides to metal oxides and/or sulfates. There is also the chloridizing roasting of an admixture of ore and salt, and segregation roasting for special cases of ore beneficiation. Depending on the ore and the subsequent treatment of the calcine by hydro or pyrometallurgical processes, the roasting is done at temperatures of 500–1000 °C, often without melting any of the constituent minerals and reaction products. In some applications, roasting is accompanied by sintering as with the roasting of zinc sulfides mixed with recycled sinter fines, or even melting as with the roasting of lead sulfides mixed with limestone.

Almost all the SO_2 generated during roasting is converted either to sulfuric acid or to elemental sulfur. For the production of elemental sulfur, the off gas should contain at least 10% SO_2. Gases containing as low as 2% SO_2 can be used in the production of sulfuric acid.

The oxidation of sulfides to metal oxides and sulfates is exothermic. The temperature in the roaster is controlled by water spray or by the moisture content of the ore. Antiquated multiple hearth furnaces for roasting are now being replaced by fluidized-bed and flash-roasting technology. In the Outokumpu sulfation process, feed consists mainly of oxide concentrates, while sulfide concentrate is used for controlling temperature and sulfating gas atmosphere only. For process engineering and technology of roasting, reference may be made to textbooks devoted to the subject (e.g., [53, 54]). Here, the discussion is confined to the physical chemistry of reactions occurring during roasting.

8.4.1 Oxidation and Sulfation Roasting

In oxidation and sulfation roasting, we are concerned with reactions of the type

$$
\begin{aligned}
2MS + 3O_2 &= 2MO + 2SO_2, \\
2MO + SO_2 + \tfrac{1}{2}O_2 &= MO \cdot MSO_4, \\
MO \cdot MSO_4 + SO_2 + \tfrac{1}{2}O_2 &= 2MSO_2.
\end{aligned}
\tag{8.30}
$$

The isothermal phase equilibria in M–S–O ternary systems, as functions of the partial pressures of SO_2 and O_2, are represented conveniently by diagrams of the type shown in Fig. 2.8 for the Pb–S–O system [55].

The isothermal bivariant equilibria defining the phase stability regions, so-called the *phase predominance areas*, are shown in Figs. 8.34 and 8.35 for the Fe–S–O and Cu–S–O systems at 600 and 800 °C. Because of negligible solid solutions in the phases considered, the isothermal phase boundaries in the $\log p_{SO_2}$ versus $\log p_{O_2}$ plots are drawn as straight lines, the slopes of which are given by the stoichiometry of the reaction equilibria involving two condensed phases. The points of intersection of the lines define the univariant equilibria, which are readily computed from the free-energy data compiled in Table 1.2.

The phase stability region for Cu_2SO_4 is small and decreases with decreasing temperature; below above 300 °C, Cu_2SO_4 is not stable [56].

Exit gases from industrial roasters contain 1%–5% O_2 and 5%–15% SO_2, the remainder being about 10% H_2O and 75% N_2; the average gas composition in the roaster is represented by the shaded circle in the phase diagram. With increasing temperature, the phase stability regions move to higher pressures of SO_2 and O_2. If the sulfate is the desired end product, the roasting is done at lower temperatures; in high temperature *dead roasting*, the metal oxide is the end product.

The sulfide ores are usually mixtures of several metal sulfides, some of which exist as sulfide solutions, or form low melting eutectic mixtures. Also, at high temperature roasting under oxidizing conditions, ferrites are formed, e.g., $CuFe_2O_4$, $ZnFe_2O_4$. Despite these complexities, however, the phase equilib-

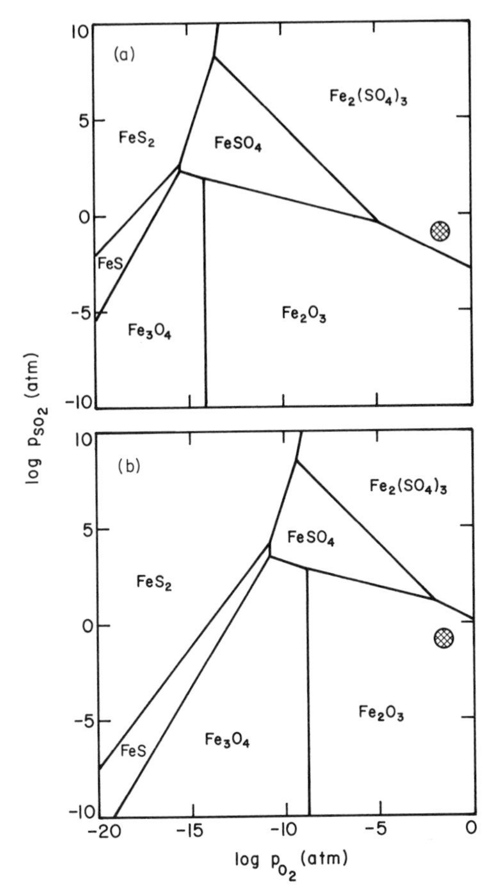

FIG. 8.34 Phase equilibria in the oxide–sulfide–sulfate region of the Fe–S–O system at (a) 600 and (b) 800 °C.

rium relations for ternary systems, such as those in Figs. 2.8, 8.34, and 8.35, are helpful in predicting the general trends in reactions during roasting of mixed sulfide ore concentrates.

In the roasting of chalcopyrite $CuFeS_2$, for example, Fe_3O_4 and Cu_2S are the primary products in the early stages of roasting. Subsequently, as Fe_3O_4 is oxidized to Fe_2O_3, the Cu_2S is oxidized in stages to Cu_2O and CuO, leading to the formation of oxysulfate $CuO \cdot CuSO_4$, and to $CuSO_4$ if the temperature is not too high. In roasting for reverberatory smelting, the calcine should not be overoxidized, otherwise an oxidizing slag forming during smelting will bring about high copper losses to the slag.

As would be expected, concentric layers of reaction products are formed during oxidation of sulfides with countercurrent diffusion of O_2 and SO_2 through the porous product layers. However, compared to the reduction of metal oxides or sulfides, the reaction products in roasting are much denser and

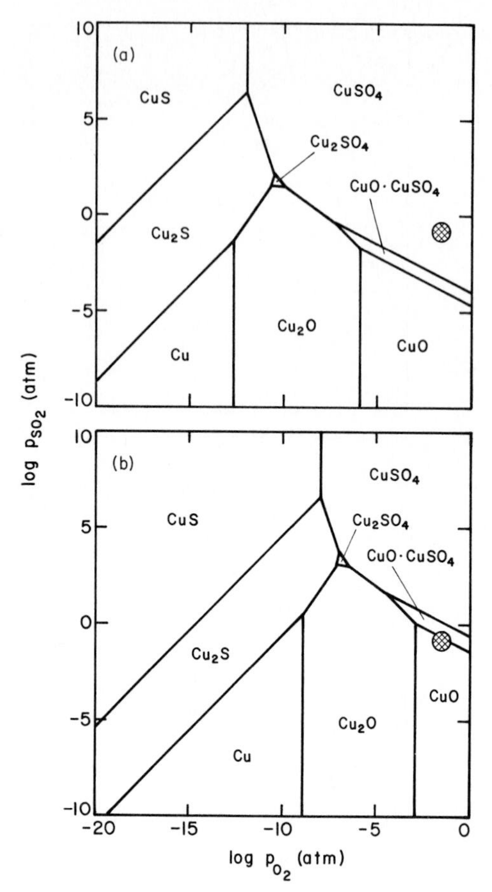

FIG. 8.35 Phase equilibria in the Cu–S–O system at (a) 600 and (b) 800 °C.

the rate of gas diffusion is correspondingly slower. When oxysulfate and sulfate product layers are formed, gas diffusion is further hindered and the oxidation of the residual sulfide becomes more sluggish.

In the oxidation of chalcopyrite, there is segregation of copper to the center and of iron to the outer region of the particle [57, 58], known as *kernel roasting.* This phenomenon is attributed to a combined thermodynamic and kinetic effect. As noted earlier, wustite and magnetite are formed first in the early stages of oxidation of chalcopyrite. Since copper oxides and sulfides have negligible solubilities in iron oxides, there is preferential diffusion of iron in the oxide phase and copper in the sulfide phase away from the chalcopyrite–wustite interface. This countercurrent diffusion results in the accumulation of copper at the center of the particle, forming chalcocite Cu_2S or digenite Cu_9S_5, while iron oxide, free of copper, accumulates on the periphery of the particle.

Migration of cations away from the sulfide–oxide interface also results in sulfur enrichment and buildup of sulfur pressure which ultimately ruptures the oxide scale and allows oxygen to diffuse into the particle. The growth of oxide films by cationic migration and their periodic disruption by built-up sulfur pressure at the sulfide–oxide interface maintains some open porosity in the oxide layer. The rate of roasting is therefore not unduly impaired by the growth of the oxide layer.

The selectivity in sulfation roasting is controlled by adjustment of the temperature and the partial pressure of oxygen and sulfur dioxide in the roast bed, as indicated by the oxide–sulfate equilibrium relations. The equilibrium partial pressures of O_2 and SO_2 are plotted in Fig. 8.36 for coexisting metal oxides and sulfates in their standard states at 800 °C. In gas compositions above the equilibrium line of a given system, the oxide will be converted to an oxysulfate or a sulfate. In sulfation roasting of oxide or sulfide ores that contain many of these elements, however, the preferential sulfation is not always possible, as was noted in the sulfation of Duluth Gabbro sulfide concentrates that contained iron, nickel, and copper sulfides [59].

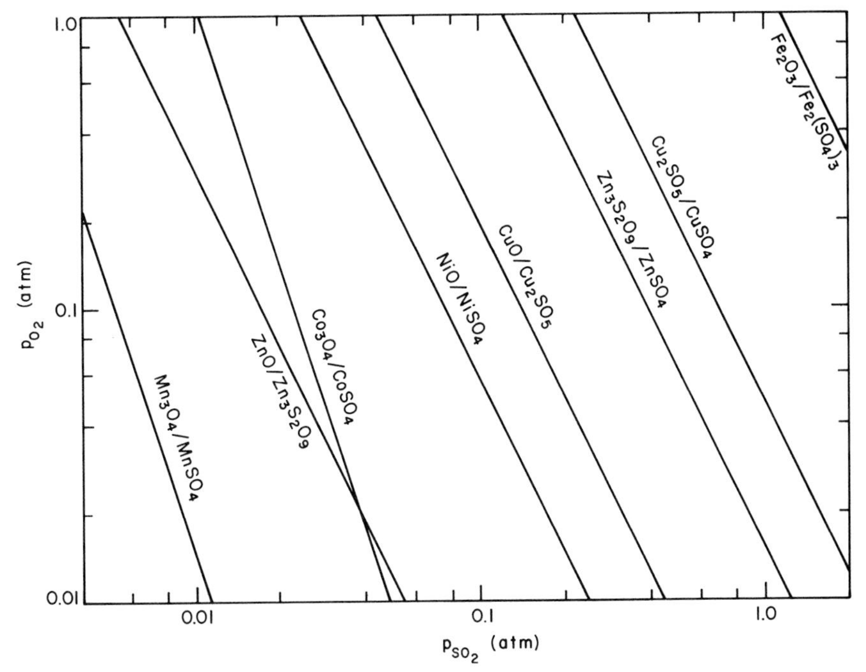

FIG. 8.36 Equilibrium partial pressures of O_2 and SO_2 for coexisting metal oxides and sulfates at 800 °C.

More striking are the results of sulfation experiments with ocean manganese nodules containing about 22% Mn, 5% Fe, 1% Ni, 0.8% Cu, and 0.15% Co as oxides [60]. For example, when the nodules were roasted at 800 °C in air that contained about 1% SO_2, manganese sulfate was formed, and additionally, some of the copper, nickel, and cobalt was converted to their respective sulfates, as found in the water–leach solution. Yet, according to the equilibrium diagram in Fig. 8.36, under the stated roasting conditions, there should have been preferential sulfation of manganese only. This nonselectivity of sulfation is due to the formation of extensive solid solution of sulfates of nickel and copper in manganese sulfate [60]. These metal sulfates, however, do not form solid solutions with calcium sulfate.

In search of alternative methods of roasting without SO_2 evolution, several experimental studies have been made of sulfation and reduction reactions. For example, when a compacted mixture of lime and powdered chalcopyrite is roasted in air at temperatures below 600 °C, about 98% of the sulfur is converted to $CaSO_4$ and copper sulfates [61]. The reaction is highly exothermic, e.g., about 474 kcal mole^{-1} $CuFeS_2$. The feasibility of such a method of roasting may be limited by the difficulty of rapid heat extraction from the reactor. In reduction roasting of a mixture of lime and chalcopyrite in hydrogen [62], metallic copper and iron are formed and most of the sulfur becomes fixed as CaS, water vapor being the only gaseous reaction product. The pentlandite, $(Fe, Ni)_9S_8$, may be reduced also by hydrogen in a similar manner [63]. The reaction is only slightly endothermic, e.g., 10–20 kcal mole^{-1} metal sulfide. However, because of the disposal problems, the by-product CaS has to be treated to produce H_2S and regenerate CaO.

8.4.2 Oxidation of CaS and MnS

Processes are being developed to remove H_2S from hot reducing gases generated by the gasification of coal. The removal of H_2S from hot reducing gases must be accomplished with little or no change in the composition and temperature of the gas. Also, the method of desulfurization should be such that the absorbent can be regenerated and recycled without much loss of reactivity, and the exhaust gases from the regeneration step be rich in H_2S and SO_2 for economical recovery of the sulfur.

Because of the differences in temperature, gas composition, and the reactions occurring in the sulfidation and regeneration steps, there are severe limitations to the choice of H_2S absorbent.

Calcined dolomite, CaO(MgO), is often advocated as a suitable material for H_2S removal from hot reducing gases. Although MgO does not participate in the reaction, its presence as a finely dispersed phase in calcined dolomite maintains open porosity in the material during sulfidation and regeneration. The calcined dolomite may be regenerated by oxidizing the sulfided dolomite, CaS(MgO), in a mixture of CO_2 and H_2O under pressure, followed by calcination:

$$CaS(MgO) + CO_2 + H_2O = CaCO_3(MgO) + H_2S, \tag{8.31a}$$

$$CaCO_3(MgO) = CaO(MgO) + CO_2. \tag{8.31b}$$

As indicated by the equilibrium constant (in atmospheres),

$$\log(p_{H_2S}/p_{CO_2}p_{H_2O}) = (5003/T) - 6.977, \tag{8.32}$$

the higher the total pressure, the lower the temperature and the higher the equilibrium concentration of H_2S in the off gas. Although the thermodynamics of the reaction is favorable, the experimental studies of Turkdogan and Olsson [64] have shown that the rate of reaction (8.31a) is inherently sluggish because of the formation of $CaCO_3$, which covers the unreacted sulfide particles. In addition, the efficiency of regeneration and H_2S absorbing capacity of the material rapidly deteriorates after a few recycles.

The reaction kinetics are more favorable when regeneration is done by oxidation roasting of the sulfided dolomite in CO_2–H_2O mixtures containing 1%–3% O_2, for which the overall reaction is

$$CaS(MgO) + \tfrac{3}{2}O_2 = CaO(MgO) + SO_2. \tag{8.33}$$

A packed-bed reactor is well suited for such a cyclic process. When roasted at temperatures of 1000–1100 °C with an ingoing CO_2–H_2O mixture containing less than 3% O_2, $CaSO_4$ does not form, and the calcined dolomite is regenerated satisfactorily for from seven to ten useful recycles, with an amount of SO_2 in the exhaust gas sufficient for recovery as sulfuric acid.

Further studies [65] have shown that the manganese oxide is a much better material for H_2S removal from hot reducing gases in a regenerative cyclic process in which the overall reactions are

H_2S removal:

$$MnO + H_2S = MnS + H_2O; \tag{8.34}$$

regeneration:

$$MnS + \tfrac{5}{3}O_2 = \tfrac{1}{3}Mn_3O_4 + SO_2, \tag{8.35a}$$

$$\tfrac{1}{3}Mn_3O_4 + \tfrac{4}{3}CO = MnO + \tfrac{4}{3}CO_2. \tag{8.35b}$$

The pellets made of a mixture of 75% MnO and 25% Al_2O_3 and sintered in air at 1200°C were found to perform well on repeated cycles of H_2S removal and regeneration. The breakthrough curve for sulfidation on the 18th cycle at 1000°C with a 3% H_2S–H_2 mixture is shown in Fig. 8.37 for a small experimental packed bed, using 148 pieces of 10-mm-diameter pellets. The concentration of H_2S in the exhaust gas at the steady state part of the curve is 159 ppm; as shown below, this is close to the value of the MnS–MnO equilibrium.

For coexisting phases MnO and MnS, the equilibrium p_{H_2O}/p_{H_2S} ratio is 0.202 at 1000 °C; since the amount of water vapor in the exhaust gas is equal to the quantity of H_2S reacted, for the ingoing 3% H_2S–H_2 gas mixture the computed equilibrium H_2S content of the exhaust gas is 148 ppm. The area under the curve in Fig. 8.37 multiplied by the gas flow rate gives the residual

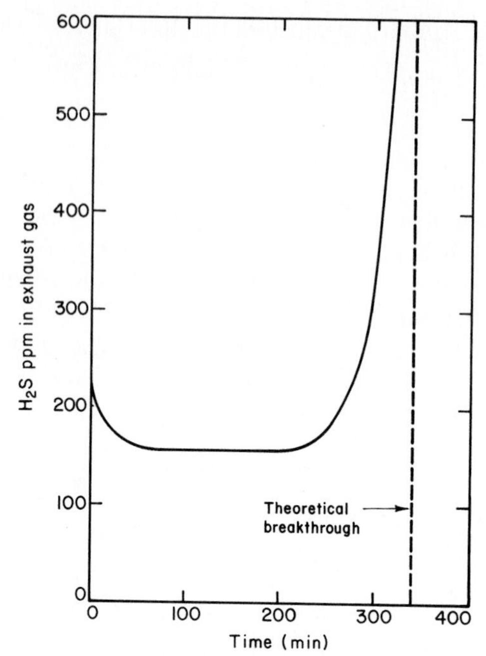

FIG. 8.37 Breakthrough curve for sulfidation of 75% MnO–25% Al_2O_3 pellets in a packed bed at 1000 °C with a mixture of 3% H_2S–H_2 flowing at 3 l min^{-1}. After Turkdogan and Olsson [65].

quantity of H_2S in the exhaust gas. This, together with the amount of MnO in the packed bed and the gas flow rate, gives the indicated theoretical time for breakthrough. For the breakthrough, defined relative to 500 ppm H_2S in the exhaust gas, this experimental packed bed would be 90% saturated with MnS, at which time the regeneration cycle may be started.

The breakthrough curve is shown in Fig. 8.38 for regeneration in an experimental packed bed with air flowing at 1 l min^{-1}. A single curve fits the data for bed temperatures of 800–1000 °C. During the early and middle stages of regeneration, elemental sulfur was seen to deposit in the exhaust lines, indicating that the oxygen in the reacting air was completely consumed. Also, the composition of the exhaust gas approached the theoretical maximum SO_2 concentration of 13.6%, which corresponds to the complete consumption of oxygen in the air.

8.4.3 Chlorination and Segregation Roasting

The halogenation of minerals will not be discussed in this chapter; however, in relation to roasting a few examples should be given of chlorination roasting.

The oxides and sulfides of residual metals in pyrite cinder are claimed as water-soluble salts by air roasting, at temperatures of 500–600 °C, the pyrite cinder mixed with about 10% NaCl. Although the reactions occurring are

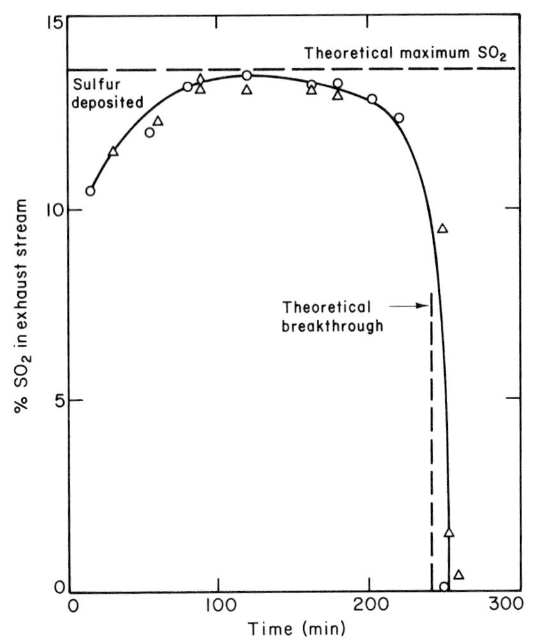

FIG. 8.38 Breakthrough curve for oxidation of sulfided MnO pellets (75/25) in a packed bed with air flowing at $1\ 1\,\mathrm{min}^{-1}$. △—cycle 7, 900 °C; ♢—cycle 9, 1000 °C; (○) cycle 17, 1000 °C. After Turkdogan and Olsson [65].

complex, resulting in off gases containing HCl, Cl_2, and SO_2, the primary reaction is of the type

$$MS + 2NaCl + 2O_2 = MCl_2 + Na_2SO_4. \tag{8.36}$$

If the volatile metal chlorides are the desired end products, calcium chloride or chlorine gas is used in air roasting at temperatures of 1000–1250 °C.

$$MS + CaCl_2 + \tfrac{3}{2}O_2 = MCl_2\,(g) + CaO + SO_2, \tag{8.37}$$

$$MS + Cl_2 + O_2 = MCl_2\,(g) + SO_2. \tag{8.38}$$

In some low-grade ores, such as laterites, the metal-bearing minerals are so finely dispersed that a pyrometallurgical treatment is necessary prior to the concentration of the metal-bearing minerals in the ore. Basically, there are two processes that have potential commercially: selective reduction and segregation roasting.

The segregation roasting involves heating of the ore mixed with salt and carbon so that the metal chlorides formed decompose on the surface of carbon particles and deposit a metallic layer. Subsequently, the metal-coated carbon particles are separated by flotation or magnetic techniques. As suggested initially by Rey [66, 67] and subsequently substantiated experimentally by many investigators, a series of consecutive reactions occur during roasting. These may be grouped into three types: hydrolysis of salt, chlorination, and decomposition of metal chlorides on carbon particles.

The hydrolysis of salt, such as NaCl or $CaCl_2$, by water vapor at roasting temperatures, is accompanied by other reactions with silica or clay matter in the ore:

$$2NaCl + H_2O + SiO_2 = Na_2SiO_3 + 2HCl, \tag{8.39a}$$

$$CaCl_2 + H_2O + SiO_2 = CaSiO_3 + 2HCl. \tag{8.39b}$$

The hydrogen chloride thus generated chlorinates some of the metal oxides to volatile metal chlorides:

$$Cu_2O + 2HCl = \tfrac{2}{3}Cu_3Cl_3\,(g) + H_2O, \tag{8.40a}$$

$$NiO + 2HCl = NiCl_2\,(g) + H_2O, \tag{8.40b}$$

$$FeO + 2HCl = FeCl_2\,(g) + H_2O. \tag{8.40c}$$

The partial pressures of metal chlorides in equilibrium with their respective oxides and the corresponding equilibrium ratios p_{HCl}^2/p_{H_2O} for two salt–H_2O–SiO_2 systems are shown in Fig. 8.39 for temperatures of 600–1000 °C. For a given temperature and an ore–salt mixture, the partial pressure of Cu_3Cl_3 is much higher than that of iron and nickel. For this reason, copper segregation is readily achieved with NaCl at temperatures of 700–750 °C as in the TORCO process [68]. For nickel segregation in, for example, garnierite ores (moderately

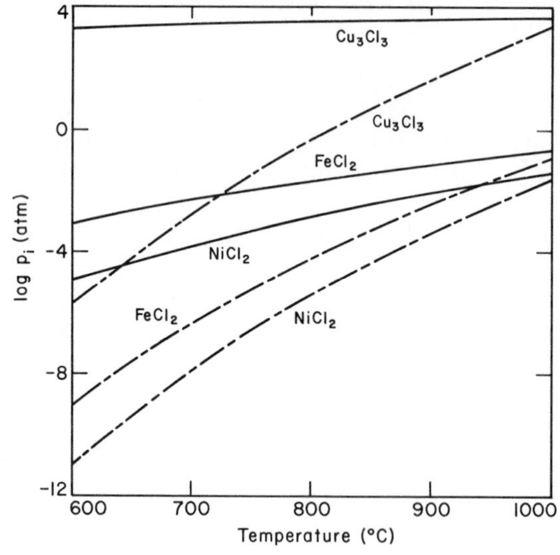

FIG. 8.39 Partial pressures of metal chlorides in equilibrium with their oxides and ratios p_{HCl}^2/p_{H_2O} for salt–silica–water vapor equilibrium. (——) for $CaCl_2$–SiO_2 equilibrium; (— — —) for NaCl–SiO_2 equilibrium.

weathered serpentines) or in nickeliferous limonitic laterites [69], the roasting temperature has to be much higher, 900–1000 °C, and $CaCl_2$ should be used.

The metal deposition on the surfaces of carbon particles is brought about presumably by the reduction of the metal chloride by hydrogen on the surface of carbon where hydrogen is regenerated by reaction with water vapor:

$$MCl_2(g) + H_2 = M + 2HCl, \tag{8.41a}$$

$$H_2O + C = CO + H_2. \tag{8.41b}$$

Since the rates of hydrolysis of salts [70] and the formation and decomposition of metal chlorides [71] are fast, the rate controlling reaction in the segregation process is believed to be the regeneration of hydrogen by reduction of water vapor with carbon on which the metal is deposited. In experiments with limonitic laterite ores, it has been found [72] that, for a given temperature and reaction time, the percentage of conversion to metallic nickel is higher for more reactive carbonaceous materials. Graphite and coke respond poorly, while a high degree of segregation is obtained with charcoal, lignite, or wood. Marcuson and Kellogg [73] also found that better results are obtained on segregation roasting of copper oxide concentrates with carbons of high reactivity.

As would be anticipated from the equilibrium relation in Fig. 8.39, in the treatment of copper ores there is little or no iron in the segregated metallic copper. With nickel-bearing ores, however, iron and nickel segregate together, forming an iron–nickel alloy deposit around the carbonaceous particles. The greater the extent of segregation, the lower is the Ni/Fe ratio in the metallic deposit [72].

A residue formed in some zinc-leaching processes contains lead sulfate and lead jarosite which cannot be charged directly to a lead smelter. Bear *et al.* [74, 75] have found that such lead sulfate-containing materials may be upgraded by segregation roasting of this material mixed with salt and carbon. The sequence of overall reactions is

$$PbSO_4 + 2HCl = PbCl_2(g) + H_2O + SO_3, \tag{8.42a}$$

$$H_2O + SO_3 + 2C = H_2S + 2CO_2, \tag{8.42b}$$

$$PbCl_2(g) + H_2S = PbS + 2HCl. \tag{8.42c}$$

As in the case of copper and nickel, the segregation of lead sulfide is more efficient with carbon of high reactivity. Also, similar to the case of reduction calcination of gypsum discussed in Section 8.2.2, if insufficient or less reactive carbon is used in the segregation process, some of the sulfur may be lost from the system as SO_2, and lead sulfate may be decomposed partly to lead oxide instead of segregating as lead sulfide. In fact, Bear and Merritt [74] have found that a high degree of segregation of lead sulfide is obtained only when the amount of carbon used is in excess of that for the stoichiometry of the overall reaction

$$PbSO_4 + 2C = PbS + 2CO_2. \tag{8.43}$$

8.5 Sintering

The brief discussion of sintering presented here is limited to the coarsening of the pore structure during reaction of gases with porous materials. Sintering by induration of compacted powder mixtures has been studied extensively in relation to many process applications in colloids, ceramics, and metals industry. Most of the theoretical and experimental studies done on sintering are mentioned in a review paper by Kuczynski [76].

In the early stages of sintering, welding occurs at points of contact of particles in the compacted powder mixture. The neck thus formed at the point of contact has a negative radius of curvature. Because the surface of the neck is under tension, the vapor pressure of the matter in the cavity of the neck is less than that over the surfaces of the adjacent particles. This state of nonequilibrium provides the driving force for the transport of matter to the neck.

Owing to the surface energy σ, the existence of tensional stress in the cavity of the neck results in local increase in the concentration of vacant lattice sites. Since in all cases we have the inequality $\sigma\Omega \ll kT$, where Ω is the atomic volume, the Thomson equation (3.75) may be simplified to give the following thermodynamic relation for local equilibrium excess vacancy concentration:

$$\Delta c/c_{\circ} = 2\sigma\Omega/kTa, \tag{8.44}$$

where Δc is the excess vacancy concentration, $c_{\circ}$ the equilibrium vacancy concentration under a flat surface, i.e., in the absence of stress, and a the radius of neck curvature. Under the concentration gradient $d\Delta c/dz$, the vacancies migrate from the neck to the free surface of the particle or to the grain boundary of the weld plane, accompanied by equivalent countercurrent migration of the atoms to the neck. With the progress of sintering, the radius of curvature of the neck decreases, and hence the total energy of the system decreases, ultimately leading to the coalescence of the neighboring particles.

There are primarily five mechanisms of mass transfer in the sintering process: vaporization, plastic flow, volume diffusion, grain boundary diffusion, and surface diffusion. Although each of these mechanisms contribute to varying degrees to the neck growth, only three of them may be in operation when there is substantial shrinkage accompanying sintering: plastic flow, volume diffusion, and grain boundary diffusion. However, as demonstrated by Brett and Seigle [77], the role of plastic flow is small, and the mass transfer by volume diffusion is by far the most important mechanism.

In a compacted powder mixture or a consolidated porous material, the pores are considered as vacancy sources and grain boundaries between adjacent particles as sources of atoms which fill the pores during the sintering process. Since each vacancy can be considered to contribute one atomic volume to the volume of the pore, as shown, for example, by van Bueren and Hornstra [78], the volume of spherical pores would decrease during sintering at the rate

$$dV_{p}/dt = -(8\pi\sigma\Omega/kT)D^{*}, \tag{8.45}$$

where D^* is the self-diffusivity of the diffusing atom. Substituting $V = \frac{4}{3}\pi r^3$, where r is the pore radius, and integrating gives for the pore radius as a function of time

$$r^3 - r_0^3 = -(6\sigma\Omega D^*/kT)t, \tag{8.46}$$

where r_0 is the initial pore radius at $t = 0$. An identical equation was derived by Kuczynski [79] for tubular pores, for which the factor 6 changes to 3. These theoretical equations have been verified experimentally by sintering twisted metal wires or compacted powder mixtures [76].

On the other hand, the pore structure that develops during reduction, calcination, or roasting becomes coarser with increasing time and temperature of reaction, a behavior which may appear to be contrary to that expected from the theoretical relation given by Eq. (8.46). This apparent anomaly is attributed to the nonuniformity of the inital pore structure that develops during the reduction or calcination reaction, followed by localized sintering that results in the disappearance of smaller pores accompanied by coarsening of larger pores.

For simplicity, let us consider a biomodal pore structure with a number n_s of small pores of volume V_s and n_l large pores of volume V_l per unit mass of the material, giving for the total pore volume V_p per unit mass

$$V_p = n_s V_s + n_l V_l. \tag{8.47}$$

Similarly, for the corresponding pore surface areas per unit mass

$$S_p = n_s S_s + n_l S_l. \tag{8.48}$$

In this bimodal packing arrangement, the sintering and shrinkage will occur mostly in the vicinity of small pores where there are a greater number of particles in contact with one another. Because of this localized sintering and shrinkage, there will be some increase in V_l, resulting in a coarser pore structure.

The coarsening of the pore structure with the disappearance of smaller pores is analogous to the Ostwald ripening of precipitated particles in a solution. That is, smaller pores disappear by diffusion of vacancies to the larger pores, while smaller particles disappear by diffusion of solutes to the larger particles. This analogy is also seen from the identity of Eq. (8.46) of opposite sign with Eq. (7.59) derived by Wagner [80] for particle growth by the Ostwald ripening effect. Equation (7.59), rewritten for molecular units,

$$r^3 - r_0{}^3 = (8\sigma\Omega^2 Dc_\circ/9kT)t \tag{8.49}$$

is readily transformed to the form of Eq. (8.46) by redefining Ω, D, and $c_\circ$ for the vacancy diffusion mechanism. That is, Ω is now the atomic or vacancy volume, $c_\circ$ is the vacancy concentration in the absence of stress; and for self-diffusion occurring by a vacancy mechanism, as discussed in Section 3.5.2, the self-diffusivity D^* is equal to $Dc_\circ\Omega$. Noting also that vacancy diffusion is accompanied by equivalent countercurrent atom diffusion to the smaller pores, we may transform Eq. (8.49) for particle growth to that for pore shrinkage,

$$r^3 - r_0{}^3 = -(8\sigma\Omega D^*/9kT)t, \tag{8.50}$$

which differs from Eq. (8.46) only by the numerical constant, a factor of about 7 smaller.

When V_s approaches zero in a finite sintering time, as indicated by Eq. (8.45), the structure will acquire an essentially stable and a uniform mazelike structure. For an ensemble of micropores, dV_s/dt is much less than dS_s/dt, and hence $dV_p/dt \ll dS_p/dt$, as borne out by the data in Figs. 8.5 and 8.6 for burnt lime. In accord with the concept of preferential sintering in the vicinity of smaller pores, the data in Fig. 8.5 for burnt lime substantiate the view that the lower the calcination temperature, the larger the value of n_sS_s; hence the smaller the grain size, and the longer the time of sintering to achieve a "stable" pore structure.

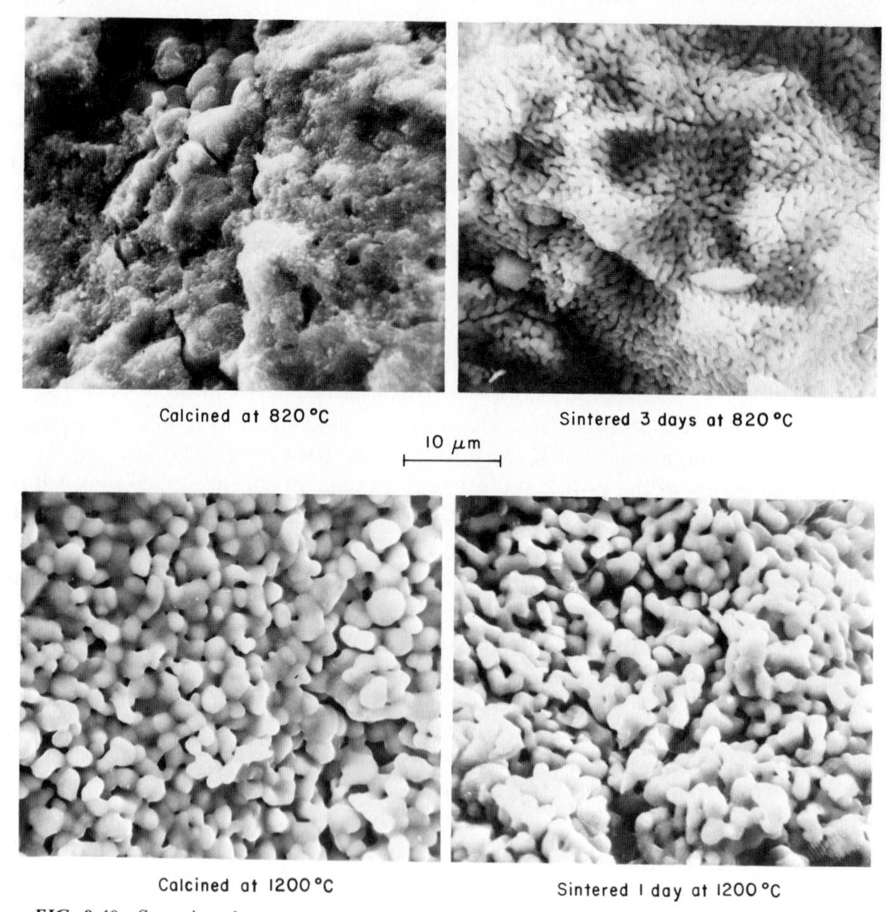

FIG. 8.40 Scanning electron micrographs of fractured surfaces of burnt lime showing the effect of calcination temperature and time of heating on the pore structure. After Turkdogan *et al.* [12].

Another implication is that the lower the temperature at which the porous reaction product is formed, the finer the pore structure even after prolonged sintering. For example, the photomicrographs in Fig. 8.40 of fracture surfaces of burnt lime as viewed under the scanning electron microscope show clearly that even after three days of heat treatment, the pore structure for calcination at 820 °C is much finer than that obtained with calcination at 1200 °C, followed by heat treatment of much shorter duration. Many other similar examples may be found in the technical literature on other porous reaction products. Therefore, the reactivity of a porous material obtained by reduction or calcination is determined primarily by the initial temperature of the reaction; subsequent sintering would have a lesser effect on the reactivity.

As indicated by Eqs. (8.46) or (8.50), the time of sintering is inversely proportional to the self-diffusivity. The diffusivity of Fe^{2+} cations in wustite is about six orders of magnitude greater than that of Ca^{2+} cations in CaO; therefore, the stabilization of the pore structure by localized sintering would take a much longer time for burnt lime than for wustite. This is in general accord with experimental observations. For example, Pepper *et al.* [81] found that the pore surface area of wustite decreases from an initial value of 2.2 $m^2\,g^{-1}$ to a stable level of 0.2 $m^2\,g^{-1}$ in about 30 min at 825 °C; we see from Fig. 8.5 that a sintering time of 200–240 hr or more is needed to stabilize the pore structure of burnt lime at 820 °C. If the pore structure is filled with a liquid reaction product, the pore structure would coarsen by the Ostwald ripening effect at a much faster rate because of faster diffusion in the liquid medium; this is well substantiated by the photomicrographs in Fig. 8.31 for iron oxide reduced at 1300 °C. The rate of local sintering, and hence the coarsening of the pore structure, may also be enhanced by introducing a suitable volatile material into the system.

References

1. E. H. Baker, *J. Chem. Soc.* p. 464 (1962).
2. A. W. D. Hills, *Inst. Min. Metall., Trans.* **76**, 241 (1967).
3. C. C. Furnas, *Ind. Eng. Chem.* **23**, 534 (1931).
4. V. J. Azbe, *Rock Prod.* **47**, No. 9, 68, 70, 98, 100 (1944).
5. V. J. Azbe, *Rock Prod.* **56**, No. 2, 100; No. 12, 111 (1953).
6. J. Y. McDonald, *Trans. Faraday Soc.* **47**, 860 (1951).
7. E. P. Hyatt, I. B. Cutler, and M. E. Wadsworth, *J. Am. Ceram. Soc.* **41**, 70 (1958).
8. C. N. Satterfield and F. Feakes, *AIChE J.* **5**, 113 (1959).
9. T. R. Ingraham and P. Marier, *Can. J. Chem. Eng.* **41**, 170 (1963).
10. W. O. Philbrook and K. Nateson, "Azbe Award No. 6, 1966." Natl. Lime Assoc., Washington, D.C., 1966.
11. A. W. D. Hills, *Chem. Eng. Sci.* **23**, 297 (1968).
12. E. T. Turkdogan, R. G. Olsson, H. A. Wriedt, and L. S. Darken, *Trans. Soc. Min. Eng. AIME* **254**, 9 (1973).
13. W. Woodside and J. H. Messmer, *J. Appl. Phys.* **32**, 1688 (1961).
14. W. D. Kingery, J. Francl, R. L. Coble, and T. Vasilos, *J. Am. Ceram. Soc.* **37**, 107 (1954).
15. H. C. Fischer, *J. Am. Ceram. Soc.* **38**, 245 (1955).

16. E. T. Turkdogan, B. B. Rice, and J. V. Vinters, *Metall. Trans.* **5**, 1527 (1974).
17. E. T. Turkdogan and B. B. Rice, *Trans. Soc. Min. Eng. AIME* **254**, 28 (1973).
18. R. C. Mullins and J. D. Hatfield, *in* "The Reaction Parameters of Lime," ASTM STP 472, p. 117. Am. Soc. Test. Mater., Philadelphia, Pennsylvania, 1970.
19. M. Jon and P. Riboud, *Rev. Metall. (Paris)* **69**, 93 (1972).
20. E. T. Turkdogan and J. V. Vinters, *Inst. Min. Metall., Trans., Sect. C* **85**, 117 (1976).
21. L. von Bogdandy and H.-J. Engell, "The Reduction of Iron Ores." Springer-Verlag, Berlin and New York, 1971.
22. J. O. Edstrom, *J. Iron Steel Inst., London* **175**, 289 (1953).
23. H. Schenck, A. Majidic, and U. Putzier, *Arch. Eisenhuettenwes.* **38**, 669 (1967).
24. T. Fuwa and S. Ban-ya, *Trans. Iron Steel Inst. Jpn.* **9**, 137 (1969).
25. O. Burghardt, H. Kostmann, and B. Grover, *Stahl Eisen* **90**, 661 (1970).
26. W. Wenzel, H. W. Gudenau, and M. Ponthenkandath, *Aufbereit. Tech.* **11**, 154, 492 (1970).
27. L. Granse, *Trans. Iron Steel Inst. Jpn.* **11** 45 (1971).
28. J. T. Moon and R. D. Walker, *Ironmaking Steelmaking* **2**, 30 (1975).
29. E. T. Turkdogan and J. V. Vinters, *Can. Metall. Q.* **12**, 9 (1973).
30. E. E. Hoffman, H. Rausch, and W. Thumm, *Stahl Eisen* **90**, 676 (1970).
31. H. vom End, K. Grebe, and S. Thomalla, *Stahl Eisen* **90**, 667 (1970).
32. H. K. Kohl and H.-J. Engell, *Arch Eisenhuettenwes.* **34**, 411 (1963).
33. E. T. Turkdogan and J. V. Vinters, *Metall. Trans.* **2**, 3175 (1971).
34. E. T. Turkdogan and J. V. Vinters, *Metall. Trans.* **3**, 1561 (1972).
35. E. T. Turkdogan, R. G. Olsson, and J. V. Vinters, *Metall. Trans.* **2**, 3189 (1971).
36. R. P. Smith, private communication, U.S. Steel Res. Lab., Monroeville, Pennsylvania (1960).
37. M. C. Udy and C. H. Lorig, *Trans. AIME* **154** 162 (1943).
38. J. Henderson, *J. Aust. Inst. Met.* **7**, 115 (1962).
39. W. M. McKewan, *Trans. Metall. Soc. AIME* **224**, 387 (1962).
40. W. M. McKewan, *in* "Steelmaking: The Chipman Conference" (J. F. Elliott and T. R. Meadowcroft, eds.), p. 141. MIT Press, Cambridge, Massachusetts, 1965.
41. A. W. D. Hills, *Metall. Trans. B* **9**, 121 (1978).
42. R. H. Spitzer, F. S. Manning, and W. O. Philbrook, *Trans. Metall. Soc. AIME* **236**, 726 (1966).
43. R. H. Tien and E. T. Turkdogan, *Metall. Trans.* **3**, 2039 (1972).
44. J. Szekely, J. W. Evans, and H. Y. Sohn, "Gas–Solid Reactions." Academic Press, New York, 1976.
45. R. J. Fruehan, *Metall. Trans. B* **8**, 279 (1977).
46. G. J. W. Kor, *Metall. Trans. B* **9**, 307 (1978).
47. L. J. Lin and Y. K. Rao, *Inst. Min. Metall., Trans., Sect. C* **84**, 76 (1975).
48. T. S. Yun, *Trans. Am. Soc. Met.* **54**, 129 (1961).
49. H. Y. Sohn and J. Szekely, *Chem. Eng. Sci.* **28**, 1789 (1973).
50. R. H. Tien and E. T. Turkdogan, *Metall. Trans. B* **8**, 305 (1977).
51. E. T. Turkdogan, *Metall. Trans. B* **9**, 163 (1978).
52. E. T. Turkdogan and J. V. Vinters, unpublished work (1976). U.S. Steel Res. Lab., Monroeville, Pennsylvania.
53. C. R. Hayward, "An Outline of Metallurgical Practice," 3rd Ed. Van Nostrand, New York, 1952.
54. J. N. Anderson and P. E. Queneau (eds.), "Pyrometallurgical Processes in Nonferrous Metallurgy." Gordon & Breach, New York, 1967.
55. H. H. Kellogg and S. K. Basu, *Trans. Metall. Soc. AIME* **218**, 70 (1960).
56. M. Nagamori and F. Habashi, *Metall. Trans.* **5**, 523 (1974).
57. T. A. Henderson, *Inst. Min. Metall., Trans.* **67**, 437 (1957–1958); **68**, 193 (1958–1959).
58. P. G. Thornhill and L. M. Pidgeon, *Trans. AIME* **209**, 989 (1957).
59. E. T. Turkdogan and B. B. Rice, U.S. Patent 3,839,013 (1974).
60. E. T. Turkdogan and J. V. Vinters, *Inst. Min. Metall., Trans., Sect. C* **86**, 59 (1977).
61. H. H. Haung and R. W. Bartlett, *Metall. Trans. B* **7**, 369 (1976).

62. F. Habashi and R. Dugdale, *Metall. Trans.* **4**, 1865 (1973).
63. I. D. Shah and P. L. Ruzzi, *Metall. Trans. B* **9**, 247 (1978).
64. E. T. Turkdogan and R. G. Olsson, *Ironmaking Steelmaking* **5**, 168 (1978).
65. E. T. Turkdogan and R. G. Olsson, *Proc. Int. Iron Steel Congr., 3rd*, p. 277. Am. Soc. Met., Metals Park, Ohio (1979).
66. M. Rey, *Rev. Metall. (Paris)* **33**, 295 (1936).
67. M. Rey, *Inst. Min. Metall., Trans., Sect. C* **76**, 101 (1967).
68. E. T. Pinkney and N. Plint, *Inst. Min. Metall., Trans., Sect. C* **76**, 114 (1967).
69. I. Iwasaki, Y. Takahasi, and H. Kahata, *Trans. Soc. Min. Eng. AIME* **238**, 308 (1966).
70. N. W. Hanf and M. J. Sole, *Inst. Min. Metall., Trans., Sect. C* **81**, 97 (1972).
71. M. I. Brittan and R. R. Liebenberg, *Inst. Min. Metall., Trans., Sect. C* **80**, 156 (1971).
72. E. T. Turkdogan and J. V. Vinters, unpublished work, (1972). U.S. Steel Res. Lab., Monroeville, Pennsylvania.
73. S. W. Marcuson and H. H. Kellogg, *Inst. Min. Metall., Trans., Sect. C* **86**, 195 (1977).
74. I. J. Bear and R. R. Merritt, *Inst. Min. Metall., Trans., Sect. C* **84**, 92 (1975).
75. I. J. Bear, R. R. Merritt, and A. G. Turnbull, *Inst. Min. Metall., Trans., Sect. C* **85**, 63 (1976).
76. G. C. Kuczynski, *Adv. Colloid Interface Sci.* **3**, 275 (1972).
77. G. J. Brett and L. Seigle, *Acta Metall.* **14**, 575 (1966).
78. H. G. van Bueren and J. Hornstra, *in* "Reactivity of Solids" (J. H. De Boer, ed.), p. 112. Elsevier, Amsterdam, 1961.
79. G. C. Kuczynski, *Acta Metall.* **4**, 58 (1956).
80. C. Wagner, *Z. Elektrochem.* **65**, 581 (1961).
81. M. W. Pepper, K. Li, and W. O. Philbrook, *Can. Metall. Q.* **15**, 201 (1976).

CHAPTER 9

PHYSICAL CHEMISTRY OF IRONMAKING AND STEELMAKING

9.1 Introduction

Historically, research and development in metallurgy, particularly that part related to the steel industry, has played an important role in advancing the frontiers of high temperature technology. Since pyrometallurgy entails the application of a wide variety of physical sciences and chemical engineering technology, this chapter is devoted to the discussion of the physical chemistry of reactions in some of the iron and steelmaking processes as examples of the application of the fundamentals of physical chemistry in high temperature technology.

The first stimulus to modern research on metallurgical reactions was a symposium on the physical chemistry of steelmaking held by the Faraday Society in London in 1925. Herty in America, Korber, Oelsen, and Schenck in Germany, and their co-workers, made earlier pioneering contributions to the physical chemistry of steelmaking. More rigorous application of the thermodynamic concepts to the study of high temperature systems and reactions of interest to pyrometallurgy did not materialize until the end of the Second World War. Notable contributions of Chipman, Darken, Richardson, and Kubaschewski marked the beginning of a new era of better understanding and better interpretation of the physical chemistry of metallurgical reactions. During the past two decades, we have seen further developments in the application of physical and computer sciences to process metallurgy, or more generally, to chemical metallurgy. It is befitting to state here that the science, technology, and engineering of process metallurgy stimulated by research in academic circles and in industrial research laboratories has played a significant role in the advancement of the technology of iron and steelmaking we have today.

Complete coverage of this subject in one chapter is not possible. Reference should be made to textbooks and review papers on the design and operation of blast furnaces [1, 2] and steelmaking processes [3, 4]. However, for the benefit of those readers who are not familiar with pyrometallurgical processes, we should at least give a brief outline of the general features of iron and steelmaking processes.

The reduction and smelting of iron ore is done mostly in the blast furnace. The burden charged at the top of the furnace consists primarily of iron ore and coke. The reducing gas carbon monoxide and the heat required for the reduction and smelting of the ore are generated at the bottom of the furnace by blowing preheated air into a bed of coke. The slag and metal accumulating as two liquid layers below the tuyere level are tapped at regular intervals.

The hot metal produced in the blast furnace is saturated with carbon and contains 0.5–1.5% Si, 0.5–1.5% Mn, 0.1–1% P, and 0.03–0.06% S as major impurities. The hot metal is charged to a steelmaking furnace together with steel scrap and burnt lime. The oxygen is blown into the metal bath to remove the impurities and decarburize the metal to the desired level, usually in the range 0.02–0.2% C. The steel produced contains as impurities about 0.1–0.5% Mn, 0.01–0.03% P, 0.01–0.02% S, and 0.02–0.08% O. The steel is then deoxidized in the ladle to the desired level in the range 30–150 ppm O. The temperature of the liquid steel at tap is in the range 1580–1620 °C; in stainless steelmaking the melt temperature is about 1700–1750 °C.

There is also the gaseous reduction of iron ore to metallic iron without smelting. This is known as the direct-reduction process. There are several direct-reduction processes: reduction of ore (i) granules in a fluidized bed, (ii) pellets in a shaft furnace, or (iii) pellets in a rotary kiln. The iron thus produced is melted in electric furnaces with steel scrap and flux to make the steel.

Until the 1960s, most of the steel was made in open hearth furnaces where carbon monoxide and later fuel oil was burned to provide the heat for refining of hot metal with hematite ore feed. With the advent of tonnage oxygen, open hearth furnaces have been replaced by oxygen steelmaking converters. However, large quantities of steel are still being made in open hearth furnaces now fitted with oxygen roof lances.

Not counting the ore preparation and coke making, there are three basic processing steps in the production of liquid steel. In ironmaking the metal is overreduced to produce carbon-saturated liquid iron, in steelmaking the metal is overoxidized, and the final stage in steelmaking is deoxidation in the ladle, where the alloying additions are also made to adjust the steel composition to the desired specification. Many attempts have been made in the past, and are still being made today, to produce liquid steel directly from the ore. With the exception of direct reduction of the ore and melting of the sponge iron in the electric furnace for small output of steel tonnage, the blast furnace–oxygen steelmaking still remains the primary route in the production of steel.

The physical chemistry of iron- and steelmaking is presented here in four sections: ironmaking, steelmaking, ladle treatment of steel, and solidification of steel. Whenever possible, an attempt is made to analyze the plant data and practical observations in terms of the thermodynamics and kinetics of gas–slag–metal reactions established from laboratory experiments.

9.2 Reactions in Ironmaking

The blast furnace may be divided into four primary reaction zones: (i) the shaft or stack, which extends about 16–18 m below the stockline, (ii) the bosh, about 3–5 m between the bottom of the shaft and the tuyeres, (iii) the tuyere or combustion zone, and (iv) the hearth, extending about 5–6 m below the tuyere level. Approximate dimensions given are for a modern blast furnace with a hearth diameter of about 12 m and an average production rate of about 7000 tons of hot metal per day. All the practical aspects of blast-furnace reactions cannot be covered adequately in this section; the present discussion is therefore confined to the primary reactions only.

9.2.1 Heat and Mass Transfer in the Stack

The unique feature of the blast-furnace stack is that, in the countercurrent flow of gas and solids, the heat transfer from gas to solids is accompanied by oxygen transfer from solids to gas. The interrelation between countercurrent heat and mass transfer between gases and solids in the stack has long been recognized, and much use has been made of the concept of heat balance conceived by Reichardt [5] in 1927. Reichardt's diagram, shown in its simplest form in Fig. 9.1a for calcined dry burden, describes the heat balance in the blast furnace and is based on the application of the first and second laws of thermodynamics. As a result of heat transfer, the enthalpy of the gas decreases with decreasing temperature of the ascending gas, as represented by the line ERA. The increase in the enthalpy of the solids with increasing temperature during descent in the stack is represented by the line SR. In the lower part of the furnace, the overall heat capacity of the burden increases because of fusion, the onset of the strongly endothermic *Boudouard reaction* ($CO_2 + C = 2CO$), and greater heat losses; hence the slope of line RC is greater than that of SR. The net result is that, as shown schematically in Fig. 9.1b, over some distance in the stack, the temperature difference between the ascending gas and descending burden reaches a minimum known as the *thermal pinch point*.

The foregoing is an oversimplified phenomenological description of Reichardt's diagram, which becomes more involved when constructed on increments of temperature as demonstrated by Ridgion [6], von Bogdandy [7], Schurmann *et al.* [8, 9], and others. Depending on the type of burden and the blast-furnace practice, the temperature level for the thermal reserve zone varies from about 900 to 1050 °C, and the length of this zone varies from about 1 to 4 m. With burdens containing hydrated ore and carbonates, additional thermal pinch points may occur at temperatures below 900 °C where the hydrates and carbonates dissociate. There is also the added complication arising from the exothermic decomposition of CO that may occur in the upper part of the stack.

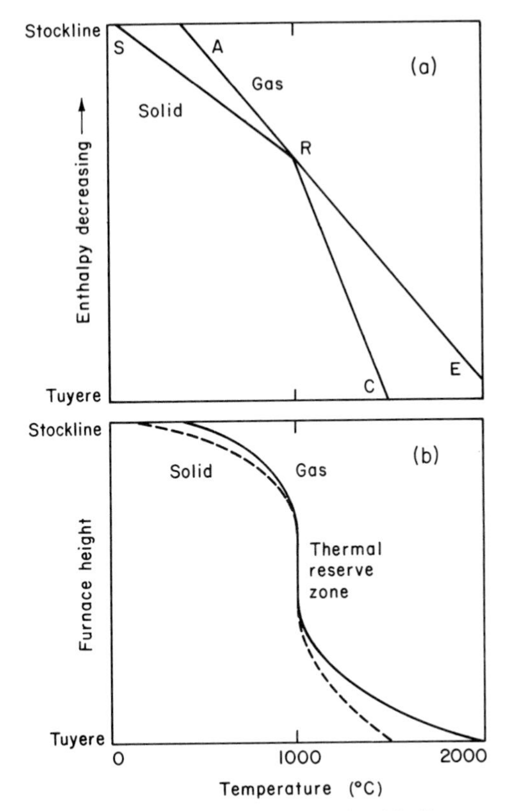

FIG. 9.1 Simplified sketch of Reichardt's diagram.

The change in the gas composition with the vertical position at half radius in the shaft, as measured, for example, by Schurmann *et al.* [9], is shown in Fig. 9.2. A sharp change in the shape of the CO curve at about 1000 °C indicates the onset of the Boudouard reaction occurring together with the CO reduction of wustite. This is also substantiated by an increase in the $CO_2 + CO$ content of the ascending gas until the temperature decreases to about 1000 °C, below which $CO_2 + CO$ concentration remains essentially unchanged at 43% ± 3% by volume up to the stockline.

The relation between the gas composition and temperature in the stack changes a great deal with the blast furnace practice. The lower curve in Fig. 9.3 is for regular (older) blast furnaces operating with acidic sinter or pellet and lump ore, and a high coke rate of about 700 to 800 kg per ton of hot metal [6–9]. The upper curve is for high pressure furnaces operating with basic sinter, oxygen-enriched high temperature air blast, and a low coke rate of about 400 kg per ton of hot metal [10]. In older-type blast furnace operations, the gas composition is reducing to wustite at all levels in the stack. In modern blast

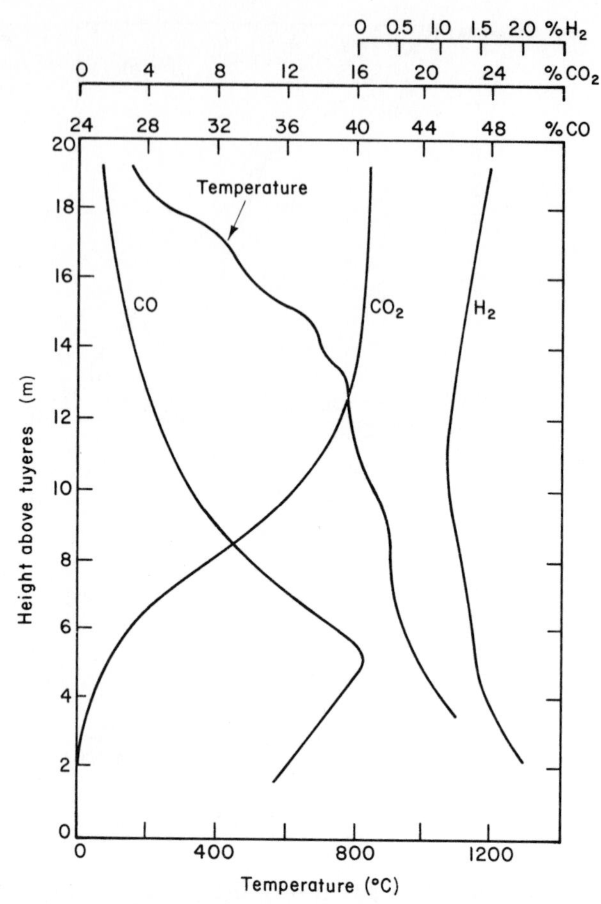

FIG. 9.2 Temperature and gas composition profiles in the blast furnace stack at the half-radius vertical plane. After Schurmann *et al.* [9].

furnace operations, the gas composition is oxidizing to iron at temperatures below the thermal pinch point of about 950 °C.

Rist and co-workers [11, 12] made detailed studies of reduction reactions by simulating the conditions in the stack, and developed a method of graphical representation of heat and mass balances. For an idealized simple case involving only CO as a reducing gas, the number of oxygen atoms per atom of iron involved in the reaction, O/Fe, is plotted in Fig. 9.4 against the number of oxygen atoms per mole of gas, O/C. Since the oxidation of CO to CO_2 occurs without a change in the total number of moles of gas, the oxygen balance is represented by a straight line. The origin of the coordinates is chosen such that the oxygen associated with the iron oxide appears on the positive side of the ordinate, and oxygen in the air blast and in other oxides (MnO, SiO_2) per gram atom of iron on the negative side of the ordinate. Rist called line AE the operating line, of

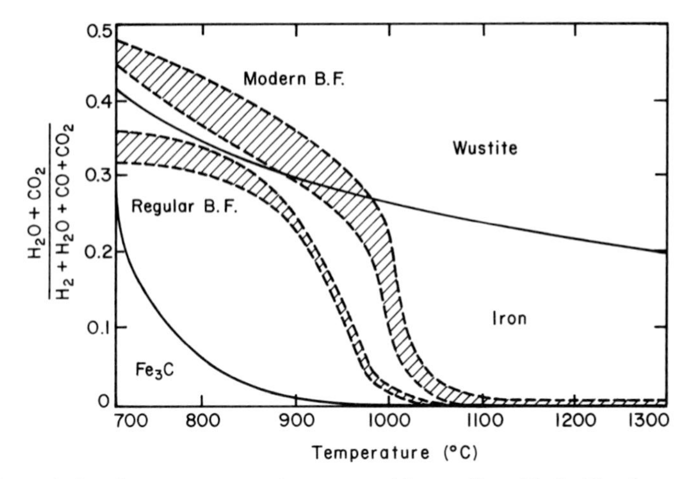

FIG. 9.3 Variations in temperature and gas composition profiles with the blast furnace practice.

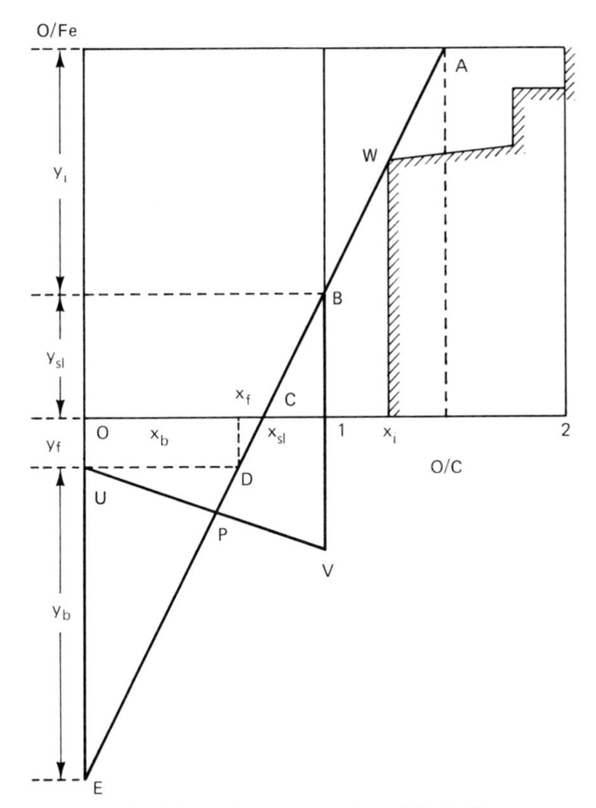

FIG. 9.4 Schematic representation of Rist's diagram.

which AB in the region $1 < x < 2$ represents the indirect reduction of iron oxides by CO,

$$y_i = \alpha x_i, \tag{9.1a}$$

where α is the proportionality factor in mole CO per gram-atom Fe. The line BE within the region $0 < x < 1$ represents the generation of CO by (i) direct reduction of FeO (solution loss) y_{sl}, (ii) reduction of other oxides y_f, and (iii) combustion at the tuyere y_b:

$$(y_{sl} + y_f + y_b) = \alpha(x_{sl} + x_f + x_b). \tag{9.1b}$$

The hatched area outlines the stability region of the iron oxides at the temperature of the thermal reserve zone. The line passing through the point W, representing the iron–wustite equilibrium, is for a maximum oxygen exchange for a given top-gas composition A, giving rise to a so-called *chemical pinch point* in the thermal reserve zone. For wustite to be reduced to iron, the line AB must lie above the pinch point W.

In the idealized blast furnace, where the thermal and chemical pinch points are assumed to coincide, the heat and mass balances are given in the following generalized form:

$$y_b q_b = y_{sl} q_{sl} + Q, \tag{9.1c}$$

where q_b is the coefficient of net heat generated in the combustion zone per mole CO, q_{sl} is the coefficient of endothermic heat of solution loss (for the overall reaction leading to reduction of wustite by carbon), and Q is the total heat requirement per unit mass of iron produced at a steady state, including heating and melting of the charge, direct reduction of nonferrous oxides, indirect reduction of wustite, heat losses, and various endothermic reactions associated with the presence of a given amount of hydrogen in the furnace per unit mass of iron. The linear relationship in Eq. (9.1c) between the two variables y_b and y_{sl} indicates that the operating line AE must pass through a fixed point P in Fig. 9.4. Equation (9.1c) is rearranged to

$$y_b/[y_{sl} + (Q/q_{sl})] = q_{sl}/q_b = \text{UE}/\text{VB} \tag{9.1d}$$

to facilitate the graphical representation shown in Fig. 9.4. Graphically, UE is a measure of the heat input in units of combustion q_b and VB is a measure of the total heat requirements in units of solution loss q_{sl}. It follows from simple geometry that the line AE intersects UV at a point P, which gives PU/PV = q_{sl}/q_b, and has the abscissa

$$x_P = q_{sl}/(q_{sl} + q_b). \tag{9.1e}$$

The abscissa of point P, x_P, is a function of the air blast characteristics, and the ordinate y_P is a function of the hot metal composition and of all the heat requirements below the thermal reserve zone per unit of iron produced. The operating line AE hinges on the pivot point P under the influence of factors affecting the amount of carbon solution loss. For example, if the operation of the blast furnace is not chemically ideal, as is often the case, the operating line AE passing through P will have a slope steeper than that shown in Fig. 9.4 for the ideal case, resulting in lower oxygen/carbon ratio in the top gas, higher values of y_{sl} and y_b, and hence a higher coke consumption in the blast furnace.

Many examples are to be found in published papers [13, 14] on the application of the Reichardt–Rist diagrams to optimize the operating variables in the blast furnace to achieve the lowest coke rate. Reference should be made also to a book by Peacey and Davenport [15] for a description of the operation of the blast furnace in terms of equations for heat and mass balances.

The mechanisms of reduction of iron oxides in hydrogen, carbon monoxide, and mixtures thereof have been discussed in Section 8.3 in relation to direct-reduction processes and the blast furnace. We should emphasize once again the importance of high temperature reducibility of iron oxide sinter or pellet, particularly when operating the blast furnace at a reduced coke rate. For obvious reasons, every effort is made to reduce the coke rate in the blast furnace. With decreasing coke rate, the volume of gas in the furnace decreases, hence more residual iron oxide is carried to the bosh region. As shown in Figs. 8.30 and 8.31, the lime/silica ratio in the sinter or pellet should be more than 0.8 to ensure rapid reduction of the residual iron oxide, before the onset of fusion of the iron and gangue in the burden.

At temperatures above 950 °C, the reduction of wustite with carbon monoxide is accompanied by the oxidation of coke with the intermediate reaction product carbon dioxide; the sum of these two reactions results in the so-called direct-reduction or carbon solution loss, which is an endothermic reaction.

Indirect reduction:

$$\mathrm{FeO + CO = Fe + CO_2;} \tag{9.2a}$$

Boudouard reaction:

$$\mathrm{C + CO_2 = 2CO;} \tag{9.2b}$$

Direct reduction:

$$\mathrm{FeO + C = Fe + CO,} \tag{9.2c}$$

for which

$$\Delta H^\circ = 35.7 \quad \text{kcal.} \tag{9.3}$$

It has long been recognized that in order to operate the blast furnace with a low coke rate the extent of direct reduction should be optimized to maintain a proper thermal balance in the furnace with a high production rate at the lowest possible coke rate. Itaya and co-workers [10] evaluated the extent of indirect and direct reduction from the material balance and the composition of gas samples taken with a probe moving with the burden in the stack. Their results are reproduced in Fig. 9.5. For both reactions, the maximum rate is obtained at temperatures of 950–1050 °C in the vicinity of the thermal reserve zone.

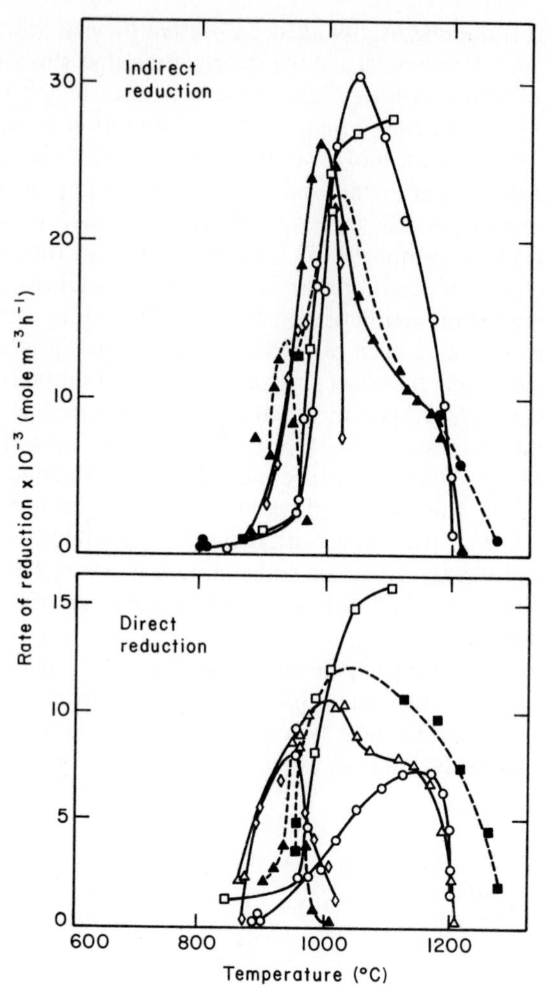

FIG. 9.5 Rates of indirect and direct reduction in the blast furnace stack. After Itaya *et al.* [10].

It should be noted in passing that the temperature, gas composition, and degree of reduction at any given level in the stack vary in a systematic manner with distance from the inner wall to the center. The CO_2 content of the gas at all levels is lower near the center and inner wall and higher at half-radius positions. Also, at a given level, the higher the CO_2 content of the gas, the lower the temperature. However, the temperature versus gas composition relation, as in Fig. 9.3, is essentially independent of the radial position.

9.2.2 Reactions in the Bosh, Tuyere, and Hearth Zones

As the burden descends in the bosh, there is rapid increase in temperature, accompanied by the reduction of more stable oxides such as MnO, SiO_2, and phosphates, and dissolution of carbon, manganese, silicon, sulfur, and phosphorus in the iron. This is followed by melting of the iron and gangue matter forming the slag. The reactions occurring in the bosh and hearth are much influenced by the volatile species forming in the tuyere–combustion zone. We shall consider here collectively only the salient features of reactions occurring in the lower part of the blast furnace.

It has long been believed that as the reduced iron begins to melt in the bosh region the metal droplets pick up silicon from the slag reduced by the carbon. The research of the past ten years brought about new thoughts on the role of silicon and sulfur-bearing volatile species on reactions in the bosh and hearth. For example, Decker and Scimar [16] have shown that silicon monoxide can be produced during the combustion of coke, and subsequently, silicon is transferred to iron droplets by reaction with silicon monoxide in the gas. Recently, Tsuchiya and co-workers [17] also demonstrated that the silicon transfer to the metal is via the formation of silicon monoxide from the coke ash in the high temperature region of the tuyere zone. They also pointed out that as the metal droplets pass through the slag layer, some of the silicon picked up earlier is oxidized by iron oxide and manganese oxide in the slag. This argument was substantiated by the silicon-concentration profile in the bosh determined from the analysis of samples taken from the quenched experimental blast furnace. The silicon content of the metal droplets reaches a maximum at the tuyere level and decreases in the slag layer.

In an earlier investigation made by Bosley *et al.* [18] on the quenched experimental blast furnace, and also in a subsequent study by Okabe *et al.* [19], it was found that the sulfur content of partially or completely reduced iron and of slag increases from the top of the bosh to the tuyere level. They also noted that while the sulfur content of the slag increases with the descent of the burden in the bosh, the sulfur content of the metal droplets reaches a maximum and then decreases as the metal droplets pass through the slag layer.

The coke ash consists of aluminum, alkali, and alkaline earth silicates containing some sulfur; in fact, coke is the primary source of sulfur input in the blast furnace. As the carbon in the coke is oxidized to carbon monoxide in the combustion zone, the following reactions will also occur:

SiO_2 (in coke ash) + C = SiO + CO,

$$\log[p_{SiO}p_{CO}/a_{SiO_2}] = -(34{,}020/T) + 16.759; \tag{9.4}$$

CaS (in coke ash) + SiO = SiS + CaO,

$$\log[p_{SiS}a_{CaO}/p_{SiO}a_{CaS}] = -(2378/T) - 0.326; \tag{9.5}$$

CaS (in coke ash) + CO = CS + CaO,

$$\log[p_{CS}a_{CaO}/p_{CO}a_{CaS}] = -(9449/T) - 0.258; \tag{9.6}$$

where the a's are the activities of CaO and CaS in the coke ash, and the p's are the equilibrium partial pressures of the volatile species in atmospheres; the equilibrium constants are derived from the free-energy data in Table 1.2.

For an activity of 0.5 for silica in the ash, the equilibrium partial pressure of SiO at the temperatures of the combustion zone, ~2000 °C, is much greater than that obtained by complete decomposition of silica in the coke ash. For coke containing, for example, 5% SiO_2, 0.4% S, and 90% C, assuming 1 atm CO (30% of the gas) and complete vaporization in the combustion zone, the partial pressures of sulfur species at 2000 °C derived from Eqs. (9.5) and (9.6) are 2.93×10^{-3} atm SiO, 0.42×10^{-3} atm SiS, and 0.13×10^{-3} atm CS. With decreasing temperature, CS becomes even less significant.

The experiments done to simulate the events in the lower part of the blast furnace have substantiated the occurrence of the following sequence of reactions [20]:

1. The formation of SiO and SiS in the combustion zone.
2. The transfer of silicon and sulfur to metal and slag droplets in the bosh.
3. The oxidation of silicon by iron and manganese oxides in the slag as the iron droplets pass through the slag layer.
4. The desulfurization of metal droplets as they pass through the slag layer.

To evaluate the extent of the reactions that may occur as the metal droplets pass through the slag layer, due consideration should be given to appropriate gas–slag–metal reaction equilibria. The numerical values given below for re-

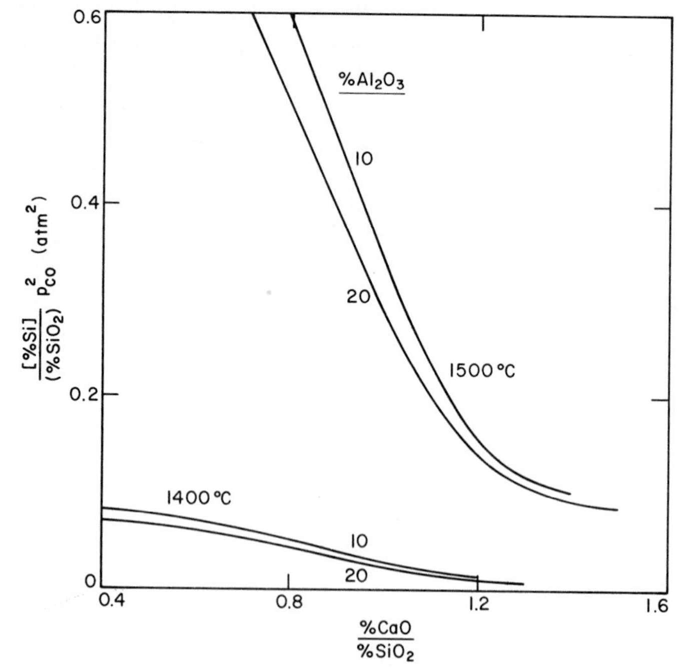

FIG. 9.6 Equilibrium distribution of silicon between metal and slag as a function of basicity of slags containing 10% and 20% Al_2O_3 and graphite-saturated iron containing less than 2% Si.

action equilibria are derived from the free-energy data in Tables 1.2 and 3.4, the activity coefficients of solutes in liquid iron in Table 3.1, and the activities of the oxides in Figs. 4.4 and 4.6. For details of the calculations, reference may be made to a critical review of blast furnace reactions presented by the author [20].

There are three reaction equilibria of primary importance:

$$SiO_2 + 2\underline{C} = \underline{Si} + 2CO, \tag{9.7}$$

$$MnO + \underline{C} = \underline{Mn} + CO, \tag{9.8}$$

$$CaO + \underline{S} + \underline{C} = CaS + CO, \tag{9.9}$$

where the underscore indicates elements dissolved in iron. The equilibrium distributions of silicon, manganese, and sulfur between graphite-saturated liquid iron and slag (CaO–MgO–Al_2O_3–SiO_2) are shown in Figs. 9.6–9.8, respectively. As shown elsewhere [20], these computed equilibrium data are in close agreement with the results of direct measurements made by numerous investigators.

Three phases are involved in reactions (9.7)–(9.9). A reaction involving three phases is not likely to approach equilibrium in a dynamic system such as a

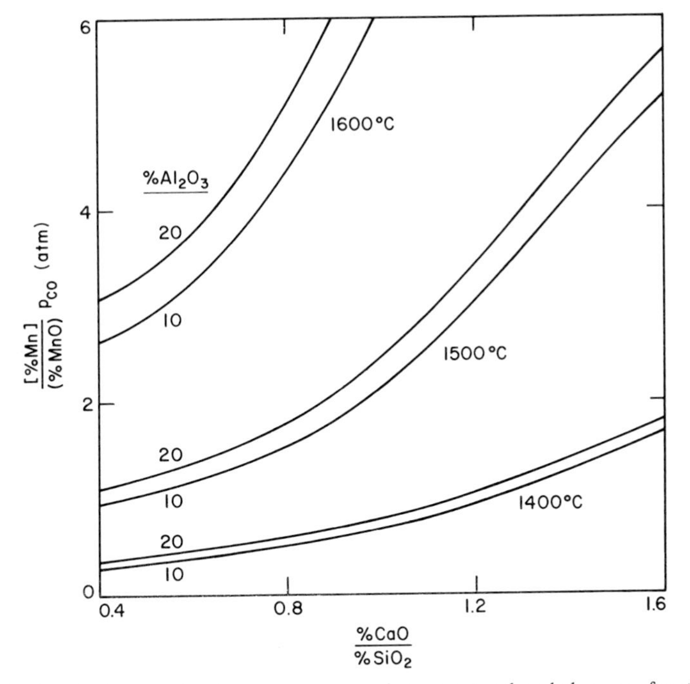

FIG. 9.7 Equilibrium distribution of manganese between metal and slag as a function of the basicity of slags containing 10% and 20% Al_2O_3 and saturated with graphite at indicated temperatures.

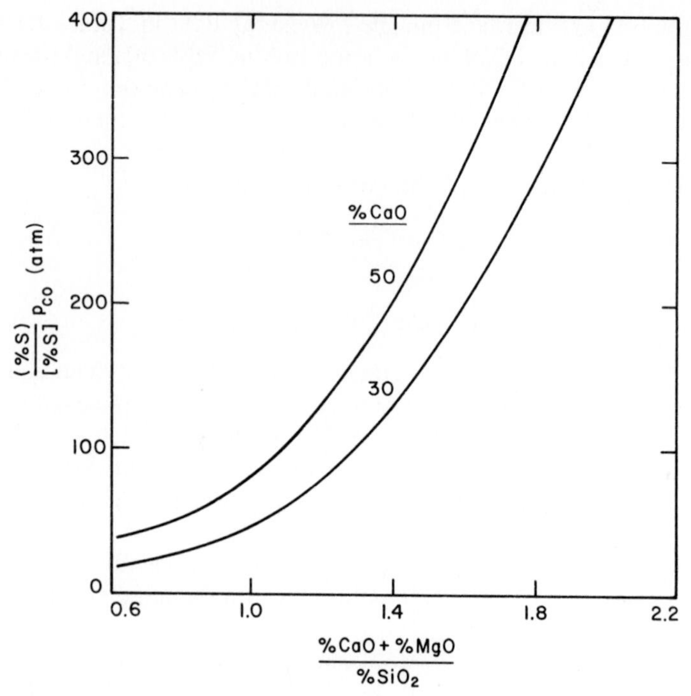

FIG. 9.8 Equilibrium distribution of sulfur between slag and metal as a function of slag basicity at 1500 °C for graphite-saturated slags containing 30% and 50% CaO.

blast furnace. We should therefore consider coupled reactions involving metal and slag only. For example, for the silicon–manganese reaction we may write

$$2MnO + \underline{Si} = SiO_2 + 2\underline{Mn}. \tag{9.10}$$

By combining the data in Figs. 9.6 and 9.7, we obtain the equilibrium relation between the product

$$k_{MS} = ([\%\,Mn]/(\%\,MnO))^2(\%\,SiO_2)/[\%\,Si] \tag{9.11}$$

and the basicity of the slag, $B = (\%\,CaO + \%\,MgO)/\%\,SiO_2$, for a given concentration of alumina in the slag. In the above and subsequent equations, the activities or concentrations of elements dissolved in iron are given in square brackets [ ] and those of oxides in the slag in round brackets ( ).

The importance of coupled reactions in the blast furnace was highlighted by Oelsen and co-workers [21, 22]. Typical examples of their results are compared in Fig. 9.9 with that calculated from the thermodynamic data. The average values of the product $(Mn/MnO)^2(SiO_2/Si)$ obtained from direct measurements of slag–metal equilibrium at graphite saturation are about 1.6 times greater than those derived from the thermodynamic data. This corresponds to an

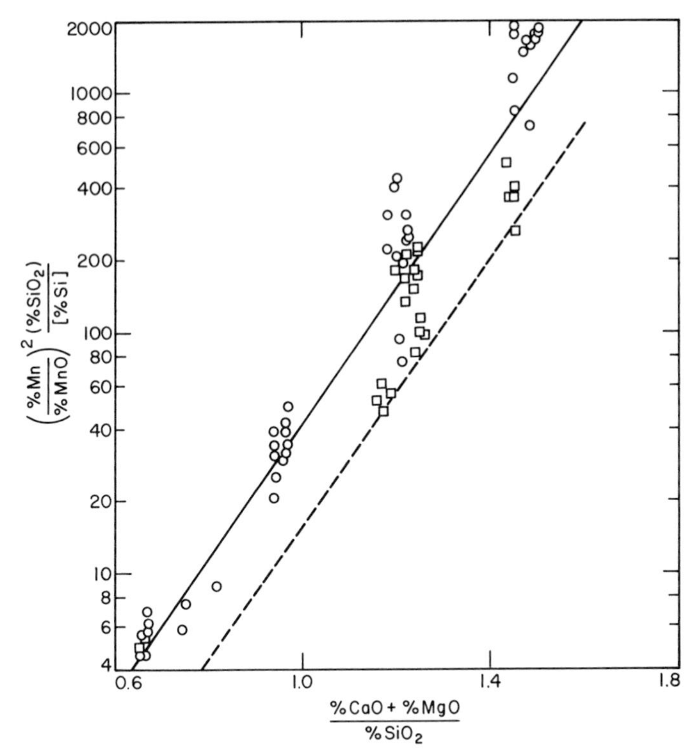

FIG. 9.9 Experimental results of Oelsen *et al.* [21, 22] for 1550 °C, compared with the relation derived from the available thermodynamic data (– – –): (○) $CaO–SiO_2–10\% \ Al_2O_3–10\% \ MgO$, (□) $CaO–SiO_2–10\% \ Al_2O_3$.

accumulated uncertainty of 1.7 kcal in the free-energy data, and is well within the limits of expected accuracy. The line drawn through the experimental points of Oelsen *et al.* is represented by the following equation:

$$\log([\% \text{ Mn}]/(\% \text{ MnO}))^2(\% \text{ SiO}_2)/[\% \text{ Si}] = 2.792B - 1.16, \qquad (9.12)$$

where $B = (\% \text{ CaO} + \% \text{ MgO})/\% \text{ SiO}_2$. The effect of temperature on this relation is almost negligible.

As discussed in Section 7.2.6, because of the electrochemical nature of slag–metal reactions, the sulfur transfer from metal to slag requiring electrons is accompanied by the oxidation of alloying elements such as manganese and silicon. For example, the sulfur reaction involving manganese may be represented by

$$\text{CaO} + \underline{\text{Mn}} + \underline{\text{S}} = \text{CaS} + \text{MnO}. \qquad (9.13)$$

By combining the data in Figs. 9.7 and 9.8, we obtain the equilibrium relation shown by the dotted line in Fig. 9.10, where typical examples of the experimental

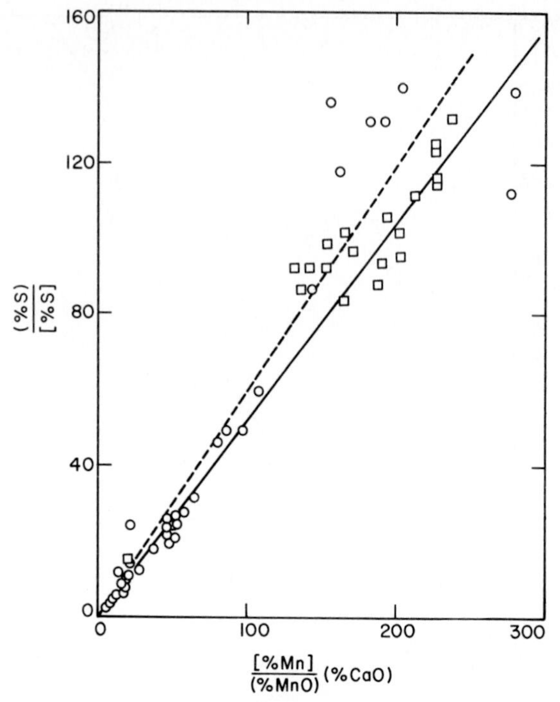

FIG. 9.10 Sulfur–manganese reaction equilibria at graphite saturation and 1550 °C measured by Oelsen *et al.* [21, 22], compared with that derived from the available thermodynamic data (– – –): (○) $CaO–SiO_2–10\%Al_2O_3–10\%$ MgO, (□) $CaO–SiO_2–10\%$ Al_2O_3.

results of Oelsen *et al.* [21, 22] for 1550 °C are included for comparison. This equilibrium relation may be represented by the equation

$$\log \frac{(\%\ \mathrm{S})}{[\%\ \mathrm{S}]} \frac{(\%\ \mathrm{MnO})}{[\%\ \mathrm{Mn}]} = \frac{9080}{T} - 5.203 + \log(\%\ \mathrm{CaO}). \tag{9.14}$$

Similarly, for the sulfur–silicon reaction

$$\mathrm{CaO} + \tfrac{1}{2}\underline{\mathrm{Si}} + \underline{\mathrm{S}} = \mathrm{CaS} + \tfrac{1}{2}\mathrm{SiO_2}, \tag{9.15}$$

the following equilibrium relation is obtained:

$$\log \frac{(\%\ \mathrm{S})}{[\%\ \mathrm{S}]} \left[\frac{(\%\ \mathrm{SiO_2})}{[\%\ \mathrm{Si}]}\right]^{1/2} = \frac{9080}{T} - 5.832 + \log(\%\ \mathrm{CaO}) + 1.396B. \tag{9.16}$$

9.2.3 Blast Furnace Plant Data

In the light of the reaction equilibria discussed in the preceding section, let us now examine some plant data to assess the state of reactions in the blast furnace. It is not uncommon to experience variations in the composition of the hot metal in consecutive tappings of the blast furnace in any given day. Despite these tap-to-tap variations, the daily averages do not change much. To evaluate the state of reactions in various blast furnaces, it is adequate to use the daily average compositions of metal and slag samples as normally recorded in the plants.

Depending on the blast furnace practice and local operating conditions, the hot-metal temperature at tap is in the range 1425–1525 °C. In most practices the slag contains 10%–15% Al_2O_3 and the slag basicity, defined by the ratio (% CaO + % MgO)/% SiO_2, varies from 1.25 to 1.55. In the combustion zone of the furnace, the gas consists of about 30%–40% CO, depending on oxygen enrichment of the air blast, and 70%–60% N_2 (other minor gaseous species need not be considered at this time). In estimating the partial pressure of CO in the bosh and hearth zone, the total pressure is assumed to be the same as the pressure of the air blast. In high-top-pressure furnaces, the blast pressure is about 4–4.5 atm absolute, and in the regular furnaces the blast pressure is about 2.5–3.0 atm absolute. In both cases the pressure differential between the bottom and the top of the furnace is about 1.5–2.0 atm. Therefore, the partial pressure of CO in the lower part of the furnace is about 0.8–0.9 atm for regular blast furnaces, and about 1.3–1.5 atm for the high-top-pressure furnaces. Normally, it takes about 8 hr for the charged material to pass through the furnace.

The manganese and silicon distribution ratios are usually in the following ranges: [Mn]/(MnO) = 0.6–4 and [Si]/(SiO_2) = 0.015–0.08. These ratios, compared with the equilibrium data in Figs. 9.6 and 9.7 for 1500 °C and 0.8–1.5 atm CO, indicate that the silicon and manganese contents of the hot metal are manyfold lower than the equilibrium values. In other words, the silicon and manganese reactions involving three phases do not approach equilibrium in the blast furnace. Another way of stating it is that the slag and metal compositions at tap are for oxygen potentials much higher than those for the C–CO equilibrium at the prevailing temperature and partial pressure of CO in the furnace hearth.

The data from five blast furnaces [20] are plotted in Fig. 9.11 for reaction (9.10); the dotted line for the slag–metal equilibrium is that given by Eq. (9.12). It is seen that even for the two-phase slag–metal reaction, the equilibrium is not reached. The extent of departure from the state of equilibrium of the silicon–manganese reaction varies from one furnace to another, the extent of departure from equilibrium being less at high basicities. However, there is a general trend for the product of the silicon and manganese ratios to increase with increasing slag basicity. In all furnaces the data points are below the equilibrium line, indicating that as the metal droplets pass through the slag layer, the direction of the reaction is

$$2MnO + \underline{Si} \rightarrow 2\underline{Mn} + SiO_2 . \tag{9.17}$$

Evidently, the level of silicon in the metal at tap is determined by the extent of oxidation of silicon in the metal droplets, and not by the carbon reduction of silica in the slag.

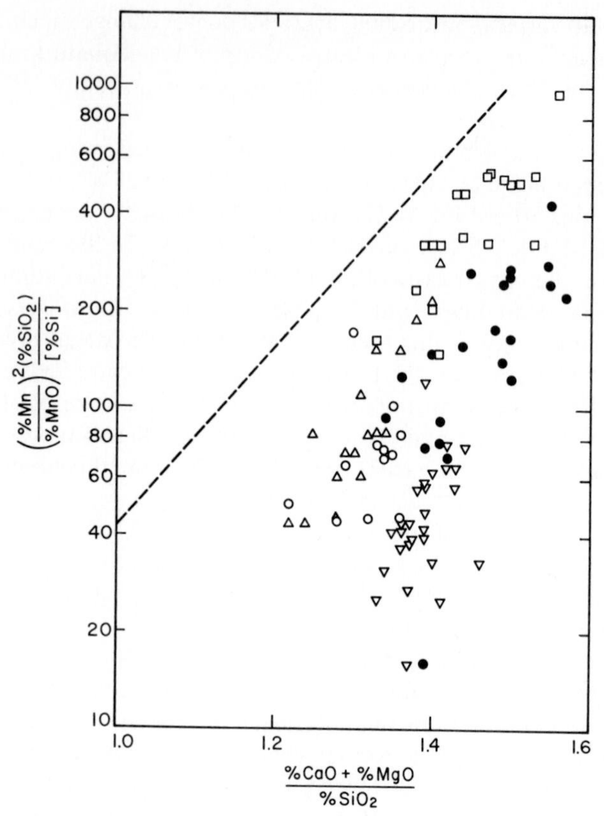

FIG. 9.11 Product of manganese and silicon distribution ratios as a function of slag basicity for daily average blast furnace data, compared with the equilibrium relation (– – –) for the silicon–manganese reaction. Steelworks: (△) Chiba BF1, (▽) Chiba BF2, (○) Homestead BF3, (○) Duquesne BF4, (□) South BF10. After Turkdogan [20].

For the sulfur reaction involving gas, metal, and slag phases, as in reaction (9.9), the equilibrium ratio (S)/[S] increases from about 120 to 220 with increasing slag basicity from 1.25 to 1.55 (Fig. 9.8). On the other hand, the sulfur distribution ratios achieved in ironmaking vary from 20 to 120. This departure from the three-phase equilibrium again indicates that the oxygen potential of the slag–metal system is higher than the value expected for the C–CO equilibrium near the furnace hearth.

The sulfur and manganese distribution ratios are compared in Fig. 9.12 with the equilibrium relation at 1500 °C for slags containing 40% CaO, which is the average lime content of most blast furnace slags. Variations in the tap temperature from one furnace to another do not account for the observed scatter in the data. However, in general, the (S)/[S] ratio increases with increasing[Mn]/(MnO) ratio. In BF2–4, the sulfur–manganese reaction is close to equilibrium. In BF1 and 10, the (S)/[S] is about half the equilibrium value.

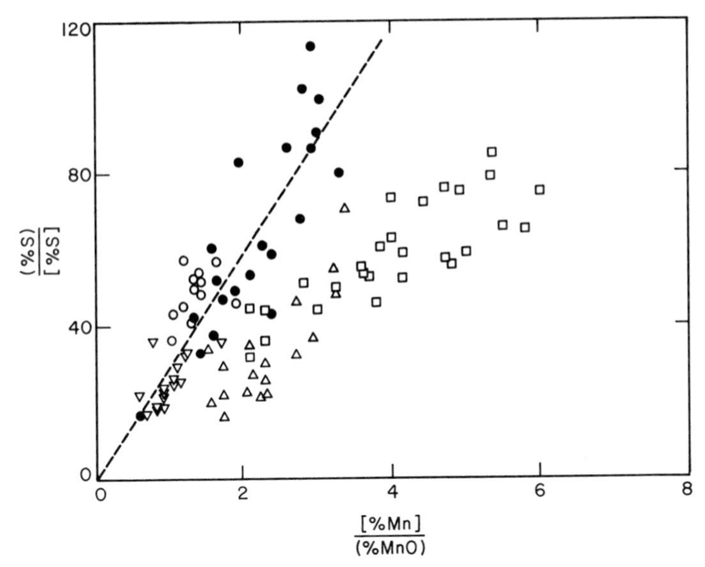

FIG. 9.12 Sulfur and manganese distribution ratios from the daily average compositions of tap samples from various blast furnaces, compared with the equilibrium relation (– – –) for 1500 °C. Steelworks: (△) Chiba BF1, (▽) Chiba BF2, (○) Homestead BF3, (●) Duquesne BF4, (□) South BF10. After Turkdogan [20].

As is indicated by the data in Fig. 9.11, the silicon–manganese reaction is out of equilibrium; had there been equilibrium, the [Mn]/(MnO) ratio would have been higher, and, in turn, there would have been more extensive desulfurization of the iron by the slag in the blast furnace hearth. In other words, the faster the rate of reaction (9.17) achieved by increasing temperature and/or MnO content of the slag, the greater would be the extent of desulfurization of iron by the slag. It is perhaps for this reason that, in general, the higher the hearth temperature, the lower the sulfur in the hot metal.

Because of fluctuations in the operating conditions in the furnace, there will be variations in the extent of reaction of metal droplets in the slag layer. Consequently, the metal droplets of varying composition collecting in the stagnant hearth will bring about the composition and temperature stratification that is common to all blast furnaces.

9.2.4 The Effect of Alkalies

The recycle of alkalies in the blast furnace and accumulation in the stack as carbonates and cyanides have been known since the 1830s. Primary reactions responsible for the recycle of alkalies and their accumulation in the stack were outlined in earlier studies made by Kinney and Guernsey [23]. A better understanding of reactions involving alkalies in the blast furnace was later developed by Richardson and Jeffes [24] through thermodynamic considerations, and the subject was recently reviewed by Abraham and Staffansson [25]. Problems associated with alkalies in blast furnaces and methods of remedy are well documented in numerous papers presented at a symposium in 1973 [26].

Depending on the raw materials and the operating practices, the alkali input to the furnace varies over a wide range from 2 to 12 kg per ton of hot metal. Because the silicates and salts of potassium are less stable than those of the sodium under reducing conditions, the potassium is the primary source of the alkali problem. The following reactions are of particular importance:

$$K_2O\,(s) + CO\,(g) = 2K\,(g) + CO_2\,(g). \tag{9.18}$$

$$KCN\,(l) + CO_2\,(g) = K\,(g) + 2CO\,(g) + \tfrac{1}{2}N_2\,(g). \tag{9.19}$$

$$KCN\,(l) = \tfrac{1}{2}(KCN)_2\,(g). \tag{9.20}$$

$$K_2CO_3\,(s,l) + CO\,(g) = 2K\,(g) + 2CO_2\,(g). \tag{9.21}$$

$$K_2SiO_3\,(s,l) + CO\,(g) = 2K\,(g) + CO_2\,(g) + SiO_2\,(s). \tag{9.22}$$

The equilibrium vapor pressures of potassium for these reactions, calculated from the thermodynamic data (Table 1.2), are given in Fig. 9.13 for temperatures of 600–1400 °C and for the corresponding CO/CO_2 ratios and total gas pressures normally encountered in modern blast furnaces. The curve for potassium cyanide is for the sum of vapor species K and $(KCN)_2$; above 900 °C the vapor pressure of liquid KCN exceeds that for reaction (9.19). The equilibrium data

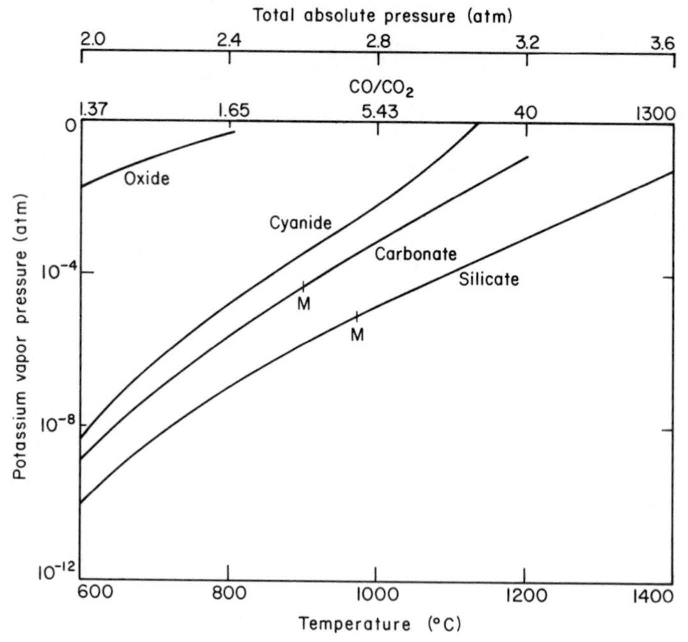

FIG. 9.13 Vapor pressure of potassium for certain pure potassium salts in equilibrium with the indicated CO–CO_2 mixtures; M is the melting point of the salt.

indicate that the alkali vapors will condense in the stack as carbonates and cyanides, but not as oxides.

The alkali silicates in the ore and coke ash will decompose at elevated temperatures in the lower part of the bosh and the combustion zone. In part the alkali vapors carried away with the ascending gas react with the slag and can therefore be removed from the furnace; part are converted to alkali carbonates and cyanides and deposited on the burden in the stack, the refractory lining; and some are carried away as dust with the furnace off gas. The cyanides and carbonates carried down with the burden readily decompose in the upper part of the bosh. This cycle of vaporization and condensation results in the accumulation of alkalies leading to scaffolding, gas channeling, furnace upsets, refractory wear, and so on. The alkali vapors are also known to catalyze the reduction of iron oxides and the oxidation of coke. Furthermore, alkali vapors cause extensive swelling and degradation of iron oxide and coke particles. All these effects are undesirable and contribute to the malfunction of the blast furnace.

A common remedy for the prevention of alkali buildup is to make periodic additions of sand or olivine (magnesium silicate) to the charge. The concentration of alkali vapors in the lower part of the stack is reduced by forming alkali silicates, which subsequently dissolve in the slag and can be removed from the furnace. Because the activity of alkali oxides in the slag decreases with increasing silica content, the lower the slag basicity, the higher the concentration of alkalies in the slag. This practice of alkali control, however, leads to a high proportion of sulfur in the hot metal. Periodic addition of calcium chloride to the burden also facilitates the removal of alkalies by the slag. Reference may be made to the papers in [26] for further reading on these practical aspects of the alkali problem.

9.3 Reactions in Steelmaking

Steelmaking is an oxidation process in which the impurities in the hot metal are oxidized by blowing oxygen into the bath. The oxides, silicates, phosphates, and sulfides forming are taken up by the slag, and the carbon is oxidized to carbon monoxide. The reactions occurring are common to all steelmaking, regardless of the type of vessel in which it is carried out, e.g., electric furnace, open hearth, oxygen top blowing, or oxygen bottom blowing.

Within the context of this book, we shall present here only the salient features of two aspects of the physical chemistry of steelmaking reactions: gas–slag–metal reaction equilibria for a chosen few reactions and the state of reactions in oxygen steelmaking assessed from studies of plant data.

9.3.1 Gas–Slag–Metal Reaction Equilibria

For the sake of brevity, the selected equilibrium data are presented here with a minimum of commentary on the studies made by numerous investigators.

In fact, most of the equilibrium data for slag–metal reactions are readily computed from the compiled thermodynamic data in Chapters 1, 3, and 4.

The equilibrium constant for the carbon–oxygen reaction in liquid iron given below is that derived by Fuwa and Chipman [27]:

$$\underline{C} + \underline{O} = CO,$$
$$K = p_{CO}/a_C a_O, \qquad \log K = (1168/T) + 2.07, \tag{9.23}$$

where p_{CO} is in atmospheres and the activities a_C and a_O are taken equivalent to [% C] and [% O], respectively, at low solute concentrations. As shown in Table 3.1, the activity coefficient of carbon in iron increases and that of oxygen decreases with increasing carbon content. The net result is that, for a given CO pressure, the product [% C][% O] decreases only slightly with increasing carbon concentration. For carbon contents below about 0.5% and at steelmaking temperatures, ~1600 °C, the equilibrium product [% C][% O] is about 0.002 for 1 atm CO. For concentrations above 0.03% C, the reaction product is essentially CO with a negligible amount of CO_2.

In a recent study of reaction (9.23) with levitated Fe–C melts in CO–CO_2 mixtures at pressures up to 70 atm, El-Kaddah and Robertson [28] found that, contrary to the previous findings, the carbon increases the activity coefficient of oxygen in liquid iron. However, their results agree with the value of [% C] [% O] = 0.002 at low concentrations of carbon.

The solubility of oxygen in liquid iron in equilibrium with liquid iron oxide at temperatures of 1530–1700 °C was first determined by Taylor and Chipman [29]. Later, Distin *et al.* [30] extended these equilibrium measurements to higher temperatures up to 1960 °C. All the data are well represented by the equation

$$Fe_xO = xFe + \underline{O},$$
$$\log\{[\%\,O]/a_{FeO}\} = -(6380/T) + 2.765, \tag{9.24}$$

where a_{FeO} is the activity of iron oxide relative to liquid Fe_xO in equilibrium with liquid iron; x increases from 0.98 at 1527 °C to 1.00 at 2000 °C. The solubility of oxygen in liquid iron in equilibrium with simple CaO–FeO–SiO_2 slags may be estimated from the activity of FeO in Fig. 4.4a and Eq. (9.24).

For a given oxygen activity, the ratio Fe^{3+}/Fe^{2+} in the slag varies with composition, e.g., increases with increasing basicity. In representing slag–metal equilibria, it is customary to use the total iron (as oxides) in the slag as % $\sum$FeO.

The compositions of most steelmaking slags are in the ranges: 50%–60% CaO, 3%–8% MgO, 3%–8% MnO, 6%–26% $\sum$ FeO, 15%–25% SiO_2, 1%–5% P_2O_5, ~1% Al_2O_3, and 0.1%–0.2% S. In some European steelmaking practices with high phosphorus hot metal, the slag may contain 20% P_2O_5 or more, but much less SiO_2. For the indicated slag compositions, the slag–metal equilibrium with respect to the carbon–oxygen reaction may be represented by the following approximate relation for 1 atm CO:

$$[\%\,C][\%\,\textstyle\sum FeO] \sim 0.25. \tag{9.25}$$

As inferred from the activities of oxides in polymeric melts presented in Section 4.1.1, the activity coefficients of oxides are relatively simple functions of the *basicity* of the slag. The definition of basicity for complex slags is somewhat arbitrary. If we assume that in steelmaking-type slags the concentrations of CaO and MgO are equivalent on molar basis, and similarly on molar basis $\frac{1}{2}P_2O_5$ is equivalent to SiO_2, in terms of weight percentages of oxides, the slag basicity B may be defined by the ratio

$$B = [\%\,CaO + 1.4(\%\,MgO)]/[\%\,SiO_2 + 0.84(\%\,P_2O_5)]. \quad (9.26)$$

The manganese exchange between metal and slag may be represented by the reaction

$$\underline{Mn} + FeO = MnO + Fe, \quad (9.27a)$$

for which the equilibrium relation for constant temperature is, in terms of weight percentages,

$$k_{Mn} = (\%\,MnO)/[\%\,Mn](\%\,\textstyle\sum FeO). \quad (9.27b)$$

In low alloy steels, the activity of manganese is proportional to its concentration. However, the ratio of the activity coefficients of the oxides $\gamma_{MnO}/\gamma_{FeO}$ increases with increasing basicity of the slag as indicated, for example, by the activity data in Figs. 4.4a and 4.4b. For this reason, the equilibrium relation k_{Mn} decreases with increasing basicity of the slag. Reevaluation of the available equilibrium data cited by Turdogan [31] gives for the equilibrium relation k_{Mn} in basic slags at steelmaking temperatures

$$k_{Mn} = 6/B. \quad (9.28)$$

That is, for a given basicity B in basic slags, the equilibrium distribution of manganese between slag and metal is directly proportional to the concentration of the iron oxide.

Because of the large difference in the free energies of formation of SiO_2 and FeO and the low activity of SiO_2 in basic slags, the silicon in iron is readily oxidized during steelmaking. At tap, the silicon content of the steel is less than 0.005%.

The sulfidation of polymeric melts was discussed in Section 4.1.4b. Analogous to reaction (4.33), we may represent the sulfur–oxygen exchange between metal and slag by the reaction

$$\underline{S} + O^{2-} = S^{2-} + \underline{O}, \quad (9.29)$$

for which the equilibrium relation may be represented by the equation

$$k_S = \{(\%\,S)/[\%\,S]\}(\%\,\textstyle\sum FeO). \quad (9.30)$$

The effect of slag composition on the activity coefficients of sulfur and iron oxide in the slag is such that the equilibrium relation k_S decreases with decreasing basicity of the slag [32]. Previous studies [33] have shown that k_S is also a

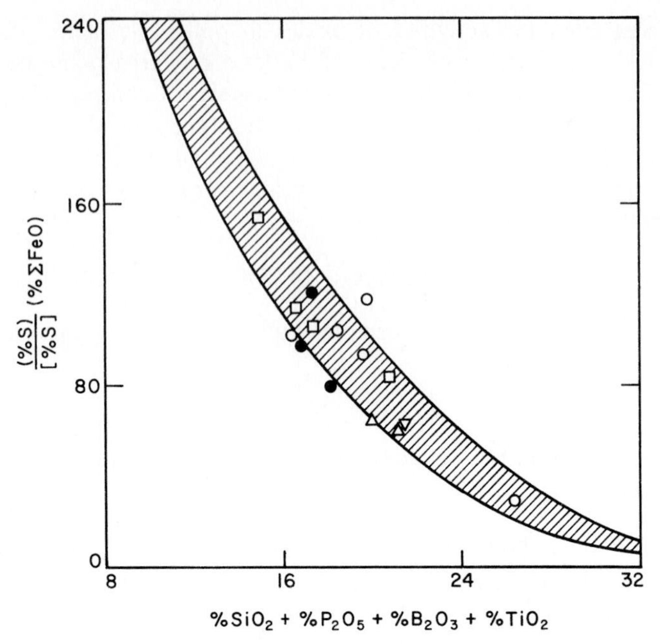

FIG. 9.14 Equilibrium relation for sulfur distribution between slag and metal, compared with plant data for open-hearth steelmaking. Flux: (○) none, (●) CaF_2, (△) B_2O_3, (▽) "Sorel," (□) MnO_2 (all Kor); shaded area from Turkdogan.

simple function of the sum $\% \, SiO_2 + \% \, P_2O_5$ in the slag. This has been confirmed subsequently by Kor [34] in a study of the effect of flux additions on the equilibrium distribution of sulfur between slag and metal.

As is seen from Kor's results in Fig. 9.14 for 1550 °C, k_S decreases with increasing sum of the concentrations of SiO_2, P_2O_5, B_2O_3, and TiO_2. The addition of 2%–5% of the fluxing agent to the slag has no perceptible effect on the sulfur distribution ratio. The shaded area is that obtained by Turkdogan [33] from an earlier study of plant data for open-hearth steelmaking for which the state of sulfur reaction was found to be essentially the same as that for the slag–metal equilibrium: The value of k_S increases slightly with increasing temperature. However, the shaded area in Fig. 9.14 gives an adequate description of the sulfur equilibrium between slag and metal for steelmaking temperatures of 1550–1650 °C.

The phosphorus reaction between metal and slag may be represented by

$$2\underline{P} + 5FeO = P_2O_5 + 5Fe,$$
$$k_P = (\% \, P_2O_5)/[\% \, P]^2(\% \textstyle\sum FeO)^5. \tag{9.31}$$

Balajiva and co-workers [35, 36] made a detailed investigation of this reaction and found that in complex steelmaking-type slags the equilibrium constant k_P

increases with decreasing temperature and increasing concentration of calcium oxide in the slag. This reaction equilibrium is well represented by the following empirical relation, which is based on the work of Balajiva and co-workers:

$$\log k_P = 10.78 \log(\% \text{CaO}) - 0.00894t - 6.245, \tag{9.32}$$

where t is temperature (°C). Subsequent equilibrium studies made by Kor [34] confirmed the validity of Eq. (9.32) representing the slag–metal equilibrium with respect to the phosphorus reaction.

In stainless steelmaking, some oxidation of chromium occurs during decarburization of Fe–Cr–Ni–C melts with oxygen blowing. The reaction of particular importance is

$$\text{Cr}_2\text{O}_3 + 3\underline{\text{C}} = 2\underline{\text{Cr}} + 3\text{CO}, \tag{9.33}$$

for which the equilibrium constant for unit activity of Cr_2O_3, i.e., for melts saturated with solid Cr_2O_3, is given by the equation

$$K = ([a_{\text{Cr}}]^2/[a_{\text{C}}]^3)p_{\text{CO}}^3, \tag{9.34}$$

where the a's are the activities of the solutes indicated by the subscripts.

Richardson and Dennis [37] investigated the state of equilibrium of iron–chromium–carbon melts, saturated with solid Cr_2O_3, with gases of known CO and CO_2 potentials. The carbon–chromium relation for this equilibrium is shown in Fig. 9.15 for 1 atm CO. For a given chromium concentration, the higher the temperature, the lower the equilibrium carbon content of the metal. For this reason, the chromium losses to the slag are minimized by decarburizing the melt with oxygen blowing at elevated temperatures. In fact, in earlier days of stainless steelmaking in electric furnaces, it was realized from practical experience that high-chromium steels could be decarburized with a minimum loss of chromium to the slag by increasing the temperature of the melt [38].

Another important feature of reaction (9.33) is that, for a given chromium concentration, the lower the partial pressure of CO in the gas the lower the equilibrium concentration of carbon. These two effects on the oxidation of chromium are shown in Fig. 9.16 for melts saturated with chromic oxide. In most current steelmaking practices, the chromium steels are decarburized by bottom blowing of argon–oxygen, or in some processes steam–oxygen, mixtures to cut down on the oxidation of chromium without excessive increase of temperature.

At the end of decarburization, the chromium in the slag is reduced back to the metal by adding ferrosilicon to the melt. Under reducing conditions, much of the Cr^{3+} [or CrO_3^{2-} in slags of low basicity as in Eq. (4.26c)] in the slag is reduced to Cr^{2+} state. Representing the total chromium oxide in the slag as CrO_x, we may write for the chromium distribution between metal and slag

$$\begin{gathered} \text{CrO}_x + x\underline{\text{Fe}} = \underline{\text{Cr}} + x\text{FeO}, \\ K_1 = ([a_{\text{Cr}}]/(a_{\text{CrO}_x}))((a_{\text{FeO}})/[a_{\text{Fe}}])^x. \end{gathered} \tag{9.35}$$

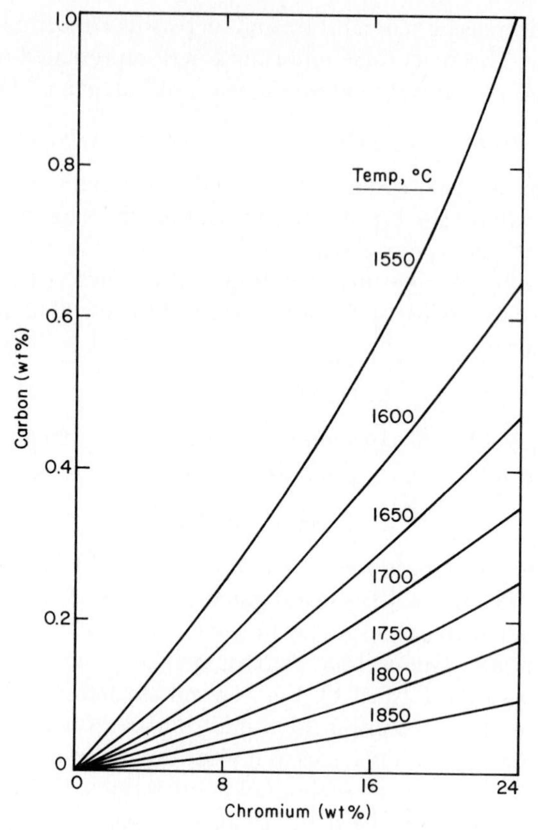

FIG. 9.15 Equilibrium concentrations of carbon and chromium in Fe–Cr–C melts in equilibrium with solid Cr_2O_3 at 1 atm CO and indicated temperatures. After Richardson and Dennis [37].

Similarly, for the chromium–silicon reaction

$$2\mathrm{CrO}_x + x\underline{\mathrm{Si}} = 2\underline{\mathrm{Cr}} + x\mathrm{SiO}_2,$$

$$K_2 = ([a_{\mathrm{Cr}}]/(a_{\mathrm{CrO}_x}))^2((a_{\mathrm{SiO}_2})/[a_{\mathrm{Si}}])^x. \tag{9.36}$$

Rankin and Biswas [39] studied these reactions by equilibrating Fe–Cr–Si alloys with SiO_2–CaO–Al_2O_3–CrO_x slags at 1600 °C. From the results, they estimated x to be about 1.07. Since the activity coefficients of the reactants in the metal and slag are not known, Rankin and Biswas used the following equations to represent the state of equilibrium:

$$k_1 = ([\%\,\mathrm{Cr}]/(\%\,\mathrm{Cr}))((\%\,\mathrm{Fe})/[\%\,\mathrm{Fe}])^{1.07}, \tag{9.37}$$

$$k_2 = ([\%\,\mathrm{Cr}]/(\%\,\mathrm{Cr}))^2(1/[\%\,\mathrm{Si}]\gamma_{\mathrm{Si}}), \tag{9.38}$$

where γ_{Si} is the activity coefficient, defined such that γ_{Si} approaches unity when % Si approaches zero; the effect of Cr on γ_{Si} was assumed to be negligible.

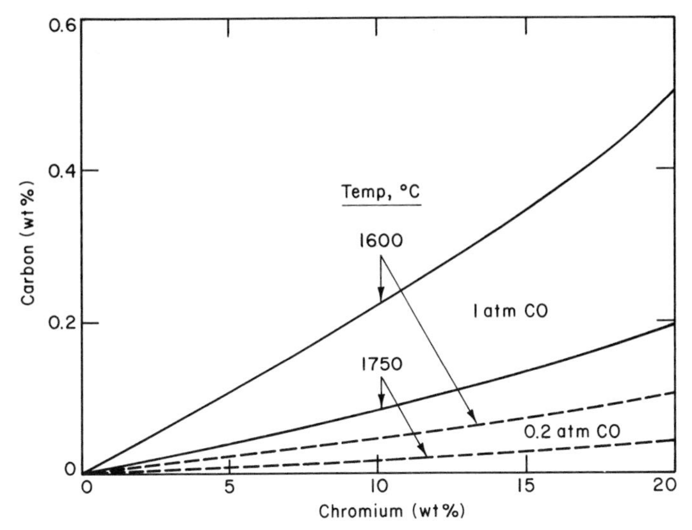

FIG. 9.16 Carbon–chromium relation in liquid steel in equilibrium with solid Cr_2O_3 at indicated temperatures and CO pressures

Variations of k_1 and k_2 with the slag composition projected on the ternary system $CaO–Al_2O_3–SiO_2$, as estimated by Rankin and Biswas from their data and those of others, are shown in Fig. 9.17 for temperatures of 1600–1640 °C. Because of the assumptions involved in the interpretation of the experimental data, the relations in Fig. 9.17 are considered approximate. Nevertheless, these data are useful in estimating the desired slag composition in the reducing stage of the process to optimize the chromium recovery from the slag with a minimum use of ferrosilicon.

As discussed in Section 4.1.1, numerous attempts have been made to interpret the thermodynamic properties of polymeric melts in terms of the concentrations of ionic species derived from assumed ionic models of the polymer structures. Flood and Grjotheim [40] and subsequently others [41, 42] proposed polymer models to represent the equilibrium relations for slag–metal exchange reactions in terms of the concentrations of ionic species in slags. Although these theoretical models are helpful in improving our understanding of the structure of polymers, their application to slag–metal reaction equilibria is not particularly rewarding when compared with the much simpler methods of representing the slag–metal equilibria presented here in generally accepted terms.

9.3.2 Plant Data for Oxygen Steelmaking

A meaningful way of evaluating the state of reactions in industrial steelmaking processes is by comparing the plant data for compositions of metal and slag samples with the equilibrium data for slag–metal reactions. We shall consider two sets of examples: steelmaking by oxygen top blowing, known as

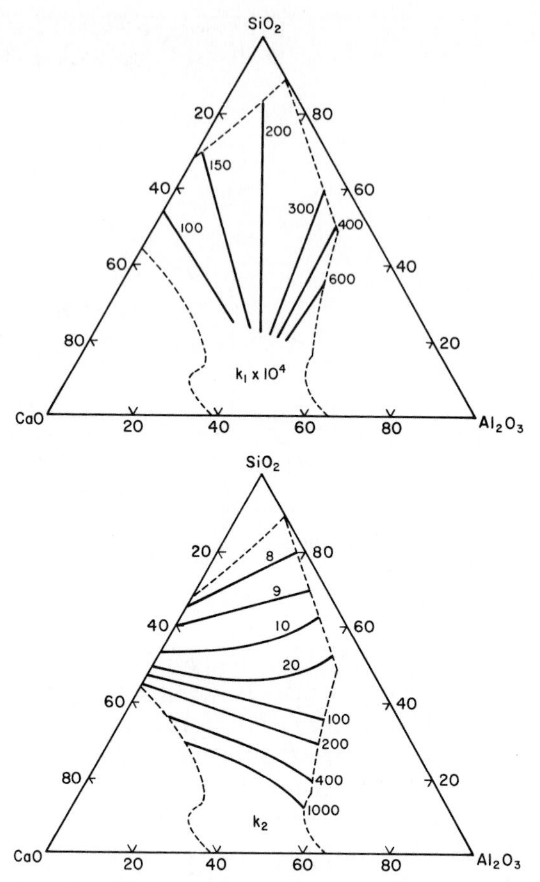

FIG. 9.17 Variations of equilibrium relations k_1 and k_2, of Eqs. (9.37) and (9.38), with the composition of CaO–Al_2O_3–SiO_2 slags. After Rankin and Biswas [39].

the LD, BOF, or BOP, and steelmaking by oxygen plus lime bottom blowing, known as the OBM or Q-BOP. The examples of the plant data given here are from the BOP and Q-BOP shops at Gary Works of U.S. Steel. The metal and slag samples taken at tap from 180-ton melts were analyzed for all the constituent elements and oxides. The slag analyses were corrected for entrapped iron particles and small amounts of undissolved free lime in the slag. Comparisons are also made with the state of reactions in the now outdated basic Bessemer steelmaking using the data compiled by Schenck [43]; in this case, refining was done by blowing air through the melt in the converter.

In all pneumatic steelmaking processes, the silicon is oxidized first in the early stages of the blowing, followed by the oxidation of carbon. The rate of decarburization is controlled by the rate of oxygen blowing which is usually in the range 650–750 $nm^3 min^{-1}$.

Below about 0.2% C, the rate of decarburization in the Q-BOP is controlled primarily by the mass transfer of carbon in the bulk metal to the gas–metal interface within the gas jet stream in the melt. From a study of plant data, Fruehan [44] obtained the following rate equation for decarburization controlled by liquid-phase mass transfer*:

$$\log[(\%\ C)_t/(\%\ C)_0] = -0.015(t - t_0), \qquad (9.39)$$

where the subscript t indicates the carbon content at time t in seconds, and the subscript zero indicates the carbon content at an earlier time t_0 when mass transfer becomes the rate controlling mechanism.

In Section 7.9.2 we discussed mass transfer to and from gas bubbles. From available information on the dynamics of gas bubbles in the swarm of bubbles in aqueous solutions, and an estimate of an average bubble size of 1 cm diameter for the swarm of bubbles in liquid steel, Eq. (7.151) was derived for the rate of mass transfer through the liquid boundary layer as a function of the superficial gas velocity u_s. Inserting in Eq. (7.151) $u_s = 250\ \mathrm{cm\ sec^{-1}}$ for the average superficial gas velocity in the Q-BOP, we obtain a rate constant of $0.25\ \mathrm{sec^{-1}}$ for a mobile bubble surface. Because of strong chemisorption of sulfur and oxygen on the bubble surface at sulfur and oxygen concentrations present in the steel, the bubble surface will be immobile so that the mass-transfer rate constant is reduced to about $0.03\ \mathrm{sec^{-1}}$. The rate constant thus estimated is of the same order of magnitude as that derived from the plant data for an oversimplified mass-transfer model.

The extent of oxidation of iron during the later stages of decarburization in the Q-BOP, when $\%\ C < 0.2$, is readily calculated from the difference between the quantity of oxygen blown and the amount of carbon oxidized, as given by Eq. (9.39). On the basis of this reasoning, Fruehan [44] suggested that by diluting oxygen with argon in the later stages of blowing low-carbon steels, excessive oxidation of iron may be suppressed without retarding the rate of decarburization at low carbon levels. Such a practice may increase the yield by as much as 2%.

Concentrations of iron oxide are lower in Q-BOP slags than in BOP slags. As is seen from the plant data in Fig. 9.18 [45], differences in iron oxide contents of BOP and Q-BOP slags are much more pronounced at low carbon concentrations. The iron oxide that forms in the tuyere zone of the Q-BOP will be reduced at least partly by carbon as the oxide droplets traverse the metal bath. On the other hand, in oxygen top blowing, there is the reverse situation of dispersion of metal droplets in the slag where decarburization and oxidation of iron occur. The iron oxide thus formed remains in the slag with little chance

* The rate constant $0.0087\ \mathrm{sec^{-1}}$ given in Fruehan's paper for pilot-plant data from 30-ton Q-BOP heats has been adjusted to $0.015\ \mathrm{sec^{-1}}$ for faster blowing rates and deeper metal bath in the commercial 180-ton Q-BOP.

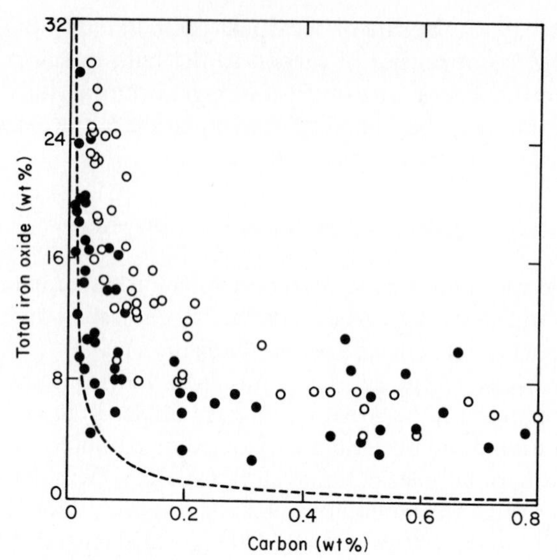

FIG. 9.18 Variation of the carbon content of the steel with the total iron oxide content of the slag in BOP (○) and Q-BOP (●) steelmaking, compared with the relation for slag–metal equilibrium (---). After Turkdogan [45].

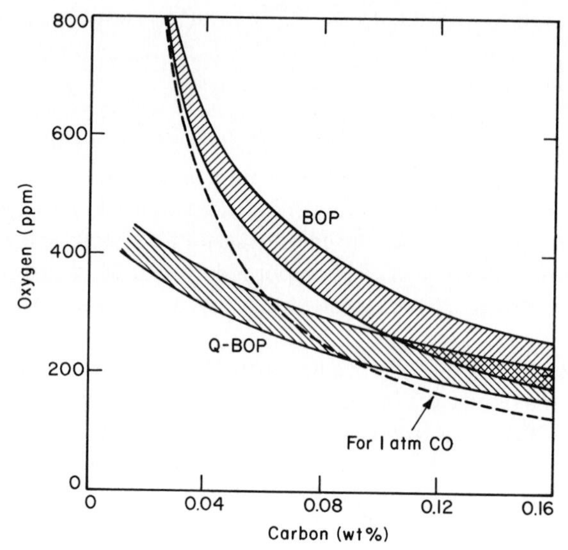

FIG. 9.19 Oxygen concentrations in BOP and Q-BOP steels, compared with values of C–O equilibrium at 1 atm CO.

of reduction by the secondary reaction with carbon. It is because of this difference in the mixing pattern of slag and metal that the iron oxide content of the Q-BOP slag is less than that of the BOP slag.

The oxygen contents of the BOP and Q-BOP steels at tap are compared in Fig. 9.19 with the equilibrium values for 1 atm CO. The nitrogen or argon is blown during rotation of the Q-BOP vessel to the tapping position; this gas purge brings about carbon deoxidation. Since the rate of this reaction is controlled by mass transfer in the melt, the higher the level of oxygen, the greater would be the extent of carbon deoxidation during turndown with an inert gas. It is for this reason that the concentration of residual oxygen in low-carbon Q-BOP steels is well below that for 1 atm CO.

It follows from the foregoing observations that lowering the partial pressure of carbon monoxide in the gas stream by dilution of the bottom-blown oxygen with argon or nitrogen will lower the residual oxygen in the steel, also lowering the extent of oxidation of the iron, and hence will increase the yield when blowing steel to exceptionally low levels of carbon.

The variation of the manganese distribution ratio between slag and metal with the total iron oxide content of the slag is shown in Fig. 9.20 for BOP, Q-BOP, and basic Bessemer steelmaking. The dotted lines are for basicities 2–4 for the slag–metal equilibrium at 1600 °C given by Eq. (9.28). In the plant data

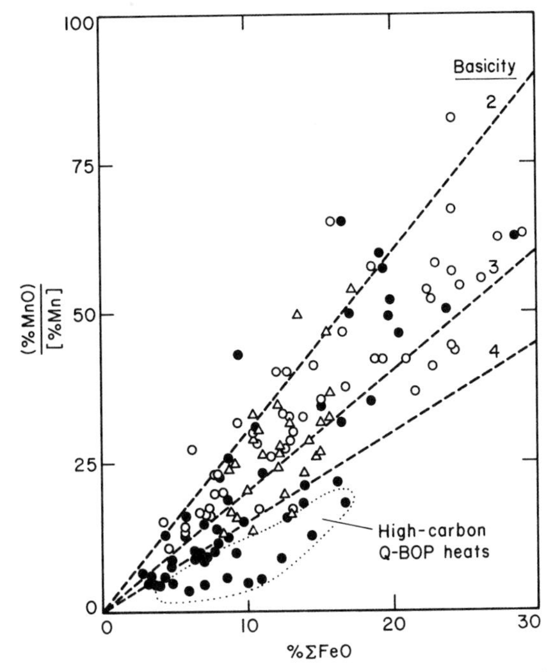

FIG. 9.20 Slag/metal manganese distribution ratio in BOP (○), Q-BOP (●), and basic Bessemer (△) processes, compared with equilibrium values for basicities of 2–4. After Turkdogan [45].

studied, the basicities are in the range 2.5–4.0. Although most of the data points are within the lines for this range of basicities, observed scatter in the data points does not relate to variations in the slag basicity. It should be noted that the plant data for basic Bessemer steelmaking are for slags containing up to 25% P_2O_5, whereas in the BOP and the Q-BOP the P_2O_5 content of the slag is 0.6%–1.2%. With the exception of high-carbon Q-BOP melts, the plant data plotted in Fig. 9.20 indicate that the state of the manganese reaction in pneumatic steelmaking is not far removed from that for the slag–metal equilibrium. This behavior is similar to that observed in the past for open-hearth steelmaking.

The data points in Fig. 9.20 for high-carbon Q-BOP melts are well below the range for the slag–metal equilibrium. To demonstrate the reason for this difference in the state of the manganese reaction, the manganese distribution ratio for slags containing 10% ± 2% total iron oxide is plotted in Fig. 9.21 against the concentration of carbon in the steel. Although the data points for the BOP are scattered and independent of the carbon content, within the indicated equilibrium range for basicities 2–4, the points for Q-BOP melts are below the equilibrium range at high carbon contents. This difference in the state of the manganese reaction, as affected by carbon, is due to the differences in the physical conditions of the metal bath in these two methods of steelmaking. As pointed out earlier, the reaction products forming in the tuyere zone of the Q-BOP are subject to a secondary reaction with carbon as the oxide particles traverse the bath; hence the MnO/Mn ratio is lower at high carbon contents. When the level of carbon drops to about 0.1%, there

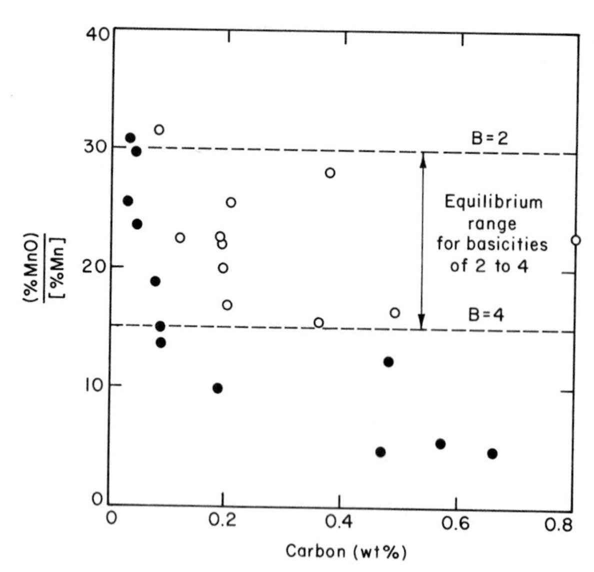

FIG. 9.21 Data for slags containing 10% ± 2% ∑FeO for BOP (○) and Q-BOP (●) melts showing the effect of carbon on the manganese ratio in Q-BOP. After Turkdogan [45].

is negligible reduction of manganese oxide by the secondary reaction, and consequently, the state of the manganese reaction becomes similar to that in the BOP and close to that for the slag–metal equilibrium.

Although the state of the manganese reaction, as represented by Eq. (9.28), is similar in the BOP and the Q-BOP at low carbon levels, the residual level of manganese in Q-BOP steel is higher than that in BOP. This is due, of course, to the iron oxide content of the slag being lower in the Q-BOP than in the BOP; and, as indicated by the equilibrium relation in Eq. (9.28), the lower the iron oxide content of the slag, the higher would be the concentration of residual manganese in the steel. The difference in the levels of manganese concentrations in these processes is further demonstrated in Fig. 9.22 by the plant data from Kawasaki Steel [46].

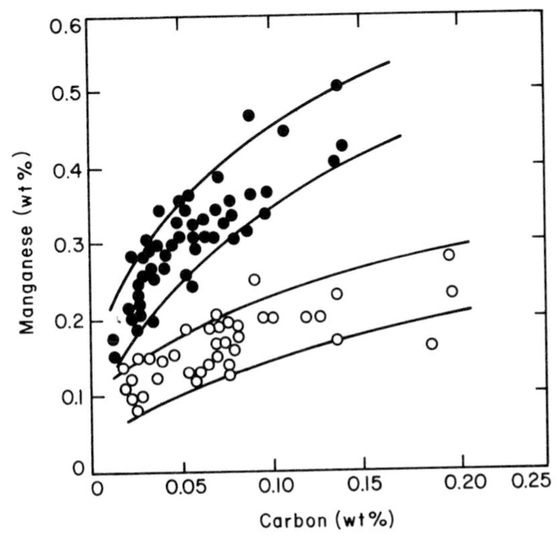

FIG. 9.22 Relation between manganese and carbon contents of steel at tap in Kawasaki Q-BOP (●) and LD (○) steelmaking [46].

To assess the state of the sulfur reaction, the values of k_S for the sulfur reaction, Eq. (9.30), obtained from the plant data are plotted in Fig. 9.23 against the sum $\%\ SiO_2 + \%\ P_2O_5$ in the slag. Most of the data points for the BOP and the Q-BOP are scattered within a band shown by dotted curves for the slag–metal equilibrium (Fig. 9.14). Much like the manganese reaction, the distribution of sulfur between slag and metal in the BOP and the Q-BOP is affected by the iron oxide content of the slag as given by the equilibrium relation. That is, the lower the iron oxide content of the slag, the higher the sulfur distribution ratio. Consequently, because the iron oxide content of the Q-BOP slag is low, the sulfur distribution ratio (S)/[S] is higher in the Q-BOP than in the BOP.

It has been known from the early days of the Bessemer and open-hearth steelmaking that the steel can be dephosphorized only under an oxidizing

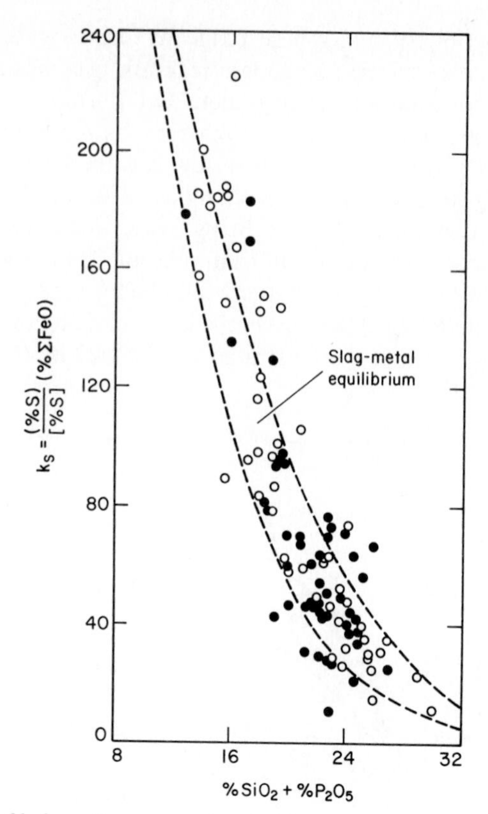

FIG. 9.23 Values of k_S for sulfur reaction from BOP (○) and Q-BOP (●) plant data, compared with the equilibrium relation for 1600 °C. After Turkdogan [45].

slag rich in lime content. Although the same is true for oxygen steelmaking processes, the state of phosphorus reaction in these processes seems to differ from the earlier methods of steelmaking. The variation of k_P, Eq. (9.31), with the concentration of calcium oxide in the slag is shown in Fig. 9.24 for BOP and Q-BOP steelmaking. The line drawn is calculated from Eq. (9.32) representing the slag–metal equilibrium. Previous studies have shown that the phosphorus distribution between slag and metal is close to equilibrium as given by Eq. (9.32) or other equivalent forms representing the slag–metal equilibrium [34, 47]. However, as is seen in Fig. 9.24, most values of k_P for the BOP and Q-BOP are 1–2 orders of magnitude higher than those for the slag–metal equilibrium; departure from the equilibrium line is more pronounced at high lime contents. This apparent nonequilibrium state is in the direction of low phosphorus in the steel; that is, in the BOP and particularly in the Q-BOP the steel is dephosphorized to levels below those expected from the composition of the slag.

More intriguing, perhaps, is the difference in the residual phosphorus levels between BOP and Q-BOP steels. This is demonstrated in Fig. 9.25, which

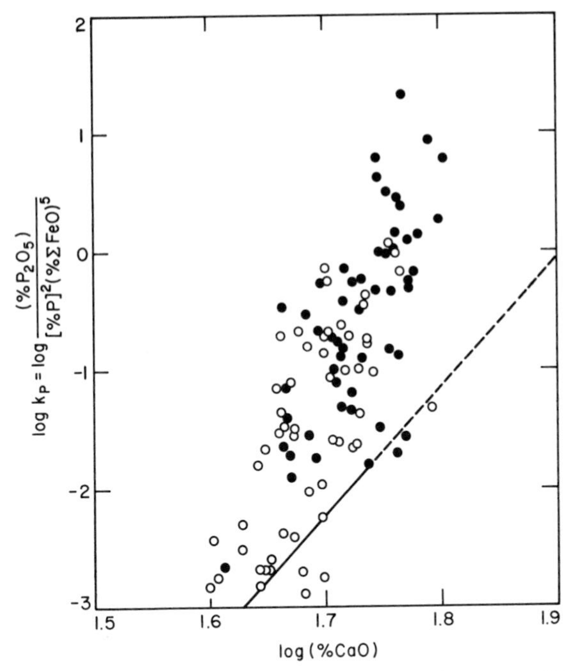

FIG. 9.24 Values of k_P for phosphorus reaction from BOP (○) and Q-BOP (●) plant data, compared with the slag–metal equilibrium relation for 1600 °C (– – –). After Turkdogan [45].

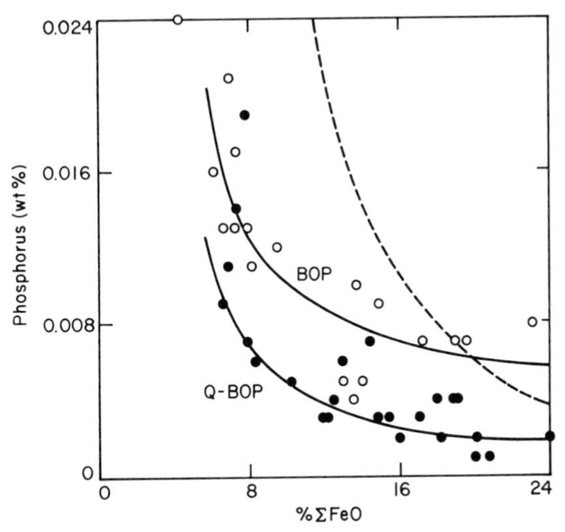

FIG. 9.25 Comparison of phosphorus concentrations in BOP (○) and Q-BOP (●) steelmaking at 1600 ± 10 °C for slags containing 52% ± 2% CaO and 0.9% ± 0.3% P_2O_5 with the equilibrium relation (– – –). After Turkdogan [45].

shows the variation of the concentration of phosphorus in the steel with the total iron oxide of slags containing 52% ± 2% CaO and 0.9% ± 0.3% P_2O_5. For iron oxide contents of 16%–24%, the average phosphorus content of steel in the BOP is about 0.007%, whereas in the Q-BOP it is about 0.002%. The dotted curve represents the relation for the slag–metal equilibrium at 1600 °C calculated from Eq. (9.32).

With oxygen plus lime bottom blowing, the oxidation reactions occur in the tuyere zone. The slag forming in the tuyere zone has a greater capacity for iron, manganese, and phosphorus than the slag that ultimately accumulates at the top of the melt. As the slag particles traverse the bath, there may be partial reversion of iron, manganese, and phosphorus back to the metal because of the secondary reaction with carbon in the bath. Approach to partial slag–metal equilibrium with respect to manganese and sulfur reactions suggests that the rates of manganese and sulfur reactions in the reduction cycle are relatively fast. By the same token, the plant data suggest that the rate of reduction of complex phosphate ions in the slag is sluggish. The phosphorus reaction therefore remains in a nonequilibrium state, in favor of low phosphorus in the steel.

The oxidation–reduction cycle is expected to occur also in the oxygen top-blowing practice. However, because of extensive dispersion of metal droplets in the foamed slag layer, there will be a faster rate of approach to the slag–metal equilibrium in the reduction phase of the cycle. Therefore, the greater the extent of metal dispersion in the slag, the faster will be the rate of approach to the slag–metal equilibrium. This will offset the nonequilibrium state of the phosphorus reaction, resulting in higher residual phosphorus in the steel. It follows from this argument that variations in the top-blowing practice (for example, lance height and foam stability of the slag) that affect the extent of metal dispersion in the slag and variations in slag temperature may also influence the level of residual phosphorus in the steel.

The Q-BOP slag is nonfoaming, and the extent of mixing of metal with the slag is much less than in the oxygen top-blowing practice. For these reasons, there is lack of opportunity for further reaction between slag and metal in the Q-BOP; consequently, the phosphorus reaction initiated in the gas jet stream containing powdered lime remains in a nonequilibrium state in the slag during the blow.

The oxidation of carbon and chromium in argon–oxygen decarburization (AOD) stainless steelmaking, by bottom blowing of argon–oxygen mixtures, may be explained also in terms of the oxidation–reduction cycle that is believed to occur in the Q-BOP discussed above. In the tuyere region of the gas jet stream, the chromium and some carbon will be oxidized independently of each other. Subsequently, as the chromic oxide particles float out of the melt, they will react with carbon, resulting in partial or complete recovery of chromium. Thus, we may consider two reaction zones.

(1) Near oxygen entrance,

$$2\underline{Cr} + \tfrac{3}{2}O_2 = Cr_2O_3, \qquad (9.40)$$

$$\underline{C} + \tfrac{1}{2}O_2 = CO. \qquad (9.41)$$

(2) In metal bath,

$$Cr_2O_3 + 3\underline{C} = 2\underline{Cr} + 3CO. \qquad (9.42)$$

The longer the residence time of chromic oxide particles in the melt, the greater the recovery of chromium on the reduction cycle and hence the lower the chromium losses to the slag. These expectations are substantiated by several independent experimental observations. For example, Fulton and Ramachandran [48] and Choulet *et al.* [49] found from experiments with 75-kg melts that during decarburization with argon–oxygen mixtures the chromium losses to the slag were low when the gas-injection rate was low.

Fruehan [50] has found from laboratory experiments that the rate of reduction of Cr_2O_3 by carbon in Fe–Cr–C alloys is controlled primarily by diffusion of carbon. On the basis of these findings, Fruehan [51] derived a rate equation from the AOD plant data for decarburization and oxidation of chromium on the assumption that local equilibrium prevails for reaction (9.42) at the bubble surface to which Cr_2O_3 particles are attached, and the rate is controlled by diffusion of carbon to the bubble surface. Calculations showed that in 11% Cr steels containing more than about 0.3% C, there is very little loss of chromium to the slag, because most of the oxygen blown is consumed in the decarburization by reaction (9.42). Therefore, in the early stages of blowing, a high O_2/Ar ratio can be used without much chromium loss to the slag. However, at carbon levels below 0.2%, the rate of decarburization and the extent of chromium oxidation increases with an increasing O_2/Ar ratio in the gas blown. Calculated rate data in Fig. 9.26 [51] are for an 80-ton AOD melt (11% Cr) at 1700 °C and a total gas blowing rate of 50 $nm^3\ min^{-1}$. Decreasing the O_2/Ar ratio increases the rate of decarburization because, as shown in Fig. 9.16, the equilibrium concentration of carbon is lowered with decreasing partial pressure of CO in the gas bubble.

Roy and Robertson [52] developed a more detailed mathematical modeling of the AOD process by including in the model joint effects of gas- and liquid-phase resistances to mass transfer inside and outside the bubbles and the change in the equilibrium concentration of carbon at the bubble surface during their ascent in the melt. For better simulation of the practical situation, they also considered the oxidation of manganese and silicon, as well as chromium, in the tuyere region. The oxidation of Cr, Mn, and Si near the jet entrance was considered to take place in proportion to the molar concentration of the solutes. In computing the rate of reduction of ascending oxide particles by carbon in the steel bath, and hence the rate of decarburization, due account was also taken of the increase in temperature of the melt during refining. The rate of oxidation

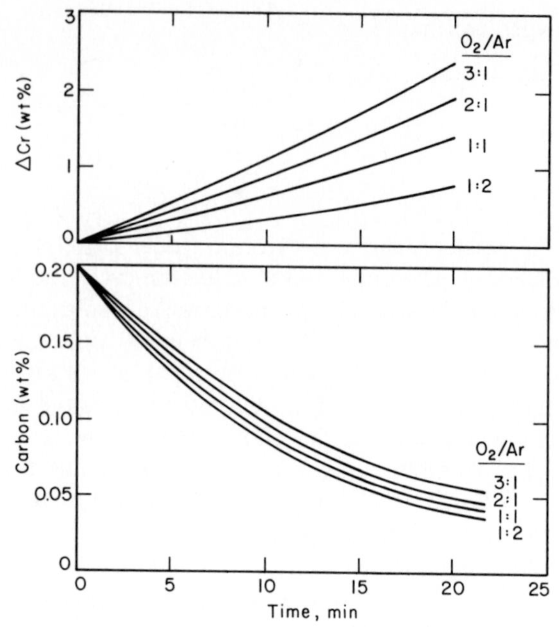

FIG. 9.26 Calculated rates of oxidation of chromium and carbon for a 409-grade stainless steel for various O_2/Ar ratios in the gas blown. After Fruehan [51].

of carbon and chromium and the temperature rise thus computed from the model were found to be in close agreement with the plant data for three-stage blowing with O_2/Ar ratios of 3:1, 2:1, and 1:2 in a 45-ton AOD vessel. Practical details of the operation of this AOD process are given in a paper by Leach *et al.* [53].

When the chromium-containing steel is decarburized to the desired low level of carbon, e.g., <0.05% C, the slag contains 20%–25% Cr_2O_3. The solubility of chromium oxide in oxidized slags is of the order of 5%. Therefore, the oxidized slag in the AOD process contains excess Cr_2O_3 as a dispersed second phase; consequently, these slags have high apparent viscosities. As pointed out earlier, the chromium is recovered from the slag by reduction with silicon in the final stages of refining. The divalent chromium formed in the reduced slag increases the solubility of chromium oxide CrO_x and lowers the slag viscosity, thus contributing to the wear of refractory lining in the slag layer.

In the light of reaction equilibria in gas–metal–slag systems and the basic concepts of rate phenomena, some general conclusions may be derived from the study of plant data for iron and steelmaking. In all these processes, we seem to encounter two or three sequences of reactions involving vaporization–oxidation–reduction cycles in the transfer of reacting elements between the metal, slag, and gas phases. Approach to complete slag–metal or gas–metal equilibrium is not to be expected for these dynamic processes. However, there are indications of approach to partial equilibrium with respect to some coupled exchange reactions, as, for example, manganese–sulfur reaction in the blast furnace and carbon–oxygen reaction in some pneumatic steelmaking processes.

The oxidation of alloying elements, or impurities in the iron in oxygen steelmaking, is often followed by reduction of the oxide reaction products by carbon in the bath. There are variations in the rates of reactions occurring on the reduction cycle leading to varying degrees of departure from the ultimate slag–metal equilibrium. For example, the oxygen potential of the slag, indicated by its iron oxide content, is invariably higher than that of the steel bath. The manganese and sulfur reactions involving iron oxide in the slag appear to approach equilibrium. Yet, in some steelmaking processes the phosphorus reaction is out of equilibrium. The rate of decarburization of steel is perhaps the simplest to interpret. At high-carbon concentrations in low-alloy steels, the rate of decarburization is determined by the rate of oxygen blowing. Below about 0.2% C, the rate is controlled primarily by the liquid-phase mass transfer of carbon to the oxidizing gas bubbles. Therefore, by controlling the composition of the oxygen-inert gas mixture blown, excessive oxidation of alloying elements and iron can be prevented.

9.4 Ladle Treatment of Liquid Steel

The ladle treatment is the final stage in steelmaking, involving the additions of alloying elements, deoxidation, desulfurization, and degassing of steel. There are many commercial processes of ladle treatment for various applications. The practical aspects of such processes will not be described here. To highlight the physical chemistry of reactions in the ladle treatment, or the so-called secondary steelmaking processes, the discussion is centered on three topics of major importance: deoxidation, desulfurization, and degassing.

9.4.1 Deoxidation

Because of the negligibly small solubility of oxygen and CO in solid iron, the liquid steel must be deoxidized to a desired low level of oxygen to insure sound castings free of blowholes and oxide inclusions. For certain applications, controlled gas evolution during solidification is necessary to produce ingots with the desired distribution of blowholes.

The choice of deoxidant depends on the desired levels of residual oxygen and deoxidant in the steel. A wide spectrum of deoxidation equilibria pertaining to the most common deoxidants is summarized in Fig. 9.27 as a log–log plot of the concentration of oxygen in solution in liquid steel against that of the added elements. In all cases, the oxygen and the alloying element in solution are in equilibrium with the appropriate gas, liquid, or solid oxide phase at 1600 °C, i.e., 1 atm CO, pure B_2O_3, pure Al_2O_3, etc. The points marked on the curves give compositions corresponding to the three-phase invariants at which two oxide phases are in equilibrium with the melt, e.g., $FeCr_2O_4$ and Cr_2O_3 at 3% Cr. Below 3% Cr chromite is the equilibrium phase, and above 3% Cr the oxide formed is Cr_2O_3.

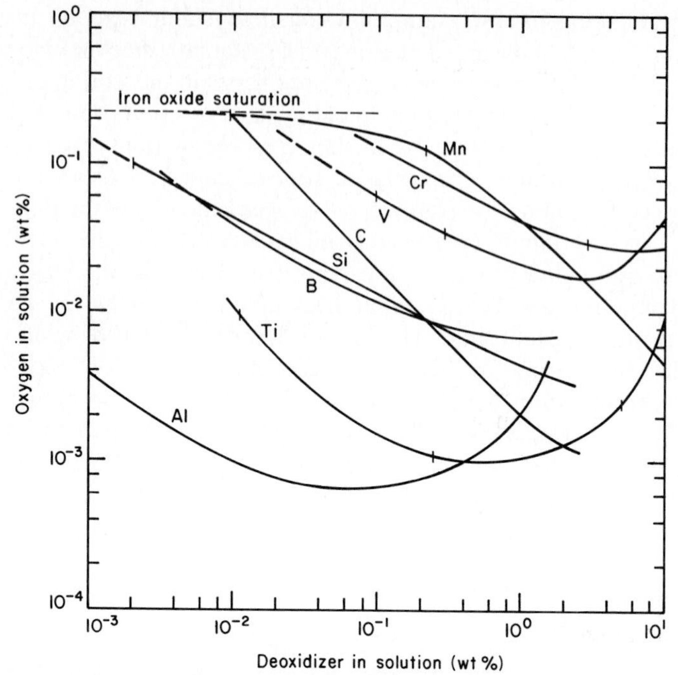

FIG. 9.27 Deoxidation equilibria in liquid iron alloys at 1600 °C.

The solubility of the deoxidation product in liquid steel is represented by

$$M_xO_y = x\underline{M} + y\underline{O},$$
$$K = [a_M]^x[a_O]^y/a_{M_xO_y}. \tag{9.43}$$

The temperature dependence of the equilibrium constant for unit activity of the deoxidation product is given in Table 9.1. The data for the Fe–Ti–O system are from the work of Suzuki and Sanbongi [54] and those for Fe–V–O from Kay and Kontopoulos [55]; the references to other data are given in a review paper [56]. The solute activities are chosen such that at infinitely dilute solutions $a_i \equiv$ wt. % i in the metal. Inserting the activity coefficients, $f_i = a_i/\%\, i$, the solubility product for unit activity of the oxide is

$$K = [\%\,M]^x[\%\,O]^y f_M{}^x f_O{}^y, \tag{9.44}$$

where f_M and f_O approach unity when % M approaches zero. The effect of alloying elements on the activity coefficient of solutes in liquid iron were given in Table 3.1. For the alloys considered in Fig. 9.27, f_M increases with increasing concentration of the solute M; however, f_O decreases with increasing % M. The net result is that a minimum may occur in the oxygen solubility at a particular solute concentration. The higher the stability of the deoxidation product, the lower the concentration of the solute at which the oxygen solubility reaches a minimum.

TABLE 9.1

Deoxidation Solubility Products in Liquid Iron

K^a	Composition range	K at 1600 °C	log K
$[a_{Al}]^2[a_O]^4$	<1 ppm Al	1.1×10^{-15}	$-(71{,}600/T) + 23.28$
$[a_{Al}]^2[a_O]^3$	>1 ppm Al	4.3×10^{-14}	$-(62{,}780/T) + 20.41$
$[a_B]^2[a_O]^3$	—	1.3×10^{-8}	
$[a_C][a_O]/p_{CO}$	<2% C	2.0×10^{-3}	$-(1168/T) - 2.07$
$[a_{Cr}]^2[a_O]^4$	<3% Cr	4.0×10^{-6}	$-(50{,}700/T) + 21.70$
$[a_{Cr}]^2[a_O]^3$	>3% Cr	1.1×10^{-4}	$-(40{,}740/T) + 17.78$
$[a_{Mn}][a_O]$	>1% Mn	5.1×10^{-2}	$-(14{,}450/T) + 6.43$
$[a_{Si}][a_O]^2$	>20 ppm Si	2.2×10^{-5}	$-(30{,}410/T) + 11.59$
$[a_{Ti}]^3[a_O]^5$	0.01%–0.25% Ti	7.9×10^{-17}	
$[a_V]^2[a_O]^4$	<0.1% V	8.3×10^{-8}	$-(48{,}060/T) + 18.61$
$[a_V]^2[a_O]^3$	>0.3% V	3.5×10^{-6}	$-(43{,}200/T) + 17.52$

[a] Activities are chosen such that $a_M \equiv \%$ M and $a_O \equiv \%$ O when % M → 0.

Although silicon is a better deoxidant than manganese, simultaneous deoxidation by these two elements gives lower residual oxygen in solution because of the reduced silica activity. The equilibrium data for simultaneous deoxidation of steel by silicon and manganese at 1600 °C are given in Fig. 9.28a [56]. The critical silicon and manganese contents of iron in equilibrium with silica-saturated manganese silicate melt are given in Fig. 9.28b for several temperatures. For compositions above a given isotherm, manganese does not participate in the reaction; instead silica is formed as the deoxidation product. In the region below the isotherm, the deoxidation product is molten manganese silicate (with little iron oxide), the composition of which is determined by the ratio $[\%\,\mathrm{Si}]/[\%\,\mathrm{Mn}]^2$ in the metal as deduced from the equilibrium constant for the reaction

$$\underline{\mathrm{Si}} + 2\mathrm{MnO} = 2\underline{\mathrm{Mn}} + \mathrm{SiO_2},$$
$$K = ([\%\,\mathrm{Mn}]^2/[\%\,\mathrm{Si}])/((a_{\mathrm{SiO_2}})/(a_{\mathrm{MnO}})^2) \tag{9.45}$$

and the activities of the oxides in the MnO–SiO_2 system (Fig. 4.1).

For special applications, the liquid steel is treated with the alloys of rare earth elements, primarily alloys of Ce and La, which are strong deoxidizers. There is much controversy on the equilibrium values of the solubility products of cerium and lanthanum oxides in liquid iron. Calculations made from what appears to be reliable thermodynamic data give, for the solubility product, at infinitely dilute solutions of Ce in liquid iron, $[\%\,\mathrm{Ce}]^2[\%\,\mathrm{O}]^3 = 3.4 \times 10^{-20}$ at 1680 °C [57]. This is about 7 orders of magnitude lower than the value measured experimentally [58]. Measured values are probably in error because of (i) a slow approach to equilibrium at very low chemical potentials of the reactants [59], (ii) incomplete separation of oxide inclusions from the melt,

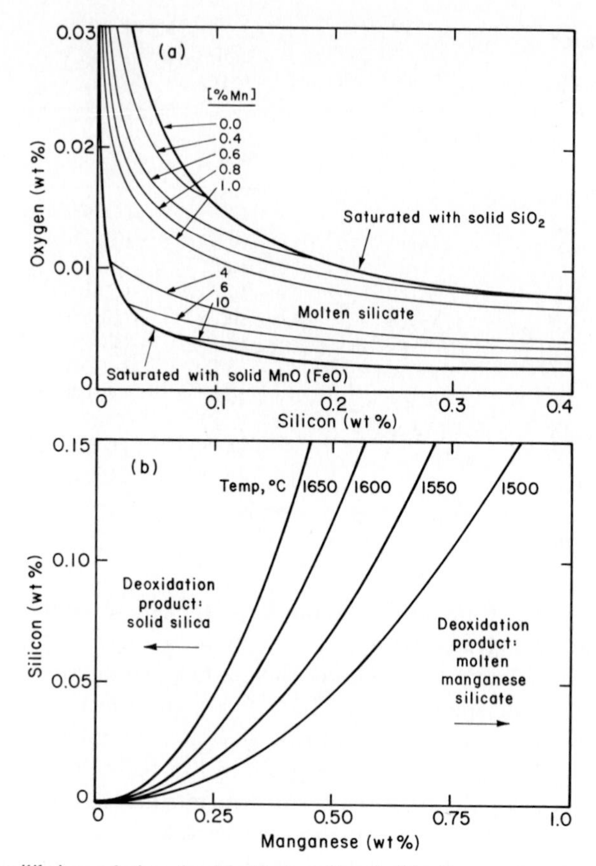

FIG. 9.28 Equilibrium relations for deoxidation of steel with silicon and manganese at 1600 °C.

and (iii) the difficulty of analyzing oxygen in iron at concentrations below a few parts per million.

The oxygen activity, or its concentration in low alloy liquid steels, is readily measured by an immersion-type oxygen sensor. The oxygen sensor is a high temperature emf–galvanic cell consisting of a lime-stabilized zirconia, ZrO_2 (CaO), as the electrolyte that is an ionic (O^{2-}) conductor, and a mixture $Cr–Cr_2O_3$ or $Mo–MoO_2$ as the reference electrode. There are numerous publications on research that led to the manufacture of oxygen sensors for use at elevated temperatures [60–62]. These cells are capable of measuring oxygen activities corresponding to concentrations in low alloy steels down to about 10 ppm. The yttria-doped thoria electrolytes may be used for lower oxygen levels of about 1 ppm.

Although elements such as aluminum interact strongly with oxygen in liquid steel and can be deoxidized to, for example 10 ppm O at 0.01% Al, in practice the total oxygen content of the steel is often much higher, e.g., 30–100 ppm. This apparent disparity is due to the kinetic effects. The rate of deoxidation of steel has been studied extensively, with particular emphasis on (i) the homogeneous nucleation of the deoxidation products in liquid steel, (ii) the progress of deoxidation reaction resulting in the growth of the size of the reaction products (inclusions), and (iii) the separation of inclusions from the melt by flotation.

The extent of supersaturation of liquid steel with oxygen and deoxidant needed for the spontaneous nucleation of the deoxidation product may be estimated from the homogeneous nucleation theory, Eq. (7.57). Turpin and Elliott [63] gave several examples of estimating the supersaturation needed for homogeneous nucleation of various types of inclusions in liquid steel. For example, for liquid iron–alumina or iron–zirconia systems, the interfacial energy is in the range 1500–2000 erg cm^{-2}, for which $K_s/K_\circ$ is of the order of 10^5–10^{10}. In fact, experimental observations of von Bogdandy *et al.* [64] show that the spontaneous nucleation of these oxides in liquid iron occurs only when $K_s/K_\circ$ is more than about 10^8. Yet, in practice, spontaneous nucleation occurs readily even though $K_s/K_\circ$ in the melt is 50 or less for homogeneous distribution of added deoxidant. The steel bath is probably never free of minute inclusions. Also, the deoxidizers added always contain inclusions which will act as nucleation sites; hence not much supersaturation, if any, is needed for the initiation of the deoxidation reaction.

The rate of deoxidation reaction is probably due to the diffusion of oxygen and/or added deoxidant solute to inclusions in the melt. Assuming an even distribution of nuclei in the melt, the diffusion-controlled rate of deoxidation may be estimated from Eq. (7.61) for an assumed number Z of nuclei per unit volume of the melt; examples of such calculations are given elsewhere [59, 65]. The inclusion counts on steel samples give values of Z in the range 10^5–10^7 particles cm^{-3}, for which the time of completion of deoxidation is of the order of a few minutes to reach the equilibrium levels of 10–20 ppm O. If the equilibrium oxygen concentration is below 0.1 ppm, as would be the case when the steel is deoxidized with rare earth elements, approach to equilibrium may be slow because of the small driving force for diffusion [59]. However, for most steelmaking practices, the diffusion-controlled rate of deoxidation is expected to be relatively fast.

What is perhaps more important is the separation of the deoxidation products from the steel bath. First, let us consider the limiting case of flotation of inclusions with terminal velocity v in accord with Stokes' law,

$$v = gd^2\,\Delta\rho/18\eta, \tag{9.46}^*$$

where g is the acceleration due to gravity, d the inclusion diameter, $\Delta\rho$ the density difference between liquid steel and inclusion, and η the viscosity of liquid steel.

* When there is slip at the particle/liquid-metal interface, the terminal velocity is 50% greater than that given by Eq. (9.46) for the case of no slip.

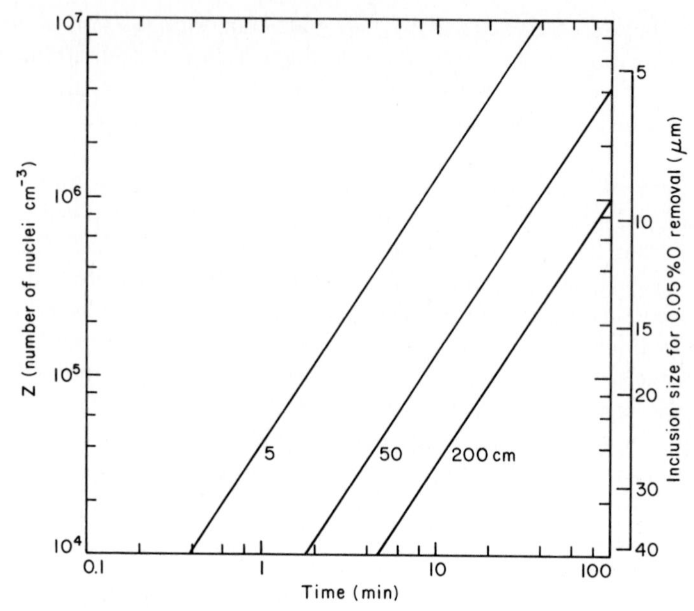

FIG. 9.29 Calculated time of flotation of inclusions in 5-, 50-, and 200-cm-deep melts as a function of inclusion size.

Let us assume that 0.05% O in steel is converted to oxide or silicate inclusions (with $\Delta\rho = 4\ \mathrm{g\,cm^{-3}}$) of number Z per unit volume of the melt. Calculated time of rise of 5, 50, and 200 cm in liquid steel is shown in Fig. 9.29. We see, for example, that 4-μm-diameter inclusions (corresponding to the removal of 0.05% O by $Z = 10^7$ nuclei cm^{-3}) would take about 40 min to rise 5 cm in the melt. In practice, however, there is sufficient stirring of liquid steel in the ladle (about 200-cm-deep melt) that the inclusions grow by collision and coalescence to sizes of 30–100 μm, resulting in faster flotation. For example, Miyashita *et al.* [66] have shown that while the rate of flotation of inclusions in unstirred melts is in accord with Stokes' law, the rate of rise is much faster in stirred melts. Experimental observations [67, 68] also indicate that in stirred melts the inclusions are trapped on the surface of the melt container.

Many experimental and plant studies have shown that the rate of separation of deoxidation products in stirred melts can be represented by an equation of the form [69]

$$C_t = C_0 \exp(-kt), \tag{9.47}$$

where C_0 is the initial concentration of inclusions, C_t the concentration of inclusions at time t, and k the apparent separation rate constant. The apparent rate constant increases with increasing intensity of stirring and depends on the means of stirring.

There are numerous methods of stirring that are applied in the ladle treatment of steel, e.g., electromagnetic, mechanical, and argon purging. Reference may be made to review papers [70] on the practical appraisals of these processes. Spurred by these industrial developments, many studies have been made of the fluidynamics of mixing in stirred melts through mathematical modeling and computer-aided calculations, as outlined in a review paper by Szekely [71].

The efficiency of coalescence of inclusions accompanying their collision depends also on the surface properties of the particles. The energy of adhesion of particles in liquids is determined by the wetting characteristics of the particle. The higher the contact angle, i.e., the poorer the wetting, the greater the force of adhesion of the particles. As shown, for example, by Bapshizmanskii *et al.* [72], the force of adhesion of alumina inclusions, with a contact angle of 140° in liquid iron, is about twice that for silica inclusions for which the contact angle is about 115°. That is, the inclusions may cluster in aluminum deoxidized steel more readily than in steel deoxidized with silicon. This has been confirmed experimentally by Torssell and Olette [73], who showed that there is faster separation of inclusions from aluminum deoxidized steels because of the cluster formation, as compared to silicon deoxidized steel.

The efficiency of aluminum usage in ladle deoxidation is about 30%–40% because the added aluminum readily floats up and is lost to the slag. The problem arises from the buoyancy effect and retarded melting of the additives. In an experimental and mathematical study, Guthrie *et al.* [74] showed that severe thermal and fluidynamic contacting resistances are encountered when buoyant objects are projected into liquid steel. The thermal effect arises from the formation of solid shell of steel around the metal praticle added to a bath of molten steel because of a rapid withdrawal of heat by the immersed object from the adjacent liquid. The depth of penetration and residence time of object in the steel bath depends on the entry velocity and size of the object. These variables, as well as the thermal properties of the added metal, determine the extent of melting that will take place before resurfacing occurs. The predictions made by Guthrie *et al.* from their mathematical analysis are summarized in Fig. 9.30 for aluminum spheres projected into a quiescent bath of liquid steel at indicated entry velocities of 5–100 m sec^{-1}.

The dotted lines show the time of residence as a function of the diameter of the aluminum sphere for indicated entry velocities. The bottom curve is for the start of melting of the aluminum inside the solid steel shell, the middle curve is for complete melting of the aluminum, and the top curve is for the start of melting of the steel shell. For example, a 2-cm-diameter aluminum sphere projected with an entry velocity of 50 m sec^{-1} will remain in the steel bath for 0.8 sec, which is the time required for complete melting of the aluminum. For a 1-cm-diameter aluminum sphere with entrance velocity of 100 m sec^{-1}, the aluminum will be completely molten and the steel shell partly melted in a residence time of 0.7 sec. These findings of Guthrie *et al.* indicate that light metals, such as aluminum, added to a quescent liquid steel bath will generally

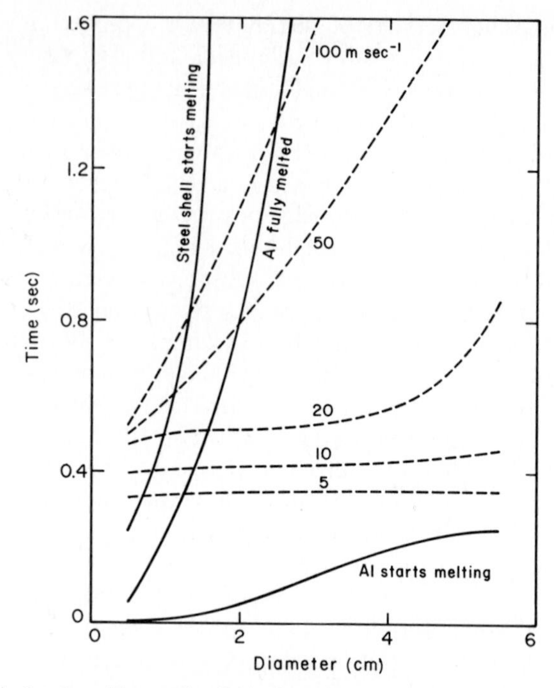

FIG. 9.30 Calculated total immersion for indicated entry velocities as a function of the diameter of the added aluminum sphere (– – –); start and completion of the melting of aluminum and start of the melting of the steel shell (———). After Guthrie *et al.* [74].

refloat before complete dissolution in the bath, even when projected with high entry velocities as with bullet shooting methods [75]. In practice, the ladle additions are made into the tapping stream, or in some cases by argon injection deep into the melt, to facilitate dissolution of additives before resurfacing.

To make sheet steel with good surface quality and deep-drawing capability, steel containing 0.30–0.35% Mn, 0.05–0.08% C, and 250–300 ppm O is cast in ingot molds with little prior deoxidation. The evolution of CO during solidification brings about the formation of a relatively pure solid iron skin next to the mold surface. When the thickness of the solidified layer (*rim*) is 35–40 mm, the liquid core is deoxidized by adding aluminum shots, while the remainder of the mold is filled with liquid steel. The efficiency of usage of aluminum is unpredictable in this process of casting rim-stabilized steel ingots because of the aforementioned difficulty of adding light solid metals into a bath of liquid steel. This problem has now been overcome in industry [76] by adding liquid aluminum into the liquid core of the solidifying ingot. Upon contact with the liquid steel, the liquid aluminum is readily assimilated and mixed with the liquid core, resulting in predictable deoxidation in the mold and even distribution of alumina inclusions in the ingot. To date, this method of deoxidation has not been extended to the ladle treatment of liquid steel.

9.4.2 Desulfurization

As indicated by the equilibrium constant k_S for reaction (9.29), the oxidizing conditions that prevail in steelmaking adversely affect the transfer of sulfur from metal to slag. The sulfur transferred to the slag during steelmaking is about one-half or one-third that present in the hot metal. Noting that the residual sulfur in most steels is in the range 0.01%–0.02%, the hot metal high in sulfur is often given a ladle treatment for desulfurization. For special applications, the residual sulfur in the steel should be less than 0.005%; this calls for ladle desulfurization of steel.

The reducing conditions in the hot metal and a high sulfur activity in the presence of carbon and silicon in the iron are most favorable for ladle desulfurization. The magnesium, calcium carbide, or soda ash are usually used in the ladle desulfurization of hot metal from about 0.04% to 0.02% S.

Although there are many variations of the methods of ladle desulfurization of steel, a feature that is common to them all is the effective deoxidation of the steel. The added desulfurizing element will also react with oxygen in the metal; therefore, the overall reaction to be considered is that involving an oxide and a sulfide,

$$\underline{S} + MO = \underline{O} + MS. \tag{9.48}$$

For the limiting case of unit activities of MO and MS, the equilibrium constant

$$K = [a_O]/[a_S] \tag{9.49}$$

may be computed from the free-energy data in Tables 1.2 and 3.4. Examples are given in Fig. 9.31 for alkaline earth sulfide–oxide and Ce_2O_2S–Ce_2O_3 equilibria. For a given level of deoxidation, the higher the equilibrium ratio a_O/a_S, the lower will be the residual sulfur in the steel. Therefore, good desulfurization is expected with barium oxide, while no desulfurization can be achieved with MgO.

In the presence of a basic oxide such as BaO or CaO, the deoxidation with, for example, aluminum is improved by the formation of aluminates, resulting in lower residual sulfur. The usual practice in ladle desulfurization is to inject alloys of calcium, e.g., Ca–Si or Ca–Ba, with argon deep into the melt after deoxidation of the steel with silicon and aluminum. The success of good ladle desulfurization depends on the precautions taken to avoid oxygen pickup from the air, slag, and the refractory lining of the ladle.

9.4.3 Degassing

Vacuum degassing of liquid steel has gone through many stages of development; pertinent information on commercial processes are well documented in comprehensive review papers by Flux [77] and others [70].

For certain critical applications, the hydrogen content of steel should be less than 3 ppm, which is about one-half or two-thirds the amount normally

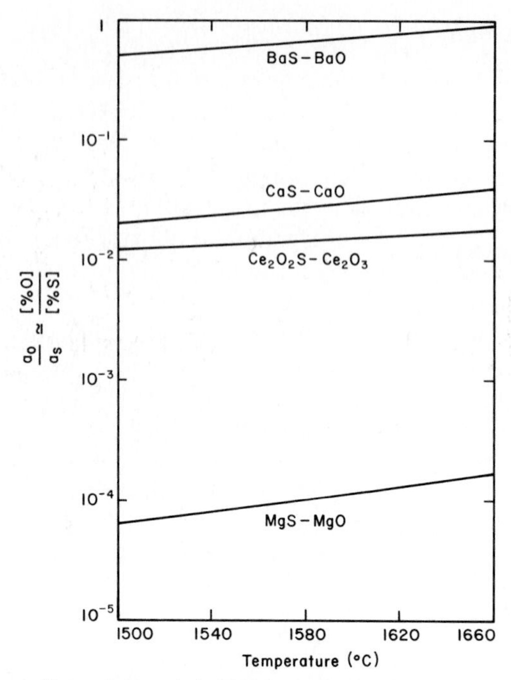

FIG. 9.31 Oxygen/sulfur activity ratio in liquid iron for the indicated sulfide–oxide equilibrium at 1600 °C.

present in the steel at tap. The primary purpose of vacuum treatment is to remove hydrogen accompanied by carbon deoxidation; evolution of CO facilitates the removal of hydrogen. The efficiency of hydrogen removal is markedly improved with argon purging of the melt during vacuum treatment, particularly when the steel has been completely deoxidized (*killed*) in a prior ladle treatment. The nitrogen cannot be removed effectively from liquid steel by vacuum treatment because of the poisoning effect of residual oxygen and sulfur on the kinetics of the nitrogen reaction, as discussed in Section 7.2.2. Reference should be made also to Section 7.9 for some physicochemical and dynamic aspects of gas evolution from liquid metals.

During vacuum treatment, there is much stirring of the melt, resulting in growth of inclusions and their rapid separation from the melt. In fact, in the manufacture of high quality special steels, the vacuum treatment is done primarily to facilitate the removal of inclusions.

9.5 Reactions during Solidification of Steel

Aside from entrapped refractory or slag particles and the products of re-oxidation which are carried into the casting mold, there are oxide, silicate, aluminate, and sulfide inclusions in the castings that originate primarily from two sources: (i) incomplete separation of the products of reactions in the ladle, which may be called the ladle inclusions, and (ii) gaseous, liquid, or solid products of reactions occurring during solidification of steel, which may be called the solification inclusions.

9.5.1 Oxide and Silicate Inclusions and Blowholes

The formation of solidification inclusions is a consequence of the microsegregation that accompanies dendritic freezing of alloys. Advances made in the understanding of the morphology of dendritic solidification are well documented in textbooks by, for example, Chalmers [78] and Flemings [79].

Microsegregation resulting from solute rejection during dendritic freezing in binary alloys has been formulated by Scheil [80] for two limiting cases. When there is complete diffusion in the solid (dendrite arms) and in the liquid phase between the dendrite arms, the concentration of solute in the liquid is given by

$$C_l = C_0/[1 + g(k - 1)] \tag{9.50}$$

where C_0 is the initial solute concentration, C_l the solute concentration in enriched liquid, k the solid/liquid solute distribution ratio, and g the volume fraction of liquid solidified. This limiting equation will apply to carbon, oxygen, hydrogen, and nitrogen in iron because of their high diffusivities.

When there is negligible diffusion in the solid and complete diffusion in the liquid phase, the solute enrichment in the liquid is given by

$$C_l = C_0(1 - g)^{k-1}. \tag{9.51}$$

This equation is more appropriate for substitutional elements in iron, e.g., manganese, silicon, and phosphorus, because of their low diffusivities in solid alloys.

For dilute solutions, the solute distribution ratio k is about 2/3 for the Fe–Mn and Fe–Si systems, 0.2 for the Fe–C system, and essentially zero for the Fe–O and Fe–S systems. For the purpose of demonstrating the consequences of microsegregation in the solidification of steel, let us make the simplifying assumption that the solute distribution ratios for low alloy steels are similar to those in the iron-base binary alloys, and not affected by the other solutes. Calculated solute content in the interdendritic liquid is shown in Fig. 9.32 as a function of percent solidification in a given volume of the liquid.

In the case considered, oxygen in solution is the only element that may react with the other solutes when their concentrations become sufficiently high in the enriched liquid. On the basis of this reasoning, Turkdogan [81] has pointed out that in calculating the solute enrichment in the entrapped freezing liquid, due account must be taken of the deoxidation reactions. Because of the complexity of the problem, certain simplifying assumptions have to be made in computing the solute enrichment controlled by solute distribution between solid and liquid phases coupled with reactions in the enriched liquid. The primary assumption is that there are sufficient nuclei present in the system,

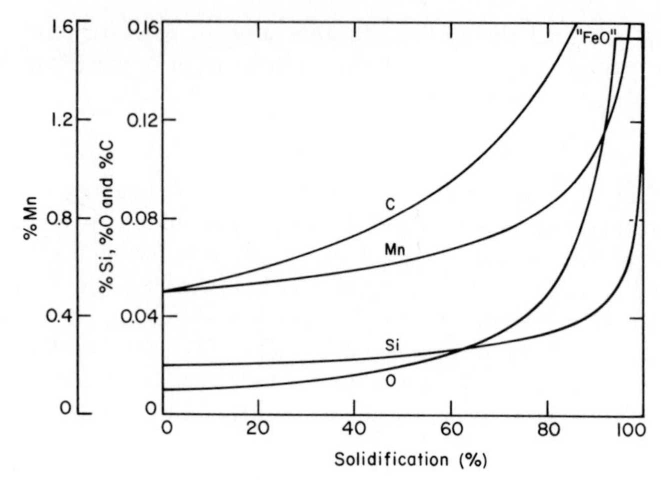

FIG. 9.32 Calculated solute enrichment in solidifying liquid steel if no reaction takes place between the solutes in the steel. After Turkdogan [81].

and therefore, little or no supersaturation is needed for the formation of the reaction products.

An example of the calculations is given in Fig. 9.33a [81], which shows the oxygen content of entrapped liquid in the solid + liquid *mushy zone* of the ingot, controlled by the Si–Mn deoxidation reaction, as a function of the local percentage of solidification. The dotted curve gives the oxygen enrichment in the absence of deoxidation reaction.

When the carbon content of the interdendritic liquid becomes sufficiently high, it will react with oxygen, forming carbon monoxide. Similarly, at sufficiently high concentrations of hydrogen and nitrogen in the enriched liquid, there will be gas evolution. This gas formation results in the displacement of liquid to neighboring interdendritic cells, and hence the formation of blowholes. The equilibrium oxygen contents of steel for the carbon–oxygen reaction at 1 atm CO at various stages of solidification for several initial carbon concentrations are superimposed in Fig. 9.33b on the curves reproduced from Fig. 9.33a. If the concentration of silicon, manganese, or other deoxidizers is sufficiently high, the oxygen in solution in the enriched liquid will be maintained at low levels during solidification, so that the carbon–oxygen reaction will not take place; thus blowholes are absent (if the hydrogen and nitrogen contents are sufficiently low). As shown in Fig. 9.33b, for 1 atm pressure of CO, a particular carbon line tangential to the Si/Mn-deoxidation curve gives the critical carbon content of steel below which blowholes will not form.

In a composition plot in Fig. 9.34, the sections of laboratory ingots that showed blowholes are indicated by open circles and the castings free of blowholes by filled circles. The critical curve derived from a theoretical analysis

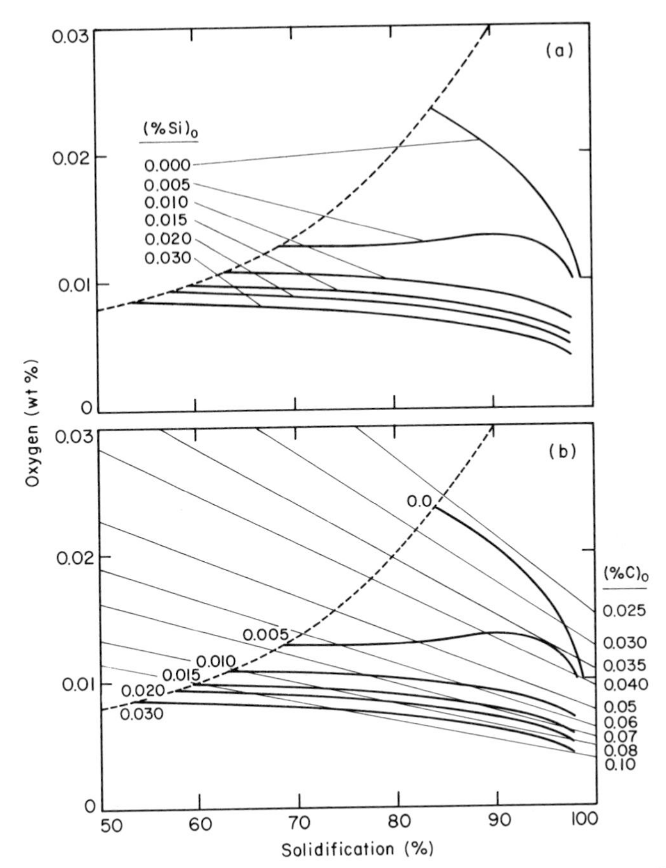

FIG. 9.33 Calculated oxygen content of interdendritic liquid controlled by (a) Si–Mn deoxidation reaction and (b) C–O reaction in the enriched liquid for initial concentrations of 0.5% Mn and 40 ppm O. After Turkdogan [81].

[81], briefly outlined here, is well supported by experimental observations [82, 83]. Furthermore, practical experience in continuous casting of low carbon and low silicon steels, deoxidized with silicon, manganese, and some aluminum, has substantiated the applicability of the critical Si–C curve (for a given manganese level) in the production of castings free of blowholes. For a given carbon concentration, the Si–C curve moves to higher silicon levels with decreasing manganese and increasing hydrogen and nitrogen concentration in the steel. The calculations made on the basis of this conceptual analysis indicate that for a given steel composition near the Si–C critical curve, when the hydrogen content of the steel increases from 2 to 5 ppm, the chances of blowhole formation are greater than when the nitrogen content increases from 20 to 50 ppm [81].

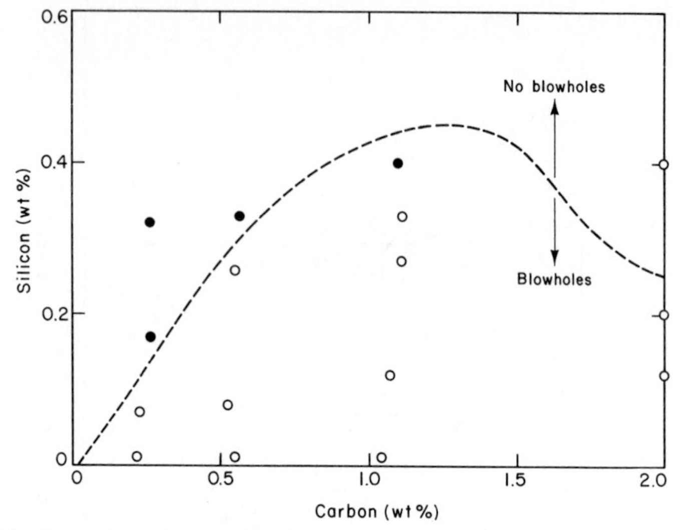

FIG. 9.34 Comparison of calculated critical concentrations of silicon and carbon for the suppression of blowholes with experimental observations in steels containing 0.5%–0.6% Mn and 25–65 ppm O. After Turkdogan and Grange [83].

9.5.2 Sulfide Inclusions

The sulfide and oxysulfide inclusions observed in castings are formed during solidification in a manner similar to that discussed above for the formation of interdendritic oxide and silicate inclusions and blowholes. The metallographic examinations made by Turkdogan and Grange [83] of as-cast and annealed samples of steels substantiated the view that the manganese sulfide inclusions in steel are indeed found in the interdendritic regions (Fig. 9.35), which ultimately become the grain boundaries in the steel. The white regions are cross sections of dendrite arms low in manganese. The pearlitic network is symptomatic of manganese enrichment in the interdendritic regions. The sulfide inclusions in the interdendritic regions are embedded in a narrow ferrite band; this is due to local depletion of manganese as a result of the precipitation of manganese sulfide in the later stages of solidification. Because of the low diffusivity of manganese in the steel and negligible dissolution of manganese sulfides in steel at elevated temperatures, the heat treatment does not much alter the original pattern of the distribution of sulfide inclusions.

Depending on the concentration of alloying elements in steel, and the level of oxygen therein, the sulfides formed during solidification may be solid sulfides or liquid oxysulfides [59, 84]. The presence of liquid oxysulfides along grain boundaries is undesirable because of their adverse effect on the high-temperature ductility of the steel. The accumulation of sulfide inclusions in interdendritic regions is responsible for the formation of streaks and stringers during hot working of the steel. The resulting banded micro-structure has an adverse

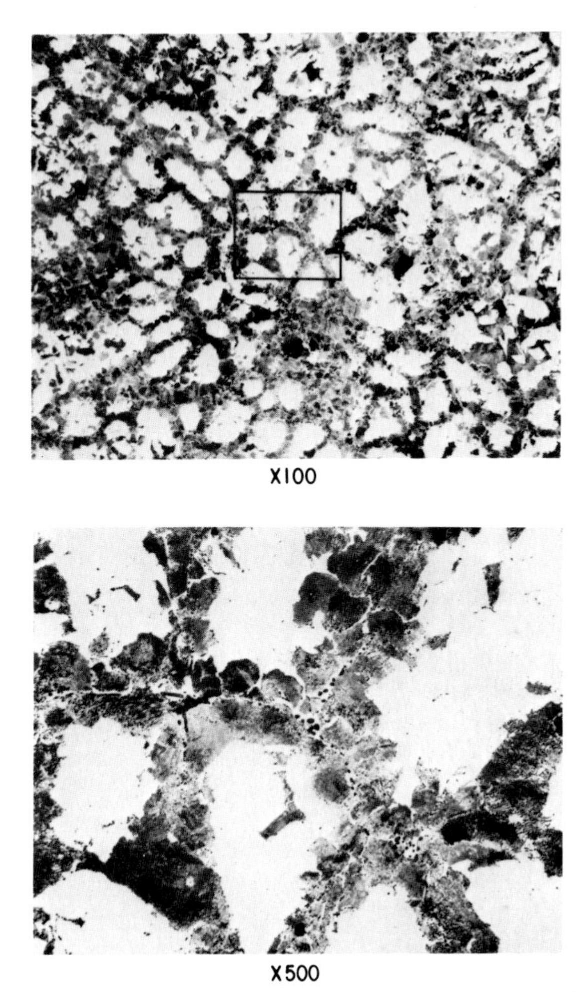

FIG. 9.35 Manganese sulfide inclusions in the interdendritic regions (pearlite network) in the as-cast condition of steel containing 1.5% Mn, 0.25% C, and 0.05% S. After Turkdogan and Grange [83].

effect on the transverse mechanical properties of the steel. Many of these problems are circumvented when steel is desulfurized to levels below 0.005% by ladle treatment with Ca–Si, Ca–Ba, or rare earth alloys [59]. For further reading on nonmetallic inclusions in steel, reference may be made to Kiessling [85].

In this condensed presentation of reactions occurring in the ladle treatment and solidification of steel, an attempt is made to draw attention to the diversity of physical sciences available for the control of these high temperature processes through a better understanding of their workings. Deoxidation is perhaps the most important reaction which directly affects the efficiency of ladle desulfurization, steel cleanliness (low inclusion count), soundness of castings (little or no blowholes), the form of sulfide inclusions, and so on. The oxide, silicate, and aluminate inclusions resulting from deoxidation have sizes that usually are below 5 μm; consequently, their separation from the melt by flotation is sluggish. Agitation of the bath with argon purging, vacuum treatment, or electromagnetic stirring enhances the inclusion growth by collision and coalescence and hence their rapid separation from the melt. Close control of deoxidation is necessary to accomplish two primary objectives: (i) to lower the residual oxygen in solution in steel to the desired level and (ii) to separate inclusions from the melt prior to casting.

Reactions occurring during solidification of steel are also of major importance in achieving casting of desired soundness and mechanical properties. As seen from the examples given, what happens during solidification may be controlled to a large extent by proper adjustment of the steel composition and the concentrations of residual hydrogen, oxygen, and sulfur.

References

1. "Blast Furnace Theory and Practice" (J. H. Strassburger, ed.). Gordon & Breach, New York, 1969.
2. L. von Bogdandy and H.-J. Engell, "The Reduction of Iron Ores." Springer-Verlag, Berlin and New York, 1971.
3. C. Bodsworth and H. B. Bell, "The Physical Chemistry of Iron and Steel Manufacture," 2nd Ed., Longmans, London, 1972.
4. "BOF Steelmaking" (R. D. Pehlke, W. F. Porter, R. F. Urban, and J. M. Gaines, eds.), Vols. I–V. AIME, New York, 1975.
5. P. Reichardt, *Arch. Eisenhuettenwes.* **1**, 77 (1927/28).
6. J. M. Ridgion, *J. Iron Steel Inst., London* **200**, 389 (1962).
7. L. von Bogdandy, *Stahl Eisen* **81**, 12 (1961).
8. E. Schurmann and D. Butler, *Arch. Eisenhuettenwes.* **35**, 475 (1964).
9. E. Schurmann, W. Zischkale, P. Ischebeck, and G. Heynert, *Stahl Eisen* **80**, 854 (1960).
10. H. Itaya, T. Fukutake, K. Okabe, and T. Nagai, *Tetsu To Hagane* **62**, 472 (1976); Engl. transl. BISI 14858.
11. A. Rist and G. Bonnivard, *Rev. Metall.* (*Paris*) **60**, 23 (1963).
12. A. Rist and N. Meysson, *Ironmaking Proc., Metall. Soc. AIME* **25**, 88 (1966).
13. C. Staib, N. Jusseau, J. Vigliengo, and J. C. Cochery, *Ironmaking Proc., Metall. Soc. AIME* **26**, 66 (1967).
14. H. Ceckler, H. N. Lander, and J. M. Uys, *in* "Blast Furnace Theory and Practice" (J. H. Strassburger, ed.), p. 785. Gordon & Breach, New York, 1969.
15. J. G. Peacey and W. G. Davenport, "The Iron Blast Furnace." Pergamon, Oxford and New York, 1979.
16. A. Decker and R. Scimar, *CNRM Metall. Rep.* **12**, 37 (1967).
17. N. Tsuchiya, M. Tokuda, and M. Ohtani, *Metall. Trans. B* **7**, 315 (1976).
18. J. J. Bosley, N. B. Melcher, and M. M. Harris, *J. Met.* **11**, 610 (1959).
19. K. Okabe *et al.*, *Kawasaki Steel Tech. Res. Lab. Rep.* May 4, pp. 1–43 (1974); English transl. BISI 13657, Nov. 27 pp. 1–11 (1974); English transl. BISI 13658.
20. E. T. Turkdogan, *Metall. Trans. B* **9**, 163 (1978).
21. W. Oelsen and H. G. Schubert, *Arch. Eisenhuettenwes.* **35**, 1039, 1115 (1964).
22. W. Oelsen, H. G. Schubert, and O. Oelsen, *Arch. Eisenhuettenwes.* **36**, 779 (1965).
23. S. P. Kinney and E. W. Guernsey, *Blast Furnace Steel Plant* p. 395 (1925).
24. F. D. Richardson and J. H. E. Jeffes, *J. Iron Steel Inst., London* **163**, 397 (1949).

25. K. P. Abraham and L. J. Staffansson, *Scand. J. Metall.* **4**, 193 (1975).
26. "Alkalies in Blast Furnaces" (N. Standish and W.-K. Lu, eds.). McMaster Univ., Hamilton, Ontario, 1973.
27. T. Fuwa and J. Chipman, *Trans. Metall. Soc. AIME* **218**, 887 (1960).
28. N. H. El-Kaddah and D. G. C. Robertson, *Metall. Trans. B* **8**, 569 (1977).
29. C. R. Taylor and J. Chipman, *Trans. AIME* **154**, 228 (1943).
30. P. A. Distin, S. G. Whiteway, and C. R. Masson, *Can. Metall. Q.* **10**, 13 (1971).
31. E. T. Turkdogan, *in* "BOF Steelmaking" (R. D. Pehlke, W. F. Porter, R. F. Urban, and J. M. Gaines, eds.), Vol. 2, p. 1. Iron Steel Soc. AIME, New York, 1975.
32. E. T. Turkdogan, *J. Iron Steel Inst., London* **179**, 147 (1955).
33. E. T. Turkdogan, *Iron Coal Trades Rev.* Dec. 16, 1471 (1955).
34. G. J. W. Kor, *Metall. Trans. B* **8**, 107 (1977).
35. K. Balajiva, A. G. Quarrell, and P. Vajragupta, *J. Iron Steel Inst., London* **153**, 115 (1946).
36. K. Balajiva and P. Vajragupta, *J. Iron Steel Inst., London* **155**, 563 (1947).
37. F. D. Richardson and W. E. Dennis, *J. Iron Steel Inst., London* **175**, 257, 264 (1953).
38. *Electr. Furn. Steel Proc.* **4**, 105 (1946).
39. W. J. Rankin and A. K. Biswas, *Inst. Min. Metall., Trans., Sect. C* **87**, 60 (1978).
40. H. Flood and K. Grjotheim, *J. Iron Steel Inst., London* **171**, 64 (1952).
41. T. Forland, *in* "Fused Salts" (B. R. Sundheim, ed.), p. 83. McGraw-Hill, New York, 1964.
42. T. Forland and K. Grjotheim, *Metall. Trans. B* **8**, 645 (1977).
43. H. Schenck, "Physical Chemistry of Steelmaking," Engl. transl. Br. Iron Steel Res. Assoc., London, 1945.
44. R. J. Fruehan, *Ironmaking Steelmaking* **3**, 33 (1976).
45. E. T. Turkdogan, *in* "Physico-Chimie et Sidérurgie," 7èmes Journées Internationales de Sidérurgie (in press).
46. M. Kawana, "Pure Oxygen Bottom Blown Converter Steelmaking Process." Kawasaki Steel. Tekkokai, Tokyo, 1978.
47. E. T. Turkdogan and J. Pearson, *J. Iron Steel Inst., London* **173**, 217 (1953).
48. J. C. Fulton and S. Ramachandran, *Electr. Furn. Steel Proc.* **30**, 43 (1972).
49. R. J. Choulet, F. S. Death, and R. N. Dokken, *Can. Metall. Q.* **10**, 129 (1971).
50. R. J. Fruehan, *Metall. Trans. B* **6**, 573 (1975).
51. R. J. Fruehan, *Ironmaking Steelmaking* **3**, 153 (1976).
52. T. D. Roy and D. G. C. Robertson, *Ironmaking Steelmaking* **5**, 198, 207 (1978).
53. J.C.C. Leach, A. Rodgers, and G. Sheehan, *Ironmaking Steelmaking* **5**, 107 (1978).
54. K. Suzuki and K. Sanbongi, *Trans. Iron Steel Inst. Jpn.* **15**, 618 (1975).
55. D.A.R. Kay and A. Kontopoulos, *in* "Chemical Metallurgy of Iron and Steel," p. 178. Iron Steel Inst., London, 1973.
56. E. T. Turkdogan, *J. Iron Steel Inst., London* **210**, 21 (1972).
57. A. Vahed and D.A.R. Kay, *Metall. Trans. B* **6**, 285 (1975).
58. R. J. Fruehan, *Metall. Trans.* **5**, 345 (1974).
59. E. T. Turkdogan, *in* "Sulfide Inclusions in Steel" (J.J. deBarbadillo and E. Snape, eds.), p. 1. ASM, Metals Park, Ohio, 1975.
60. G. R. Fitterer, *J. Met.* **18**, 961 (1966); **19**, 92 (1967).
61. R. J. Fruehan, L. J. Martonik, and E. T. Turkdogan, *Trans. Metall. Soc. AIME* **245**, 1501 (1969).
62. C. Gattelier, K. Torssell, M. Olette, M. Meysson, M. Chastant, A. Rist, and P. Vicens, *Rev. Metall.* (*Paris*) **66**, 673 (1969).
63. M. L. Turpin and J. F. Elliott, *J. Iron Steel Inst., London* **204**, 217 (1966).
64. L. von Bogdandy, W. Meyer, and I. N. Stranski, *Arch. Eisenhuettenwes.* **32**, 451 (1961).
65. E. T. Turkdogan, *J. Iron Steel Inst., London* **204**, 914 (1966).
66. Y. Miyashita, K. Nishikawa, T. Kawawa, and M. Ohkubo, *Jpn–USSR Jt. Symp. Phys. Chem. Metall. Processes, 2nd* p. 101. Iron Steel Inst. Jpn., Tokyo, 1969.
67. N. Lindskog and H. Sandberg, *Scand. J. Metall.* **2**, 71 (1973).

68. T. A. Engh and N. Lindskog, *Scand. J. Metall.* **4**, 49 (1975).
69. H. Sandberg, T. Engh, J. Andersson, and R. Olsson, *Jernkontorets Ann.* **155**, 201 (1971).
70. "Secondary Steelmaking." Met. Soc., London, 1978.
71. J. Szekely, *in* "Physico-Chimie et Sidérurgie," 7èmes Journées Internationales de Sidérurgie (in press).
72. V. I. Bapshizmanskii, N. Bakhman, and Y. V. Dmishriev, *Izv. Vuz-Chern. Met.* **12**, 42 (1969); Henry Brutcher Engl. transl. 7830.
73. K. Torssell and M. Olette, *Rev. Metall.* (*Paris*) **66**, 813 (1969).
74. R.I.L. Guthrie, L. Gourtsoyannis, and H. Henein, *Can. Metall. Q.* **15**, 145 (1976).
75. T. Tanoue, Y. Umeda, H. Ichikawa, and T. Aoki, *Sumitomo Search* May 19, p. 74 (1973).
76. M. A. Orehoski, J. W. Bales, and D. A. Venseret, *NOH-BOS Conf. Proc. AIME* **57**, 386 (1974).
77. J. H. Flux, *in* "Vacuum Degassing of Steel," p. 1. Iron Steel Inst., London, 1965.
78. B. Chalmers, "Principles of Solidification." Wiley, New York, 1964.
79. M. C. Flemings, "Solidification Processing." McGraw-Hill, New York, 1974.
80. E. Scheil, *Z. Metallkd.* **34**, 70 (1942).
81. E. T. Turkdogan, *Trans. Metall. Soc. AIME* **233**, 2100 (1965).
82. E. T. Turkdogan, *J. Met.* **19**, 38 (1967).
83. E. T. Turkdogan and R. A. Grange, *J. Iron Steel Inst., London* **208**, 482 (1970).
84. J. C. Yarwood, M. C. Flemings, and J. F. Elliott, *Metall. Trans.* **2**, 2573 (1971).
85. R. Kiessling, "Non-Metallic Inclusions in Steel," Parts I–IV. Met. Soc., London, 1978.

CHAPTER 10

REACTIONS IN ENERGY-RELATED PROCESSES AND HOT CORROSION

10.1 Introduction

In previous chapters we discussed several gas–solid and gas–liquid reactions related to calcination, roasting, reduction, smelting, and metal refining processes. Selected examples of other gas–solid reactions at elevated temperatures presented in this chapter are those related to energy conversion and hot corrosion of metal alloys in process environment. Processes related to energy conversion are in a development state both in concept and application; also, only limited aspects of such processes are pertinent to high temperature technology. Therefore, a comprehensive coverage of this subject within the context of this book is not possible. Instead, we shall confine our deliberations to energy-related reactions and systems at elevated temperatures.

10.2 Thermodynamic Limitation to Energy Conversion—Decomposition of Water

Estimation of the thermodynamic limitation is prerequisite to the evaluation of the practical feasibility of any process. By considering the specific process of decomposition of water to hydrogen and oxygen, Funk and Reinstrom [1] derived a thermodynamic argument from the Gibbs free-energy function and the Carnot cycle to evaluate the thermodynamics of multistep processes.

In the processes considered, the net reaction is the decomposition of water vapor to hydrogen and oxygen.

$$H_2O \rightarrow H_2 + \tfrac{1}{2}O_2. \tag{10.1}$$

For the reversible process at constant temperature and pressure, the useful work required is

$$w = \Delta F = \Delta H - T\,\Delta S \tag{10.2}$$

and the heat (thermal energy) required is

$$q = T\,\Delta S, \tag{10.3}$$

where ΔF, ΔH, and ΔS are changes in the Gibbs free energy, the enthalpy, and the entropy for reaction (10.1). The useful work required for the reaction must be produced from heat in an engine of some sort. Therefore, the total heat requirement is

$$Q = (w/\xi) + q, \qquad Q = (\Delta F/\xi) + T\,\Delta S, \tag{10.4}$$

where ξ is the efficiency of converting heat to work. The thermodynamic efficiency of the process may be defined by the ratio

$$\eta = \Delta H/Q, \tag{10.5}$$

which is sometimes called *a figure of merit.*

Dissociation of water vapor at room temperature, for example, involves an enthalpy change of 57.8 kcal mole^{-1}, of which about 95% has to be supplied as useful work in an electrolysis cell. Also, although the voltage efficiency is high, e.g., 80%–100%, the total energy requirement for electrolysis is high, because ξ is about 0.3 for generating electricity from heat. As indicated by Eq. (10.4), the theoretical useful work requirement may be reduced by operating the process at elevated temperatures.

Now, let us consider reaction (10.1) occurring in a two-step process operating reversibly, isothermally, and isobarically at high and low temperatures T_H and T_C. With the simplifying assumption that the reactants and products have equal heat capacities, Funk and Reinstrom derived the following equation for the useful work and heat requirements:

$$w_H = w_C - (T_H - T_C)\Delta\tilde{S}^\circ = \Delta F^\circ - (T_H - T_C)\Delta\tilde{S}^\circ, \tag{10.6}$$

$$q_H = q_C + (T_H - T_C)\Delta\tilde{S}^\circ = T_H\Delta\tilde{S}^\circ, \tag{10.7}$$

where $\Delta\tilde{S}^\circ$ is the average entropy change for reaction (10.1) over a wide temperature, and $w_C = \Delta F^\circ$ is the Gibbs free-energy change for reaction (10.1) at the reference temperature $T_C = 298\,^\circ\text{K}$. For the limiting case of zero work requirement, i.e., no work except expansion against constant pressure, which is equivalent to conducting each reaction at equilibrium, $\Delta\tilde{H}^\circ$ is equal to $(T_H - T_C)\Delta\tilde{S}^\circ$, and η of Eq. (10.5) equals the ideal Carnot efficiency:

$$\eta = (T_H - T_C)/T_H. \tag{10.8}$$

We see from Eq. (10.6) that since ΔF° and $\Delta\tilde{S}^\circ$ are fixed, the useful work requirement may be reduced only by increasing T_H. Inserting $\Delta F^\circ = 54.64$ kcal per mole water vapor at 298 °K and $\Delta\tilde{S}^\circ = 13.35$ cal mole^{-1} deg^{-1}, T_H has to be 4118 °C for the thermal decomposition of water vapor with zero work requirement. Because $\Delta\tilde{S}^\circ$ term in Eq. (10.6) is constant and the practical upper limit for T_H is in the range 900–1100 °C, the theoretical work requirement cannot be reduced to zero. That is, the thermochemical decomposition of water is not possible by a two-step process.

Funk and Reinstrom then extended their thermodynamic argument to a multistep process involving a sequence of I reactions for which the total work and heat requirements are given by the summations

$$w = \sum_{i=1}^{i=I} [\Delta F_{C(i)} - (T_{H(i)} - T_C)\,\Delta\tilde{S}^\circ_{(i)}], \tag{10.9}$$

$$q = \sum_{i=1}^{i=I} [q_{C(i)} + (T_{H(i)} - T_C)\,\Delta\tilde{S}^\circ_{(i)}]. \tag{10.10}$$

Since the equilibrium is assumed, the temperature T_H for each reaction i is given by the ratio $\Delta\tilde{H}^\circ_{(i)}/\Delta\tilde{S}^\circ_{(i)}$. Noting that the summations on $\Delta F_{C(i)}$ and $q_{C(i)}$ in a sequence of reactions are equal to ΔF° and $T_C\,\Delta\tilde{S}^\circ$, respectively, for reaction (10.1) at 298 °K, we may write

$$w = \Delta F^\circ - \sum_{i=1}^{i=I} (T_{H(i)} - T_C)\,\Delta\tilde{S}^\circ_{(i)}, \tag{10.11}$$

$$q = \sum_{i=1}^{i=I} T_{H(i)}\,\Delta\tilde{S}^\circ_{(i)}. \tag{10.12}$$

In a sequence of I reactions in a multistep process, there are J reactions that have positive entropy change and L reactions with negative entropy change. It follows from Eq. (10.11) that to minimize the useful work required in a multistep process, the reactions with positive entropy change should operate at a higher temperature T_H, while reactions with negative entropy change at a lower temperature T_C.

$$w_H = \Delta F^\circ - \sum_{j=1}^{j=J} (T_H - T_C)\,\Delta\tilde{S}^\circ_{(j)} - \sum_{l=1}^{l=L} (T_C - T_C)\,\Delta\tilde{S}^\circ_{(l)}. \tag{10.13}$$

Since the last term is zero, the sum of positive entropy changes for J reactions at T_H for zero work requirement is

$$\sum_{j=1}^{j=J} \Delta\tilde{S}^\circ_{(j)} = \frac{\Delta F^\circ}{T_H - T_C}. \tag{10.14}$$

Also, it should be noted that the total entropy change in a sequence of I reactions is equal to $\Delta\tilde{S}^\circ$ for the net reaction (10.1):

$$\sum_{i=1}^{i=I} \Delta S_{(i)} = \sum_{j=1}^{j=J} \Delta\tilde{S}^\circ_{(j)} + \sum_{l=1}^{l=L} \Delta\tilde{S}^\circ_{(l)} = \Delta\tilde{S}^\circ. \tag{10.15}$$

Based on this thermodynamic argument (which was further discussed subsequently by Abraham and Schreiner [2]) leading to Eqs. (10.14) and (10.15), Funk *et al.* [3] evaluated multistep thermochemical processes for the production of hydrogen and oxygen from water. Reference may be made also to a review paper by Chao [4] on processes for thermochemical decomposition of water.

Because of increasing demand for hydrogen in new process developments for the production of hydrocarbons from coal, and anticipated greater demand for hydrogen as the primary source of clean energy of the future, much research has been done during the past decade on novel methods of production of hydrogen by thermochemical processes, using nuclear energy (HTGR) as the source of heat. Of the numerous conceptual processes that are being investigated [3, 4], three examples are given in Table 10.1 of multistep processes for thermochemical decomposition of water.

TABLE 10.1

Examples of Thermochemical Processes for Decomposition of Water Vapor to Hydrogen and Oxygen by a Sequence of Reactions

Reaction	Temperature (°C)
I. Iron chloride–iron oxide process[a]	
$3H_2O + 3Cl_2 \rightarrow 6HCl + \frac{3}{2}O_2$	700
$3Fe_2O_3 + 18HCl \rightarrow 6FeCl_3 + 9H_2O$	120
$6FeCl_3 \rightarrow 6FeCl_2 + 3Cl_2$	420
$6FeCl_2 + 8H_2O \rightarrow 2Fe_3O_4 + 12HCl + 2H_2$	650
$2Fe_3O_4 + \frac{1}{2}O_2 \rightarrow 3Fe_2O_3$	350
II. Steam–iron–CO_2 process[b]	
$Fe + H_2O \rightarrow FeO + H_2$	550 to 950 (for the four reactions bracketed)
$3FeO + H_2O \rightarrow Fe_3O_4 + H_2$	
$Fe_3O_4 + CO \rightarrow 3FeO + CO_2$	
$FeO + CO \rightarrow Fe + CO_2$	
$2CO_2 \rightarrow 2CO + O_2$	315
III. Manganese–sodium oxide process[c]	
$Mn_2O_3 + 4NaOH \rightarrow 2(Na_2O \cdot MnO_2) + H_2O + H_2$	800
$2(Na_2O \cdot MnO_2) + nH_2O \rightarrow 4NaOH(aq) + 2MnO_2$	100
$2MnO_2 \rightarrow Mn_2O_3 + \frac{1}{2}O_2$	600

[a] See, for example, Hardy [5].
[b] See, for example, von Fredersdorff [6].
[c] See, for example, Marchetti [7].

It is well to bear in mind that the calculated thermodynamic efficiency, or figure of merit, gives an optimistic evaluation of a multistep process on the assumption of equilibria in sequence of reactions and takes no account of kinetic effects, imcomplete heat and mass transfer, and the energy required to transport materials to and from the reactors. Nevertheless, the thermodynamic argument developed by Funk and Reinstrom is of value in searching for materials which may be suitable for intermediate reactions in a multistep process.

10.3 Coal Conversion

Principles of conversion of coal to liquid, gaseous hydrocarbons, and chemicals were conceived and applied on an industrial scale first in Germany in the mid-1920s. In the years following the World War II, there was little further industrial development in coal conversion. This situation has changed much during the past decade. Limited world reserves of natural petroleum and gas, increasing demand for these commodities, and stringent regulations on emission controls in recent years brought about greater urgency to research into the production of liquid hydrocarbons, synthetic natural gas, and energy for generating electricity from coal, which is the world's major resource of carbonaceous raw material. Vast strides have been made in the 1970s in exploring alternative methods of conversion of coal to chemical feedstocks and low sulfur fuels by processes that are economically feasible and meet most regulations on environmental pollution.

Depending on the fixed carbon content and energy value, coals are classified into four primary ranks: anthracite, bituminous, subbituminous, and lignite. The weight ratio of carbon to hydrogen, on a dry basis, increases with increasing coal rank; from about 12 for lignite to about 20 for bituminous. For methane, the C/H ratio is 3, for liquid hydrocarbons the ratio is in the range 4.5 for propane, and it is 10 for the heavier oils. The addition of hydrogen and, in some cases, the partial removal of carbon are therefore necessary in converting coal to liquid or gaseous hydrocarbons.

Depending on the primary crude product, the conversion of coal may be divided into three major process areas: gasification, carbonization (pyrolysis), and liquefaction. There is much overlap, however, between gasification and pyrolysis, since both of these processes occur together in some coal conversion processes. In addition in some processes the liquid hydrocarbons are produced from synthesis gas subsequent to coal gasification. The diagram in Fig. 10.1, reproduced from a review paper by Bodle *et al.* [8], gives a general outline of primary processes of coal conversion to chemicals and clean fuels. Although much developmental work is being done on conversion processes to produce clean fuels for electric power plants, pipeline gas, hot reducing gas and formed coke for metallurgical applications, and chemicals, there are only a few commercial plants in operation at present, primarily for power generation. Also, because of the proprietary nature of the development work, there is limited technical information available for a comprehensive discussion of the high temperature aspects of the physical chemistry of coal conversion processes.

10.3.1 Gasification

Depending on the type of coal used and the method of gasification, the gas produced is a low-, medium-, or high-energy gas. To produce a medium-energy gas of 2000–5000 kcal nm^{-3}, the coal is reacted with a mixture of steam and oxygen; air or oxygen-enriched air is used together with steam for a low-energy

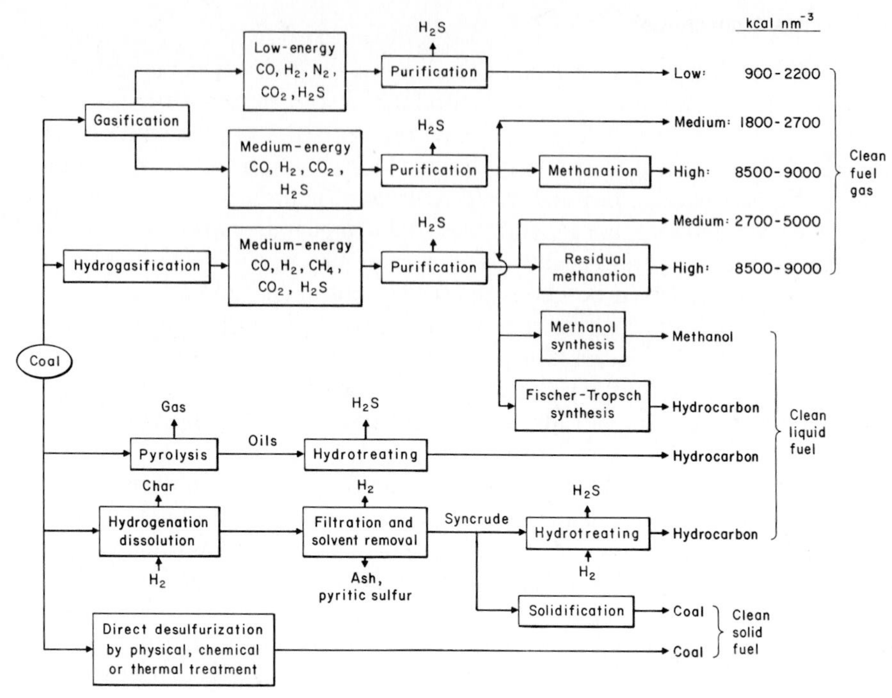

FIG. 10.1 Primary processes of coal conversion. After Bodle *et al.* [8].

gas of 900–2000 kcal nm^{-3}. The primary reactions and the corresponding heats of reactions are

$$C + \tfrac{1}{2}O_2 = CO, \qquad \Delta\tilde{H}^\circ = -27.34 \quad \text{kcal}, \tag{10.16}$$

$$C + H_2O = CO + H_2, \qquad \Delta\tilde{H}^\circ = 31.81 \quad \text{kcal}. \tag{10.17}$$

A high-energy gas of 8000–9000 kcal nm^{-3} is manufactured by one of two ways. If a medium-energy gas is to be used to produce methanol, liquid hydrocarbons, or a high-energy gas as a pipeline gas, the H_2/CO molar ratio is adjusted to 2–3 by reacting the crude gas (after the removal of tars and entrained particulate matter) with steam. Following the water–gas shift reaction,

$$CO + H_2O = CO_2 + H_2, \qquad \Delta\tilde{H}^\circ = -8.00 \quad \text{kcal} \tag{10.18}$$

the carbon dioxide and sulfurous gases (H_2S, COS) are removed by acid–gas cleaning; the gas is then methanated in the presence of a catalyst:

$$CO + 3H_2 = H_2O + CH_4, \qquad \Delta\tilde{H}^\circ = -53.57 \quad \text{kcal}. \tag{10.19}$$

Another method of producing a high-energy gas is by direct methanation of coal with hydrogen under pressure; this is known as hydrogasification:

$$C + 2H_2 = CH_4, \qquad \Delta\tilde{H}^\circ = -21.76 \quad \text{kcal}. \tag{10.20}$$

Although reaction (10.19) is more exothermic than reaction (10.20), the heat generated at relatively low temperatures of methanation of carbon monoxide (250–300 °C) is of little value. In hydrogasification of coal with hydrogen and steam at temperatures of 650–930 °C, the heat generated by reaction (10.20) is utilized in the endothermic reaction of steam with carbon, reaction (10.17). Therefore, the product of a high-energy gas by hydrogasification has an inherent thermochemical advantage over the catalytic methanation of synthesis gas produced by gasifying coal with oxygen and steam. For example, the theoretical thermal efficiency of hydrogasification is about 70%, as compared with about 57% for methanation.

As dictated by the equilibrium for reaction (10.20), the hydrogasification must be done at high pressures. Also, there are many advantages to gasification with oxygen and steam at elevated pressures. For example, (i) high gasification rates per unit volume of the reactor under pressure reduces the percentages of heat losses; (ii) savings incurred in compressing only the oxygen instead of compressing all the product gas; (iii) lower gas velocities employed at elevated pressures reduce the elutration of particulate matter; (iv) increased residence time of gas in the reactor at high pressures and reduced velocities increases the efficiency of waste-heat recovery and cooling of gas streams; and (v) pressure enhances hydrogenation reactions resulting in higher concentrations of methane, hence increasing the energy value of the gas.

There are also disadvantages of using elevated pressures in some gasifiers. For example, (i) in nonslagging gasifiers operating at moderate temperatures, CO-rich gas cannot be produced under pressure, because of unfavorable equilibrium conditions; (ii) if the gas is to be used in ammonia synthesis, the methane has to be removed, therefore the methane level should be kept low by gasification at atmospheric pressure; and (iii) noncaking characteristics of some coals are adversely affected by pressure.

Although the equilibrium conditions are not realized in coal gasification, the effects of temperature and pressure on the composition of the product gas are in general accord with the trends predicted by assuming gas equilibria involving CO, CO_2, H_2, H_2O, and CH_4. That is, the concentrations of H_2 and CO increase and those of H_2O, CO_2, and CH_4 decrease with increasing temperature or decreasing pressure.

10.3.1a Gasifiers for steam–oxygen (or air) gasification. Basically, there are four types of gasifiers operating either at atmospheric or at elevated pressures: (1) fixed- or moving-bed reactor, (2) fluidized-bed reactor, (3) entrained reactor, and (4) molten bath reactor. There are numerous review articles (in [8–10]) describing the basic features of gasifiers that are operated commercially as well as gasifiers that are in an advanced development state. Here we shall give a brief outline of some of these gasifiers as typical examples.

1. *Fixed-bed reactor.* Lurgi gasifier [11] is a shaft-type reactor; coal, sized within the range 2–4 cm diameter, is charged at the top and coal ash is removed

from the bottom through a rotating grate. A mixture of steam and oxygen is blown through the bottom grate and the crude gas is withdrawn from the top. This process is suitable for noncaking, low swelling coals with an ash-fusion temperature greater than 1150 °C. The reactor operates at pressures of 25–30 atm. As the coal descends in the reactor by gravity, the coal is heated up and dried, followed by initial devolatilization. At temperatures of 600–850 °C, devolatilization is accompanied by gasification in the countercurrent flow of steam and oxygen. The residence time of coal in the reactor is about 1 hr. Most of the oxygen is consumed in the combustion zone, the height of which is about 5–10 times the diameter of the coal particles. Hot CO_2 and H_2O emerging from the combustion zone gasify the coal, producing a CO- and H_2-rich crude gas. The steam to oxygen ratio is in the range 4–8 kg(steam) $nm^{-3}(O_2)$, depending on coal rank and the desired H_2/CO ratio in crude gas, usually in the range 1.5–2.5. Also, the higher the steam/oxygen ratio, the lower the temperature in the combustion zone and the higher the H_2/CO in the crude gas. For a given steam/oxygen ratio, the higher the reactivity of the coal, the higher the H_2/CO ratio. Crude gas leaving the gasifier at temperatures of 350–600 °C contains tar, oil, naphtha, phenols, ammonia, and coal and ash dust. After removal of tar, oil, and dust by quenching in recycle oil, the gas is passed through a catalytic shift converter where the light oils in the gas are also desulfurized. The gas is then processed in several stages to produce the desired end product and recover the by-products.

2. *Fluidized-bed reactor.* The Winkler process, conceived in 1922, was the first fluidized-bed gasifier used in industry. The Winkler process is still in use to produce a low- and medium-energy gas from a wide variety of coals [12]. The run-of-mine coal, crushed to a particle size range below about 8 mm and dried, is injected with steam at the bottom of the reactor to fluidize the coal and to cool large ash particles discharging from the bottom of the reactor. A mixture of steam and oxygen (or air) is injected at different levels into the fluidized bed to gasify the coal at about atmospheric pressure (higher pressures in some installations). The heavier ash particles fall down the fluidized bed to the bottom of the reactor into the ash discharge unit. The lighter ash particles, about 70% of the ash, are carried away with the product gas. Depending on coal rank and steam/oxygen ratio in the gas injected, the gasifier temperature is in the range 850–1050 °C. The gas in the freeboard above the fluidized bed is cooled by a radiant boiler section to stabilize the gas temperature below the fusion point of ash. The crude gas leaving the reactor is cooled in a waste–heat–recovery unit, and after the removal of fly ash, passed through a shift converter, purified, and further processed to produce, for example, methanol, ammonia, or pipeline-quality gas.

3. *Entrained reactor.* The Koppers–Totzek process, in commercial operation since the early 1950s, is based on the concept of an entrained reactor where there is partial combustion of a carbonaceous feed in suspension with oxygen and steam to produce a medium-energy gas rich in CO and H_2 with a negligible

amount of methane [13]. All ranks of coal, petroleum coke, char, and liquid hydrocarbons can be used as feedstock. Depending on rank, the coal is dried to 2%–8% moisture and pulverized (about 70% below 200 mesh size). Prepared coal powder is premixed with steam and injected together with oxygen through burners on opposite ends of a refractory-lined combustion chamber. The gas temperature is about 1500 °C and the pressure in the chamber is slightly over atmospheric pressure. Ash in the coal is liquified in the high temperature zone of the flame. About 60% of the molten slag falls into a slag quench tank and is recovered as granular slag. The remainder of the slag and all the unreacted carbon are entrained in the exit gas. The crude gas is passed through the waste heat boiler to generate high-pressure steam. The crude gas is then treated by conventional methods to remove entrained solids. The heating value of the product gas is equivalent to 70%–80% of the calorific value of the coal; additional energy recovered as steam from the available sensible heat is about 15% of the calorific value of the coal.

4. *Molten bath reactor.* Molten sodium carbonate is the reaction medium in the Kellogg molten salt gasifier [14]. Crushed coal and recycled sodium carbonate are blown together with steam and oxygen into a molten bath of sodium carbonate. Because of the catalytic effect of the molten salt, all the tar in coal is gasified and the product gas contains large amounts of methane. To counteract buildup of coal ash, a bleed stream of molten salt is treated with water to remove ash by dissolving sodium carbonate. Filtered solution is then carbonated to precipitate sodium bicarbonate which, after filtering, is converted to sodium carbonate by calcination and recycled to the gasifier. The crude gas leaving the gasifier at a temperature of about 950 °C is passed through a heat recovery section and any entrained salt is recovered. The gas is then treated in the usual way to produce pipeline-quality gas.

10.3.1b Gasifiers for hydrogasification. Fluidized-bed reactors are used in hydrogasification of coal with steam and hydrogen at elevated pressures and temperatures of 650–1500 °C to produce a medium-energy gas. The gasification is done in two consecutive stages to take advantage of the initial rapid rate of reaction of hydrogen with coal to form methane and to maximize the use of the heat generated by reaction (10.20) for the heat needed in reaction (10.17). The most active fraction of coal is hydrogasified at lower temperatures to form methane, while the less active fraction is gasified at higher temperatures to generate hydrogen for hydrogasification. Of the several gasifiers that are in an advanced state of development, we shall cite two types of gasifiers as examples.

1. *IGT-Hygas.* In the Hygas process developed by Institute of Coal Technology (IGT) [15], the coal is partially gasified in hydrogen-rich gas in a three-stage reactor. The hydrogen-rich gas for gasification is generated in another reactor where char from the gasifier is reacted with steam. Crushed, dried, and pretreated (if necessary) coal is slurried with a light oil and fed into the first (top) stage of a three-stage reactor where the pressure is maintained at about

70 atm. The light oil vaporizes in the first stage of the reactor. In the second stage, the coal is hydrogasified at about 650 °C by hot gases exiting from the third (bottom) stage of the reactor. The third stage is the high temperature zone (950–1000 °C) of the reactor where coal char is further gasified in steam and hydrogen. The hydrogen-rich gas fed into the gasifier is generated by one of the following three alternative processes, using the char discharged from the bottom of the gasifier: (i) steam–oxygen gasifier, (ii) steam–iron gasifier, and (iii) electrothermal gasifier. The crude gas produced by the Hygas process contains primarily 15%–25% CH_4, 8%–20% CO, 8%–20% CO_2, 20%–25% H_2, and 18%–30% H_2O. The light oil is recovered by quenching the exit gas which is then treated in the usual way to produce pipeline gas.

2. *Bi-Gas.* In the Bi-Gas process, developed by Bituminous Coal Research, Inc. [16], the pulverized coal is injected with steam into the gas space in the upper stage of the reactor, where the coal is rapidly heated to about 950 °C by hot hydrogen-rich gas from the lower section of the reactor. In a short residence time of a few seconds, the coal is devolatilized and gasified, forming methane and char. Entrained char, separated from the crude gas in a cyclone, is returned to the second (lower) stage of the gasifier. In the second stage, the char is gasified at about 1500 °C with preheated oxygen (~650 °C) and steam to generate hydrogen-rich gas for hydrogasification in the first stage of the reactor. The reactor operates at pressures of 70–100 atm. All the ash in the char is melted during the high temperature gasification; molten slag is continuously drained from the bottom of the gasifier into a quench tank. The crude gas containing about 8% CH_4 and $H_2/CO \sim 0.5$ is cleaned and treated by conventional means.

10.3.1c Gasifiers for power generation. In a conventional power station, coal is burned with air at atmospheric pressure to raise superheated high pressure steam (540 °C and 160 atm); the electricity is produced by the expansion of steam in turbogenerators. In a two-turbine system with reheating of steam between the high pressure and low pressure turbines and preheating of the recirculating water, the highest efficiency attainable is about 38%.

In a combined gas-and-steam process, coal is partially combusted with air at elevated pressures to produce fuel gas and the electricity is produced in two parallel systems. The steam raised from partial combustion of coal in boilers operates turbogenerators. In the second part of the system, the fuel gas is completely combusted and expanded in a gas turbine. The overall efficiency in the combined-cycle power generation is about 51%.

A combined-cycle power plant in operation at present is based on Lurgi pressure gasifiers [17]. Currently, various types of fluidized-bed gasifiers are being developed for combined-cycle power generation from coal. The present state of the development has been reviewed recently by, for example, Robson [18].

10.3.2 Carbonization

Conventional coke making is the oldest method of pyrolysis or carbonization of coal. Only limited range of noncaking and low sulfur coals are suitable for the production of high quality metallurgical coke. Also, costly and energy intensive process modifications have to be made for close control of emissions from coke ovens. Processes are now being developed for the production of high quality metallurgical coke from high sulfur, high ash coals by carbonization of coal to a low sulfur char, coupled with hydrogenation of coal to produce aromatics, other chemicals, gaseous and liquid fuels with low energy consumption, and simpler and better emission control.

As an example, let us consider the process development unit (PDU) studies of U. S. Steel Corporation on the CLEAN COKE process [19, 20]. The process is based on a combination of fluidized-bed carbonization/desulfurization with hydrogenation/liquefaction to convert high sulfur coal to low sulfur metallurgical coke, chemical feedstocks, and to a lesser extent, liquid and gaseous fuels. The run-of-mine coal is beneficiated and divided into two feed portions: a sized fraction for fluidized-bed carbonization and a fine fraction for hydrogenation. The carbonization char is mixed with a heavy oil by-product and processed to make a low sulfur metallurgical formed coke (or formcoke) in the form of pellets. Product gases from all operations are combined and processed to produce hydrogen for the hydrogenation operation, fuel, ethylene, propylene, sulfur, and ammonia. Here we shall discuss briefly only the carbonization part of the process to produce low sulfur char.

Sized coal (0.2–3 mm) is dried and preoxidized at 180–200 °C. This mild surface oxidation of the coal diminishes its caking tendency sufficiently to prevent agglomeration in first-stage carbonization at 425–450 °C [21]. About 70% of the carbonization tar and 15% of the total gas is produced in first-stage carbonization. Cleaned, desulfurized recycle gas contains 71% CH_4, 13% C_2H_6, 2% H_2, 10% CO, 4% hydrocarbons, and about 100 ppm H_2S. In first-stage carbonization at about 440 °C and 13 atm, coal containing 35% volatile matter and 2.2% S is converted in about 20 min to semichar containing 20% volatile matter and 1.7% S. Completion of devolatilization accompanied by desulfurization is done in second-stage carbonization at 760 °C and 13 atm. In a residence time of 120 min, the char produced contains about 0.4% S.

The results obtained on the rate of devolatilization and desulfurization of Illinois coal char from the PDU tests have been substantiated by bench-scale experiments done by Kor [22]. An example of his results given in Fig. 10.2 shows the effect of composition of H_2–CH_4 mixtures on the rate of devolatilization and desulfurization of char at 800 °C and 5 atm pressure. The char, made by heating Illinois coal in hydrogen at 600 °C, contained 0.75% S, of which 0.25% S was in the form of pyrrhotite, the remainder being as organic sulfur. Studies by Kor have shown that the removal of pyrrhotite sulfur during carbonization is much more sluggish than the removal of organic sulfur. In

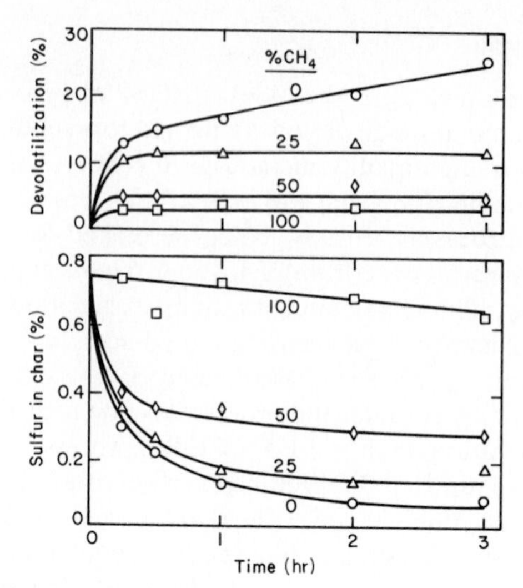

FIG. 10.2 Devolatilization and desulfurization of Illinois coal char (prepared at 600 °C) in H_2–CH_4 mixtures at 5 atm and 800 °C. After Kor [22], in "Coal Desulfurization—Chemical and Physical Methods."

fact, the desulfurization achieved during carbonization is closely associated with devolatilization. The data in Fig. 10.3 are for carbonization of coal char in hydrogen at pressures of 1–10 atm and temperatures of 600–1000 °C. The extent of desulfurization is a direct function of devolatilization, independent of temperature and pressure. The higher the temperature and higher the pressure, the faster the rate of devolatilization accompanied by desulfurization.

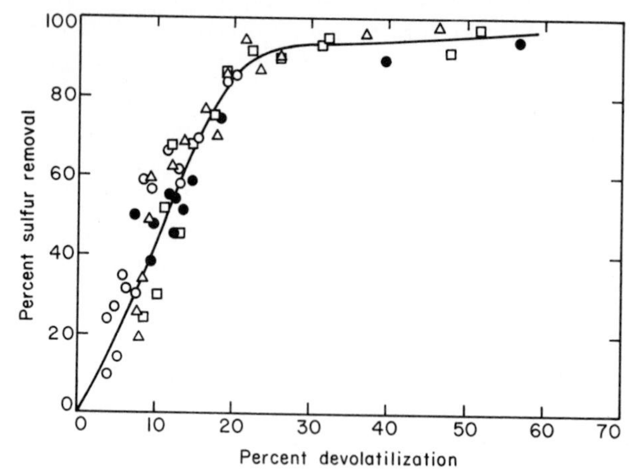

FIG. 10.3 Interrelation between desulfurization and devolatilization of coal char in hydrogen at pressures of 1–10 atm and temperatures of (○) 600, (△) 800, (□) 900, and (●) 1000 °C. After Kor [22], in "Coal Desulfurization—Chemical and Physical Methods."

10.3.3 Liquefaction

Conversion of coal by liquefaction is outside the scope of this chapter. However, we might mention in passing that liquefaction of coal may be divided into three process areas: catalytic conversion of synthesis gas, solvent extraction of coal, and gaseous extraction of coal.

The concept of gaseous extraction is based on the solvent capability of compressed supercritical gas and *retrograde condensation* of the extract upon reduction of pressure. When coal is treated with compressed supercritical gas at a temperature slightly above the critical temperature, coal constituents pass into the gas phase, leaving behind low-volatile char and ash. When the gas is withdrawn to a vessel at atmospheric pressure, the coal extract precipitates as a liquid hydrocarbon. For example, Whitehead and Williams [23] have shown that a coal containing 38% volatile matter, when treated with toluene at 350 °C and 100 atm, yielded a liquid extract amounting to about 20% by weight of dry coal. The percentage of liquid extract decreases with decreasing percentage of volatile matter in the coal. However, yield with gas extraction is lower than that can be obtained by using liquid solvents such as anthracene oil.

At present, liquid hydrocarbons are produced from synthesis gas by the Fischer–Tropsch process. In this process, the synthesis gas with an H_2/CO ratio of about 2 is reacted in the presence of a catalyst at pressures of 25–30 atm and temperatures of 220–250 °C to produce liquid hydrocarbons.

10.3.4 Desulfurization during Gasification of Coal and Combustion of Fuels

In coal conversion processes and combustion of fuels, the sulfur is removed from the liquid or gaseous products at some stage of the operation to control H_2S or SO_2 emissions in stack gases. There are numerous methods of removing sulfurous gases from fuel gases and combustion products by generative–cyclic processes at ambient temperatures. In some practical applications, however, desulfurization at elevated temperatures is necessary or economically more attractive. Most of the current ideas on hot desulfurization in coal gasification or fuel combustion are in experimental or pilot-plant development state.

As discussed in Section 8.4.2, experimental studies have shown that hot reducing gases can be desulfurized either by calcined dolomite or manganous oxide (preferably in the form of porous sintered MnO–Al_2O_3 pellets) in a packed-bed reactor. By an appropriate oxidizing roasting of the sulfides, the oxide sorbents can be regenerated with the by-product of SO_2-rich gas suitable for sulfur recovery. We have shown in Section 10.3.3 that about 90% of the sulfur in coal can be removed during carbonization in a recycled gas mixture of H_2–CH_4–CO at temperatures of 425–760 °C.

There are also methods of *in situ* desulfurization during gasification of coal. In the Westinghouse PDU fluidized-bed gasifier for a combined-cycle power plant, crushed dolomite is charged to the fluidized-bed gasifier operating at temperatures of 800–900 °C and pressures of 10–15 atm [24]. The gas velocity in the gasifier is so adjusted that the product gas is removed from the top of the gasifier, while spent dolomite is withdrawn from the bottom of the gasifier. The resulting calcium sulfide is treated for disposal or the sorbent is regenerated for return to the process, and sulfur is recovered. In this process, about 80% of the sulfur in coal is absorbed by dolomite.

In the CO_2-acceptor process [25], the calcined dolomite is added to the gasifier to react with CO_2 that is evolved together with H_2, CO, and CH_4 during gasification of coal with steam at about 830 °C and 10–20 atm. The heat generated by the formation of methane and reaction of CaO with CO_2 balances the endothermic heat of reaction of coal with steam. The residual char and spent dolomite ($CaCO_3 \cdot MgO$) is transferred to the regenerator, where it is burned with air to supply the heat required for calcination of spent dolomite. The residual ash, containing about 5% carbon, is elutriated from the acceptor bed and carried out of the regenerator with the flue gas. This ash contains about 80% of the sulfur in the coal charged to the gasifier.

In an experimental study, Jonke *et al.* [26] have shown that effective desulfurization can be achieved by dolomite or limestone during combustion of coal in a fluidized-bed combustor. During combustion of coal with excess air in a fluidized bed, dolomite or limestone added with coal readily decomposes and reacts with SO_2 forming calcium sulfate. The extent of sulfur retention by limestone increases with increasing Ca/S ratio in the charge. With a Ca/S molar ratio of 3 to 4, the highest sulfur retention of 92%–98% is achieved at the fluidized-bed temperatures of 790–820 °C. A decrease in SO_2 removal at lower temperatures may be attributed to slow rate of calcination and sulfation of the limestone. At higher temperatures, the calcium sulfate becomes less stable, hence the extent of sulfur retention decreases at high bed temperatures.

10.3.5 Emission of NO_x in Fuel Combustion

Nitric oxide NO and nitrogen dioxide NO_2 are formed during combustion of fuels, and after release to the atmosphere, more NO is oxidized to NO_2, which is an air pollutant and participates in smog formation. A mixture of NO and NO_2 as pollutant in the atmosphere is usually referred to as NO_x. A great deal of research has been done in recent years on the control of NO_x exhaust emissions from automobile engines, gas turbines, and fuel combustion in power plants. For in-depth reading on this subject, reference may be made to papers published in *AIChE Symposium Series* [27–29]. Only a brief summary is made here of factors affecting NO_x emission in fluidized-bed combustion.

Basically, there are two mechanisms of NO formation: (i) thermal fixation (or thermal NO) formed by reaction of N_2 and O_2 in air during the combustion of fuel, and (ii) fuel NO formed by conversion of organic nitrogen in fuel to NO. Amount of thermal NO formed increases with increasing oxygen/fuel ratio and increasing bed temperature. Because of the plug–flow-type gas concentration distribution, more NO is formed near the coal inlet point, where the oxygen concentration is high. This is partly counteracted in staged combustion with secondary air injection above the fluidized bed.

Most coals contain 1%–2% nitrogen which is tied up in aromatic rings such as pyridine (C_5H_5N), picoline (C_6H_7N), quinoline (C_9H_7N), and nicotine ($C_{10}H_{14}N_2$). As coal is devolatilized and gasified, intermediate gaseous species of nitrogen are formed, e.g., NH_3, CN, HCN, the combination of which is designated by NX. The reaction of NX with O or OH in the combustion zone leads to the formation of NO.

$$NX + O = NO + X, \qquad NX + OH = NO + HX. \tag{10.21}$$

However, the intermediate NXs are not completely converted to NO because of competing recombination reactions which produce N_2:

$$NX + NO = N_2 + XO. \tag{10.22}$$

This reaction takes place under reducing conditions, i.e., in fuel-rich zones in the flame. For this reason, two-stage combustion gives lower NO_x emissions. For example, 10% secondary air injection lowers NO_x emission by about 20%. For combustion with about 3% excess oxygen, typical NO_x values are in the range 400–1200 g NO_2 per 1000 kcal fired, corresponding to 200–600 ppm (vol) of NO_x in furnace exhaust.

The nitric oxide, forming either by fixation of atmospheric nitrogen or by conversion of fuel nitrogen, may be reduced by reaction with carbon,

$$2NO + C = N_2 + CO_2, \tag{10.23}$$

for which the thermodynamics are favorable. Therefore, improved mass transfer between the gas phase and the char particles, as achieved in pressurized–fluidized combustion, would favor reduction of NO_x emission.

10.4 Hot Corrosion in the Process Environment

Construction materials used in, for example, coal gasifiers, gas reformers, petroleum refineries, gas turbines, jet engines, and nuclear reactors are exposed to hostile process environments at elevated temperatures and pressures. A great deal of research effort has gone into the development of special alloys and superalloys to withstand corrosive and erosive environments without loss of mechanical strength at elevated temperatures. Selection of the primary constituents of the alloys or superalloys resistant to hot corrosion is based on the fulfillment of certain essential requirements: (i) high solidus temperature, (ii) high toughness, ductility, creep, and stress rupture resistance at elevated temperatures, (iii) no low melting or volatile compound formation, and (iv) ability to sustain a surface oxide layer in which diffusion is slow. The refractory metals such as tungsten and molybdenum are not suitable as base metals because of the formation of volatile oxides and sulfides. Of the transition metals, those with bcc structure are not suitable because of low toughness and ductility compared to the alloys with fcc structure. The chromium in the alloy forms a protective oxide layer in many corrosive environments. With these requirements in mind, the superalloys resistant to hot corrosion are based on austenitic Ni–Cr–Fe and Ni–Cr–Co alloys to which a few percentages of other alloying elements are added, e.g., Al, Ti, Mo, Ta, Nb, Si; the choice of these additives depends on the nature of the corrosive or erosive attack and the high temperature strength required of the alloy.

In this section, we shall give a few specific examples of the type of hot corrosion encountered in industrial processes, combustion engines, and nuclear reactors.

10.4.1 Carbon Deposition from Gases

There are several modes of formation of solid carbon from carbonaceous gas mixtures, either intentionally as in the production of carbon black and pyrolytic graphite, for example, or inadvertently, as carbon deposition in reheating of reformed gases, thermal cracking of hydrocarbons, and in gas-cooled nuclear reactors with graphite as the moderating and structural material in the core. There are two necessary conditions for the formation of solid carbon from carbonaceous gases: (i) the thermodynamic activity of carbon in the gas, relative to graphite, must be greater than unity and (ii) the temperature must be high enough and/or a catalyst must be present to activate the reaction.

10.4.1a Carbon activities in gas mixtures. The activity of carbon in a carbonaceous gas mixture depends on the state of reactions between various species, i.e., whether or not there is complete gas equilibrium at the reheating temperature and pressure. Let us consider as, an example, reheating of a reformed gas containing CO, CO_2, H_2O, and CH_4. If there is complete gas equilibrium at the reheating temperature and pressure, the activity of carbon in the equilibrated gas mixture must satisfy three basic reaction equilibria:

$$2CO = C + CO_2, \qquad K_1 = a_C p_{CO_2}/p_{CO}^2; \tag{10.24}$$

$$H_2 + CO = C + H_2O, \qquad K_2 = a_C p_{H_2O}/p_{H_2} p_{CO}; \tag{10.25}$$

$$CH_4 = C + 2H_2, \qquad K_3 = a_C p_{H_2}^2/p_{CH_4}. \tag{10.26}$$

At the temperatures and pressures of interest, the partial pressures of O, H, and other hydrocarbon species are small enough to be neglected.

From the mass balance for C, O, and H, we may write the following equalities:

$$\frac{\sum C}{\sum O} = \left(\frac{p_{CO_2} + p_{CO} + p_{CH_4}}{2p_{CO_2} + p_{CO} + p_{H_2O}}\right)_i = \left(\frac{p_{CO_2} + p_{CO} + p_{CH_4}}{2p_{CO_2} + p_{CO} + p_{H_2O}}\right)_e, \tag{10.27}$$

$$\frac{\sum C}{\sum H} = \left(\frac{p_{CO_2} + p_{CO} + p_{CH_4}}{2p_{H_2O} + 2p_{H_2} + 4p_{CH_4}}\right)_i = \left(\frac{p_{CO_2} + p_{CO} + p_{CH_4}}{2p_{H_2O} + 2p_{H_2} + 4p_{CH_4}}\right)_e, \tag{10.28}$$

where i and e indicate partial pressures in the ingoing and equilibrated gas mixtures, respectively. For the total pressure P of the equilibrated gas mixture, we have the equation

$$p_{CO} + p_{CO_2} + p_{H_2} + p_{H_2O} + p_{CH_4} = P. \tag{10.29}$$

For known values of the equilibrium constants K_1, K_2, and K_3 and the total pressure P, we can compute for any given inlet gas composition the partial pressures of gaseous species and the activity of carbon in the equilibrated gas at the reheating temperature by simultaneous solution of Eqs. (10.24)–(10.29).

Examples are given in Fig. 10.4 of carbon activities, relative to graphite, for complete gas equilibrium at 4 atm total pressure and indicated temperatures for an initial gas mixture containing 73% H_2, 18% CO, 8% CO_2, 1% CH_4 to which 0.2%–6.0% H_2O has been added. When the carbon activity in the equilibrated gas exceeds unity, the carbon deposition is imminent. With the addition of 0.2% H_2O to this gas mixture, there should be no carbon deposition in the equilibrated gas mixture at temperatures below 512 and above 720 °C. Yet,

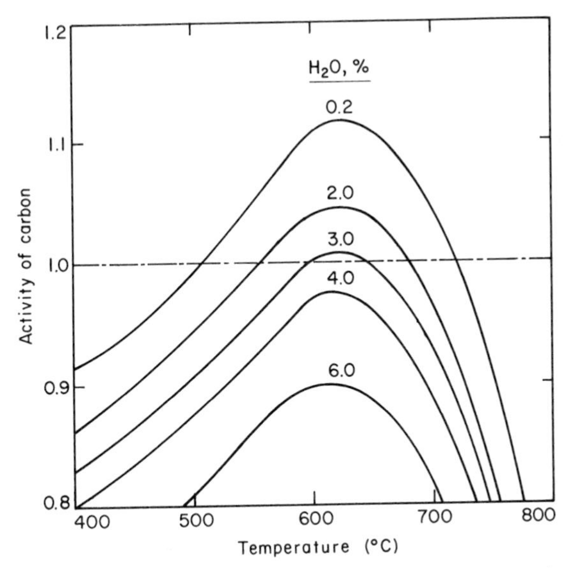

FIG. 10.4 Calculated activity of carbon for complete gas equilibrium at 4 atm and indicated temperatures for reformed natural gas containing 73% H_2, 18% CO, 8% CO_2, and 1% CH_4 to which an indicated amount of H_2O is added.

when iron oxides were reduced in a similar gas mixture, carbon deposition was observed at all temperatures below 1000 °C [30], indicating lack of complete gas equilibrium. These experimental findings suggest that carbon deposition may occur during reheating of reformed natural gas in industrial operations, even when the calculated carbon activity is less than unity in the gas mixture under equilibrium conditions.

The activity of carbon cannot be calculated for a nonequilibrated gas mixture. However, for the purpose of identifying the reactions that may be responsible for carbon deposition, we may calculate carbon activities for individual reactions (10.24)–(10.26) on the assumption that there is no change in the composition of the inlet gas upon heating. As is seen from the example given in Fig. 10.5, the carbon activities for individual reactions differ considerably from those calculated for complete gas equilibrium. For complete gas equilibrium, the carbon activity is below unity at temperatures below 540 and above 700 °C, while for the individual reactions the carbon activity is below unity only within the temperature range 880–1060 °C. Therefore, in the nonequilibrated gas, there can be no carbon deposition only within the temperature range 880–1060 °C. The temperature at which carbon deposition may start depends on the relative rates of reactions (10.24) and (10.25): below 730 °C if reaction (10.24) is fast, and below 880 °C if reaction (10.25) is fast.

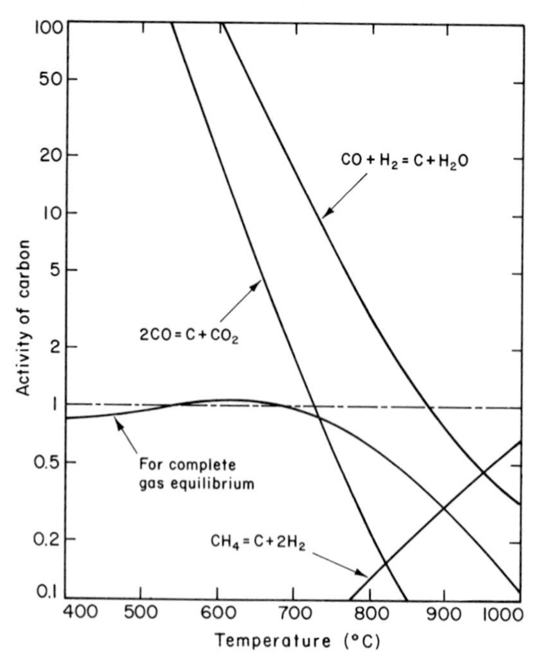

FIG. 10.5 Calculated activity of carbon for complete gas equilibrium at 4 atm, compared with the activity of carbon for individual reactions in a reformed inlet gas containing 2% H_2O.

On the basis of the foregoing reasoning, we may compute a diagram of % H_2O in the inlet reformed gas versus temperature, as in Fig. 10.6, to delineate the regions where (I) carbon will not deposit, (II) carbon may deposit, and (III) carbon will deposit. With increasing pressure, regions II and III move to higher temperatures and I to higher concentrations of H_2O, and chances of carbon deposition become greater. From the equilibrium constant for reaction (10.25), we can also compute the critical p_{H_2O}/p_{CO} ratio for possible carbon deposition at any temperature and partial pressure of H_2 in the inlet gas. The possibility of carbon deposition in region II depends much on the catalytic behavior of the inner surface of the reheater tubes.

10.4.1b Catalytic decomposition of carbon monoxide. The catalytic decomposition of carbon monoxide has been the subject of much study, because of the occurrence of carbon deposition in many processes, e.g., blast furnaces, the Stelling process, the Fischer–Tropsch process, nuclear reactors, and so on. References to early work on this subject are listed in an annotated bibliography by Donald [31]. The consensus is that iron, cobalt, and nickel are the most effective catalysts for the decomposition of carbon monoxide. Kehrer and Leidheiser [32] reported virtually no catalytic effect of copper, silver, chromium, molybdenum, palladium, and rhenium.

In a comprehensive study of carbon deposition in CO–H_2 mixtures at temperatures of 450–700 °C, Walker *et al.* [33] observed that the C/H atom ratio in the deposit increased from 10 to 50 with increasing reaction temperature and increasing CO/H_2 ratio in the gas. They also found that the properties of carbon deposit were affected by the amount accumulated on the catalyst, e.g., with

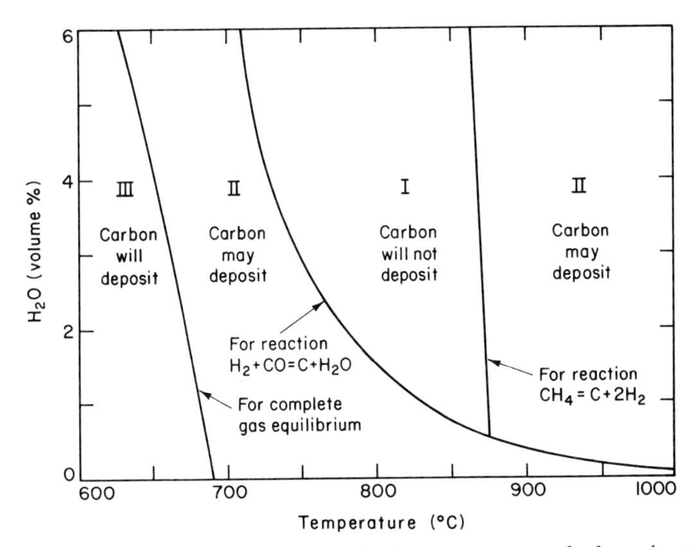

FIG. 10.6 Regions of carbon deposition at reheating temperatures of reformed natural gas at 1 atm containing indicated amounts of H_2O.

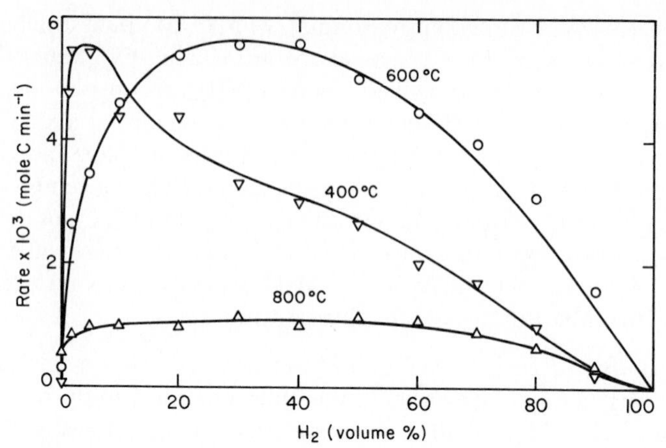

FIG. 10.7 Effect of gas composition (H_2–CO) on rate of carbon deposition on 660-mg porous iron granules at atmospheric pressure and indicated temperatures. After Olsson and Turkdogan [35].

increasing thickness of the carbon layer, the crystallinity became poorer, the surface area increased, the electrical conductivity decreased, and the C/H_2 ratio increased. Another important finding was that carbon deposition ultimately ceased when most of the iron was converted to cementite. Upon hydrogen treatment, cementite decomposed to iron and graphite, but the regenerated iron lost most of its reactivity as a catalyst. Similar observations were made by Turkdogan and Vinters [30] in subsequent studies of the catalytic decomposition of carbon monoxide in the presence of porous iron.

It is generally agreed that iron, not cementite, catalyzes the decomposition of carbon monoxide. The chemisorption of H_2 and CO on the surface of iron is believed to approach equilibrium rapidly. The carbon formed by reactions (10.24) and (10.25) migrates across the surface to a nucleating center where cementite and free carbon are deposited. Based on x-ray diffraction and electron microscopic studies, Ruston *et al.* [34] suggested that although the decomposition of carbon monoxide was catalyzed by iron, an iron carbide Fe_7C_3 formed as an intermediate step in the decomposition of the activated reaction product to graphite.

As is seen from the experimental results of Olsson and Turkdogan [35] in Fig. 10.7, the rate of carbon deposition on an iron catalyst in CO–H_2 mixtures is a complex function of temperature and gas composition. From the measured rate of carbon deposition and the measured CO_2/H_2O in the exhaust gas, the rates of reactions (10.24) and (10.25) were determined. These measured rates are shown in Fig. 10.8 as a function of gas composition at 400, 600, and 800 °C. The hydrogen appears to have a dual role: at low concentrations the hydrogen catalyzes reaction (10.24), and at high concentrations of hydrogen reaction (10.25) contributes directly to carbon deposition.

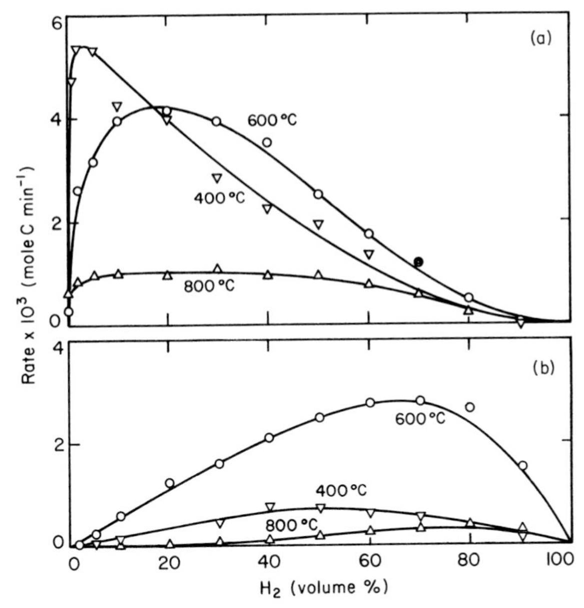

FIG. 10.8 Effect of gas composition (H_2–CO) on rate of carbon deposition on 660-mg porous iron granules at 1 atm (a) by reaction $2CO \rightarrow C + CO_2$ and (b) by reaction $CO + H_2 \rightarrow C + H_2O$. After Olsson and Turkdogan [35].

Although H_2O is expected to retard or inhibit carbon deposition by virtue of reverse reaction (10.25), it has been found [35] that an addition of 0.5%–2% H_2O to CO–CO_2 mixtures enhances the rate of carbon deposition at 400–600 °C. A similar catalytic effect of H_2O has been observed for carbon deposition from methane in the presence of nickel [36]. We might mention in passing that the rate of carbon formation by the disproportionation of carbon monoxide is much faster than the pyrolysis of methane under similar conditions [37].

Ever since the first observations made by Byrom [38] and soon after by Carpenter and Smith [39], it has become generally recognized that sulfur in the gas retards or inhibits the catalytic action of iron in the decomposition of carbon monoxide. This is due to the coating of iron surface with cementite, which does not decompose readily in the presence of sulfur. In fact, in the Stelling process [40] the iron ore is reduced to cementite in a CO–CO_2 atmosphere without the formation of free carbon, probably because of the presence of sulfur in the system.

The nitrogen-bearing gaseous species such as NH_3 and $(CN)_2$ are also known to retard the decomposition of carbon monoxide [40–42]. Again, this may be attributed to extended metastability of cementite in the presence of nitrogen. The retarding effect of sulfur and nitrogen on carbon deposition is similar to sluggish graphitization of steel (by decomposition of cementite) in the presence of sulfur and nitrogen in solution in steel.

10.4.1c Metal dusting. Catastrophic deterioration of alloys when heated in carbonaceous gases is a consequence of carbon deposition, which ultimately leads to the so-called *metal dusting*, a phenomenon that has been known for a century. Fischer and Bahr [43] appear to be the first to demonstrate that the disintegration and intimate mixing of the iron catalyst in the carbon deposit is brought about mechanically. They decomposed carbon monoxide on an iron–copper catalyst and found that carbon deposit, well separated from the main mass of the catalyst, contained copper as well as iron. Although copper did not participate in the reaction, the substrate of the alloy was disrupted by volume expansion accompanying the formation of cementite, and both copper and iron were intermixed with carbon deposit.

This phenomenon has been demonstrated by another experimental technique [30]. About 0.7 g of porous iron granules placed in a platinum basket were held in a stream of 80% H_2 + 18% CO + 2% CO_2 mixture at 600 °C. About 8.5 g carbon accumulated in the basket to a height of 5 cm was analyzed for iron at several levels; there was essentially uniform distribution of finely dispersed iron as cementite in carbon deposit.

The surface pitting on a test sample of HK40 alloy steel (20% Ni, 25% Cr, 1.5% Si, and 0.4% C) resulting from metal dusting is shown in Fig. 10.9 [44].

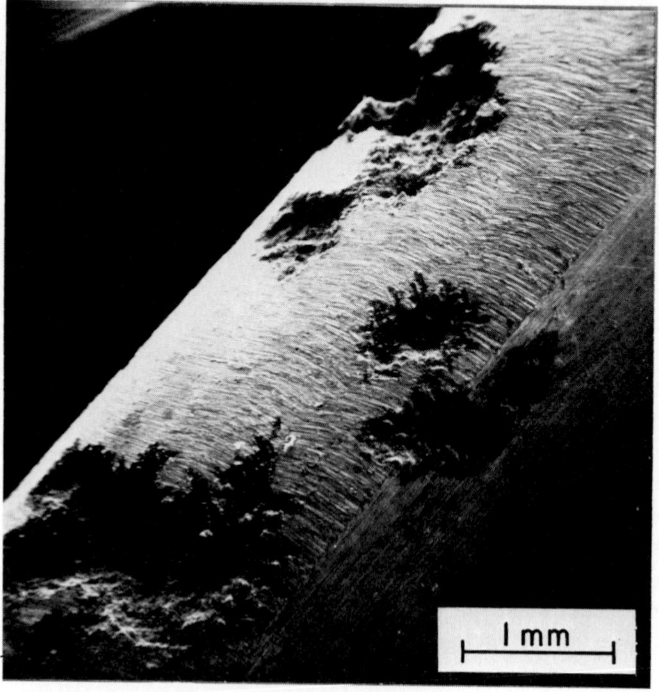

FIG. 10.9 Corrosion pits on HK40 alloy after 3-day exposure to reformed natural gas at 4 atm and 700 °C. After Turkdogan and Olsson [44].

This sample was exposed to a stream of simulated reformed natural gas at 4 atm and 700 °C for three days. To permit a better view of the pits, soot deposited on the sample was brushed away. As is seen from Fig. 10.10, the x-ray scan spectrum of the accumulated soot is almost identical to that obtained for the unreacted HK40 steel. This observation once again demonstrates that the metallic substrate is disrupted and dispersed in carbon deposit.

The metal dispersed in carbon deposit has a high chemical reactivity as a catalyst, as noted, for example, in hydrogenation of carbon deposited on nickel particles [37, 45].

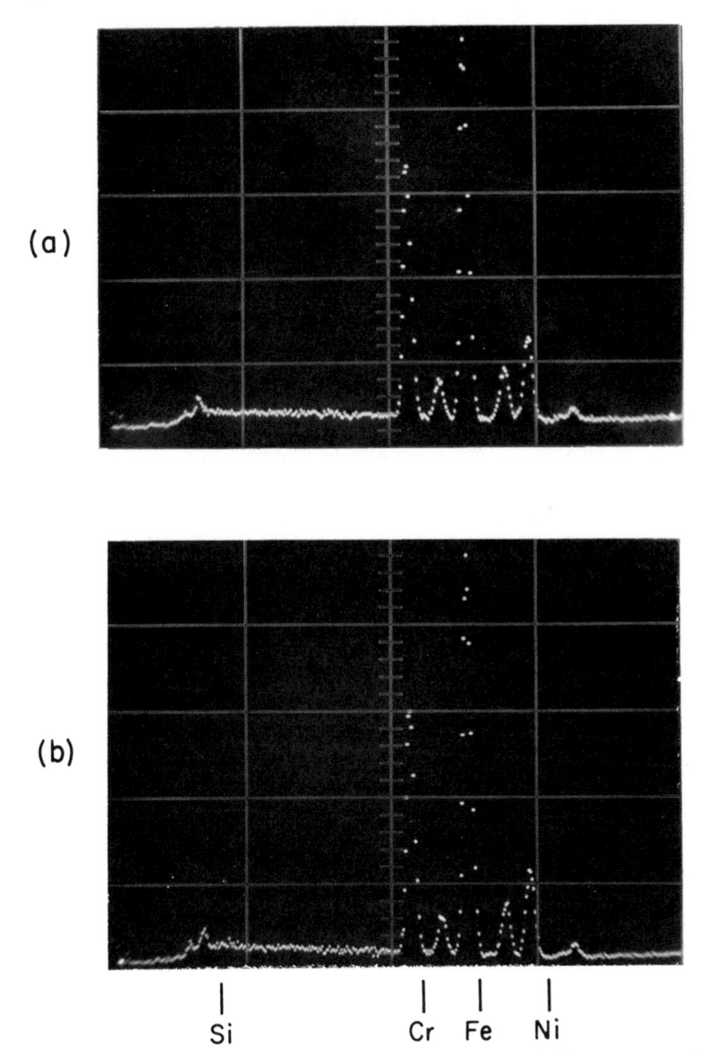

FIG. 10.10 X-ray scan spectra of (a) unreacted HK40 alloy and (b) soot accumulated during 3-day exposure to reformed natural gas at 4 atm and 700 °C. After Turkdogan and Olsson [44].

Equipment exposed to carburizing–oxidizing environments are usually made from heat-resistant austenitic Ni–Cr–Fe alloys containing small percentages of C, Si Mn, Al, and Ti, such as *Inconel* 600 (Ni/Cr = 75/15) and *Incoloy* 800 (Ni/Cr = 32/20). For these alloys to be resistant to metal dusting, the oxygen potential of the carburizing gas should be sufficiently high to sustain a continuous dense layer of Cr_2O_3 on the surface of the metal. As shown by Schaas and Grabke [46], for example, when Incoloy 800 was held at 1000 °C in a mixture of CO–H_2–H_2O with $a_C = 1$ and $p_{H_2O}/p_{H_2} = 10^{-3}$, the metal surface became coated with a dense layer of Cr_2O_3 through which carbon could not diffuse, and hence there was no carburization of the metallic substrate and no surface pitting. However, in high temperature creep tests under an applied stress, cracks developed in the Cr_2O_3 surface layer and at grain boundaries of the metal that led to the onset of carburization and ultimate rupture of the metal in the creep test.

When cracks are developed in the surface oxide layer during service, due to the temperature cycle and applied stress, a carbide subscale begins to form. Because of the formation of chromium-rich carbides, M_7C_3 and $M_{23}C_6$, the matrix in the subscale and around the grain boundaries becomes depleted of chromium, and consequently, a new protective layer of Cr_2O_3 may not form. The strength and ductility decreases because of the formation of a chromium-depleted ferritic Fe–Ni matrix. Also, the Fe–Ni matrix exposed to the carburizing gas via the surface cracks catalyzes the disproportionation of the carbon monoxide leading to metal dusting and ultimate failure of the metal. Reference may be made to papers by, for example, Hochman [47] and Schueler [48] for numerous examples of hot corrosion of stainless steels and superalloys in carburizing–oxidizing environments. It should be noted also that the alloys in the as-cast condition have lower resistance to metal dusting because of enhanced carburization of the substrate by gas diffusion through the porous cast structure. Whenever possible, wrought alloys should be used.

With gases of low oxygen activity, e.g., when $p_{H_2O}/p_{H_2} < 10^{-5}$, a protective oxide layer Cr_2O_3 will not form on the Ni–Cr–Fe or Ni–Cr–Co alloys. For such applications, the aluminum is added to the steel to form a protective surface layer of alumina. Aluminide-coated alloys give better service in applications at higher temperatures. Useful background information on the behavior of aluminide-coated superalloys is given in review papers by Lindblad [49] and Goward [50].

10.4.2 Oxidation

Only a brief presentation was made in Sections 7.6.2 and 7.6.3 of the parabolic rate law of oxidation of pure metals, and the simple form of internal oxidation of binary alloys. For background reading on the physical chemistry of oxidation of metals and alloys, reference may be made to textbooks by Kubaschewski and Hopkins [51] and by Hauffe [52]. Much of what we know about the theories of oxidation of metals and alloys is attributed to the teachings of Carl Wagner, over a period of four decades, on the thermodynamics of point defects and mechanisms of diffusion in semiconductors, the basic concepts of scale and subscale formation, interface stability, and so on.

In keeping with the general theme of Section 10.4, we shall consider here a specific case of oxidation pertinent to dispersion-strengthened alloys. First, we should perhaps give a simple example of preventing catastrophic oxidation of iron–aluminum alloys. As estimated from Eq. (7.105), in a low alloy steel containing more than about 0.1% Al, in air oxidation the aluminum should be preferentially oxidized to form a layer of alumina on the surface of the metal. However, the oxygen diffusing through the thin layer of alumina will react with the aluminum-depleted iron substrate and form an iron oxide subscale, accompanied by internal oxidation of aluminum. Because of preferential diffusion of iron in wustite, outward growth of iron oxide disrupts the initial thin film of alumina, resulting in the growth of iron oxide nodules on the outer surface, as observed, for example, by Boggs [53]. Wagner [54] pointed out that this sequence of events leading to catastrophic oxidation may be alleviated by the addition of a third alloying element which has affinity for oxygen between those for iron and aluminum, such as chromium or silicon. The third alloying element acts as a getter for oxygen during the initial stages of oxidation and prevents diffusion of oxygen atoms into the interior of the alloy followed by internal oxidation of aluminum which would preclude the formation of a protective surface layer of alumina. This reasoning of Wagner would account for excellent oxidation resistance of Fe–Cr–Al alloys [55]. Also, as shown by Boggs [56], an addition of about 2% Si to iron containing 2%–4% Al greatly improves the oxidation resistance of the metal.

Stimulated by demands for alloys of high strength in oxidizing atmospheres at elevated temperatures, intensified research efforts in the 1960s resulted in the development of dispersion-strengthened alloys [57]. The Ni–20% Cr-2 vol% ThO_2 alloy, supplied under the trade name TD-NiCr, has received much attention because of its combination of high strength, due to the presence of finely dispersed thoria particles (submicron size), and oxidation resistance at temperatures of 900–1200 °C. The oxidation behavior of TD-NiCr or other dispersoid-containing alloys has been investigated by many; studies by Giggins and Pettit [58] and Stringer *et al.* [59] are particularly enlightening. The characteristic oxidation behavior of TD-NiCr compared to the Ni–20Cr and Ni–30Cr alloys are summarized below.

1. Dispersoids promote selective oxidation of chromium, and a scale composed essentially of Cr_2O_3 is formed with only small amounts of spinel. Initially, both NiO and Cr_2O_3 are formed; however, because of the nucleating effect of dispersoids, there is a rapid lateral growth process to form a continuous layer of Cr_2O_3 which prevents further growth of NiO on the surface. Ultimately, NiO is converted to spinel $NiCr_2O_4$. In the Ni–20Cr alloy, a larger amount of NiO is formed before the Cr_2O_3 layer becomes continuous.

2. The rate of oxidation of TD-NiCr alloy is 15–20 times slower than that of Ni–30Cr [58]. After the initial stages, oxidation reaction in TD-NiCr occurs at the metal–scale interface, while with the Ni–30Cr alloy there is outward growth of the scale. Slow rate of oxidation and inward growth of the Cr_2O_3 scale is attributed by Stringer *et al.* to two independent diffusion processes: (i) The dispersoids retard the diffusion of chromium in Cr_2O_3 and (ii) the fine grain size of the Cr_2O_3 layer, induced by dispersoids, enhances grain-boundary diffusion of oxygen; the latter is responsible for inward growth of the scale. The chromic oxide is a cation deficient *p*-type semiconductor which also contains some Frenkel defects, i.e., interstitial plus vacancy pairs. Michels [60] suggested that, similar to the behavior of TD-NiCr, the oxidation resistance of Ni–20Cr alloys containing dispersed Y_2O_3, La_2O_3, or Al_2O_3 may be attributed to slow diffusion of chromium in chromic oxide because of fillage of cation vacant sites with trivalent ions Y^{3+}, La^{3+}, or Al^{3+}. Since thorium is capable of acquiring several valence states in compounds including thoria, Michels suggested that the concentration of cation vacancy in Cr_2O_3 is reduced in the presence of dispersed ThO_2, thus accounting for the slow rate of scaling.

3. There is little or no spalling and strong adhesion of scale on Ni–20Cr alloys containing dispersoids; also, the scale is resistant to spalling when subjected to temperature cycle. Although no explanation could be given for this beneficial effect, Stringer *et al.* [59] suggested that the improved adhesion may be associated with the inward growth of the Cr_2O_3 layer.

4. The formation of a protective layer of Cr_2O_3 on chromium alloys is offset by its oxidation to the volatile species CrO_3,

$$Cr_2O_3(s) + \tfrac{3}{2}O_2(g) = 2CrO_3, \qquad (10.30)$$

for which the temperature dependence of the equilibrium constant at unit activity of Cr_2O_3 is approximately [61]

$$\log(p^2_{CrO_3}/p^{3/2}_{O_2}) \simeq -(27{,}200/T) + 5.32, \qquad (10.31)$$

where the partial pressures are in atmospheres. The chromium losses due to the formation of CrO_3 increase with increasing temperature, partial pressure of oxygen, and gas velocity. Therefore, for given conditions of oxidation, the overall rate of consumption of chromium is determined by an interplay of the parabolic rate of scale thickening due to oxidation of chromium and the linear rate of scale thinning due to vaporization of CrO_3. Ultimately, the scale acquires a limiting thickness X_l when the parabolic rate of thickening is equal to the linear rate of thinning,

$$X_l = k_p/2k_v, \qquad (10.32)$$

where k_p is the parabolic rate constant for oxidation and k_v is the volatilization rate constant for given temperature, oxygen pressure, and gas velocity. Using the rate equation derived by Tedmon [62] for this oxidation/vaporization process, Giggins and Pettit [58] showed that, because the values of k_p for

TD-NiCr alloy are smaller than those for Ni–30Cr alloy, during oxidation in essentially static oxygen (free convection only), the chromium consumption rate for TD-NiCr alloy is smaller than that for Ni–30Cr alloy for oxidation times up to 1000 hr. Beyond this exposure time, when the scale acquires the limiting thickness, rates of chromium consumption are the same for both alloys.

The chromium loss due to CrO_3 volatilization is not too severe in static or slowly moving gas environments. The rate constant for reaction (10.30) in essentially static oxygen at 0.1 atm has been determined by Hagel [63]:

$$k_v = 0.214 \exp[(-48{,}800 \pm 3000)/RT] \quad g(Cr_2O_3)\,cm^{-2}\,sec^{-1}. \tag{10.33}$$

Because of high temperature strength and excellent oxidation resistance, TD-NiCr alloy has been considered for use in high performance jet engines and as part of a thermal protection cover for the space shuttle. The sonic and hypersonic gas velocities encountered in such applications greatly enhance material loss due to vaporization. For example, Lowell and Sanders [64] have found that the rate of material loss due to vaporization in a 1 atm, Mach-1 turbine gas stream at 1200 °C was about two orders of magnitude greater than that occurring in static tests. Also, the rate of oxidation of Cr_2O_3 to CrO_3 is greatly enhanced in the presence of atomic oxygen, as would be expected from the comparison of the equilibrium constant in Eq. (10.31) with the equilibrium relation

$$\begin{gathered} Cr_2O_3(s) + 3O(g) = 2CrO_3(g), \\ \log(p_{CrO_3}^2/p_O{}^3) \simeq (12{,}400/T) - 4.94. \end{gathered} \tag{10.34}$$

Since a large fraction of molecular oxygen in the boundary layer is dissociated into atomic oxygen by the shock wave [65], the loss of material due to vaporization will be appreciable at hypersonic velocities.

Tenney *et al.* [66] investigated the oxidation behavior of TD-NiCr alloy under simulated reentry conditions for the space shuttle. Specimens were oxidized in the arc-heater system [67] under the conditions of 1200 °C, Mach-5 air stream velocity ($\equiv$0.9 g sec^{-1} mass flow rate), 1070 cal g^{-1} stream enthalpy, 3.5 cal cm^{-2} sec^{-1} cold-wall heating rate, and a stagnant pressure of 0.0066 atm. A boundary-layer thickness of 360 μm was estimated for this flow condition. Based on microscopic examinations, microprobe analyses, and x-ray diffraction measurements, Tenney *et al.* made the following observations on the dynamic oxidation behavior of TD-NiCr alloy. A thin film of Cr_2O_3 on the sample readily vaporizes in the initial stages of exposure to the air stream at 1200 °C, followed by the formation of NiO, an intermediate duplex oxide, and an inner Cr_2O_3 layer. The inner Cr_2O_3 subscale blocks the supply of Ni to the outer NiO layer with the result that the thickness of the NiO layer decreases with time, due to vaporization. There is also vaporization of Cr_2O_3 from regions not covered by NiO. The growth of porous NiO nodules and their transport on the surface in the flow direction is caused by a vaporization–condensation process promoted by the high gas velocity and the presence of atomic oxygen in the boundary layer. The amount of alloy consumed during dynamic oxidation is about 20 times larger than that in static tests.

Whittenberger [68] studied the oxidation behavior of thin-gage (0.037 cm) TD-NiCr alloys and superalloys (Hastelloy X and Haynes Alloy 188) in multiple reentry conditions simulated through cyclic plasma-arc-tunnel exposure. The exposure consisted of 100 cycles of 10-min duration at temperatures of 980–1200 °C in a Mach-4.6 airstream with an impact pressure of 0.008 atm. It was found that the mechanical properties of the alloys at ambient and elevated temperatures were not much affected after cyclic exposure to severe oxidation under simulated reentry conditions.

10.4.3 Sulfidation

Hot corrosion is particularly serious in gases containing sulfur-bearing species that are encountered in most process environments and gas turbines. The extent and mechanism of hot corrosion by sulfidation of alloys depend on the oxygen potential of the environment; alkalies also have a large effect on corrosion by oxidation and sulfidation.

10.4.3a Sulfidation in the absence of oxides. In applications such as coal gasification and processing of hydrocarbons, we are concerned with sulfidation of reactor walls under reducing conditions. The sulfur activities ($\equiv p_{H_2S}/p_{H_2}$) are given in Fig. 10.11 for selected metal–metal sulfide equilibria up to eutectic melting temperatures. The sulfides of base metals Fe, Co, and Ni have relatively low eutectic melting temperatures as compared to 1530 °C for the Cr–CrS eutectic. The resistance of base metals to sulfidation is increased by alloying with chromium. The resistance to sulfidation is improved also by the addition of, for example, aluminum. For general guidance, the dotted curves are estimated sulfur activities for base metal containing about 1% Al, in equilibrium with Al_2S_3, or about 30% Cr, in equilibrium with CrS. Because of the formation of sulfide solid solutions and spinels, e.g., *pentlandite*, $(Fe, Ni)_9S_8$, and *daubreelite*, $FeCr_2S_4$, the sulfur activities for metal–metal sulfide equilibria in binary alloys are much more complex than those given by the dotted curves in Fig. 10.11.

Extensive and repeated studies have been made of the sulfidation of chromium alloys with iron [69, 70], cobalt [71, 72], and nickel [73, 74], just to name a few. At temperatures below the melting points of the sulfides, the growth of the sulfide scale obeys the parabolic rate law for pure metals, binary, and multi-component alloys. In the early stages of sulfidation in H_2S–H_2 mixtures, the reaction is controlled by the rate of dissociation of H_2S on the surface of the sulfide exposed to the gas phase, as, for example, in the sulfidation of iron [75]. This aspect of the sulfidation reaction will not be discussed here.

The marker experiments have shown that the scale grows outwards; that is, the alloying elements reacting with the sulfide MS at the inner scale–metal interface,

$$M = M^{2+} + 2\varepsilon, \tag{10.35}$$

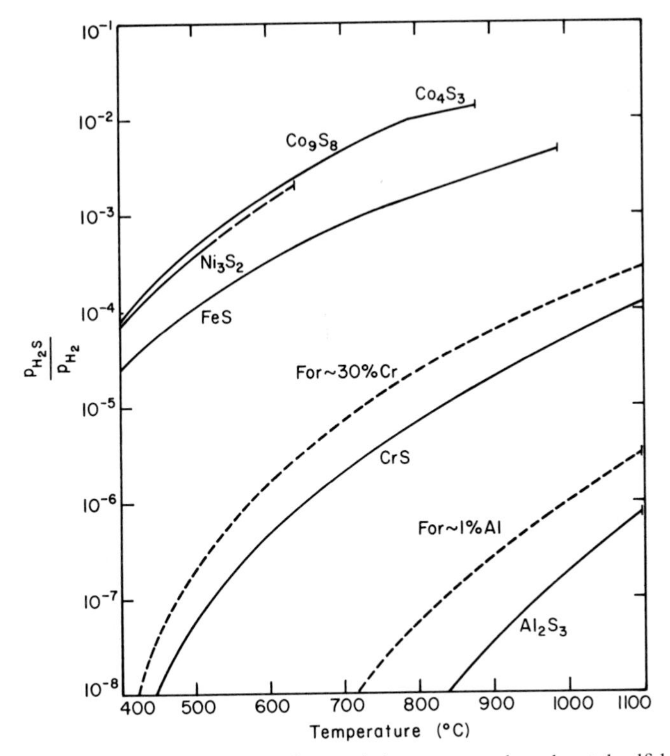

FIG. 10.11 Equilibrium p_{H_2S}/p_{H_2} ratios for coexisting pure metals and metal sulfides as a function of temperature up to the eutectic invariant; dotted curves are approximate for Fe–30% Cr and Fe–1% Al alloys.

diffuse across the scale via the cation vacant sites and react with sulfur in the gas at the scale–gas surface:

$$M^{2+} + \tfrac{1}{2}S_2 + 2\varepsilon = MS. \tag{10.36}$$

The plastic flow accompanying the growth of the scale maintains good contact between the scale and the retreating surface of the metal core. When the plastic deformation is impaired as in thick scales, corners of flat samples, or surface of wires, the cracks develop at the interface; this leads to local dissociation of the scale [reverse of reaction (10.36)], followed by inward growth of the inner scale by diffusion of sulfur, and the development of a porous inner layer.

Depending on the concentration of chromium, there are three modes of scale formation in the sulfidation of chromium alloys with iron, cobalt, and nickel. The variation of the parabolic rate constant with the composition of Ni–Cr alloys, determined by Mrowec *et al.* [73], is shown in Fig. 10.12; similar trends have been observed with Fe–Cr and Co–Cr alloys. For chromium concentrations below about 2%, region I, the scale consists of a single phase $M_{1-x}S$ with some Cr^{3+} in solution. In the range 2%–40% or 70% Cr (depending on the

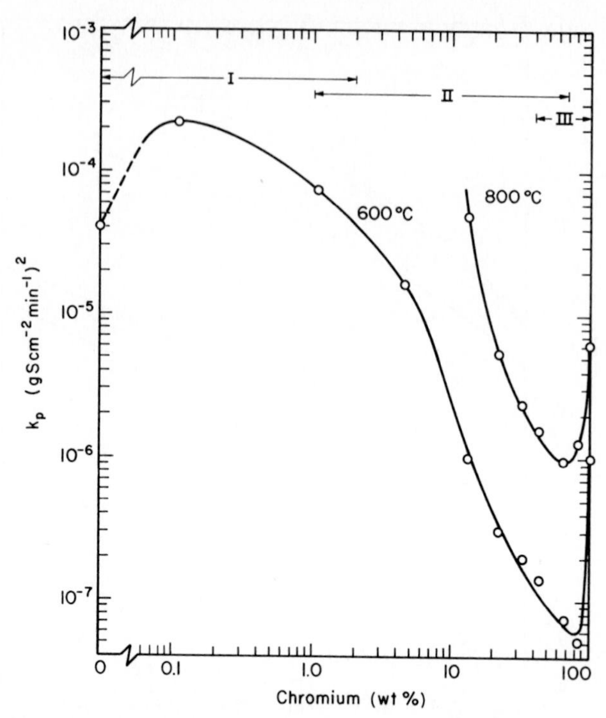

FIG. 10.12 Variation of the parabolic rate constant for oxidation and sulfidation with the composition of Ni–Cr alloys investigated by Mrowec *et al.* [73], at indicated temperatures and 1 atm S_2.

system and temperature), region II, the scale consists of two sulfide layers: the outer layer is $M_{1-x}S$ with Cr^{3+} in solution, and the inner layer is Cr_2S_3 with M^{2+} in solution. Above 40% or 70% Cr, region III, the scale consists of Cr_2S_3 (or Cr_3S_4, CrS, depending on the system, temperature, and gas composition). There are, of course, transition regions between these three types of scales. At low concentrations of chromium in region II, the inner layer has a duplex structure consisting of $M_{1-x}S$ with dispersed particles of Cr_2S_3 or spinel. At higher chromium concentrations the duplex layer becomes Cr_2S_3 with dispersed particles of $M_{1-x}S$, and above about 10% Cr the inner layer becomes a single phase Cr_2S_3. The thickness ratio of the layers $Cr_2S_3/M_{1-x}S$ increases with increasing chromium concentration, and since the diffusion of M^{2+} in Cr_2S_3 is slower than in $M_{1-x}S$, the parabolic rate of sulfidation decreases with increasing concentration of chromium in the alloy.

As postulated originally by Wagner [76] and proven experimentally time and again, in p-type semiconductors, as in the oxides and sulfides of chromium, iron, cobalt, and nickel, there are vacancies in the cation sublattice and the corresponding electron holes. The electroneutrality is maintained by fillage of some cation sites with the cations of higher valence to an amount equal to the vacancy concentration. The dissolution of another cation of higher valence in a p-type semiconductor increases the cation–vacancy concentration; conversely, doping of the lattice with a cation of lower valence has the reverse effect.

For the reasons given above, the addition of a small amount of chromium increases the diffusivity of M^{2+} in Cr-doped $M_{1-x}S$, hence increasing the rate of sulfidation. Further increase in chromium concentration decreases the nickel activity gradient in the scale, hence decreasing the rate of sulfidation. These two competing effects give rise to a maximum rate of sulfidation at low chromium contents. In region II the rate of sulfidation is controlled by diffusion in the Cr_2S_3 layer, because the cation diffusion in M^{2+}-doped Cr_2S_3 is lower than in Cr^{3+}-doped $M_{1-x}S$. Also, because of increase in the ratio $Cr_2S_3/M_{1-x}S$ with increasing concentration of chromium, the rate of sulfidation is lower at higher concentrations of chromium in the alloy. In region III the rate of sulfidation increases with increasing chromium concentration because of a decrease in M^{2+} doping of Cr_2S_3 and a reduction in nickel enrichment at the receding metal surface. Therefore, the lowest rate of sulfidation occurs when the chromium concentration is just enough for the formation of a single-phase Cr_2S_3 scale.

Unequal rates of transfer of the alloying elements across the metal–scale interface may bring about appreciable changes in alloy composition near the interface. Consequently, in later stages of sulfidation the mode of scale formation may change from region I to II or from region II to III. Romeo *et al.* [74] have shown that the rate of sulfidation of Ni–20Cr alloy in H_2–H_2S mixtures at 700 °C is independent of the sulfur activity (H_2S/H_2) in the gas ($\equiv 10^{-2}$–10^{-7} atm S_2) and the parabolic rate constant agrees well with that measured by Mrowec *et al.* [73] at 1 atm S_2. This is as would be expected for the rate controlled by diffusion in the inner sulfide layer Cr_2S_3, where the activity gradient is determined by the alloy composition, and that for the $Ni_{1-x}S/Cr_2S_3$ equilibrium; these activities are independent of the activity of sulfur in the gas. However, in regions I and III for single sulfide scale, the rate of sulfidation increases with increasing sulfur activity of the gas, as indicated by Wagner's rate constant, Eq. (7.101).

Additions of less noble elements, e.g., Al, Mo, Zr, and rare earths, increase the resistance of M–Cr alloys to sulfidation and suppress the formation of liquid sulfides. In ternary and multicomponent alloys of chromium, the scale consists of several layers of sulfides; the inner layer usually consists of Cr_2S_3 (or Cr_3S_4, CrS) with dispersed sulfide of the third alloying element. There are large variations in the retardation effect of the added third element on the rate of sulfidation. For example, an addition of 5% Al lowers the rate of sulfidation of Fe–20% Cr alloy by a factor of about 5 [77], while the rate for Co–25% Cr alloy is reduced by a factor of about 300 [78]. The rate of sulfidation of Ni–20% Cr alloy is halfed by an addition of 5% Mo [79].

10.4.3b Sulfidation under oxidizing conditions. When a pure metal or an alloy is exposed to an oxidizing gas containing SO_2, sulfidation occurs along with oxidation such that the scale formed consists of an outer oxide layer and an inner sulfide layer with an intermediate layer of oxide containing sulfide dispersoids. Depending on temperature, gas composition, and alloy composition, some internal oxidation and sulfidation may also occur. The oxidation–sulfidation process is characterized by countercurrent diffusion in the scale: (i) outward diffusion of metal via the cation vacant sites and oxidation at the gas–scale surface and (ii) inward diffusion of sulfur and sulfidation at the metal–scale interface. There is also sulfur transfer to the metal–scale interface via the cracks and fissures in the scale. In fact, we see from the examples cited by Birks [80] that both mechanisms of sulfur transfer have been noted to contribute to the sulfidation of alloys accompanying oxidation.

The metal sulfides are sparingly soluble in metal oxides even at elevated temperatures, e.g., 0.01%–0.03% S. However, the lattice diffusion of sulfur in metal oxides is relatively fast and similar to the metal diffusivity in the oxide via the cation vacant sites. Examples are given in Table 10.2 of measured sulfur diffusivities in selected oxides at 1000 °C. Numerous studies [84, 85] have shown that despite low sulfide solubilities in oxides, the rates of oxidation of alloys are enhanced in the presence of sulfides, which are believed to increase cation migration in the oxide scale.

TABLE 10.2

Sulfur Diffusivities in Selected Oxides at 1000 °C

Oxide	Diffusivity ($cm^2\ sec^{-1}$)	Reference
Wustite	3.1×10^{-8}	81
CoO	6.3×10^{-13}	82
NiO	4.7×10^{-13}	82
Cr_2O_3	1.0×10^{-10}	83

The lattice diffusion of sulfur across the oxide scale will occur only when the sulfur potential at the gas–scale interface is higher than that at the metal–scale interface. The critical gas composition for the sulfidation to occur may be determined by considering two reaction equilibria:

gas equilibrium at the gas–scale interface:

$$\tfrac{1}{2}S_2 + O_2 = SO_2; \tag{10.37}$$

sulfidation at the metal–scale interface:

$$\tfrac{1}{2}S_2 + M = MS. \tag{10.38}$$

The sulfur potentials of the sulfides of cobalt, nickel, iron, and chromium (in equilibrium with respective pure metals) are given in Fig. 10.13. The bottom broken line is estimated for Al_2S_3 in equilibrium with Fe–1% Al alloy. The dotted lines are sulfur potentials for indicated p_{SO_2}/p_{O_2} ratios in the gas phase.

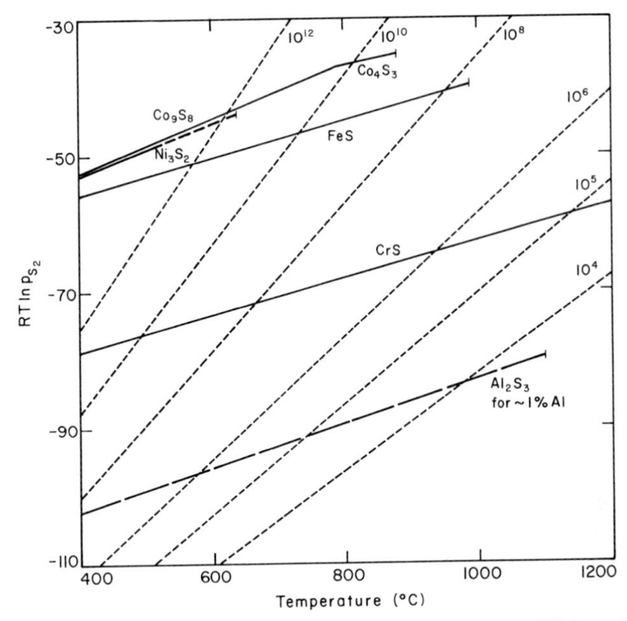

FIG. 10.13 Sulfur potentials for indicated pure metal–metal sulfide equilibria and for base metal (iron) containing about 1% Al, compared with those for indicated (p_{SO_2}/p_{O_2}) ratios (– – –) in the gas phase .

From such an equilibrium diagram, we can read off the range of gas compositions and temperatures where oxidation can or cannot be accompanied by sulfidation. This is, of course, for the case of lattice diffusion of sulfur in the scale assumed to be free of cracks and fissures, precluding molecular diffusion SO_2. For example, in a gas mixture with p_{SO_2}/p_{O_2} equal to 10^6, the sulfidation at the metal–scale interface can occur only at temperatures above 575 °C for Fe–1% Al alloy and above 935 °C for chromium. Whether or not sulfidation can occur at a given temperature and SO_2 concentration depends on the oxygen activity (CO_2/CO or H_2O/H_2 ratio) in the gas. The lower the oxygen activity, the greater are the chances of sulfidation, particularly in oxidation-resistant alloys containing aluminum or other elements which have strong affinity for sulfur. In other words, a stable oxide coating on an alloy for protection against oxidation is no guaranty for the suppression of hot corrosion by sulfidation at low oxygen activities. For example, as shown by McKee and Romeo [86], Ni–10% Cr alloy resistant to sulfidation in an N_2–2% SO_2 mixture at 1000 °C is severely corroded, due to sulfidation, when 1% CH_4 is added to the gas mixture.

10.4.4 Hot Corrosion Enhanced by Alkalies

Deposition of fuel ash on metals and alloys is known to cause hot corrosion, which is particularly serious with Na_2SO_4 as the principal constituent of the ash. The sodium sulfate forms during combustion of fuel containing salt and sulfur and condenses on metal parts exposed to the combustion gases [87]. The superalloys used for turbine blades of aircraft jet engines are also susceptible to hot corrosion by condensed Na_2SO_4, which forms from reaction of sulfur in the fuel and sea salt ingested inadvertently into the engine. Many studies have been made of hot corrosion of superalloys, under accelerated testing conditions, to understand the mechanism of hot corrosion and identify the constituent alloying elements which enhance hot corrosion. Of the various mechanisms postulated by many investigators, those proposed by Goebel *et al.* [88–90] from thermodynamic and electrochemical considerations are perhaps most appealing.

The mechanism of hot corrosion enhanced by Na_2SO_4 proposed by Goebel *et al.* [88, 89] is based on their oxidation experiments using alloys of nickel with Al, Cr, Mo, W, V, and combinations thereof, as in superalloy B-1900 (Ni–10% Co–8% Cr–6% Al–1% Ti–6% Mo–4.3% Ta–0.015% B–0.1% C–0.075% Zr). The alloy was coated with 0.5-mg Na_2SO_4 per cm^2 of the sample and oxidized in 1-atm oxygen at 1000 °C. In the absence of Na_2SO_4, all the alloys had a similar resistance to oxidation. In the presence of Na_2SO_4, two types of enhanced oxidation behavior were observed: (i) Na_2SO_4-induced accelerated oxidation and (ii) Na_2SO_4-induced catastrophic oxidation. The catastrophic oxidation occurred only with alloys containing molybdenum, tungsten, or vanadium. Accelerated oxidation occurring with Ni–Al, Ni–Cr, and Ni–Cr–Al alloys was not self-sustaining and depended on the amount of Na_2SO_4 on the sample. Whether accelerated or catastrophic, the enhanced oxidation is triggered by the dissolution of the protective oxide layers in molten Na_2SO_4 deposited on the surface.

The change in the oxide ion activity of molten Na_2SO_4 caused by the dissolution of the oxidation products is the key to the mechanism of enhanced oxidation. In oxide dissolution consuming oxide ions,

$$MO + O^{2-} = MO_2^{2-} \tag{10.39a}$$

or

$$MO + SO_4^{2-} = MO_2^{2-} + SO_3(g), \tag{10.39b}$$

Na_2SO_4-base flux becomes acidic and SO_3 is evolved. In oxide dissolution resulting in donation of oxide ions,

$$MO = M^{2+} + O^{2-} \tag{10.40a}$$

or

$$MO + SO_3(g) = M^{2+} + SO_4^{2-}, \tag{10.40b}$$

Na_2SO_4-base flux becomes basic and the equilibrium SO_3 pressure decreases. Goebel *et al.* have found that SO_2 is evolved when MoO_3, WO_3, V_2O_5, or Cr_2O_3 is dissolved in molten Na_2SO_4 and the oxide ion activity decreased as indicated by the emf-cell measurements, the effect being greatest with MoO_3 dissolution and least with Cr_2O_3 dissolution. Although NiO and Al_2O_3 are sparingly soluble in molten Na_2SO_4, a change in the oxide ion activity of the salt increases the solubility of NiO and Al_2O_3, either as MO_2^{2-} in acidic flux or M^{2+} in basic flux.

The oxidation of alloying elements, molybdenum, tungsten, or vanadium in nickel–aluminum alloys in the presence of Na_2SO_4 on the surface, decreases the oxide ion activity of the flux by virtue of reaction (10.39) and dissolves the protective layers NiO and Al_2O_3. Away from the flux–metal interface, cations Ni^{2+} and Al^{3+} react with the molybdate anions MoO_4^{2-}, resulting in the liberation of volatile MoO_3, and precipitation of NiO and Al_2O_3 particles near the flux–gas surface.

$$Ni^{2+} + MoO_4^{2-} = NiO(s) + MoO_3(g), \qquad 2Al^{3+} + 3MoO_4^{2-} = Al_2O_3(s) + 3MoO_3(g). \quad (10.41)$$

Similar reactions occur with the oxidation of tungsten and vanadium in the alloy. This oxidation process is catastrophic or self-sustaining because, once initiated, elements such as Mo, W, and V in the alloy insure the continued presence of the acidic flux and prevent the formation of a protective oxide layer on the alloy.

Although the dissolution of Cr_2O_3 in molten Na_2SO_4 lowers the oxide ion activity, the flux is not sufficiently acidic to dissolve Al_2O_3, so that catastrophic oxidation does not occur with Ni–Cr–Al alloys and instead there is accelerated oxidation in the presence of Na_2SO_4. The accelerated oxidation is initiated by the transfer of sulfur from Na_2SO_4 to the alloy:

$$SO_4^{2-} + \tfrac{8}{3}Al = \tfrac{1}{3}Al_2S_3 + Al_2O_3 + O^{2-}. \quad (10.42)$$

An increase in flux basicity accompanying this reaction leads to the dissolution of the protective Al_2O_3 layer:

$$Al_2O_3 + O^{2-} = 2AlO_2^{2-}. \quad (10.43)$$

The accelerated oxidation continues until all the sulfur in the flux is converted to Al_2S_3. Unless Na_2SO_4 on the surface is replenished continuously, the accelerated oxidation is not self-sustaining.

The addition of chromium to Ni–Al alloys improves the resistance to accelerated oxidation in the presence of Na_2SO_4. In this case, Cr_2S_3 forms and the solution of Cr_2O_3 in the flux does not lower the oxide ion activity enough to dissolve the inner Al_2O_3 layer. Also, the dissolution of Cr_2O_3 in the flux counteracts the increase in flux basicity by the formation of Cr_2S_3 from reaction of Na_2SO_4 with the chromium in the alloy, thus minimizing attack on the Al_2O_3 layer.

The foregoing brief outline of the mechanisms of catastrophic oxidation proposed by Goebel *et al.* need not be limited to the presence of Na_2SO_4 on the alloy surface. For the onset of self-sustaining catastrophic oxidation, three conditions are necessary and sufficient: (i) single application of a low melting nonvolatile flux, (ii) oxidation and dissolution of alloying elements that render the flux acidic, and (iii) vaporization of the oxides of the alloying elements at the gas–flux surface. The dissolution of MoO_3 or WO_3 at the metal–flux interface and their vaporization at the gas–flux surface provide the basic mechanism for continual destruction of the inner Al_2O_3 layer and deposition of porous nodules of Al_2O_3 on the surface, with flux acting as a reaction and transfer medium. The incubation period for the onset of catastrophic oxidation varies from a few hours to a few hundred hours, depending on temperature, gas composition, and alloy composition.

The superalloys based on nickel–chromium, cobalt–chromium, or TD-NiCr are resistant to hot corrosion to a moderate degree only. Further improvements on resistance to hot corrosion cannot be made by major changes in the alloy composition without impairing the ductility, toughness, and creep strength of the alloy. The problem has been overcome to a large extent by applying special metal or alloy coatings on superalloys. The diffusion–aluminide coatings that have been in use for many years are now being replaced with FeCrAlY, NiCrAlY, or CoCrAlY alloy coatings, the latter being more resistant to hot corrosion induced by salt or Na_2SO_4. For information on current developments of alloy coatings resistant to catastrophic oxidation and thermal fatigue under conditions of thermal cycling, reference may be made to a review paper such as that by Goward [50].

10.4.5 Corrosion of Alloys by Molten Fluorides

The chemical attack of molten salts on stainless steel or superalloys is another example of hot corrosion pertinent to molten salt breeder reactors that are being developed, based on the concept of using fluorides of fissile materials UF_4, PuF_3) dissolved in a mixture of lithium fluoride, beryllium fluoride, and thorium fluoride as fuel [91]. The fuel salt also acts as a heat-exchanger medium with hot and cold legs of the loop at temperatures of about 700 and 560 °C, respectively. The molten fluorides are also considered for use as unfueled heat-transfer media. With these possible applications in mind, extensive studies have been made of the attack of molten fluorides on construction materials such as types 304, 316, 410, 430, and 446 stainless steel and numerous superalloys [92].

The characteristic feature of the attack of molten fluorides on Fe–Ni–Cr alloys is that there is preferential reaction of chromium in the alloy with either impurities in the salt,

$$\underline{Cr} + FeF_2 \text{ (or } NiF_2) = CrF_2 + \underline{Fe} \text{ (or Ni)}, \tag{10.44}$$

$$\underline{Cr} + 2HF = CrF_2 + H_2, \tag{10.45}$$

or constituents of the salt,

$$\underline{Cr} + 2UF_4 = CrF_2 + 2UF_3. \tag{10.46}$$

The preferential attack on chromium is as would be expected from measured, or calculated, electrode potentials in molten fluorides; e.g., in LiF–BeF_2 melts at 500 °C, the electrode potential is −0.789 V for Cr^{2+}, −0.204 V for Fe^{2+} and 0 for Ni^{2+} [93].

The onset of a concentration gradient resulting from chromium depletion in the surface layer causes the diffusion of chromium from the bulk metal toward the surface, leaving behind a zone enriched with vacancies, The vacancies precipitate at grain boundaries and inclusions, leading to the development of a porous substrate.

Since reaction (10.46) is endothermic, the forward and reverse reactions provide the mechanism of transfer of chromium from the walls of the hot leg to the cold leg of the circulating loop. The increase in chromium concentration of the salt continues until the rate of dissolution of chromium from the walls of the hot leg equals the rate of deposition on the walls of the cold leg. When this steady state condition is achieved, the composition of the salt remains unchanged and the transfer of chromium from hot to cold parts of the loop continues at a slow rate, determined by diffusion of chromium to the surface of the alloy.

Some HF is formed in molten fluoride mixtures by reaction with H_2O or other hydrogen-containing impurities present in the salt at the start of the loop operation. Since reaction (10.45) is exothermic, there will be material removal from the cold zone and deposition in the hot zone of the loop until the impurities generating HF have been consumed.

Extensive work done in Oak Ridge National Laboratory on hot corrosion of stainless steel and superalloys has been reviewed by Koger [94]; the following are a few examples of the experimental findings. Most stainless steels and Inconel 600 alloys are susceptible to attack by molten fluorides containing UF_4. For example, a molten salt, of the composition 0.5NaF–0.46ZrF_4–0.04 UF_4, flowing in an Inconel 600 tube at 815 °C, picked up 765 ppm Cr in a period of 1000 hr, while there was little change in the nickel and iron concentrations which were in the range 10–50 ppm present in the salt as impurities. Appreciable chromium losses from the alloy were observed after 1000-hr exposure; even at a depth of 500 μm, the chromium content was 11.3% as compared to the original 14.5% Cr.

Of the various superalloys tested, Hastelloy N, of the composition Ni–15% Mo–7% Cr–5% Fe, was found to have a better resistance to hot corrosion by molten fluorides containing UF_4. After an exposure time of 10,000 hr to a molten salt mixture 0.62LiF–0.37BeF_2–0.01UF_4 at 705 °C, a zone of chromium depletion, i.e., void formation, extended only to a depth of 12 μm. After an exposure time of 9.2 years at 675 °C, the depth of chromium depletion and void formation in Hastelloy N was about 60 μm.

10.4.6 Corrosion of Alloys by Liquid Sodium

Liquid metals are used as the working fluid in numerous areas of advanced technology as, for example, in space power systems which utilize thermoelectric and thermionic converters in conjunction with nuclear heat sources and in liquid metal fast breeder reactors (LMFBR) to remove the fission heat produced in the fuel elements. Because of these and other critical applications, the hot corrosion of container walls by liquid metals has received much attention during the past decade, as evidenced from the papers published in the proceeding of a symposium on corrosion by liquid metals [95]. As another example of hot corrosion, a brief description is given below of the corrosion of stainless steels and nickel-base alloys by liquid sodium in LMFBR loops.

The reactors are designed with two separate sodium-coolant loops. In the primary loop, the sodium circulates between the core and the intermediate heat exchanger, where the heat is transferred to the sodium circulating in the secondary loop. The heat sink in the secondary loop is the steam generator. The highest rate of corrosion occurs in the high temperature region (~700–750 °C) of the intermediate heat exchanger where the material is removed from both the outside and inside of the tubes by primary and secondary sodium. In the low temperature region of the loop, there is material deposition on both sides of the tube. Although the material deposition is not great enough to cause flow blockage, it may affect the efficiency of heat transfer. Failure of the tube in the steam generator portion would be catastrophic because of the violent interaction of liquid sodium with water.

Extensive studies have been made of the corrosion of metals and alloys by liquid sodium in several research establishments. Much of the early work is documented in [95–97] as well as in a review paper by Weeks and Isaacs [98] and in a book by Olander [99] on the fundamental aspects of nuclear reactor-fuel elements.

The mechanism of material transfer from hot to cold parts of the loop is controlled by the interplay of the solubility of alloy constituents in liquid sodium and the effect of oxygen therein, alloy composition, the velocity of fluid flow, and temperature. First, let us consider the solubility of oxygen and the primary alloying elements in liquid sodium.

Since oxygen in liquid sodium has a marked effect on the corrosion of alloys, many measurements have been made of the solubility of Na_2O in liquid sodium. Measurements made by Minushkin and Kissel [100] using an emf cell and earlier measurements using conventional sampling and analytical techniques give for the solubility of oxygen in liquid sodium, in equilibrium with sodium oxide,

$$\mathrm{Na_2O(s)} = 2\mathrm{Na(l)} + \underline{\mathrm{O}},$$
$$\log[\mathrm{ppm\ O}] = -(2600/T) + 6.46. \tag{10.47}$$

The temperature dependence of the solubilities of chromium and iron [101] and nickel [102] in liquid sodium is represented by the following equations:

$$\log[\mathrm{ppm\ Cr}] = -(9010/T) + 9.35, \tag{10.48}$$

$$\log[\mathrm{ppm\ Fe}] = -(4310/T) + 5.16, \tag{10.49}$$

$$\log[\mathrm{ppm\ Ni}] = -(1230/T) + 1.55. \tag{10.50}$$

The oxygen interacts strongly with chromium in liquid sodium because of the formation of sodium chromite, $NaCrO_2$. The chromium solubility represented by Eq. (10.48) is for liquid sodium containing less than 3 ppm O when no sodium chromite is formed. At sodium oxide activities below about 0.1, the solubility of iron is independent of the concentration of oxygen in sodium. At higher activities of sodium oxide, the iron solubility increases with increasing concentration of oxygen in sodium. This behavior has been attributed to the formation of di-sodium ferrite, $(Na_2O)_2 \cdot FeO$ [100, 103]. The oxygen in solution has no effect on the solubility of nickel in sodium.

When the oxygen content of liquid sodium is less than about 10 ppm, there is preferential dissolution of chromium and nickel in sodium in the early stages of contact of stainless steel with liquid sodium. The extent of surface depletion of chromium and nickel increases with decreasing concentration of oxygen in sodium. This change in the surface composition of stainless steel leads to a phase transformation from austenite to ferrite in the surface layers because austenite is not stable below about 15% Cr and 8% Ni. The rate of dissolution of steel in the high temperature portion of the loop decreases with increasing time of loop operation, reaching a steady state rate after several weeks at about 600 °C, or about 300 hr at temperatures above 700 °C.

At oxygen contents above about 10 ppm, the sodium chromite forms on the surface of the stainless steel in the initial stages of the loop operation. However, because of the higher stability of chromite at lower temperatures, the initial chromite deposit in the high temperature region is gradually removed by circulating sodium, and ultimately, a new steady state rate of corrosion is established. Romano and Klamut (cited in a review paper by Weeks and Isaacs [98]) found that the steady state rate of corrosion of stainless steel increases almost linearly with increasing concentration of oxygen in sodium; e.g., at 760 °C the rate for type 304 stainless steel is 25 $\mu m\ yr^{-1}$ at about 4 ppm O and increases to 75 $\mu m\ yr^{-1}$ at about 15 ppm O.

The rate of corrosion of nickel-base alloys, such as Inconels and Hastelloy N, is not affected by the concentration of oxygen in sodium up to about 500 ppm O [104]. As nickel in the alloy is replaced by iron, the rate of corrosion increases; this increase becomes more pronounced with increasing temperature and oxygen content of sodium.

In stainless steel loops, the deposition of the constituent elements in the cooler areas of the loop are segregated. As the liquid sodium is cooled from the maximum circuit temperature, the chromium deposits first, followed by nickel, manganese, and iron. However, deposits rich in iron and silicon are usually dispersed in sodium and tend to collect in areas where there is change in flow direction and in magnetic fields of electromagnetic pumps and flow meters. The carbon leached out from the hotter areas is deposited in chromium segregated areas because of strong Cr–C interaction.

The rate of corrosion of nickel-base alloys increases with increasing velocity of sodium flow in the loop, in accord with what is to be expected for the rate controlled by liquid-phase mass transfer in the boundary layer [104]. In the case of stainless steel, the rate of corrosion becomes insensitive to fluid velocity beyond about 7 m sec^{-1}.

Several mechanisms have been proposed to explain the steady state rate of corrosion of stainless steel by liquid sodium at high fluid velocities [98]. Although much speculation is involved in the proposed rate mechanisms, Eq. (10.51), derived by Weeks and Isaacs [98], does fit the available experimental data from LMFBR tests on the rate of corrosion of stainless steel by liquid sodium at high fluid velocities and for oxygen contents above 3 ppm:

$$\mathscr{R} = \frac{2 \times 10^{13} T}{x_{\mathrm{Fe}}} \left[\left(\frac{[\mathrm{O}]}{[\mathrm{O}]_s} \right)^{2x_{\mathrm{Fe}}} \exp\left(-\frac{50{,}300}{RT} \right) \right] \left[1 - \varepsilon \left(\frac{L}{d} \right)^2 \right] \quad \mu\mathrm{m\,yr}^{-1} \qquad (10.51)$$

where x_{Fe} is the atom fraction of iron in the alloy; [O] the oxygen concentration in ppm and $[\mathrm{O}]_s$ that for saturation with Na_2O, as given by Eq. (10.47); d the tube diameter; L the downstream position; and ε the downstream coefficient. The downstream coefficient ε has been estimated to be about 5×10^{-5} at 750 °C and about 10^{-6} at 700 °C.

The stainless steel cladding of the fuel element is about 0.6–0.8 mm thick. Considering that the core has a life of about 3 years, the thinning of the cladding due to corrosion by sodium is relatively small, particularly at oxygen contents below about 10 ppm. The electrochemical oxygen meter utilizing a high purity ThO_2–15% Y_2O_3 electrolyte is capable of measuring oxygen in solution in liquid sodium at concentrations as low as 1 ppm O. By employing such measuring devices, it is possible to monitor continuously the oxygen content of sodium in the loop at levels low enough to minimize the damage of the alloy tubing by hot corrosion.

References

1. J. E. Funk and R. M. Reinstrom, *Ind. Eng. Chem., Process Des. Dev.* **5**(3), 336 (1966).
2. B. M. Abraham and F. Schreiner, *Ind. Eng. Chem., Fundam.* **13**(4), 305 (1974).
3. J. E. Funk, W. Conger, and R. Carty, *in* "Hydrogen Energy" (T. N. Veziroglu, ed.), p. 457. Plenum, New York, 1975.
4. R. E. Chao, *Ind. Eng. Chem., Prod. Res. Dev.* **13**(2), 94 (1974).
5. C. Hardy, Euratom Rep. EUR 49581. Euratom, Ispra, Italy (December 1973).
6. C. G. von Fredersdorff, "Conceptual Process for Hydrogen and Oxygen Production from Nuclear Decomposition of Carbon Dioxide," Memorandum to Project S-128 Sponsors' Committee. Inst. Gas Technol., Chicago, Illinois (October 1959).
7. C. Marchetti, Euratom Rep. No. 1, EUR 4776e (December 1970); No. 2, EUR 4955e (December 1971); No. 3, EUR-CIS-35-73e (December 1972). Euratom, Ispra, Italy.
8. W. W. Bodle, K. C. Vyas, and A. T. Talwalkar, *in* "Clean Fuels From Coal Symposium II" (J. W. White, R. Sprague, and W. McGrew, eds.), p. 53. Inst. Gas Technol., Chicago, Illinois, 1975.
9. "Energy Technology Handbook" (D. M. Considine, ed.). McGraw-Hill, New York, 1977.
10. "Kirk-Othmer Encyclopedia of Chemical Technology," 2nd Ed. Wiley (Interscience), New York, 1966.

11. P. F. H. Rudolph, *in* "Energy Technology Handbook" (D. M. Considine, ed.), pp. 1–188. McGraw-Hill, New York, 1977.
12. I. N. Banchik, *in* "Clean Fuels From Coal Symposium II" (J. W. White, R. Sprague, and W. McGrew, eds.), p. 359. Inst. Gas Technol., Chicago, Illinois, 1975.
13. J. F. Farnsworth and D. M. Mitsak, *in* "Clean Fuels From Coal Symposium II" (J. W. White, R. Sprague, and W. McGrew, eds.), p. 107. Inst. Gas Technol., Chicago, Illinois, 1975.
14. A. E. Cover, W. C. Schreiner, and G. T. Skaperdas, *Chem. Eng. Prog.* **69**(3), 32 (1973).
15. J. Huebler, F. C. Schora, and B. S. Lee, *in* "Energy Technology Handbook" (D. M. Considine, ed.), p. 1.201. McGraw-Hill, New York, 1977.
16. R. J. Grace, *in* "Energy Technology Handbook" (D. M. Considine, ed.), p. 1.212. McGraw-Hill, New York, 1977.
17. K. Bund, K. A. Henney, and K. H. Krieb, *World Energy Conf., 8th, Bucharest* p. 2.3 (1971).
18. B. Robson, *Energy Res.* **1**, 157 (1977).
19. K. A. Schowalter and N. S. Boodman, *Chem. Eng. Prog.* **70**(6), 76 (1974).
20. N. S. Boodman, T. F. Johnson, and K. C. Krupinski, *in* "Coal Desulfurization—Chemical and Physical Methods" (T. D. Wheelock, ed.), ACS Symposium Series No. 64, p. 248. Am. Chem. Soc., Washington, D.C., 1977.
21. R. J. Gray and K. C. Krupinski, *in* "Coal Agglomeration and Conversion Symposium," p. 37. Coal Res. Bur., West Virginia Univ., Morgantown, 1975.
22. G. J. W. Kor, *in* "Coal Desulfurization—Chemical and Physical Methods" (T. D. Wheelock, ed.), ACS Symposium Series, No. 64, p. 221. Am. Chem. Soc., Washington, D.C., 1977.
23. J. C. Whitehead and D. F. Williams, *J. Inst. Fuel* **48**, 182 (1975).
24. D. M. Archer, D. L. Keairns, and E. J. Vidt, *Energy Commun.* **1**(2), 115 (1975).
25. C. E. Fink, J. D. Sudbury, and G. P. Curran, *Nat. Meet. AIChE, 77th, Pittsburgh, 1974.*
26. A. A. Jonke, G. J. Vogel, E. L. Carls, D. Ramaswami, L. Anastasia, R. Jarry, and M. Haas, *AIChE Symp. Ser.* **68**, 241 (1972).
27. "Air Pollution and Its Control," *AIChE Symp. Ser.* **68** (1972).
28. "Air: II. Control of NO_x and SO_x Emissions," *AIChE Symp. Ser.* **71** (1975).
29. "Control and Dispersion of Air Pollutants: Emphasis on NO_x and Particulate Emissions," *AIChE Symp. Ser.* **74** (1978).
30. E. T. Turkdogan and J. V. Vinters, *Metall. Trans.* **5**, 11 (1974).
31. H. J. Donald, "An Annotated Bibliography." Mellon Inst. Ind. Res., Pittsburgh, Pennsylvania, 1956.
32. J. V. Kehrer and H. Leidheiser, *J. Phys. Chem.* **58**, 550 (1954).
33. P. L. Walker, J. F. Rakszawski, and G. R. Imperial, *J. Phys. Chem.* **63**, 133, 140 (1959).
34. W. R. Ruston, M. Warzee, J. Hennaut, and J. Waty, *Carbon* **7**, 47 (1969).
35. R. G. Olsson and E. T. Turkdogan, *Metall. Trans.* **5**, 21 (1974).
36. J. Macak, P. Knizek, and J. Malecha, *Carbon* **16**, 111 (1978).
37. E. R. Gilliland and P. Harriott, *Ind. Eng. Chem.* **46**, 2195 (1954).
38. T. H. Byrom, *J. Iron Steel Inst., London* **92**, 106 (1915).
39. H. C. H. Carpenter and C. C. Smith, *J. Iron Steel Inst., London* **96**, 139 (1918).
40. O. Stelling, *J. Met.* **10**, 290 (1958).
41. H. Schenck and W. Maachlanka, *Arch. Eisenhuettenwes.* **31**, 271 (1960).
42. W. Baukloh and G. Henke, *Metallurgia* **19**, 463 (1940).
43. F. Fischer and H. A. Bahr, *Gesammelte Abh. Kennt. Kohle* **8**, 225 (1928).
44. E. T. Turkdogan and R. G. Olsson, unpublished work, U.S. Steel Res. Lab., Monroeville, Pennsylvania (1976).
45. Y. Nishiyama and Y. Tamai, *Carbon* **14**, 13 (1974).
46. A. Schaas and H. J. Grabke, *Oxid. Met.* **12**(5), 387 (1978).
47. R. F. Hochman, *in* "Properties of High Temperature Alloys" (Z. A. Foroulis and F. S. Pettit, eds.), p. 715. Electrochem. Soc., Princeton, New Jersey, 1976.
48. R. Schueler, *Hydrocarbon Process.* **51**(8), 73 (1972).

49. N. R. Lindblad, *Oxid. Met.* **1**(1), 143 (1969).
50. G. W. Goward, *in* "Properties of High Temperature Alloys" (Z. A. Foroulis and F. S. Pettit, eds.), p. 806. Electrochem. Soc., Princeton, New Jersey, 1976.
51. O. Kubaschewski and B. E. Hopkins, "Oxidation of Metals and Alloys." Buttersworth, London, 1962.
52. K. Hauffe, "Oxidation of Metals." Plenum, New York, 1965.
53. W. E. Boggs, *J. Electrochem. Soc.* **118**, 906 (1971).
54. C. Wagner, *Corros. Sci.* **5**, 751 (1965).
55. E. Scheil and K. Kiwit, *Arch. Eisenhuettenwes.* **9**, 405 (1935–1936).
56. W. E. Boggs, *Oxid. Met.* **10**(4), 277 (1976).
57. "Oxide Dispersion Strengthening" (G. S. Ansell and T. D. Cooper, eds.), Metall. Soc. Conf., AIME, 1966, Vol. 47. Gordon & Breach, New York, 1968.
58. C. S. Giggins and F. S. Pettit, *Metall. Trans.* **2**, 1071 (1971).
59. J. Stringer, R. A. Wilcox, and R. I. Jaffee, *Oxid. Met.* **5**(1), 11 (1972).
60. H. T. Michels, *Metall. Trans. A* **7**, 379 (1976).
61. M. W. Chase, J. L. Curnutt, H. Prophet, R. A. McDonald, and A. N. Syverud, "JANAF Thermochemical Tables, 1975 Supplement." Natl. Bur. Stand., Washington, D.C., 1975.
62. C. S. Tedmon, *J. Electrochem. Soc.* **113**, 766 (1966).
63. W. C. Hagel, *Trans. Am. Soc. Met.* **56**, 583 (1963).
64. C. E. Lowell and W. A. Sanders, MACH-1 Oxidation of Thoriated Nickel Chromium at 1204°C (2200°F) NASA TN-D-6562 (1971).
65. W. P. Gilbreath, *Prog. Astronaut. Aeronaut.* **31**, 127 (1972).
66. D. R. Tenney, C. T. Young, and H. W. Herring, *Metall. Trans.* **5**, 1001 (1974); *Metall. Trans. A* **6**, 2253 (1975).
67. "Instructions and Operating Procedures for Aerotherm 100 KW Constrictor Arc Heater System," Aerotherm Corp., Tech. Narrative Rep. No. 1M-68-1. NASA Contract No. NASI-7560 (1968).
68. J. D. Whittenberger, *Metall. Trans. A* **9**, 1327 (1978).
69. S. Mrowec, T. Walec, and T. Werber, *Oxid. Met.* **1**, 93 (1969).
70. T. Narita and K. Nishida, *Oxid. Met.* **6**, 157, 181 (1973).
71. A. Davin, D. Coutsouradis, and L. Habraken, *Cobalt* **35**, 69 (1967).
72. T. Biegun, A. Bruckman, and S. Mrowec, *Oxid. Met.* **12**, 157 (1978).
73. S. Mrowec, T. Werber, and M. Zastawnik, *Corros. Sci.* **6**, 47 (1966).
74. G. Romeo, W. Smeltzer, and J. Kirkaldy, *J. Electrochem. Soc.* **118**, 740, 1336 (1971).
75. W. L. Worrell and E. T. Turkdogan, *Trans. Metall. Soc. AIME* **242**, 1673 (1968).
76. C. Wagner, *Z. Phys. Chem.* **22**, 181 (1933); **32**, 447 (1936).
77. S. Mrowec, *in* "Properties of High Temperature Alloys" (Z. A. Foroulis and F. S. Pettit, eds.), p. 413. Electrochem. Soc., Princeton, New Jersey, 1976.
78. S. Verma, D. Whittle, and J. Stringer, *Corros. Sci.* **12**, 454 (1972).
79. D. J. Young, W. W. Smeltzer, and J. S. Kirkaldy, *Metall. Trans. A* **6**, 1205 (1975).
80. N. Birks, *in* "Properties of High Temperature Alloys" (Z. A. Foroulis and F. S. Pettit, eds.), p. 215. Electrochem. Soc., Princeton, New Jersey, 1976.
81. G. J. W. Kor, *Metall. Trans.* **3**, 2343 (1972).
82. R. H. Chang, W. Stewart, and J. B. Wagner, *Reactiv. Solids, Proc. Int. Symp., 7th, Bristol* p. 231 (1972).
83. A. U. Seybolt, *Trans. Metall. Soc. AIME* **242**, 752 (1968).
84. M. R. Wootton and N. Birks, *Corros. Sci.* **12**, 829 (1972).
85. G. Romeo and H. S. Spacil, *in* "High Temperature Gas–Metal Reactions in Mixed Environments" (S. A. Jansson and Z. A. Foroulis, eds.), p. 299. Metall. Soc. AIME, New York, 1973.
86. D. W. McKee and G. Romeo, *Metall. Trans.* **5**, 1127 (1974).
87. E. L. Simons, G. V. Browning, and H. A. Liebhafsky, *Corrosion* **11**, 505t (1955).
88. J. A. Goebel and F. S. Pettit, *Metall. Trans.* **1**, 1943 (1970).
89. J. A. Goebel, F. S. Pettit, and G. W. Goward, *Metall. Trans.* **4**, 261 (1973).

90. J. A. Goebel and F. S. Pettit, *in* "Metal–Slag–Gas Reactions and Processes" (Z. A. Foroulis and W. W. Smeltzer, eds.), p. 693. Electrochem. Soc., Princeton, New Jersey, 1975.
91. M. W. Rosenthal, P. N. Haubenreich, and R. B. Briggs, "Development Status of Molten Salt Breeder Reactors," USAEC Rep. ORNL-4812. Oak Ridge Natl. Lab. (1972).
92. G. M. Adamson, R. S. Crouse, and W. D. Manly, "Interim Report on Corrosion by Alkali-Metal Flourides," AECORNL-2337. Oak Ridge Natl. Lab. (March 1959); "Interim Report on Corrosion by Zirconium Base Flourides," ORNL-2338 (January 1961).
93. C. F. Baes, *in* "Thermodynamics," Proc. Symp., Vol. I, p. 409. IAEA, Vienna, 1966.
94. J. W. Koger, *Adv. Corros. Sci. Technol.* **4**, 245 (1974).
95. "Corrosion by Liquid Metals" (J. E. Draley and J. R. Weeks, eds.). Plenum, New York, 1970.
96. *IAEA Symp. Alkali Met. Coolants—Corros. Stud. Syst. Oper. Experience Vienna, 1966.*
97. *Proc. Int. Conf. Sodium Technol. Large Fast React. Des.* ANL-7520, Part I. Argonne Natl. Lab. (1968).
98. J. R. Weeks and H. S. Isaacs, *Adv. Corros. Sci. Technol.*, **3**, 1 (1973).
99. D. R. Olander, "Fundamental Aspects of Nuclear Reactor Fuel Elements." Tech. Inf. Cent., Energy Res. Dev. Adm., Washington, D.C., 1976.
100. B. Minushkin and G. Kissel, *in* "Corrosion by Liquid Metals" (J. E. Draley and J. R. Weeks, eds.), p. 515. Plenum, New York, 1970.
101. R. M. Singer, A. H. Fleitman, J. R. Weeks, and H. S. Isaacs, *in* "Corrosion by Liquid Metals" (J. E. Draley and J. R. Weeks, eds.), p. 561. Plenum, New York, 1970.
102. R. M. Singer and J. R. Weeks, *Proc. Int. Conf. Sodium Technol. Large Fast React. Des.* ANL-7520, Part I, p. 309. Argonne Natl. Lab. (1968).
103. C. Tyzack, *in* "The Alkali Metals," Spec. Publ. No. 22, p. 236. Chem. Soc., London, 1967.
104. A. W. Thorley and C. Tyzack, *in* "Alkali Metal Coolants," p. 97. IAEA, Vienna, 1967.

INDEX

S

T